Oats:
Chemistry
and
Technology

Edited by
Francis H. Webster

John Stuart Research Laboratories
The Quaker Oats Company
Barrington, Illinois

Published by the
American Association of Cereal Chemists, Inc.
St. Paul, Minnesota, USA

Cover photograph by Steve Kronmiller.

Library of Congress Catalog Card Number: 86-071926
International Standard Book Number: 0-913250-30-9

©1986 by the American Association of Cereal Chemists, Inc.

Reference in this volume to a company or product name by personnel of the U.S.
Department of Agriculture is intended for explicit description only and does
not imply approval or recommendation of the product by the U.S. Department of
Agriculture to the exclusion of others that may be suitable.

Printed in the United States of America

American Association of Cereal Chemists, Inc.
3340 Pilot Knob Road
St. Paul, Minnesota 55121, USA

CONTRIBUTORS

James W. Anderson, V. A. Medical Center, University of Kentucky, Lexington, Kentucky

A. Chris Brinegar, U.S. Department of Agriculture-Agricultural Research Service, Department of Agronomy, University of Wisconsin, Madison, Wisconsin

Vernon D. Burrows, Cereal Section, Plant Research Centre, Agriculture Canada, Ottawa, Ontario

Wen-Ju Lin Chen, V. A. Medical Center, University of Kentucky, Lexington, Kentucky

F. W. Collins, Food Research Centre, Research Branch, Agriculture Canada, Ottawa, Ontario

Edward Commers, Carter-Day Company, Minneapolis, Minnesota

Donald Deane, Carter-Day Company, Minneapolis, Minnesota

R. G. Fulcher, Cereal Section, Plant Research Centre, Agriculture Canada, Ottawa, Ontario

Menard G. Haydanek, Jr., The Quaker Oats Company, Barrington, Illinois

H. David Hurt, The Quaker Oats Company, Barrington, Illinois

Haines B. Lockhart, The Quaker Oats Company, Barrington, Illinois

Linda A. MacArthur-Grant, Department of Cereal Science and Food Technology, North Dakota State University, Fargo, North Dakota

Robert J. McGorrin, The Quaker Oats Company, Barrington, Illinois

David Paton, Food Research Centre, Research Branch, Agriculture Canada, Ottawa, Ontario

David M. Peterson, U.S. Department of Agriculture-Agricultural Research Service, Department of Agronomy, University of Wisconsin, Madison, Wisconsin

Donald J. Schrickel, The Quaker Oats Company, Chicago, Illinois

Francis H. Webster, The Quaker Oats Company, Barrington, Illinois

Peter J. Wood, Food Research Center, Research Branch, Agriculture Canada, Ottawa, Ontario

Vernon L. Youngs, U.S. Department of Agriculture-Agricultural Research Service, Department of Cereal Chemistry and Technology, North Dakota State University, Fargo, North Dakota

PREFACE

The oat monograph is the newest in a series on key cereal grains published by the American Association of Cereal Chemists. This publication is intended to provide cereal chemists, students, and industrial processors with an in-depth and authoritative reference on oat chemistry and technology. The information in it represents the efforts of leading North American oat researchers. The individual chapters have a strong technical focus based on each contributor's experiences in his or her respective area. The result is the most complete text ever published on oats.

This monograph has several unique features that are new to the cereal monograph series. The full-color fluorescent micrographs illustrating oat structure and component compartmentalization complement the individual chapters on specific components. Additionally, separate contributions on flavor chemistry, phenol chemistry, and dietary fiber add new dimensions.

I would like to thank each author for his or her cooperation. The assistance and support of the Quaker Oats Company and its technical staff are greatly appreciated.

F. H. Webster

CONTENTS

Oats:
Chemistry
and
Technology

CHAPTER 1

OATS PRODUCTION, VALUE, AND USE

DONALD J. SCHRICKEL
Quaker Oats Company
Chicago, Illinois

I. INTRODUCTION

The origin of oats can be traced back to about 2000 B.C. Archaeological discoveries reveal that oats originated in the Middle East, particularly the areas surrounding the Mediterranean Sea. Some of the first evidence of oats was found in Egypt and among the ancient lake dwellers of what is now Switzerland.

The oat species include *Avena abyssinica, A. byzantina, A. fatua, A. nuda, A. sativa, A. strigosa,* and others. The species most cultivated in the world are *A. sativa* and *A. byzantina.* More than 75% of the total cultivated world oats production area is sown to cultivars of *A. sativa,* and most of the remaining cultivated oats are *A. byzantina.* A small acreage is devoted to *A. strigosa* and other species (Coffman, 1961).

In recent years, many planned genetic crosses have been made between *A. sativa* and *A. byzantina,* so that many of the commonly grown cultivars— particularly in the southern part of the United States and in much of South America—are cultivars developed from these two species. *A. sativa* is referred to by growers as *white* oats, or in some cases as *yellow* oats, and *A. byzantina* is commonly referred to as *red* oats. Because of the breeding programs that have been used in recent years, the color of the oat grain is becoming more like that of *A. sativa* and less like that of *A. byzantina.* Even in Argentina, where *A. byzantina* has been the most popular oat species over the years, the producers now prefer oats of the *A. sativa* type. For example, in the major production areas of central Argentina, more than half of the oats are now *A. sativa* types. In the United States, most of the spring oats are *A. sativa,* whereas the winter oats grown in the southeast and southwest originated as *A. byzantina* but, through breeding programs, have become more like *A. sativa* in appearance (Coffman, 1977).

II. PRODUCTION

A. In the World

The world production of oats for grain has been maintained at the level of 50 million metric tons since the early 1960s. Recent data on world production, yield, and areas of production are shown in Tables I–III. The world production statistics reflect the fact that oats are primarily a cool-season crop. A large proportion of world oat production occurs in the Northern Hemisphere between the latitudes of 35 and 50° (Fig. 1).

The USSR's leadership in 1984 oats production is particularly dramatic when viewed graphically (Fig. 2). Outstanding yields per hectare in 1984 are noteworthy in the United Kingdom, Sweden, and West Germany (Fig. 3). These yields are attributable to good cultivars, a long growing season, and excellent cultural management.

Studies by the Quaker Oats Company (P. F. Sisson, *personal communication*, 1985) project a decline in world oats production into 1990, although the USSR will continue to increase its production (Table IV).

Thus, it becomes apparent that the share of world oats production in the USSR is increasing—from 26% in 1975 to 33% in 1984—and, according to projections, will reach 37% in 1990. Meanwhile, the share produced by the United States is decreasing—from 19% in 1975 to 15% in 1984 and projected to be 12% in 1990. In many parts of the world, oats are grown for multipurpose use: i.e., for pasture, forage (hay, silage, "green-chop"), and grain. In addition, the oat straw remaining after grain harvest often is removed from the field for use as hay and animal bedding (Schrickel and Scantland, 1981).

B. In the United States

Although oats production is declining in the United States, it continues to be an important crop, particularly in the north central states of Iowa, Minnesota, North Dakota, South Dakota, and Wisconsin. These five states produce about

Fig. 1. Areas in the world where oats are produced. (Courtesy Quaker Oats Co., Chicago)

65% of the oats harvested for grain each year. The U.S. production of oats by state for 1982–1984 is shown in Table V. One should note, however, that there is a large difference between the area planted and the area harvested for grain,

TABLE I
World Oat Production (1,000 t)[a]

Country	1975	1976	1977	1978	1979	1980	1981	1982	1983	1984
USSR	12,495	18,113	18,407	18,507	15,200	15,544	15,000	15,500	17,500	15,200
United States	9,275	7,838	10,930	8,443	7,646	6,659	7,396	8,602	6,926	6,858
West Germany	3,445	2,497	2,714	4,049	3,697	3,249	3,200	3,777	2,489	3,054
Poland	2,920	2,695	2,552	2,492	2,186	2,245	2,731	2,608	2,376	2,750
Canada	4,480	4,831	4,303	3,621	2,978	3,028	3,188	3,637	2,773	2,700
Sweden	1,321	1,251	1,416	1,550	1,524	1,567	1,816	1,663	1,268	1,990
France	1,948	1,431	1,928	2,203	1,845	1,927	1,774	1,802	1,455	1,752
People's Republic of China	1,549	1,540	1,500	1,500	1,800	1,800	1,700	1,668	1,650	1,680
Finland	1,450	1,573	1,022	1,082	1,283	1,258	1,008	1,320	1,407	1,350
Australia	1,141	1,072	990	1,763	1,411	1,128	1,617	848	2,269	1,300
Spain	609	528	418	553	456	680	445	443	470	800
East Germany	780	506	411	595	532	582	598	848	500	608
United Kingdom	795	765	790	706	542	601	620	575	460	550
Argentina	433	530	570	676	522	433	339	637	593	530
Others	4,457	3,736	2,643	3,826	3,872	3,850	3,916	3,961	3,582	4,208
World	47,098	48,906	51,594	51,566	45,494	44,551	45,348	47,889	45,718	45,330

[a]Source: USDA Foreign Agricultural Service (1984).

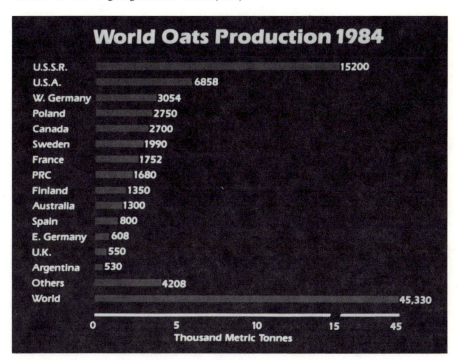

Fig. 2. Production of oats in selected countries, 1984. (Courtesy Quaker Oats Co., Chicago)

especially in the southern United States. In these regions, oats are grown primarily as a pasture forage crop. For example, Texas ranks fifth in the United States in terms of area planted to oats but only eighth in area harvested for grain, since approximately 60% of the area is pastured or harvested for forage.

TABLE II
World Oat Yields (t/ha)[a]

Country	1975	1976	1977	1978	1979	1980	1981	1982	1983	1984
United Kingdom	3.43	3.26	4.05	3.88	3.99	4.06	4.31	4.42	4.18	5.00
Sweden	2.85	2.78	3.09	3.42	3.33	3.47	3.83	3.49	3.14	4.68
West Germany	3.74	2.92	3.42	4.16	4.02	3.80	3.88	4.25	3.41	4.59
France	2.97	2.19	3.08	3.61	3.42	3.61	3.54	3.48	3.30	4.01
East Germany	3.21	2.66	2.69	3.89	3.91	3.75	3.48	3.89	3.07	3.80
Finland	2.53	2.85	2.45	2.43	2.84	2.81	2.32	2.88	3.13	3.11
Poland	2.26	2.42	2.33	2.42	2.00	2.25	2.36	2.40	2.28	2.75
United States	1.76	1.64	2.00	1.87	1.95	1.90	1.94	2.07	1.89	2.10
Canada	1.86	2.01	2.02	1.98	1.93	2.00	2.04	2.26	1.98	1.89
Spain	1.33	1.16	1.03	1.25	1.05	1.48	.96	1.00	1.01	1.67
Argentina	1.28	1.38	1.33	1.35	1.27	1.24	1.14	1.56	1.43	1.33
USSR	1.03	1.61	1.41	1.53	1.24	1.32	1.20	1.35	1.40	1.30
People's Republic of China	1.06	1.03	1.00	1.00	1.20	1.20	1.21	1.19	1.18	1.20
Australia	1.15	1.08	.92	1.30	1.26	1.03	1.16	.70	1.30	1.13
Others	1.56	1.35	1.53	1.58	1.59	1.65	1.66	1.61	1.52	1.75
World	1.56	1.71	1.71	1.81	1.66	1.71	1.65	1.78	1.68	1.78

[a]Source: USDA Foreign Agricultural Service (1984).

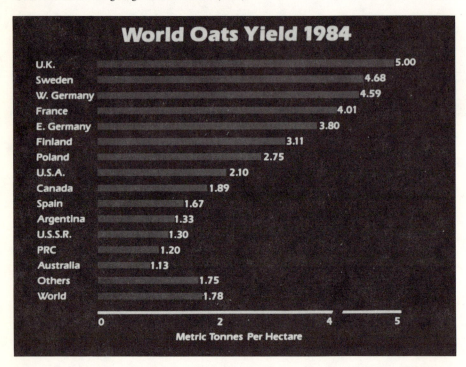

Fig. 3. Yields of oats produced in selected countries, 1984. (Courtesy Quaker Oats Co., Chicago)

TABLE III
World Oats—Area Harvested (1,000 ha)[a]

Country	1975	1976	1977	1978	1979	1980	1981	1982	1983	1984
USSR	12,107	11,269	13,026	12,097	12,239	11,770	12,470	11,489	12,500	11,700
United States	5,261	4,775	5,463	4,503	3,918	3,503	3,807	4,151	3,673	3,273
Canada	2,414	2,409	2,132	1,828	1,541	1,515	1,561	1,612	1,400	1,426
People's Republic of China	1,464	1,499	1,500	1,500	1,500	1,500	1,400	1,400	1,400	1,400
Australia	988	995	1,076	1,359	1,123	1,093	1,388	1,212	1,742	1,150
Poland	1,291	1,115	1,097	1,030	1,094	997	1,156	1,086	1,042	1,000
West Germany	920	855	793	973	919	856	825	888	729	666
Spain	457	455	405	442	436	458	464	442	466	479
France	655	652	625	611	540	534	501	518	441	437
Finland	572	551	417	446	451	448	434	459	449	434
Sweden	464	450	458	453	457	452	474	477	404	425
Argentina	338	383	430	500	410	350	298	408	414	400
East Germany	243	190	153	153	136	155	172	218	163	160
United Kingdom	232	235	195	182	136	148	144	130	110	110
Others	2,864	2,765	2,380	2,480	2,442	2,327	2,363	2,466	2,360	2,405
World	30,270	28,598	30,150	28,506	27,342	26,106	27,457	26,956	27,293	25,465

[a]Source: USDA Foreign Agricultural Service (1984).

TABLE IV
Oat Production (1,000 t)—Projections to 1990[a]

Area	Actual Production		Projection to 1990
	1975	1984	
World	47,098	45,330	42,500
USSR	12,495	15,200	16,000
United States	9,275	6,858	5,300

[a]Source: Quaker Oats Co.

III. CULTURAL PRACTICES

A. sativa, the spring or northern white oat, is the most popular species grown in the world and in the United States. In the Northern Hemisphere, spring oats are planted in April or May and harvested in July or August. Well-drained soils are preferred, and the seedbed should be reasonably well prepared through plowing or disking. Oats are drill-planted to a depth of 3.8–6.8 cm at a rate of approximately 90 kg/ha. Oats produce well with only moderate amounts of fertilizer. Fertilizer application rates of about 50 kg of nitrogen, 24 kg of phosphate, and 17 kg of potash per hectare would be typical. Ideally, the fertilizer and seed are put down at the same time and at about the same depth.

The oat cultivars planted differ considerably from state to state and even within states because of environmental differences and disease situations in specific growing areas. Agricultural extension services and crop improvement associations are able to recommend the most suitable cultivars for particular circumstances.

Weed control is important in obtaining maximum oat yields. Broadleaf weeds can be controlled easily with the recommended amounts of appropriate herbicides. Since oats are often used as a nurse crop for establishing small-seeded legumes, it is important that the herbicides be chosen with this in mind.

Harvest usually begins three months after planting, when the moisture content of the grain is 14% or less. Two methods of harvest are practiced: direct combining or windrowing followed several days later by combining of the grain from the windrow. The latter method allows the grain to dry uniformly and the straw to move easily through the combine. Windrowing or cutting oats and placing them in a swath also removes some of the risk of lodging from high winds or storms, which may occur when the grain is near maturity.

The winter or cultivated red oat, *A. byzantina*, is grown in latitudes between 20 and 40°. Winter oats are grown in much the same manner as spring oats, but the timing of production steps is quite different. Winter oats are planted in the fall of

TABLE V

Area, Yield, and Production of Oats in the United States, 1982–1984[a]

State	Area Harvested (1,000 ha)			Yield (t)			Production (1,000 t)		
	1982	1983	1984	1982	1983	1984	1982	1983	1984
Ala.	16	16	12	1.86	1.76	1.72	30	28	21
Ark.	13	20	11	2.51	2.58	2.51	34	52	28
Calif.	20	18	20	2.40	2.33	2.47	49	42	50
Colo.	16	17	20	1.86	2.04	1.97	30	35	40
Ga.	36	34	24	2.19	2.19	1.97	80	75	48
Idaho	19	19	18	2.47	2.72	2.44	46	53	43
Ill.	81	85	67	2.12	2.15	2.33	171	183	156
Ind.	43	32	32	2.29	2.04	2.22	98	66	72
Iowa	385	304	300	2.04	1.83	2.29	786	555	687
Kans.	70	43	49	1.69	1.72	1.90	117	73	92
Ky.	3	3	2	1.58	1.58	1.51	4	4	4
Maine	18	15	16	2.15	2.22	2.01	38	34	33
Md.	6	6	6	2.08	2.01	2.04	13	11	12
Mich.	182	121	142	2.26	1.86	2.22	412	226	315
Minn.	619	547	486	2.29	2.04	2.33	1,421	1,117	1,132
Mo.	32	22	13	1.51	1.69	1.72	48	37	23
Mont.	61	49	43	1.83	1.58	1.33	111	77	56
Nebr.	186	126	121	2.08	1.58	1.79	387	198	218
N.J.	3	2	2	1.94	1.83	2.01	5	4	5
N.Y.	113	81	73	2.33	2.04	2.08	264	165	152
N.C.	30	28	28	2.12	2.01	2.08	64	57	57
N. Dak.	425	510	397	1.90	1.81	1.83	808	924	725
Ohio	138	97	89	2.47	2.29	2.15	341	223	192
Okla.	36	32	32	1.36	1.76	1.65	50	57	53
Oreg.	34	30	30	2.69	2.87	3.16	93	87	96
Pa.	136	121	113	2.12	1.94	2.04	287	235	232
S.C.	20	16	16	2.01	1.90	2.08	41	31	34
S. Dak.	862	668	628	2.08	1.72	2.01	1,793	1,150	1,260
Tenn.	3	3	2	1.69	1.58	1.69	5	4	3
Tex.	117	202	101	1.33	1.72	1.25	156	348	127
Utah	6	6	5	2.44	2.44	2.40	15	14	13
Va.	7	9	5	1.72	1.79	1.69	12	16	8
Wash.	12	13	12	2.22	2.26	2.44	27	30	30
W. Va.	4	4	3	1.83	1.86	1.83	7	7	6
Wis.	377	344	340	1.90	1.90	2.29	715	654	780
Wyo.	22	28	28	1.97	1.76	1.65	44	49	47
U.S.	4,153	3,673	3,289	2.07	1.89	2.08	8,602	6,923	6,850

[a]Source: USDA Crop Reporting Board (1984).

the year, or between September and October in the Northern Hemisphere. Before the onset of winter (January-February), the oats do not progress beyond the juvenile leaf stage and therefore are often used as a pasture for livestock. The following spring the livestock are usually removed to allow the plant to produce a panicle or grain head. Alternatively, the fields may be plowed in the spring and replanted to other crops. Harvesting of grain takes place in May and June in the same fashion as spring oats. At this point, producers often follow immediately with a second crop, such as soybeans.

In the Southern Hemisphere, oats are grown primarily as a forage crop. The economic benefit to the cattle industry is considerable, particularly in countries such as Argentina and Uruguay. After having been grazed once or twice during the fall and winter (June, July, and August primarily), the oats are permitted to develop to the grain production stage and are harvested for grain in December. The grain is then used for animal feed or sold into commercial channels for milling.

The primary oat diseases in the United States in order of economic importance are yellow dwarf virus or redleaf, leaf or crown rust (*Puccinia coronata*), septoria (*Septoria avenae*), stem rust (*Puccinia graminis*), halo blight (*Pseudomonas coronafaciens*), loose smut (*Ustilago avenae*), and covered smut (*U. kolleri*). Through the efforts of oat breeders over the last 50 years, reasonably good genetic tolerance or resistance to each of these diseases among the several varieties of oats has been developed. However, new races of oat pathogens constantly occur in nature and the plant breeder is challenged to continue a program of selective resistance development within the genetic pool at all times.

IV. ECONOMIC IMPORTANCE

A. In the World

Throughout the world, oats are grown as a multipurpose crop. They are used as a nurse crop to establish small-seeded legumes and grasses and are a major pasture crop in many of the cattle-raising areas of the world—for example, in Texas in the United States and in Argentina and Uruguay in South America. Oat straw is often used for livestock bedding or in maintenance rations for cattle. More than half of the oat grain never leaves the farm where it is produced. Additional benefits derived from growing oats include the breaking of cycles of soilborne insects and diseases. Most of the uses and benefits of oats described above are difficult to evaluate from the standpoint of economic value. Consequently, when net returns per unit area of production (hectare or acre) of oats are compared with those of grains that are marketed entirely for cash, oats often suffer in the comparison.

Seventy-eight percent of the world production is used for livestock feed, 18% for human food, and the remaining 4% for industrial, seed, and export.

B. In the United States

In 1984, approximately 6.8 million metric tons of oats were harvested as grain from 3.3 million hectares in the United States. However, an additional 1.7 million hectares were planted, most of which were used as pasture. The oats grain was valued at 807 million dollars, or an average price of $117.80 per metric ton (Table VI). Growers sold 39% of the 1984 oats crop, using the remaining 4.2 million metric tons on the farm as feed. Oats ranked eighth in value of production in 1984, as compared with the other major feed and food grains (Table VII).

The economic importance of the nongrain portion of the oat crop has been estimated by the Quaker Oats Company. In 1984, the value of oats as grain accounted for only about 60% of the total crop value. The value of the straw and pasture each accounted for 17%, and forage the remaining 6% (Table VIII). In areas where oat straw is used for livestock bedding, the value of straw may be as great as 35% of the total crop value.

TABLE VI
Area, Yield, Production, Disposition, and Value of Oats in the United States, 1982–1984[a]

Year	Area (million ha) Planted	Harvested	Yield (t/ha)	Production (1,000 t)	Used on Farms (1,000 t)	Sold (1,000 t)	Average Price ($/t)	Grain Value (million $)
1982	5.65	4.15	2.07	8,602	5,333	3,269	101.96	877
1983	8.21	3.67	1.89	6,923	4,225	2,698	115.05	796
1984	5.00	3.29	2.08	6,850	4,175	2,675	117.80	807

[a]Source: USDA Economic and Statistics Service (1984).

TABLE VII
Value of Production: Major U.S. Feed and Food Grain Crops in 1984[a]

Crop	Value (million $)
Corn	20,425
Soybeans	10,979
Wheat	8,773
Grain sorghum	3,697
Barley	1,372
Rice	1,128
Peanuts	1,120
Oats	807[b]
Sunflower	412
Rye	62
Total	48,775

[a]Source: Quaker Oats Co.
[b]Grain value only.

TABLE VIII
Value of U.S. Oats in 1984[a]

Disposition	Value (million $)	Percentage of Value
Grain	807	60
Straw	228	17
Pasture	228	17
Forage	82	6
Total	1,345	100

[a]Source: Quaker Oats Co.

Over the three-year period (1982–1983 to 1984–1985), feed accounted for an average of 81% of the total oat utilization in the United States. Oats used as food accounted for 11% of the total, seed accounted for 7%, and exports absorbed the remaining 1%. Oats grain imports have become increasingly important as the higher values for the U.S. dollar encouraged foreign producers to sell to U.S. users. Imports of oats during the past two crop years have averaged 420,000 t, which is equal to 6% of U.S. production and 10 times the long-term average for imports.

V. FOREIGN TRADE

More than a million metric tons of oats move in foreign trade each year. Major exporters are Sweden, France, Canada, East Germany, and Argentina. Major importers are the United States, Switzerland, West Germany, East Germany, Japan, Italy, and Belgium. The fact that East Germany is both a major importer and an exporter may be explained by the fact that this country could be moving oats from the Western nations through East Germany into the Soviet Union or other Eastern European nations.

Table IX lists the foreign trade in oats among selected countries.

VI. NUTRITIONAL VALUE OF OATS FOR HUMANS

In the processing of oats for human food, the hulls are removed and the interior portion or groat is generally consumed as a whole grain. As consumed, oats contain the largest quantity of protein of any of the naturally occurring cereal grains and have a protein quality that compares favorably with the protein of other cereal grains. Oats have ample carbohydrate of a high digestibility, except for fiber, which has desirable physiological properties. Oats have the highest level of fat of any of the cereal grains, and the fat has a favorable ratio of polyunsaturated to saturated fats. Many of the vitamins and minerals of oats are found in the bran and germ. Since oats are generally consumed as a whole grain, these nutrients are not lost in processing.[1]

VII. HANDLING AND DISTRIBUTION IN THE UNITED STATES

Of the oats produced in the United States, 60–65% are fed on the farm where they are produced. The remaining 35–40% move into the marketplace. Oats coming into the marketplace generally are accumulated at country elevators near the point of production and moved by truck or rail into terminals. In some cases, producers may sell direct to feed or food processors and bypass the elevator system; similarly, many country elevators today are selling direct to processors and bypassing the terminal elevators. However, the terminal elevator is important in the short-term storage and merchandising of oats from producer to the ultimate user or into export channels. Some 726,000–871,000 t are purchased and milled into human food products. This human food portion currently represents 9–10% of the total U.S. production.

[1] H. B. Lockhart. 1982. Nutrition of Oats. Address to Int. Oats Conf., Am. Oat Workers, University Park, PA.

VIII. BUYING OATS OF MILLING QUALITY

The primary goal in purchasing oats for milling is to obtain grain that will assure a product of high quality. The secondary goal is to purchase oats that give a good milling yield. The best oats produced will yield 100 lb (45.36 kg) of oat groats from 160 lb (72.58 kg) of country-run oats. Because of varietal factors and poor cultural practices, some oats are so low in quality that it would take 200 lb (90.72 kg) to produce 100 lb of oat groats. To the producer and grain handler, this generally means a test weight of 38 lb/bu (48.9 kg/hl) sound count of 96% or better, and foreign material of 3% or less. The grain must be cool and sweet, with no insect infestation nor the presence of any contaminants. Other grains in the oats, such as wheat and barley, should not exceed 1.5%. Barley is especially objectionable because it does not hull in the oat huller; the hull breaks loose in the rollers, however, and slivers remain in the finished product. The moisture of the grain must be 13% or less.

TABLE IX
Foreign Trade in Oats by Selected Countries (1,000 t)[a]

Country	1979	1980	1981	1982	1983
	Imports				
United States	11	16	13	27	277
Switzerland	145	129	115	128	148
West Germany	109	107	84	92	96
East Germany	200	111	37	33	82
Japan	207	177	114	95	74
Italy	118	118	58	89	64
Belgium-Luxembourg	71	60	52	54	61
Poland	110	125	40	26	39
USSR	109	258	223	42	33
United Kingdom	66	25	6	2	28
Peru	4	12	5	8	24
Denmark	22	13	11	5	23
Netherlands	13	17	30	27	22
Brazil	40	24	18	4	18
Ecuador	22	11	12	25	15
Finland	0	0	25	190	0
Other	367	226	242	356	214
World total	1,447	1,327	978	942	1,088
	Exports				
Sweden	291	282	275	331	341
France	352	292	187	218	265
Canada	63	220	176	138	245
Australia	196	153	290	472	196
East Germany	122	125	121	128	103
Finland	13	0	0	0	65
Netherlands	66	39	31	29	39
United States	35	76	95	17	11
West Germany	13	10	12	1	8
USSR	16	8	41	7	4
Norway	0	0	14	14	0
Other	198	412	1,199	603	409
World total	1,364	1,617	2,441	1,958	1,685

[a]Source: FAO (1983).

All of these factors, except test weight in years of extreme climatic conditions, can be controlled through very careful practices by the producer and the grain handlers. The extra care that has to be given by the farmer and the elevators costs time and money. The marketplace recognizes this by paying a premium price for milling-quality oats over ordinary feed-grade oats. This premium is variable from week to week as determined by the opposing forces of supply and demand, ranging from 5–20 cents per bushel (14.52 kg) (Donald J. Adair, Quaker Oats Co., *personal communication, 1984*).

IX. SUMMARY

A. sativa and *A. byzantina* are the two most widely grown of the oats species. Because of genetic crossings between these species, many current cultivars are a combination of the original white or yellow color of *A. sativa* and the red color of *A. byzantina.*

World production has continued at approximately 50 million metric tons since the early 1960s. Most oats are grown in cool climates, and the USSR is the largest producer. Oats production in the United States has declined about 30% since the early 1960s; the leading production area is the north central region. Foreign trade of oats exceeds 1 million metric tons per year.

Oats are grown for both grain and forage needs. Cultural practices include good seedbed preparation, proper fertilization, use of high-quality adapted seed, control of weeds and diseases, and good harvest and storage practices. The economic importance of oats includes not only the value of the grain, but also its value as pasture and forage for feed and straw for animal bedding.

Oats provide one of the most highly nutritious cereal grains for humans, with protein quantity and quality that are the best of all the cereals. Preliminary research indicates that the fiber content of oats can be highly advantageous in the human diet, and extensive work is underway to support this hypothesis.

LITERATURE CITED

COFFMAN, F. A. 1961. Oats and oats improvement. Am. Soc. Agron., Madison, Wis.

COFFMAN, F. A. 1977. Oat history—identification and classification. U.S. Dep. Agric. Tech. Bull. 1516. 352 pp.

FAO. 1983. Trade Yearbook. Trade in Agricultural Products—Oats. Vol. 37, pp. 127-128. Food and Agric. Org. of the U.N., Rome.

SCHRICKEL, D. J., and SCANTLAND, D. A. 1981. Oats. Pages 51-59 in: CRC Handbook of Biosolar Resources. CRC Press, Boca Raton, Fla.

USDA. Crop Reporting Board. 1984. Annual crop summary for 1984. U.S. Dep. Agric., Washington, DC.

USDA. 1984. Economic and Statistics Service. Production, disposition, value. U.S. Dep. Agric., Washington, DC.

USDA. Foreign Agricultural Service. Production Estimates Division—Grain, 1984. U.S. Dep. Agric., Washington, DC.

CHAPTER 2

BREEDING OATS FOR FOOD AND FEED: CONVENTIONAL AND NEW TECHNIQUES AND MATERIALS

VERNON D. BURROWS
Plant Research Centre
Agriculture Canada
Ottawa, Ontario, Canada

I. INTRODUCTION

Some of the breeding work directed toward improving the image and usefulness of oats both as a food and feed grain is described in this chapter. The declining area devoted to world oats production, especially in North America, is testimony enough that oats have not kept pace with other small grains in bringing sufficient monetary returns per hectare. In many respects, oats are an old-fashioned, wholesome crop that has not been groomed by breeders or food processors for international trade or for modern industrial feed and food uses. Oats have been anchored to the farm because of their thick, bulky hull, and their nutritious kernel has lacked the identified functional quality in food processing that would allow them access to high-volume modern food markets.

In the past era of surplus grains in North America, the qualities of wheat flours that allow their use in bread, pasta, or pastry products, or those of malting barley useful in making beer, have ensured a growing use in the marketplace. In contrast, a popular functional quality has not been identified for oats, and merchandisers have had to rely upon the relative cheapness and superior nutritional quality of the crop for sales. Utilization of food oats has been restricted almost exclusively to products made from rolled oat flakes or whole oat flour because conventional wheat-milling equipment cannot be used efficiently to produce refined, white oat flours of low ash content. Food processors have satisfied the relatively small market demand for rolled oats and whole grain flours by selecting high-quality shipments of oats from the feed trade. All the remaining oats have been used as feed, and the presence of the hull on the kernel reduced the energy content of the grain to less than that of corn or barley. Increasing numbers of farmers have been producing cash crops such as barley and maize for energy, and oats have often been relegated to a utility role on the farm, being grown on the marginal soils often associated with poor

13

drainage and lower fertility. Farmers like to grow oats, and many of the attributes of the crop will be discussed later. It is sufficient to say now, however, that more research must be done on the crop to prevent the loss of this valuable species.

The era of surplus grains will ultimately come to an end, and emphasis must be placed upon planning for greater cereal production to meet the needs of future populations. Increased oat production can be attained on both higher- and lower-class soils by providing the necessary monetary incentives to producers to combine good management practices with the growing of modern, productive cultivars. Farmers in many countries have recognized the useful role that oats can play in providing feed and nutritious grain on relatively poor land. As part of an overall breeding and production strategy, breeders could attempt to exploit this feature and breed better cultivars for the millions of hectares of marginal soils that are available in the world. This is not to say that oats should only be improved for less productive soils, or that breeders should abandon the needs of the feed industry; but increased emphasis should also be placed on the production and utilization of oats for food, possibly even nutritious naked-seeded oats (*Avena nuda* L.) that can be transported over greater distances as economically as wheat. The growing of oats could be revived if new markets were established in response to either a recognition of nutritional merit or any newly discovered, functional quality characteristic of the protein, starch, oil, bran, or gum components arising from a fractionated kernel.

In writing this chapter, I have refrained from compiling an exhaustive review of the breeding literature dealing with the ecological, pest, and production problems facing oat culture in the world; the specific breeding techniques that most breeders follow to combine useful traits; and the taxonomic, cytological, and adaptational relationships that exist among the many different oat species. Excellent publications are already available on these subjects (Coffman, 1961; Rajhathy and Thomas, 1974; Baum, 1977; Simmonds, 1979; Youngs et al, 1982). Instead, I have placed emphasis upon trying to develop breeding strategies both to raise the yield potential of spring-type oats and to make them a more useful and profitable crop to grow. This is not an attempt to minimize the importance of existing breeding strategies or programs at various institutions around the world for feed oats, nor does it minimize the importance of providing new cultivars with appropriate genes to protect them against the damaging effects of common diseases and pests. The problems of lodging and diseases will be discussed later in a general way because they both affect kernel quality and yield.

Oat breeders have mainly been concerned in the past with the production aspects of the crop. Emphasis has been placed on breeding for improved yield, winter hardiness, lodging and disease resistance, proper maturity, thinness of hull, seed size, hectoliter or bushel weight, plant height, and, more recently, protein content. Since the grain has not been selected extensively for functional properties, as has been done with wheat, the genes governing other functional characteristics probably reside somewhere in the genetic reservoirs of oats. The amount of genetic information used in present-day commercial cultivars represents only a very small portion of that residing in world oat collections. Oats are a very diverse crop. Many thousands of samples of different diploid, tetraploid, and hexaploid wild species, and many thousands of the more commonly used hexaploid genetic stocks and cultivars, are available in gene banks for exploitation. The amount of genetic variation that exists in these collections is not known, but the collection itself is used by breeders as a source of new or specific gene combinations. Many breeders are eager to select for specific quality characteristics when they are identified by food scientists because improved food quality will almost certainly lead to an increase in value of the crop. Unless some rather dramatic changes are made in the characteristics and uses of oats to increase value and net returns per hectare, interest in oats on the farm will dwindle and oats will cease to be a major cereal crop.

II. BREEDING STRATEGY TO INCREASE YIELD POTENTIAL

The yield of an oat crop is determined by the number of kernels produced per unit area of land multiplied by the average seed weight. The yield potential varies with the cultivar and the environmental conditions at the time of floret differentiation and grain filling. Even cultivars with the same yield potential may achieve this by different mechanisms. For example, some cultivars produce many tillers with few seeds per tiller, whereas others produce few tillers bearing panicles with many seeds. Since seeding rate can control tiller number to some extent, the number and size of seeds per panicle appear to be very important in affecting yield potential. The highest yield potential is obtained in winter rather than in spring-type oats. Winter oats are sown and germinate in autumn; they tiller profusely before slowing or stopping growth during winter, and then they resume growth in springtime. The high yield potential is made possible by the development of many tillers under the shorter photoperiods and cool, moist conditions of autumn and of large panicles with many florets under the cool, moist conditions of springtime. Even though winter oats are the preferred choice when considering high yield potential, they are not hardy enough to be grown successfully in many northern areas, where they are destroyed by winter conditions.

To obtain maximum yields of spring-sown oats in northern regions, the crop should be sown as early as possible to take advantage of abundant moisture and cool temperatures, which are conducive to the development of tillers and large flowering panicles. In Canada, however, farmers often fail to sow oats early enough because they are busy sowing crops such as wheat, barley, and rapeseed (Canola), which are of higher cash value. They sow oats late in the season, thus severely reducing yield potential because fewer florets are differentiated in response to warmer growing temperatures. Yield potential is reduced even further because the land that is used is often being sown for second, third, or fourth consecutive years in the rotation, and its nutrient and moisture status is often deficient. However, producers argue that they cannot take the time to sow oats earlier because of economic considerations. Their oat requirements for feeding livestock are frequently met by sowing more hectares.

In some years, the seeding date may be quite late because of prolonged wet weather, which poses the added problem that late-planted crops often are damaged by fungal and viral diseases. Because of these practices, official statistics on oats productivity commonly reflect crop management rather than yield potential.

This problem of reduced yield potential in oats has led me to attempt to breed a different type of oat crop whereby the producer would sow dormant spring-type seed in the autumn and have that dormant seed overwinter as a seed but germinate in early spring to take advantage of spring rains and cool temperatures. I have termed this class of oats *dormoats,* and it is intended that they be grown in climates with severe winters and short growing seasons (Burrows, 1970).

A. Dormoat Breeding

Dormoats are an experimental oat crop bred at the Plant Research Centre, Agriculture Canada, from hybrids between *A. fatua* L. (wild oat) and commercially grown cultivars of *A. sativa* L. All or a portion of the seed dormancy characteristics of wild oats have been combined with the desirable agronomic characteristics of spring-type oats. Dormoat seed and plants have been selected to resemble the oat cultivars of commercial importance, and the seed has been selected so that it will not shatter like the seed of wild oats.

Many years of selection and many crosses and backcrosses were necessary to derive strains comparable in grain and plant quality to normal spring cultivars. The initial wild oats collected in the region of Regina, Saskatchewan, Canada, when combined with the cultivar Clintland yielded an extensive array of offspring of poor seed quality, very weak straw, and high susceptibility to disease. Only after several years of growing bulk populations of the hybrid, and selecting for dormancy and spring emergence, were desirable plants that no longer shattered at maturity isolated. The dormoat breeding program is now quite large and diversified, and populations and strains of high-protein, high-fat, forage-type, daylength-insensitive, dwarf and semidwarf, naked-seeded dormoats have been synthesized for experimentation (Fig. 1).

The procedure to breed dormoats that has been followed in recent years has been to cross the most promising dormoat strains with advanced strains or cultivars of *A. sativa* that possess one or more desirable traits. The F_1 is grown in a growth room, and the F_2 and F_3 generations are grown as bulk populations in the field. The F_3 plants are cut and threshed with a small-plot combine, and all of the seed is dropped on the stubble, where the seeds and stubble are incorporated into the soil with tillage equipment. All nondormant grains and the seeds of those showing a very short primary dormancy germinate in autumn and are destroyed by tillage or by freezing temperatures during winter. In spring, a sparse dormoat population usually emerges and produces seed. These plants are again harvested but instead of scattering all of the seed on the soil for the second round of selection for dormancy, only one-quarter to one-half of the seed is replanted in autumn. The dormoat stand the following spring is usually more dense and in some hybrids may equal a commercial sowing. Panicle selections are then made, and these are planted unthreshed in hills in autumn. Those showing the highest emergence of seedlings the following springtime are retained. Seeds from these hills are selected for visual attractiveness and chemical quality, and the best strains are evaluated in following years for agronomic qualities in rows, plots, and replicated yield trials.

The program is now at the stage where the yields of the best spring-sown dormoat strains approach, and in some cases equal, the yields of the best conventional spring oat cultivars. When autumn-sown dormoats were first tested

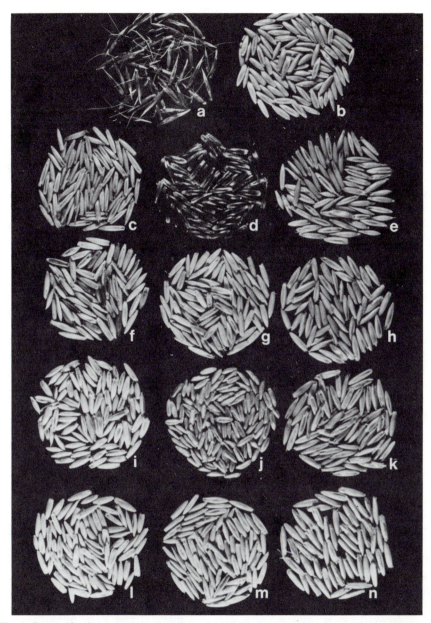

Fig. 1. Progressive improvement of seed quality of dormoat strains as a result of crossing and backcrossing (one to six times) to selected parental cultivars of *Avena sativa*. (a) *A. fatua* wild oat parent and donor of seed dormancy genes; (b) Clintland parent, first donor of desirable agronomic genes; (c) D-124, one of the first nonshattering, yellow, but small-seeded strains; (d) B-47, another selection with black hulls from the same hybrid; (e) DC14 from the first backcross (Roxton); (f) a dormoat bulk population segregating for seed color from the second (Gemini); (g) OA455 from the third (OA123-44); (h) OA422 from the fourth (Scott); (i) DB78-28-1, a long-peduncled dwarf from the fifth (OT207); (j) DB79-3, a naked-seeded dormoat from the fifth (3932-16); (k) FTDO-1107, a forage-type dormoat (Foothill) from the fifth; (l) HPDO-2082, a high-protein strain from the sixth (Hinoat); (m) HFDO-864, a high-fat-content dormoat from the sixth (CI4492); (n) DIDO-119, a daylength-insensitive dormoat from the sixth (OA309) backcross to *A. sativa*.

for yield potential, they outyielded the same nondormant, spring-sown dormoats sown at the normal time by as much as 10–25% (Burrows, 1970). If wet weather conditions force very late spring planting, the autumn-sown dormoats even outperform the standard commercial cultivars. It is hoped that the yield potential of dormoats can be raised through breeding and that, when this is accomplished, the new autumn-sown dormoats will consistently outyield the best commercial oats by 10–25%.

The advantages of sowing dormoats are not restricted to improving yield potential. Sowing dormoat seeds in autumn is convenient for growers, and better use is made of labor and machinery resources. Since some of the dormoat seeds of the best strains fail to remain dormant in autumn, they grow and can be grazed by livestock. Tillage late in the season just before winter incorporates fertilizer, destroys winter annual weeds, and aerates the dormoat seed. The land is not worked in spring because the oats emerge normally before it is possible to put heavy equipment on the wet land, and the lack of tillage helps reduce the loss of valuable soil moisture, especially in dry years. The seed quality of autumn-sown dormoats is most often superior to that of spring-sown oats because the plants usually flower earlier in the season (Fig. 2) and thus tend to avoid diseases and the high temperatures common at flowering time. The plants are free of smut (*Ustilago avenae* (Pers.) Rostr.) because the spores do not overwinter on dormoat seed in soil in Canada, and the incidence of crown rust (*Puccinia coronata* Cda. f. sp. *avenae* Erikss. and Henn.) and stem rust (*P. graminis* Pers. f. sp. *avenae* Erikss. and Henn.) is very much less than on the same strains sown in springtime. Barley yellow dwarf virus (BYDV) is not as common in dormoats as

Fig. 2. Difference in heading time in 1966 of dormoat-26 plots sown with dormant seed in the autumn of 1965 (left) and with nondormant seed in the spring of 1966 (right).

in spring-sown oats because viruliferous aphids are not as prevalent in early spring to infect the succulent dormoat seedlings. By the time aphids are plentiful, they prefer to feed upon the young, succulent, spring-sown oats rather than the older, less tender dormoats. Lodging also is less because the lower internodes are shorter and stiffer in response to growing at the lower temperatures of early springtime, and the straw is not weakened as much by disease-inducing organisms. If wet conditions occur at harvesttime, dormoat kernels fail to sprout in the standing crop or in the swath.

The ability to grow dormoats successfully without using chemical seed protectants or possibly herbicides is particularly attractive for economic and health reasons. Dormoat seed does not need to be treated with fungicides because, like the wild oat itself, it has been selected to resist attack from microorganisms in soil. Even if some seed loss occurs, especially to kernels damaged in threshing, such losses can be overcome by excessive tillering of surviving plants in springtime. Experience at Ottawa has shown that when good stands of dormoats emerge in springtime, they do not need to be sprayed for the control of annual weeds. This will probably not be an advantage in all areas because it depends upon the germination characteristics of the different species of annual weeds present in the soil.

When this breeding project began, it was hoped that strains of dormoat seeds could be bred to act in unison by remaining completely dormant in autumn, holding the dormancy overwinter, and breaking dormancy and growing luxuriously in springtime. Although significant advances have been made toward this objective and superior strains such as line PGR8658 (Plant Gene Resources of Canada Office, Agriculture Canada, Ottawa, K1A 0C6) have been isolated, it seems clear now that a seed management step must be introduced into the protocol to synchronize the seeds so that they are all in the same physiological state as they enter and emerge from winter.

The duration of primary dormancy for some strains is measured in days, whereas for other strains it is measured in months. Those strains with short primary dormancy are of major interest because they probably do not present the same weed hazard as the strains with a long after-ripening requirement. However, if the seeds lose their primary dormancy before an adequate level of secondary dormancy is induced after they are sown in autumn, germination begins and the seedlings are destroyed by winter.

At least three main physiological processes must be satisfied in the seed before seedlings will emerge from soil in springtime. First, freshly harvested seed must pass through an after-ripening period to overcome primary dormancy. Its completion is a prerequisite to germination. Second, sometime during or just after the completion of the after-ripening period, the seed must be induced into a type of secondary dormancy either naturally in soil or artificially before planting. This is necessary so the seed will not germinate in autumn. Third, this dormancy must be released during winter or early spring to permit spring germination. At present, it is not known whether these required changes within the seed are under hormonal control or are due to structural changes in cell membranes within the embryo and endosperm tissues.

Investigations at Ottawa are now centered on finding an efficient method to induce a proper level of secondary dormancy. Exposure of the moistened seed to a heat stress ($+30-40°$ C) or to a reduced oxygen supply (Sawhney and Naylor,

1980) before planting show promise, but sowing freshly harvested seed immediately after harvest produces the best, but still inadequate, spring emergence for many strains. What role winter conditions play in the breaking of dormancy is not clear, but it may be related to the drying effects of freezing. Experience has shown that the moisture content of seed has a profound effect on its ability to germinate, and drying imbibed dormant seed improves its ability to germinate at a later date. This is also true of parental wild oats.

The weediness characteristic of dormoats could present a seed contamination problem to some growers. It is almost certain that some dormoats will lie dormant in soil for one or more years before germinating. Pedigree seed growers and farmers growing quality crops such as wheat or malting barley may consider the risks of contamination too great to grow dormoats even though they use wild-oat herbicides. Growers of feed grains, hay, or plant silage may consider volunteer oats of no importance in their operation and may welcome the opportunity to grow a higher yielding dormoat.

I have not yet conducted inheritance studies on seed dormancy, but Jana et al (1979) have reported that at least three genes control the primary dormancy duration in six *A. fatua* hybrids. They concluded that a gene E confers early after-ripening and is partially dominant. Two other genes, L_1 and L_2, confer slow after-ripening and are also partially dominant. L_1 is epistatic to E, whereas L_2 has only a mild suppressing effect in the presence of E. Jana et al (1979) concluded that the heritability of dormancy of primary seeds in four wild-oat populations was approximately 50% under field conditions.

The potential advantages of dormoats over conventional spring-type oats are many, but the crop remains experimental. Commercialization could come rapidly if a practical seed management protocol were formulated to synchronize seed germination behavior so that complete stands of oats could emerge from soil in springtime.

B. Breeding Oats for Intensive Management

Even though emphasis in this chapter is on breeding superior oats for marginal soils, it is obvious that if grain yields can be increased significantly by intensive management, both covered and naked-seeded oats would compete with other cereals for productive land. The extent to which yields would have to be increased to provide an economic return to the grower would depend on many factors, including the value of land and the price and productivity of the cereals. High yields of other cereal grains have been obtained by breeding and growing dwarf and semidwarf cultivars. Dwarf oats have been reported (Warburton, 1919; Derick, 1930; Florell, 1931; Litzenberger, 1949), but they were either too dwarfed or genetically unstable because of chromosomal irregularities (Goulden, 1926; Huskins and Hearne, 1933). Dwarfing in oats has been reported to be caused by a single recessive gene (Warburton, 1919; Derick, 1930; Litzenberger, 1949), whereas in other plant material, dwarfing is caused by a dominant gene (Stanton, 1923b).

Brown et al (1980) reported the isolation of a vigorous new dwarf line of oats, OT207, from progeny resulting from irradiation of OT184 with fast neutrons. They reported that OT207 possesses 42 chromosomes and that meiosis and mitosis were normal. Dwarfing is conditioned by a single dominant gene designated *Dw*6, and this conclusion has been confirmed by me (*unpublished data*). *Dw*6 represents a major discovery, but OT207 possesses three undesirable traits: in most environments, the peduncle is too short to extend the entire panicle outside the leaf sheath, which causes sterility of the trapped, lower-panicle florets; the crop matures later than standard commercial, tall cultivars; and the seed size is too small to be attractive to growers.

The effect of the *Dw*6 gene is to increase tillering and reduce the internode length of the culms. Panicle size is smaller than that of OT184 because of a shortening in pedicel lengths; the seeds are also short and stubby compared with those of OT184. The physiological mechanisms by which this gene operates are not known, but the extra tillering might be interpreted to be the result of a reduction of apical dominance caused by either a reduction of plant hormone synthesis at the apex or a reduction of basal hormone transport in each tiller. Kolb and Marshall (1984) found fewer cells in the peduncle of OT207 dwarf plants than in tall plants, but it is not known whether the mitotic activity of an entire dwarf plant has been reduced by the dwarfing gene. A reduction in mitotic activity of each tiller may be compensated for by mitotic activity in more tillers. It is tempting to propose that the dwarfing gene may act by reducing basal plant hormone transport in each tiller, resulting in a reduction of intercalary meristematic divisions and a release of apical dominance over the basal tiller bud. At Ottawa, I have found that spraying dwarf plants immediately before panicle emergence with a single application of gibberellic acid (GA$_3$ at 50–100 ppm) caused an elongation of the uppermost internode and peduncle, resulting in a complete emergence of the panicle, high floret fertility, and elongation of the developing seeds. The rapid growth response probably indicated that the applied GA$_3$ acted by elongating preformed cells rather than by stimulating new cell divisions. Field spraying of dwarf oats with GA$_3$ solution could provide a means of attaining maximum grain yields, but such a procedure is expensive in terms of the costs of GA$_3$, petroleum, equipment, and labor.

Another genetic solution would involve the discovery of a modifier gene or genes that would be almost completely quiescent throughout early plant development and permit full expression of *Dw*6, but that would become operative later when the peduncle was beginning to elongate and stimulate it to project the panicle outside the leaf sheath. The modifier would thus mimic the effect of a late application of GA$_3$. The isolation of such a gene in rice has been reported by Rutger and Carnaham (1981).

I have made several hundred hybrids between OT207 and commercial oat cultivars of American, Canadian, and European origin in search of such a modifier gene, but none of the dwarf progeny from any of these hybrids displayed the desired long peduncles. Some of the dwarf progeny derived from hybrids with tall cultivars such as Lodi, Roxton, and a Canadian naked-seeded genetic stock 3932-16 were taller than OT207, but they could not be classified as having long peduncles. Success was achieved, however, when OT207 was crossed with a very tall dormoat strain DC739, which probably obtained the long-peduncled trait from *A. fatua* (Fig. 3). Presumably, the long-peduncled trait had been eliminated from standard-height commercial oat cultivars because it undoubtedly led to excessive plant height and susceptibility to lodging.

This trait is easy to transfer and identify in selections, and long-peduncled, dwarf, naked-seeded, daylength-sensitive and daylength-insensitive (see next section) oats and dormoats have been developed. Comparison of the growth characteristics of two daylength-insensitive dwarf lines bred at Ottawa and grown at the Agriculture Canada winter (November to April) oat nursery at

Fig. 3. Representative panicles of dwarf oat plants with long (left), intermediate (center), and short (right) peduncles.

Brawley, California, is illustrated in Fig. 4. PGR12646 is a short-peduncled strain, like OT207, and PGR12647 is a long-peduncled strain. The panicle length of PGR12646 was 34.5 cm, whereas for PGR12647 it was 38.5 cm. The height of the peduncle node above the soil was 48.2 cm for PGR12646 and 49.2 cm for PGR12647. The most spectacular difference between the two strains was in peduncle length, which was 25.9 cm for PGR12646 and 51.2 cm for PGR12647. These results suggest that growth and differentiation of the panicle and stem internodes are only slightly affected by the presence of the long-peduncled trait during the early stages of plant growth; later in the development of PGR12647 plants, however, either $Dw6$ is "turned off" and mitotic activity in the intercalary meristem is allowed to proceed normally, or excess growth hormone (possibly even a different growth hormone) is synthesized, resulting in a marked increase of peduncle length.

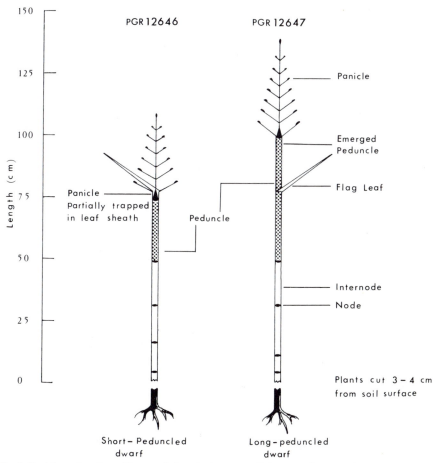

Fig. 4. Panicle, peduncle, and internode lengths (average of 36 stems) of short (PGR12646) and long (PGR12647) peduncled, daylength-insensitive, dwarf-type oat cultivars grown from early November to early April (1981–1982) at Brawley, California.

I determined that the long-peduncled trait was controlled by a single recessive gene designated *pl*1. The frequency distribution of three F_2 populations derived from three hybrids between PGR12646 and PGR12647 grown in California was decidedly bimodal, with approximately three-quarters of each population being short-peduncled and one-quarter long-peduncled. The F_3 lines derived from these F_2 plants were grown in Ottawa, and each population segregated in a ratio of 1:2:1, as illustrated in Table I. In some breeding populations, the desired long-peduncled segregate occurs less frequently than one in four. Presumably, other genes present in some parents affect the expression of the peduncle extender gene *pl*1.

Marshall and Murphy (1983) have described a new, partially dominant gene *Dw*7 and postulated that it may be useful in developing semidwarf, lodging-resistant cultivars. *Dw*7 is believed to have resulted from a mutation in the winter oat line NC2469. The new dwarf, NC2469-3, resembles the cultivar Scotland Club in having a short, stout culm; a compact, club-type panicle; occasional dead spikelets because of weak, twisted pedicels; and occasional spikelets with an absent glume. The authors state that the dwarf had excellent lodging resistance even when fertilized with high levels of nitrogen but that the grain yields of dwarf-type lines derived from various crosses were below those of current taller cultivars. The main weaknesses of the dwarfs are fewer seeds per panicle, low grain test weight, and a tendency to shatter grain during ripening.

The breeding of high-yielding, lodging-resistant semidwarfs is not a simple matter of just including the dwarfing and long-peduncled genes into commercial cultivars. Often, grain yields are lower and straw quality is inferior to that of tall cultivars. Dwarfs without disease resistance genes also appear to be more susceptible to pathogens such as rust and septoria than tall susceptible plants, possibly because growth in length of the straw is inhibited but photosynthesis proceeds and substrates may accumulate in the tissues, providing a better medium for the growth of pathogens. There also appears to be greater moisture condensation on the leaves of rather densely packed dwarf plants, promoting an increase of spore germination that results in an increase of infection sites in the leaves. Breeders must thus concentrate on breeding high-yielding dwarfs with good-quality straw and with appropriate resistance genes.

The discovery and distribution of *Dw*6, *Dw*7, and *pl*1 genes have stimulated several breeders to begin development of new semidwarf oats. Steady progress has been made in raising the yields of the new long-peduncled semidwarfs at

TABLE I
Segregation of F_3 Lines Derived from Three Separate Hybrids Made Between a Short (PGR12646) and a Long (PGR12647) Peduncled Dwarf Oat Parent

Population	Short Peduncle		Segregating Short and Long Peduncle		Long Peduncle		Total	χ^2
	Observed	Expected	Observed	Expected	Observed	Expected		
1	44.00	41.75	78.00	83.50	45.00	41.75	167.00	0.73[a]
2	49.00	48.00	94.00	96.00	49.00	48.00	192.00	0.08
3	65.00	65.50	129.00	131.00	68.00	65.50	262.00	0.13
Total	158.00	155.25	301.00	310.50	162.00	155.25	621.00	0.63

[a] All chi-square values not significant.

Ottawa, and several of them are being evaluated in official tests. Marshall and Kolb (1983) and Marshall et al (1983) have registered Pennline 116, which possesses *Dw7* and two semidwarf germ plasm lines, Pennlo and Pennline 6571. They suggest that all three releases may be useful as parents to produce cultivars highly resistant to lodging.

C. Breeding of Daylength-Insensitive Oats

One of the great achievements of international programs to breed superior wheat and rice cultivars for the different areas of the world has been the isolation and use of genes governing daylength insensitivity. Daylength-insensitive (DI) cultivars develop and flower normally over a wide range of photoperiods and thus can be grown fairly successfully at all latitudes if conditions of temperature, rainfall, soil fertility, and the activity of biotic agents permit. Such cultivars are not strictly insensitive, but they are less so than sensitive cultivars to photoperiod as it affects the number of days from seeding to panicle emergence. Because plants with DI genes flower normally at different latitudes, it is easier to transfer or exchange cultivars and breeding stocks from one area of the world to another. These genes also make it possible for the breeder to grow and process two breeding nurseries per year: one at home in summer and another at a different latitude in winter. Although one can accomplish the same thing with daylength-sensitive (DS) cultivars by growing plants both north and south of the equator in different seasons, or by using artificial field lighting to correct the local photoperiod, the added costs of transportation and lighting equipment may tend to limit the size of such breeding nurseries. In addition, plants possessing DI genes may prove more useful in solving specific production problems or creating new cropping possibilities not possible with DS cultivars in northern regions.

In 1965, I began a search for DI oat genes and chose as a source an early flowering sample of *A. byzantina* Koch. (CAV2700) collected by the Canada-Wales Oat Collecting Expedition from Bodrum, Turkey (Baum et al, 1975). The trait was transferred into DS Canadian cultivars, which permitted the growing of an oat winter nursery from November to late April at the USDA Imperial Valley Station, Brawley, California, without using artificial lights to shorten the night period to promote flowering.

Plants of DS cultivars grown at Brawley displayed excessive tillering, very thick stems, very wide leaves, extreme plant height, delayed flowering, and a very abnormal elongated panicle with very low floret fertility (Fig. 5). The degree of abnormality varied with the cultivar or strain grown, with some even failing to display any flowers. In contrast, DI plants flowered normally. They also tended to tiller well and produced very robust stems and wide leaves (Fig. 5).

A single major dominant (designated *Di1*) gene appears to condition the DI reaction in oats derived from CAV2700, although it is sometimes difficult to demonstrate a 3:1 DI-DS ratio in spaced F_2 plants in California because of the disturbing effects of BYDV on flowering expression. Classification of two F_2 populations derived from crossing the homozygous DS strain DC 739 with DI homozygous strains OA400-2 and OA405-5 are illustrated in Table II. The hybrid DC 739 × OA400-2 data are in support of single-gene involvement, whereas the data from DC 739 × OA405-5 do not confirm a single-gene model. However, F_2 seed of DC 739 × OA405-5 was also grown in Canada under long

TABLE II
Segregation of F_2 Populations of Hybrids from Crosses of Daylength-Sensitive (DC739)
and Daylength-Insensitive (OA400-2, OA405-5) Parents

Sensitive Female Parent (DC739)	Insensitive Male Parent					
	OA400-2			OA405-5		
	Insensitive	Sensitive	Total	Insensitive	Sensitive	Total
Observed	1,820.00	597.00	2,417	2,052	624	2,676
Expected	1,812.75	604.25	2,417	2,007	669	2,676
χ^2	0.03	0.09	0.12	1.01	3.03	4.04
P			0.90–.95			0.02–.05

days and returned to California the following year, where F_3 lines were scored for photoperiodic reaction (Table III). The F_3 data are in agreement with a single dominant gene conditioning daylength insensitivity. Experience derived from processing several hundred other hybrids between DI and DS parents has demonstrated that about one-quarter of F_2 progeny fail to flower in California, although in some hybrids other modifier genes may be operating.

Many of the improved strains of DI oats failed to yield as much grain as standard DS cultivars when grown in the north. This was not surprising because

Fig. 5. Contrasting flowering behavior of daylength-sensitive (A) and daylength-insensitive (B) oat strains grown from early November to early April at Brawley, California. Both strains flower normally at Ottawa.

TABLE III
Segregation of F$_3$ Lines of Hybrid from Crosses of Daylength-Sensitive (DC739)
and Daylength-Insensitive (OA405-5) Parents

	Class			
	Insensitive	Segregating	Sensitive	Total
Observed	238.00	424.00	233.00	895
Expected	223.75	447.50	223.75	895
χ^2	0.91	1.23	0.38	
P				0.20–.30

CAV2700 did not possess good agronomic characteristics, and many lines displayed seed shattering. By repeated crossing to improved DS cultivars and selection of improved types in both Canada and California, yield levels advanced; but evidently DI oats flowered too early in the north to attain yields equal to those of the best DS cultivars. Emphasis was then placed upon selecting for late maturity in Canada, and this had the effect of raising both grain yield and seed size, presumably because of a longer filling period. Some of the more advanced strains have performed well in standardized yield trials in the United States and Canada, and some show good milling quality. The first DI cultivar, Donald (PGR8643, CI9361), was licensed in Canada in 1982 for commercial production and is proving to be an excellent oat for milling purposes. It is anticipated that yield levels will even be higher in Canada when rust resistance is incorporated into these DI strains. The large seed size of many improved DI strains has been transferred to both dwarf and naked-seeded oats, which has improved their yields.

High kernel weight and early maturity of DI oats may be of real benefit to farmers who grow mixtures of oats and barley. The maturity of DI oats matches the maturity of barley better than the maturity of DS cultivars, and the heavy seed weight of DI oats is closer to that of barley, making it easier to regulate wind velocity in the combine to prevent losses of oats at harvest. Mixtures of oats and barley are grown in eastern Canada for feed purposes, and total yields of mixed grain are reported to be higher because of better utilization of land and better control of fungal diseases (Zavitz, 1927; Clark, 1980).

DI oats may also prove useful for late seeding when land is double-cropped in one season. Normally, spring oats are sown in early May in Ottawa and harvested in mid-August, with first frosts arriving in late September (a 144-day frost-free period). As stated earlier, dormoats get an early start and mature in late July or early August. Harvest of this crop followed by immediate planting of DI oats alone or in combination with dormoats would allow sufficient time to produce a crop of DI oats for forage in autumn before killing frosts. If properly managed, the dormoat seed would not emerge until the following spring. DI oats will flower properly under the shorter daylengths experienced this late in the season, whereas DS cultivars will not flower before frost. This may prove to be an important role for DI oats by helping farmers in northern regions better utilize the full growing season and increase net returns per hectare.

Another benefit derived from the establishment of the California winter oat nursery has been the opportunity to select segregating plant populations for resistance to BYDV. Oats are space planted in fertile soil in early November in California and are irrigated, causing them to tiller profusely. Viruliferous aphids feed on the leaves of the tender tillers that are produced all winter, and the virus multiplies in each plant. Virus infection results in the production of red-leaf symptoms and reduction in root growth, which becomes very obvious when irrigation water is withheld in late March to ripen the crop. Susceptible plants that have been able to grow with reduced root systems simply die, whereas resistant plants mature normally. The strains of BYDV present in California may not exactly match those present in the northern United States or Canada, but oats selected in this nursery display a great deal of resistance in the province of Ontario and in specific regions of western Canada.

It is still too early to determine whether DI will mean as much to oat improvement as it has to wheat and rice improvement. Research and breeding work with DI in oats is in its infancy, but already significant advances have been made, possible unique uses are being explored, and seed samples have been distributed to interested cooperators around the world.

D. Breeding of Naked-Seeded Oats

The oat hull (lemma and palea) that covers the groat constitutes a very serious constraint in the resurgence of oats as a major industrialized crop. The hull is high in crude fiber, and thus it both lowers the intrinsic energy content of the grain for feeding purposes and decreases its bulk density so that covered oats are uneconomical to store and transport over long distances. However, the covering hull is a useful envelope to keep the groat clean for food purposes and to protect the groat from mechanical damage during threshing. Pulverized hull material also serves as an important source of crude fiber in the formulation of certain feeds and as a raw material for the chemical industry in the production of furfural. The thick, lignified, tightly adhering hulls are usually separated mechanically from the groat by specialized dehulling machines located in processing factories.

Cultivars vary considerably in percentage of hull, even when the plants are grown under ideal conditions, but the variation is even greater when environmental conditions affect the extent to which the groat is filled. Hull values of 23–28% are common in some areas of western Canada, for example, but values of 28–34% are common in eastern Canada. Breeders use hull percentage values, determined in the laboratory, to calculate groat yields as part of the strain selection process. Food manufacturers use these data but they also evaluate cultivars, or oat samples, by the weight of covered oats required to produce 45.36 kg (100 lb) of groats. This weight is usually more than that calculated from the hull percentage value because of groat losses incurred during dehulling and cleaning. Food processors are anxious to purchase oat cultivars that dehull efficiently. Breeding efforts are directed toward reducing the hull percentage to as low a value as possible without making the hull so thin and membranous that a substantial portion of the kernels are dehulled during threshing. The embryo of a dehulled groat is often damaged or the kernel broken if cylinder speeds are too high and clearance tolerances set too narrowly in the combine harvester at the time of threshing. These kernels represent a loss for seed purposes because of low germination and a loss for food purposes because the seed develops enzymatic rancidity when damaged.

The proper use of hull-less genes offers the possibility of producing oats that can be threshed completely free of hulls on the farm. These genes affect the development of the spikelet so that the lemma and palea of each floret are very thin and membranous and the spikelet is multiflorous, containing as many as four to eight florets rather than the usual two or three florets in standard covered-seeded types of oats. Upon gentle threshing, all or most of the groats in each spikelet are separated from hulls (Fig. 6). Temperature has been found (Lawes and Bowland, 1974) to play a role in the expression of the hull-less genes, with warm temperatures favoring and cool temperatures reducing full expression. The genetic constitution of the cultivar also plays a role in expression (Lawes and Bowland, 1974), so breeders were advised by Moule (1972) to select segregating progeny for maximum expression of nakedness.

The inheritance of the naked-seeded condition has been reported by Moule (1972), Jenkins and Hanson (1976), and others. Nakedness is incompletely dominant, and panicles of F_1 plants from crosses between naked and covered cultivars are structural mosaics, with the spikelets uppermost on the panicle being multiflorous and those at the base being of the normal *A. sativa* type. Both

Fig. 6. Naked-seeded oats of improved large-seeded strain OA504-6 grown at Ottawa in 1979. Many hairs (trichomes) were removed from the surface of each kernel during cleaning and handling of the grain.

authors propose a principal or "switch" gene N, which if present in the homozygous condition results in naked kernels. The presence of nn leads to complete suppression of the multiflorous condition that is present in early ontogeny, and all plants are covered. Moule proposed two modifier genes, N_1 and N_2, and Jenkins and Hanson accepted Moule's basic model but proposed the involvement of a third modifier gene to explain the segregation of nakedness when the switch gene N is in the heterozygous condition (Nn). The third gene, N_3, is proposed to operate in the opposite direction to the modifiers N_1 and N_2; when homozygous and dominant (N_3N_3), it results in the differentiation of covered grain. When absent (n_3n_3), covered grain will be produced if a single dose of either N_1 or N_2 is in association; but if four ($N_1N_1N_2N_2$), three ($N_1N_1N_2n_2$ or $N_1n_1N_2N_2$), or two ($N_1n_1N_2n_2$, $N_1N_1n_2n_2$, or $n_1n_1N_2N_2$) doses of the modifiers are present, the kernels are naked. This model results in an F_2 segregation ratio of naked to covered grain of 35:29, which was a close fit to the observed frequencies in hybrids between the covered cultivar, Maris Oberon, and two naked experimental strains and three naked cultivars.

Naked oats originated in China, where they have been grown for centuries (Stanton, 1923a), but several constraints have been identified to explain why they have failed to gain acceptance in other parts of the world. Groat yields have been too low; expression of the naked condition has not been complete; seed size is too small; trichomes (hairs) on the surface of the groat are liberated during threshing and handling and act as skin and respiratory irritants to operators; storage of groats too high in moisture content can result in a loss of germination capacity, heating, development of rancidity, increased mycofloral growth, and a higher incidence of mites and insects. Markets for widespread use of naked oats have not developed because interested users have not been guaranteed continuity of supply and producers do not have a market demand to satisfy.

Many of the problems associated with the proper storage of naked oats can be more readily solved today than in previous decades. Most storage problems have been found to be related to the moisture content of the groat and the extent to which the groat was damaged in the threshing operation (Welch, 1977; Sinha et al, 1979a, 1979b). The availability of this knowledge and the presence of superior grain-drying facilities and harvesting equipment on the farm that can be properly adjusted to prevent groat damage and protect the operators from dust should make it relatively easy to store and handle naked oats. Breeders can help overcome some of these problems by producing cultivars that are of the proper maturity for the area of production to minimize moisture content of grain at harvest; whose groats are relatively free of trichomes to increase hectoliter weights (closer packing of seeds) and reduce the amount of grain dust when handling the seed; and whose groats have recessed embryos to minimize threshing damage. Breeding for proper maturity is a relatively simple matter, but breeding for "bald" groats and groats with completely recessed embryos will be somewhat more difficult. Variation does exist in the amount of surface hairs on oat groats of strains catalogued in the international oat collections. Seed of accessions CN17825, CN17826, CN17827, CN17828, and CN17829 have been identified by me as having few or short surface trichomes, and they are being used as parents in the naked oat breeding program at Ottawa. A flotation technique has been published by Comeau and Dubuc (1977) to separate groats with different amounts of surface trichomes. Variation also exists in the shape of the kernel: many of the groats of older cultivars are slightly crescent shaped with elevated embryos, making them vulnerable to damage, whereas the seeds of some of the more modern cultivars lie flat on a plane surface. Their embryos do not protrude to the same degree and thus are not as vulnerable to damage (Fig. 6).

Variation in seed size is more of a problem in naked cultivars than in covered cultivars because of the greater number of florets per spikelet. The basic shape of the multiflorous spikelet is pyramidal, and higher ranking seeds at the top of the spikelet present special problems. If they are naked seeded, they are quite small and are often lost during threshing. Unfortunately, many of them retain their hulls after threshing and are difficult to remove during cleaning of naked-seeded samples. Attempts by many breeders to prevent the formation of the small kernels by combining the naked-seeded trait with the normal *A. sativa*-type spikelet have failed because the multiflorous and naked traits are probably conditioned by the same genes. I have been able to solve the problem of preventing the formation of small seeds in another manner, by breeding for very large-seeded naked oats. Hybrids between very large-seeded oat strains, derived from the DI breeding program, and normal naked oats have yielded some large-seeded, naked strains (Table IV) that are virtually free of covered seeds and produce usually one, two, or three groats per spikelet. The spikelet morphology remains multiflorous, but the smaller florets at the top abort early in development, thus permitting all nutrients to be translocated to the developing primary, secondary, or tertiary kernels.

An analysis was made of the size and frequency of these kernels taken from the primary, secondary, and tertiary florets of panicles on one of the large-seeded naked strains, OA504-5-21, grown at Ottawa in 1981. The average kernel weight of all the grain in the sample was 37.4 mg, but the primary, secondary, and tertiary groats averaged 44.2, 34.0, and 23.5 mg, respectively, and these grains comprised 48.6, 37.2, and 14.2%, respectively, of the total number of seeds in the sample. The high percentage of large groats (primary plus secondary) in the sample should prove to be an attractive feature to growers and food processors because large seeds contribute to increased harvestable yields, smaller losses when cleaning and grading, uniformity of sample, and improved movement of air through the sample during grain drying and storage. Even though the grain yield levels of a few of these large-seeded naked oats appear satisfactory (Table

TABLE IV

Groat Yield and Kernel Weight of Terra (Naked-Seeded) and Scott (Covered) Cultivars and of Four Large and Naked-Seeded Strains of Daylength-Insensitive Origin[a]

Cultivar	Yield (kg/ha)			Kernel Weight (mg/seed)		
	Ottawa	Inkerman	Avg.	Ottawa	Inkerman	Avg.
Terra	2,061	2,130	2,096	22.8	19.2	21.0
Scott[b]	2,497	2,839	2,668	23.6	20.9	22.2
OA504-5	2,870	3,113	2,991	37.0	31.0	34.0
OA504-6	2,552	2,739	2,646	41.6	42.3	41.9
OA507-2	2,287	2,791	2,539	41.2	29.2	35.2
OA507-3	1,848	2,339	2,094	44.2	31.8	38.0

[a]Strains were OA504-5, OA504-6, OA507-2, and OA507-3. Oats were grown in replicated field trials at Ottawa and Inkerman, Ontario, Canada, in 1979.
[b]Groat yield of Scott calculated using an experimentally determined hull percentage value of 30.1.

IV), a more complete appraisal of the value of breeding for fewer but larger kernels in multiflorous naked oats will have to be delayed until the genes controlling this trait are incorporated into many different genetic backgrounds and improved strains are evaluated under a wider range of environmental conditions.

A distinctive naked oat with an exaggerated multiflorous spikelet possessing up to 10–12 florets (PGR11621) has been isolated in the Ottawa breeding program (Fig. 7). The lineage of this oat is complex, genes having been contributed by strains of *A. nuda, A. strigosa* Schreb., *A. byzantina, A. fatua,* and *A. sativa* over the past 60 years of breeding at the research station. PGR11621 was selected in 1980 as a long-peduncled, naked-seeded dwarf because of its unique spikelet, which is typically pyramidal at the tip but having a basal portion that tends to resemble a short spike; the seeds taken from this region tend to be fairly uniform in size. I have called this spikelet a "chevron-type" spikelet because of its similarity in appearance to the chevron pattern of a military badge. The chevron spikelet is not confined to long-peduncled dwarfs but has been found on semidwarfs and tall oat plants (Fig. 8), and it is expressed under both glasshouse or field conditions. The discovery of this oat is too recent to predict the impact such a spikelet will have on the yield of naked oats, but its isolation is probably indicative that new forms of naked oats will arise as the genetic base and amount of breeding effort are increased for this crop.

Several investigators (Christison and Bell, 1980; Cave and Burrows, 1985; Myer et al, 1985; Morris and Burrows, 1986) have concluded that naked oats can replace all or a large proportion of the corn, wheat, and soybean meal in poultry and pig rations. Naked oats are a high-energy, high-protein, nutritious cereal

Fig. 7. Spikelet development of a typical covered oat (left) and an exaggerated, multifloretted, naked-seeded oat (right). This new spikelet form has been called a "chevron-type" spikelet by the author.

grain. The metabolizable energy content of the oat groat is comparable to that of corn (apparent metabolizable energy: 3,240 kcal/kg, Christison and Bell, 1980; true metabolizable energy: 3,475 vs. 3,365 kcal/kg, Sibbald, 1983; 3,370 vs. 3,350 kcal/kg, Dale and Fuller, 1985) largely because of the relatively high groat lipid content. Naked oat cultivars such as Tibor (Burrows, 1986; Morris and Burrows, 1986) and the experimental cultivar Coker 82-30 (Myer et al, 1985) have been reported to have protein contents of 16.8–19.5%, which is in the correct range for pig and poultry grower rations. L. Poste, Food Research Centre, Agriculture Canada, Ottawa, compared the quality of meat (longissimus dorsi muscle) taken from pigs fed a standard grower corn-soybean ration with that taken from pigs fed Tibor oats as the sole source of energy and protein (*personal communication*). Experienced panelists concluded that the meat from the oat-fed pigs possessed a significantly higher flavor intensity, tenderness, and juiciness.

The encouraging results obtained in the feeding of oat groats to domestic animals have motivated breeders to attempt to solve some of the agronomic

Fig. 8. Chevron-type, naked-seeded, long-peduncled dwarf (a), semidwarf (b and c), and tall (d) oats grown under glasshouse conditions at Ottawa (1981).

problems associated with naked oats. Cultivars such as Terra and Tibor in Canada, Rhiannon in the United Kingdom, Manu and Caesar in Germany, and Nuprime in France represent improvements in groat yields over older cultivars. Progress has been sporadic because of the lack of sustained effort due to the dearth of recognized, well-regulated markets. In North America, in the past 20–30 years, there has been little incentive to build a marketing system for naked oats because both Canada and the United States were in a surplus grain position. However, as the demand for cereal-based products for food and specialized feed increases, producers will probably make a greater effort to utilize the more marginal lands for food production. The naked oat would be a great candidate for this purpose. I believe that the possibility of supplementing world grain products with a nutritious, transportable crop derived from the more marginal lands, or from good lands where oats are used effectively in crop rotations, is a compelling argument for putting renewed effort into the breeding of naked oats.

III. PROTECTION OF HIGH YIELD POTENTIAL

In all breeding programs, a proper strategy must be formulated and executed to protect new cultivars from factors such as lodging and from agents such as diseases and pests that prevent cultivars from attaining their yield potential. Lodging is a problem encountered in most oat-growing areas, whereas the spectrum of disease and the frequency of the biological races or strains of these diseases vary in the different oat production regions. Many of these problems can be controlled by suitable crop and insect management practices, including the use of chemicals, but agriculturalists concerned with costs of production and pollution of the environment prefer a genetic approach to protection whenever possible. The type of breeding program followed will depend upon many factors, including the number and expressivity of resistance genes, state of primitiveness of the plants serving as donors of resistance, availability of appropriate selection techniques, and severity of the problem in the area where the crop is to be grown.

There are new trends in crop protection, as there are in any area of crop improvement. Thus the philosophical approaches and the techniques used to breed for resistance are continually being modified. Some breeders spend most of their time breeding cultivars to combat a specific disease, whereas others are forced to deal with several diseases at one time. The greater the number of factors that the breeder has to consider, the lower the probability of combining all the required useful genes into one cultivar. Some breeders attempt to combine useful resistance genes with other desirable genes at the beginning of a program by choosing appropriate parents, making the hybrids, growing large populations, and attempting to select superior plants with the required resistance from the segregating progeny. This procedure often works best when the original parents possess good agronomic characteristics. When working with poor donor parents, such as wild species, it is virtually impossible to combine outstanding agronomic performance with resistance in a single step. In this instance, resistance is best transmitted through a series of intermediate plant types in separate breeding programs, so that eventually the resistance genes can be combined with other desirable genes in cultivars of commercial quality. Any suitable pedigree or backcrossing program will accomplish this result if selection techniques are efficient and chromosomal relationships permit the transfer.

A. Lodging Resistance

Lodging, or the falling down of the plant during development, can result in severe yield losses and a reduction in seed quality. The most severe lodging occurs within 25 days after anthesis. This early lodging results in poor floret fertility and poor floret filling. The seed cannot usually be used for food purposes because hull percentage values are too high and the groats are shrunken and often dark in color because of fungal growth. Later lodging caused by stem breakage commonly results in seed loss due to shattering and discoloration of kernels.

Oat lodging may or may not be caused by a single factor. In the absence of diseases that weaken the stems, lodging may be a physiological disorder resulting from hypernutrition, which causes excessive growth and weak stems. Care must be exercised when fertilizing tall oat cultivars because they respond to added nutrients, especially nitrogen, and grow very leafy and tall. A leaf canopy is formed rapidly, which shades the lower regions of the plant and causes excessive elongation of intercalary meristems. The crown internodes are also elongated, and coronal roots are weakly developed. The tissues above the ground become soft, and the crop lodges when subjected to the driving forces of wind and weight of rainwater. Diseases such as the rusts and septoria, which affect the integrity of the stem tissues, can also contribute to severe lodging; breeding for resistance thus not only reduces losses caused by parasitism but improves the harvestability of the crop.

Selection for lodging resistance is usually done in the more advanced stages of breeding programs where there is enough seed to grow replicated plots at one or more locations. Single-spaced plants are more resistant to lodging because their lower regions receive enough sunlight during development to produce firm straw and condensed internodes. All too often, breeders discover that most of their highest-yielding selections are susceptible to lodging, and consequently the cultivars cannot be released to growers. In practice, the breeder has to make compromises when combining high yield with lodging resistance. Since lodging susceptibility is very visible to growers, resistance usually has top priority even over grain yield.

I have developed a seedling lodging test to aid in the identification of lodging resistance lines. The test can be conducted in a growth room or in the field. A total of 585 seeds of a cultivar or experimental strain are planted in a 30-cm-diameter clay pot or in a 30-cm-diameter hill in the field, and the seedlings are allowed to grow for two to three weeks after emergence. The lodging reaction of the seedlings is determined after watering the plant from above with the aid of a watering can or after a rain. Growth of the seedlings in the clump is rapid; a canopy of leaf tissue is developed early because of the high seedling density; and the seedlings in the interior of the clump are etiolated and bend over when watered (Fig. 9). Lodging-susceptible cultivars reach this stage earlier than resistant cultivars, which indicates that susceptibility to lodging is related to a greater inherent relative seedling growth rate. The correlation coefficient between seedling lodging reaction and the adult lodging reaction of the same 29 tall-oat cultivars grown in plots was $r = 0.76***$ (Table V).

There probably will always be difficulty in combining high levels of lodging resistance with high grain yields, and this is why breeders have an interest in breeding high-yielding, lodging-resistant dwarfs and semidwarfs.

TABLE V
Comparison Between Lodging Reaction of Adult Plants of Oat Cultivars Grown in Field Plots and of Seedlings Grown in Densely Sown Clumps in the Field[a]

Cultivar	Field Plots[b]	Seedlings	Cultivar	Field Plots[b]	Seedlings
Dasix	9.0	5.0	Shield	4.0	3.0
Old Island Black	9.0	6.5	Russell	4.0	4.0
Don de Dieu	9.0	7.0	OA18-21	3.5	3.0
Markton	8.5	7.5	OA26-5	3.5	5.5
Victory	7.4	7.5	OA65-3	3.4	3.5
Roxton	7.3	7.5	Milford	3.2	6.0
Fundy	6.1	7.0	Vigor	3.0	2.0
OA77-1	6.0	6.0	Waubay	2.8	1.5
Dorval	6.0	7.0	AO2-6	2.7	3.0
Rodney	5.6	3.0	OA53-1	2.5	2.0
Minor	5.4	2.5	3932-16	2.4	1.5
Gopher	5.3	5.5	Clintland 60	1.7	2.5
Angus	5.1	5.0	QO23-5	1.5	1.5
OA69-4	5.1	7.5	Stormont	1.3	1.0
Cartier	4.9	2.5			

[a] Correlation coefficient = 0.76***.
[b] 1 = erect, 9 = lodged.

B. Disease Resistance

RUSTS

Several different strategies have been followed by breeders to reduce the damage caused by both stem and crown rust. In the past, major genes for resistance were located in plants in international oat collections and then were transferred to commercial cultivars to offer host protection. Each gene conditioned resistance to a specific group of rust races, and breeders were content to work with single genes to obtain resistance to those races prevalent in their region. These resistant cultivars were soon attacked by new races that arose in nature through mutation or recombination, and breeding for resistance has thus become a never-ending struggle for the breeder.

One of the strategies devised to solve the quick breakdown in cultivar resistance was to incorporate two different genes offering resistance to the same

Fig. 9. Seedling lodging reaction of 15 different lodging-resistant (R) cultivars and 15 different lodging-susceptible (S) cultivars of oats grown for 21 days under field conditions at Ottawa.

prevalent race (gene pyramiding). In this system, when a new virulent race arises from a prevalent race in nature, it is prevented from multiplying on the host because of the other resistant gene. A variation on this theme is to breed multiline cultivars, in which different lines of the cultivar carry different genes for rust resistance. In an epidemic, only some lines are susceptible to the prevalent race, and this prevents a massive buildup of inoculum. Another method of reducing the amount of inoculum is to utilize "slow rusting." When oats are inoculated with the rust organism, a certain time interval is required for the rust to grow and produce uredospores. Some cultivars have a combination of minor and possibly major genes that permits them to slow down the infection cycle and limit the production of spores. This slowing down of rust development limits the amount of inoculum in the field. There is growing evidence that slow rusting is effective against a broad spectrum of races and thus that the resistance is more lasting.

SEPTORIA

Septoria is a very serious oat disease, especially in areas of adequate rainfall. It is caused by the organism *Leptosphaeria avenaria* Weber f. sp. *avenaria,* and both grain yield and seed quality are reduced by infection. The fungus overwinters on oat stubble, and ascospores are ejected from the perithecia on the straw in springtime and travel by wind to young oat leaves. Infection occurs, pycnidia are formed, and macrospores are splashed by rain to other sites on the same plant or to adjacent plants, forming new infection sites. Growth of the fungus beneath the leaf sheath results in "black stem," which leads to stem breakage when the oat plant ripens. Resistance to septoria is not nearly as available in common cultivars as rust resistance, and the resistance that has been located in other *Avena* species is difficult to transfer to *A. sativa* (R. V. Clark, *personal communication*). Damage to the host occurs mainly during the grain filling and ripening process. Late-maturing strains are often rated as resistant, as septoria appears to be a disease mainly affecting senescing tissues.I have found it difficult to combine a high level of resistance into early-maturing cultivars, and this is a particularly serious problem with the early-maturing DI strains. CN 17823 is a septoria-resistant, midseason strain bred by me and selected for resistance by R. V. Clark at the Ottawa Research Centre (*personal communication*). Reference is made in this chapter to oat strains that show differential ripening of grain and straw. These lines may prove useful in breeding for septoria resistance because, when the grain is ripe, the fungus has not yet consumed and weakened the unripe straw.

BARLEY YELLOW DWARF VIRUS

Barley yellow dwarf virus (BYDV) is a serious worldwide disease of oats. It is a member of the luteovirus group that is carried and transmitted by several species of aphids. When the oat plant becomes infected, a growth reduction occurs in the host, which ultimately leads to fewer tillers, reduced root mass, and smaller panicles and kernels. The leaves become short, erect, stiffened, and leathery, and a red or yellow color develops, usually beginning at the tip and margins of the leaf and progressing toward the base. Breeding for resistance is thought to be the only effective way to combat the virus.

Genes conditioning resistance to BYDV have been located in several different cultivars of *A. sativa,* and Brown and Jedlinski (1978) have made available 13 resistant germ plasm lines to oat breeders. Jedlinski et al (1977) have reported that tolerance to BYDV was expressed by a reduction in the replication of BYDV in the host. Selection of segregating plant material for resistance is sometimes complicated by poor symptom development. I have found that growing infected plants under irrigation in California and then withdrawing the water toward maturity quickly identifies resistant plants because they have better root systems than susceptible plants and they ripen normally. Susceptible plants wilt and die very quickly and can be identified easily.

SMUT

Smut can be a serious oat disease, especially for naked-seeded cultivars, but breeders have many genes for resistance at their disposal. Breeding for resistance is a relatively simple matter, involving the proper choice of parents, inoculation of seed in the segregating populations with races of smut common to the area of production, and selection for the absence of infection.

IV. BREEDING FOR FORAGE QUALITY

The vegetative portion of oats is valuable to many producers as a source of pasture, cut green forage, and straw bedding material for animal production. In fact, many animal producers in the world grow oats only for these purposes because they recognize the value of oat forage in the production of milk and meat. Even in areas of the world where land is highly priced and cannot economically support grain production, seed is imported and sown to produce annual forage. The cultivars used for this purpose must be capable of producing a maximum amount of highly digestible vegetation, and the plants must show high regrowth characteristics when cut one or two times during the growing period. Since the crop may have to grow over a long period of time, resistance to diseases that reduce vegetative growth, such as BYDV, is very important.

In some regions, farmers want the option of using oats for either grain or forage. Where the winter oat is adapted, such as in Texas, as much as 75% of the area sown is used for pasture or cut green for feed. In the spring oat areas, the ratio is often reversed. In all areas, straw residues left after harvest are either incorporated into the soil or collected and used as bedding for livestock.

A. Dual-Purpose Oats

Cultivars used for forage in grain-growing areas must produce a high yield of dry matter per hectare but be fairly resistant to lodging, whereas pedigree seed growers producing grain for sale, and farmers producing their own seed, want the crop to resist lodging and produce good yields of grain. Up to a certain yield level, grain yield and vegetative yield are correlated (Takeda and Frey, 1979); but when attempts are made to combine maximum grain yield with maximum forage yield, difficulties arise. Compromises have to be made in the breeding of dual-purpose oats for producers who want to delay a decision of whether to cut the oats for forage or leave the crop for grain and straw. Dual-purpose oats are defined here as oats that produce the highest yield of dry matter of good digestibility combined with grain yields equal to at least some of the grain-type cultivars commonly grown in the area. The cultivar Foothill, licensed and released by

Agriculture Canada in 1977, is an example of this oat type. Interest in Foothill is growing, but farmers still have a tendency to use cultivars that have been bred for maximum grain yield for forage purposes. This is understandable because animal producers often do not know what their annual forage requirements are going to be until they know the productivity of their perennial grasses. A producer who is short of perennial hay land would be well advised to grow dual-purpose oats. Oat breeders in Canada are watching this development closely, especially now that there appears to be a trend toward the use of small grains for silage purposes in cooler areas where maize is not well adapted.

B. Potential Value of Cultivars Exhibiting Differential Ripening of Grain and Straw

Oat straw is sometimes used as a low-grade feed for overwintering beef cattle. Ripe straw contains a low protein level (2–5% at Ottawa), but Meyer and McMullen (*personal communication*) reported an average protein value of 8.4% in the straw of 28 different cultivars they monitored over a three-year period at North Dakota State University. Straw from the DI cultivar PGR8642 (CI9255) developed at Ottawa averaged 10.1% protein, whereas the U.S. cultivar Spear averaged 7.3% protein. PGR8642 grains had a low protein concentration of 15.7%, and the high-protein cultivar Otee had 20.1% protein. Meyer and McMullen found examples of cultivars that displayed straw protein-grain protein relationships of high-high, high-low, low-high, and low-low. This and the differences in grain yield indicate that cultivars differ greatly in the amount of nitrogen absorbed from soil, the amount of dilution of protein that takes place with carbohydrates during grain filling, and the redistribution of nitrogen in the plant toward maturity.

In general, oat breeders select oats with grain and straw that ripen simultaneously to facilitate mechanical harvesting. Strains that produce fully ripened grain while the straw and leaves remain partially green are usually discarded from breeding populations. Breeders believe that farmers would not want to grow such cultivars because they would have to wait for the straw to ripen before direct combining. The fact that differential ripening occurs would indicate that if the straw were allowed to ripen, nitrogen in the straw could not move to the grain but would remain in the straw because protein synthesis in the grain would already be complete. The fate of this nitrogen would depend on how the crop was managed and on the physiological events taking place in the straw during ripening. If the crop were swathed immediately, then dried and threshed or put into a silo, the protein would be trapped in the straw. This could be attractive to farmers who use this more nutritious straw to overwinter beef cattle. If the straw were allowed to ripen completely before harvest, I believe that as ripening proceeded the nitrogen-containing compounds in the partially green straw would be degraded, with an ultimate loss of simple nitrogen compounds to the atmosphere. We know that fully ripened straws are traditionally low or very low in nitrogen content. If this technique to increase the nutritive value of straw is accepted by producers, then the total value of the grain plus straw could produce sufficient monetary returns per hectare to warrant the growing of oats in some areas.

V. BREEDING FOR GRAIN QUALITY

A. Seed Morphology

Controversy exists between producers and processors over the most desirable size for oat kernels. Producers want kernels to be large and plump, and this specification is as important as yielding ability to many growers. The argument is made that an increase in seed size will increase harvestable yields and that larger and heavier seeds are easier to separate from weed seeds such as the wild oat, *A. fatua*. In areas where crops of oats and barley are grown in mixture for feed, the large-seeded oats match the seed size of barley, which aids adjustment of the combine harvester. In contrast, food processors prefer average-sized oats with groats that can be easily separated from barley seed contaminants. Since the amount of oats used for food purposes is now small, breeders should probably pay more attention to producers and hope that processors can select high-grade oats free of barley from the total crop.

Large seeds are easy to separate from small seeds in segregating populations by sieving, and the trait displayed a heritability value of 71% in a hybrid between Kent and Kelsey (R. J. Baker, University of Saskatchewan, *personal communication*). It is difficult to combine high yield with large seed size, although advances are being made in our DI program in naked (Table IV) and covered (cv. Donald) grain. I have noted in some hybrids that, when extremely large seed sizes are bred, the incidence of "bosom" oats often increases. Large seeds also commonly produce low hectoliter weights because of poor packing in measuring containers, and this conflicts with the desire of both producers and processors to have oats of high hectoliter weight, a worldwide measure of grain quality. Breeders may have to increase the density of large-seeded oats to overcome the problem, although what effect this would have in groat composition and quality is not known.

Hull color is an important factor to some producers because it affects the marketing of the grain as seed. Seed color is often, but not always, an emotional issue, and the arguments put forward by producers in defense of whether a kernel should be white, yellow, tan, or red are open to question. In North America, black, gray, or dark brown hulls are not wanted because it is difficult to detect wild oat contaminants that display the same colors, and the presence of the occasional dark-colored hull in rolled oats is easily detected by consumers and confused with insect skeletons or rodent excrement.

Processors of oats want the hull percentage to be as low as possible; a value of 20% is often quoted as ideal but is not often realized.

B. Protein Concentration

The protein concentration and the quality of the protein contribute greatly to the high nutritional quality of oats (Chapters 7 and 10). There is a wide variation (15–30%) in the concentration of protein in different oat cultivars (Robbins et al, 1971) and species (Frey, 1951; Briggle et al, 1975), and this has motivated breeders to attempt to produce adaptable, high-yielding, high-protein cultivars. Initially it was hoped that a high protein concentration in the groat would help raise the status and usefulness of the crop both as a food and a feed grain, but the lower yield potential of high-protein cultivars has reduced enthusiasm for this breeding objective. The problem of the negative relationships between yield and protein concentration is also shared by other cereal and pulse crops.

One of the unique features of oats is that, as protein levels are increased through breeding, the relatively good amino acid profile of the protein is only slightly changed. The prolamine fraction of the protein, which is deficient in essential amino acids such as lysine, is low in oats (approximately 12%) and does not increase appreciably with increase in total protein. This is in contrast to crops such as the millets, sorghum, maize, wheat, rye, and barley, in which the prolamine content is normally 40–60% of the total protein and a preferential increase in this fraction results in a lowering of nutritional quality.

Several high-protein oat cultivars have been bred in North America—e.g., Hinoat in Canada and Dal, Otee, Lancer, Brooks, Madison, and Goodland in the United States. The availability of these cultivars has stimulated a considerable amount of research, and some commercial interest, but a major shift to the growing of high-protein cultivars has not occurred. The lower yield potential and the reluctance of food processors to pay large enough premiums to producers to grow cultivars like Hinoat have prevented widespread use of the cultivar. Dal was a leading oat in Wisconsin for many years because it was agronomically superior to other available cultivars and also had high protein concentration. It thus represented an exception to the rule of a negative relationship between yield and protein concentration for the Wisconsin area.

The protein content of oat groats, like that of other cereal grains, is affected greatly by such environmental factors as the available nitrogen level in soils and moisture conditions during the growing season. In some years, where growth of the crop is limited by drought, a groat protein value of 14.5% ($N \times 6.25$) may be considered low, whereas in years of higher grain productivity resulting from adequate moisture, such a value may be considered average or even high. Experience has shown, however, that the relative ranking of very high- or very low-protein cultivars grown in plots or large fields is maintained in both circumstances, even though the absolute protein levels may fluctuate considerably. These fluctuations among small samples of grain taken from various regions of a breeding nursery make it difficult to select desirable plants with confidence. Check cultivars should be grown at frequent intervals throughout the nursery to establish the range of protein that can be expected from oats growing on that soil in that year. In Canada, the cultivars Hinoat and Rodney are often used as standards for high and low protein values, respectively.

Selection for protein content in grain usually begins in the $F_4–F_6$ generation. In the past, most protein analyses were performed in duplicate on 1-g samples of finely ground oats using the macro-Kjeldahl or the Udy colorimetric procedures. In recent years, near infrared procedures have become more popular for screening populations of breeding lines for protein content because the method is quite rapid and easy to perform. Seed from spaced plants also can be analyzed, as long as the grain yield exceeds 6 g. This amount provides enough groats for analysis plus sufficient residue seed for planting the next generation. The availability of improved micro techniques to measure protein concentration has made it possible to assay half-seeds for protein concentration to select for high- or low-protein strains. For this analysis, the oat groat is cut transversely into two portions: the embryo-containing half is germinated to grow a new oat plant, whereas the endosperm half is analyzed for protein concentration. Several years ago, I used this technique for screening for protein concentration; the relationship of the protein concentration of the two halves to each other and to the whole groat was determined (Dumas method) for 29 different oat cultivars. The average protein concentrations of the whole groat, embryo half, and endosperm half were 20.42, 20.92, and 19.65% (N × 6.25), respectively. In all 29 cultivars, the embryo half had the highest protein concentration, followed by the whole groat and then by the endosperm half of the kernel. The correlation coefficient between the whole groat and the embryo half was 0.97***; that of the whole groat and the endosperm half was 0.96***; and that of embryo half and the endosperm half was 0.94***. These results indicated that the protein concentration was fairly uniform throughout the kernel and that the protein concentration of the endosperm half of the kernel is a reliable estimate of the protein in the whole seed.

The half-seed protein prediction test was then evaluated in the breeding programs using primary seeds of two (OA123 and OA130) F_8 populations of oats of interspecific (*A. sativa* × *A. strigosa*[2]) origin. The endosperm half of each of 200 seeds from each population was assayed for protein concentration, and the embryo half of each seed was germinated and grown to an adult plant in the field. The seeds from each plant were assayed for protein concentration, using a macro-Kjeldahl procedure. The correlation coefficients obtained between the protein concentration of the half-kernel and the seed derived from the plant grown from the embryo half of the kernel were 0.52*** for the OA123 population and 0.36*** for the OA130 population. These coefficients are significant, indicating that the half-seed protein prediction test may be useful to identify seeds differing widely in protein concentration.

The nutritional or functional value given to oat protein in the future by human nutritionists, feed formulators, and farmers will largely determine the strategy that oat breeders follow to breed new cultivars for grain protein concentration and yield. If oat protein is found to have some important functional property in the preparation of processed foods, or if it becomes a cheaper substitute for some expensive or imported protein source, this will improve the value of the crop and greater emphasis will be placed upon breeding for higher protein concentration. Anderson et al (1978) reported that the increased protein and amino acid levels in Hinoat could be regarded as having potentially significant economic and nutritional merit when fed to swine.

If genetically controlled mechanisms can be found to make grain yield and protein concentration more independent, then emphasis will be placed upon breeding high-yielding, high-protein cultivars. If the situation remains as it is at present, where producers are paid for oats on a tonnage basis, then breeders will probably concentrate on breeding oats for yield and broad adaptability and the concentration of protein will probably remain the same or decline.

High-protein parental material has been located in cultivars of *A. sativa* and in wild species such as *A. sterilis* L., *A. maroccana* Gdgr., *A. murphyi* Ladiz., and *A. strigosa*. Frey (1977) has published an excellent literature review dealing with oat protein and concluded that the inheritance of grain percentage in crosses among strains of cultivated oats and between *A. sativa* and *A. sterilis* seems to be polygenic, with partial dominance for low protein content in F_1, but largely additive gene action in F_2 and later generations. In interspecific crosses, one or more chromosomes carry major loci or groups of closely linked minor loci for groat protein percentage. Frey is of the opinion that, as a breeding strategy, attempts to raise the protein content of commercial cultivars can best be accomplished by utilizing genetic variation in *A. sativa*. For the long run, Frey believes that special breeding procedures should be used to utilize genes in *A. sterilis*.

The very high protein content and large seed size of *A. magna* (*A. maroccana*) and *A. murphyi* (Ladizinsky and Fainstein, 1977a) will only be exploited when a greater degree of introgression of the genes from these tetraploid species into *A. sativa* can be achieved (Ladizinsky and Fainstein, 1977b). *A. strigosa* has already been used by Zillinsky (*personal communication*) and by me in the breeding of the high-protein oat, Hinoat; but because of outcrossing in the nursery to the sterile progeny arising from the hybrid *A. sativa*/*A. strigosa*[2], it is impossible to document the origin of the high-protein gene or genes.

High priority should be given to determining the physiological mechanism causing the negative relationship between yield and protein in all cereals, but it is especially important in oats because of the superior biological value of its protein.

C. Lipid Concentration

Oat groats contain higher concentrations of hexane-extractable lipids than wheat, rye, or barley, and the concentration varies from approximately 3 to 11% in different cultivars (Brown and Craddock, 1972). Some breeders are attempting to produce high-yielding, high-lipid cultivars to overcome the criticism that covered oat grains are low in energy for feeding purposes. The high content coupled with the soft texture of the groat has made it difficult to prepare refined oat flours efficiently. The oil is of good fatty acid composition (Chapter 8) and thus contributes to the good nutritional qualities of oat groats. Youngs and Forsberg (1979) and Gullord (1980) reported no significant correlation between oil and protein content. Lipid content in oat groats is highly heritable (Baker and McKenzie, 1972; Brown and Aryeetey, 1973; Brown et al, 1974; Frey and Hammond, 1975; Frey et al, 1975). Youngs and Püskülcü (1976) reported broad-sense heritabilities of 91.4, 98.6, and 95.7% for palmitic, oleic, and linoleic acids, respectively. It should then be a rather straightforward program to breed both high- and low-lipid cultivars to satisfy specific needs.

D. Starch, Gums, Minerals, and Vitamin Content

I am not aware of any breeding programs designed to alter the kind or the amount of starch, dietary fiber (including β-glucan gums), minerals, or vitamins in oats. Variation in the amount of one or all of these ingredients has been noted, especially in high-protein cultivars (Chapter 3), and presumably this variation is caused by both genetic and environmental factors. If oats gain importance as human food, a search for genes that will control or modify these traits may be warranted.

VI. SUMMARY

The area devoted to the production of oats in the world is declining, primarily because the oat is not yet a cereal grain for international trade and its role has been as a feed grain consumed largely on the farm. Its relatively thick hull has lowered its bulk density, so that it is uneconomical to transport, and lowered the intrinsic energy of the kernel, so that it is not a high-energy feed like maize or barley. The monetary returns received by producers for oat grains have not been sufficient to justify sowing the crop; rather, oats are grown because the crop has assumed a general utility role on the farm, being sown on the poorer land sites for pasture, forage, straw for bedding, and as a nurse or companion crop to establish legumes.

A new and more important role for oats could arise in the future as the demand for food becomes greater in the world and the importance of fiber in the diet is more widely recognized. Except as a hot breakfast cereal, oats have barely penetrated the human food markets because the kernel itself and the whole-grain flour lack identified, desirable functional properties. The kernel is also too soft in texture and too high in oil to permit efficient use of conventional wheat-milling equipment to make refined oat flours of low ash content. The superior nutritional qualities of oats have not been exploited on a large scale. The fact that a nutritious cereal grain like oats serves a useful role in crop rotations and can be produced on the more marginal soils of the world will probably form an attractive strategy for feeding greater numbers of people in the future.

The genetic base for improving oats for more extensive and more specialized food and feed uses is very large and is contained in many different wild and cultivated species or cultivars. The immediate exploitation of hull-less genes to uncouple oats from farm use and the employment of DI genes to uncouple oats from specific latitudes and photoperiods will make it an international crop and one that is more easily processed for food. The options of breeding cultivars with different levels of protein, starch, lipid, and gums and of changing seed shape and size are important in meeting future demands of food processors as they learn how to adapt oats to more extensive human use.

To make the crop more productive, an attempt is being made in Canada to raise yield potential by breeding the experimental crop dormoats and semidwarf oats for growing under intensive management. Dormant seed of dormoats is planted in the autumn, remains dormant over winter, and germinates in early spring to tiller well and differentiate large panicles bearing many florets. Semidwarf oats and especially semidwarf, naked-seeded oats are intended for growing on productive land, possibly under irrigation. Breeding strategies to protect oats from the most serious diseases such as rusts, septoria, smuts, and BYDV are already in place at many institutions in the world.

The oat crop, as we know it today, will have to change if it is to remain a major crop. Increased oat utilization in human diets seems logical because the groats are nutritious and tasty. Oats also appear to have a unique dietary function in the treatment of diabetic and hypercholesteremic individuals (Chapter 11). They are also economical to purchase and do not necessarily have to compete with well-established food grains for the same land. Breeders are eager and equipped to combine forces with producers and food scientists to change the morphology and composition of oats to facilitate the gradual shift of this feed grain to a food role.

ACKNOWLEDGMENT

This chapter is Contribution No. 796A of the Plant Research Centre, Research Branch, Agriculture Canada, Ottawa, Canada, K1A 0C6.

LITERATURE CITED

ANDERSON, D. M., BELL, J. M., and CHRISTISON, G. I. 1978. Evaluation of a high-protein cultivar of oats (Hinoats) as a feed for swine. Can. J. Anim. Sci. 58:87-96.

BAKER, R. J., and McKENZIE, R. I. H. 1972. Heritability of oil content in oats *Avena sativa* L. Crop Sci. 12:201-202.

BAUM, B. R. 1977. Oats: Wild and Cultivated. Biosystematics Research Institute, Canada Dept. Agric., Research Branch. Monogr. 14. Printing and Publishing Supply and Services, Ottawa, Canada. 463 pp.

BAUM, B. R., RAJHATHY, T., MARTENS, J. W., and THOMAS, H. 1975. Wild oat gene pool. Can. Dep. Agric. Res. Branch Publ. 1475.

BRIGGLE, L. W., SMITH, R. T., POMERANZ, Y., and ROBBINS, G. S. 1975. Protein concentration and amino acid composition of *Avena sterilis* L. groats. Crop Sci. 15:547-550.

BROWN, C. M., and ARYEETEY, A. N. 1973. Maternal control of oil content in oats (*Avena sativa* L.). Crop Sci. 13:120-121.

BROWN, C. M., ARYEETEY, A. N., and DUBEY, S. N. 1974. Inheritance and combining ability for oil content in oats (*Avena sativa* L.). Crop Sci. 14:67-69.

BROWN, C. M., and CRADDOCK, J. C. 1972. Oil content and groat weight of entries in the world oat collection. Crop Sci. 12:514-515.

BROWN, C. M., and JEDLINSKI, H. 1978. Registration of 13 germplasm lines of oats. Crop Sci. 18:1098.

BROWN, P. D., McKENZIE, R. I. H., and MIKAELSEN, K. 1980. Agronomic, genetic and cytologic evaluation of a vigorous new semi-dwarf oat. Crop Sci. 20:303-306.

BURROWS, V. D. 1970. Yield and disease-escape potential of fall-sown oats possessing seed dormancy. Can. J. Plant Sci. 50:371-377.

BURROWS, V. D. 1986. Tibor oats. Can. J. Plant Sci. In press.

CAVE, N. A., and BURROWS, V. D. 1985. Naked oats in feeding the broiler chicken. Poult. Sci. 64:771-773.

CHRISTISON, G. I., and BELL, J. M. 1980. Evaluation of Terra, a new cultivar of naked oats (*Avena nuda*) when fed to young pigs and chicks. Can. J. Anim. Sci. 60:465-471.

CLARK, R. V. 1980. Comparison of spot blotch severity in barley grown in pure stand and in mixtures with oats. Can. J. Plant Pathol. 2:37-38.

COFFMAN, F. A. 1961. Oats and Oat Improvement. F. A. Coffman, ed. Am. Soc. Agron., Madison, WI. 650 pp.

COMEAU, A., and DUBUC, J. P. 1977. A flotation method to separate out oat groats possessing different degrees of surface hairs. Can. J. Plant Sci. 57:397-399.

DALE, N. M., and FULLER, H. L. 1985. Energetic content of yellow corn and several alternate energy sources. (Abstr.) Poult. Sci. 64:S13.

DERICK, R. A. 1930. A new dwarf oat. Sci. Agric. 10:539-542.

FLORELL, V. H. 1931. Inheritance of type of floret separation and other characters in interspecific crosses in oats. J. Agric. Res. 43:365-386.

FREY, K. J. 1951. The relation between alcohol-soluble and total nitrogen content in oats. Cereal Chem. 28:506-509.

FREY, K. J. 1977. Protein of oats. Z. Pflanzenzuecht. 78:185-215.

FREY, K. J, and HAMMOND, E. G. 1975. Genetics, characteristics, and utilization of oil in caryopses of oat species. J. Am. Oil Chem. Soc. 52:358-362.

FREY, K. J., HAMMOND, E. G., and LAWRENCE, P. K. 1975. Inheritance of oil percentage in interspecific crosses of hexaploid oats. Crop Sci. 15:94-95.

GOULDEN, C. H. 1926. A genetic and cytological study of dwarfing in wheat and oats. Minn. Agric. Exp. Stn. Tech. Bull. 33.

GULLORD, M. 1980. Oil and protein content and its relation to other characters in oats. Acta Agric. Scand. 30:216-218.

HUSKINS, C. L., and HEARNE, E. M. 1933. Meiosis in asynaptic dwarf oats and wheat. J. R. Microsc. Soc. 53:109-117.

JANA, S., ACHARYA, S. N., and NAYLOR, J. M. 1979. Dormancy studies in seeds of *Avena fatua*. X. On the inheritance of germination behaviour. Can. J. Bot. 57:1663-1667.

JEDLINSKI, H., ROCHOW, W. F., and BROWN, C. M. 1977. Tolerance to barley yellow dwarf virus in oats. Phytopathology 67:1408-1411.

JENKINS, G., and HANSON, P. R. 1976. The genetics of naked oats (*Avena nuda* L.). Euphytica 25:167-174.

KOLB, F. L., and MARSHALL, H. G. 1984. Peduncle elongation in dwarf and normal height oats. Crop Sci. 24:699-703.

LADIZINSKY, G., and FAINSTEIN, R. 1977a. Domestication of the protein-rich tetraploid wild oats *Avena magna* and *Avena murphyi*. Euphytica 26:221-223.

LADIZINSKY, G., and FAINSTEIN, R. 1977b. Introgression between the cultivated hexaploid oat *Avena sativa* and the tetraploid *Avena magna* and *Avena murphyi*. Can. J. Genet. Cytol. 19:59-66.

LAWES, D. A., and BOLAND, P. 1974. Effect of temperature on the expression of the naked grain character in oats. Euphytica 23:101-104.

LITZENBERGER, S. C. 1949. Inheritance of resistance to specific races of crown and stem rust, to Helminthosporium blight, and to certain agronomic characters of oats. Iowa Agric. Exp. Stn. Res. Bull. 370.

MARSHALL, H. G., and KOLB, F. L. 1983. Registration of Pennline 116 oat germplasm. Crop Sci. 23:190.

MARSHALL, H. G., KOLB, F. L., and FRANK, J. A. 1983. Registration of Pennlo and Pennline 6571 oat germplasm lines. Crop Sci. 23:404.

MARSHALL, H. G., and MURPHY, C. F. 1983. Inheritance of dwarfness in three oat crosses and relationship of height to panicle and culm length. Crop Sci. 21:335-338.

MORRIS, J. R., and BURROWS, V. D. 1986. Naked oats in grower-finisher pig diets. Can. J. Anim. Sci. In press.

MOULE, C. 1972. Contribution a l'étude de l'héredité du caractère 'grain nu' chez l'avoine cultiveé. Ann. Amelior. Plant. 22:335-361.

MYER, R. O., BARNETT, R. D., and WALKER, W. R. 1985. Evaluation of hullless oats (*Avena nuda* L.) in diets for young swine. Nutr. Rep. Int. 32:1273-1277.

RAJHATHY, T., and THOMAS, H. 1974. Cytogenetics of oats (*Avena* L.). Genet. Soc. Can. Misc. Publ. 2. 90 pp.

ROBBINS, G. S., POMERANZ, Y., and BRIGGLE, L. W. 1971. Amino acid composition of oat groats. J. Agric. Food Chem. 19:536-539.

RUTGER, J. N., and CARNAHAM, H. L. 1981. A fourth genetic element to facilitate hybrid cereal production—a recessive tall in rice. Crop Sci. 21:373-376.

SAWHNEY, R., and NAYLOR, J. M. 1980. Dormancy studies in seed of *Avena fatua*. XII. Influence of temperature on germination behaviour of nondormant families. Can. J. Bot. 58:578-581.

SIBBALD, I. R. 1983. The TME system of feed evaluation. Agric. Can. Res. Branch Publ. 1983-2OE.

SIMMONDS, N. W. 1979. Principles of Crop Production. Longman, New York. 408 pp.

SINHA, R. N., WALLACE, H. A. H., MILLS, J. T., and McKENZIE, R. I. H. 1979a. Storability of farm-stored hulless oats in Manitoba. Can. J. Plant Sci. 59:949-957.

SINHA, R. N., WHITE, N. D. G., WALLACE, H. A. H., and McKENZIE, R. I. H. 1979b. Effect of moisture content on viability and infection of hulless Terra oats in storage. Can. J. Plant Sci. 59:911-916.

STANTON, T. R. 1923a. Naked oats. J. Hered. 14:177-183.

STANTON, T. R. 1923b. Prolific and other dwarf oats. J. Hered. 14:301-305.

TAKEDA, K., and FREY, K. J. 1979. Protein yield and its relationship to other traits in backcross populations from an *Avena sativa* × *A. sterilis* cross. Crop Sci. 19:623-628.

WARBURTON, C. W. 1919. The occurrence of dwarfness in oats. J. Am. Soc. Agron. 11:72-76.

WELCH, R. W. 1977. The development of rancidity in husked and naked oats after storage under various conditions. J. Sci. Food Agric. 28:269-274.

YOUNGS, V. L., and FORSBERG, R. A. 1979. Protein-oil relationships in oats. Crop Sci. 19:798-802.

YOUNGS, V. L., PETERSON, D. M., and BROWN, C. M. 1982. Oats. Pages 49-105 in: Advances in Cereal Science and Technology, Vol. V. Y. Pomeranz, ed. Am. Assoc. Cereal Chem., St. Paul, MN.

YOUNGS, V. L., and PÜSKÜLCÜ, H. 1976. Variation in fatty acid composition of oat groats from different cultivars. Crop Sci. 16:881-883.

ZAVITZ, C. A. 1927. Forty years' experiments with grain crops. Ont. Dep. Agric. Bull. 337:1-98.

CHAPTER 3

MORPHOLOGICAL AND CHEMICAL ORGANIZATION OF THE OAT KERNEL

R. G. FULCHER
Cereal Section
Plant Research Centre
Agriculture Canada
Ottawa, Ontario, Canada

I. INTRODUCTION

The primary function of a mature, viable oat kernel is to generate a complete, new vegetative plant in favorable soil conditions. To do this, it must: 1) include in its organization quiescent vegetative tissue (the embryo), which will generate new shoots and roots during germination; 2) provide sufficient stores of the nutrients necessary for the first several days of growth while young leaves and roots establish photosynthetic and absorptive function; 3) provide mechanisms for releasing these nutrients from their usual polymeric storage state and transport them as small, soluble molecules to the germinating embryo; and 4) protect itself from the rigors of hostile environmental conditions such as drought, freezing temperatures, and a vast array of potentially damaging soil microorganisms.

Not surprisingly, each of these (and other) functions is the mandate of separate and distinct tissues, which in turn suggests an impressive degree of structural and chemical compartmentalization within the kernel. Indeed, each tissue contains specific combinations of a wide variety of biochemical constituents, which may include the major reserves of storage and structural carbohydrates, proteins, and lipids, to various amines, lignins, waxes, nucleic acids, phytin, vitamins, and enzymes. Moreover, these constituents vary considerably in concentration or chemical and morphological form depending on their genetic background and the environmental conditions in which the plants were grown. In short, the cereal grain is differentiated *morphologically* (structurally) into components that also display considerable *chemical* variation. Because specific tissue and cultivar properties ultimately determine the nutritional and processing characteristics of cereals, efforts to improve or modify any of these characteristics will be well served by a sound understanding of the physical and chemical interactions within the kernel.

The oat kernel is as complex as other cereal grains, all of which, as members of the Gramineae (the grass family), are organized according to a similar structural pattern. But whereas the cereals (e.g., wheat, oats, barley, rice, and corn) contain similar tissues with equivalent physiological functions, in many ways the oat is chemically and structurally unique. For example, it has been recognized for many decades that oat protein provides an amino acid balance superior to that of wheat, barley, rye, and corn. In addition, the groats (dehulled kernels) of several recent oat cultivars contain over 20% total protein. Beyond the protein-related features of oats, however, only sporadic interest has been shown in the many other components of the kernel. A large percentage of oat starch, for example, is contained in compound granules similar to those of rice, quite unlike the simple granules of wheat, barley, or corn. Recent studies by Paton (1977, 1979, 1981) and MacArthur and D'Appolonia (1979) have renewed interest in oat starch by defining several of its intriguing physical properties, and Wood et al (1977, 1978) have provided considerable data on the β-glucans and other nonstarchy oat polysaccharides. Until Wood's recent publications, there had been little emphasis on the fact that oat β-glucans (gums) exhibit viscosities in solution that rival or exceed those of other biological hydrocolloids. Even oat lipids, which have a favorable fatty acid balance and stability (Sahasrabudhe, 1979), may exceed the concentrations of wheat and barley lipids by twofold to fivefold in some cultivars. Furthermore, lipid concentration in the groat is not directly correlated with yield (Youngs et al, 1982). Thus, considerable potential exists for raising total energy levels (by increasing lipid content) in a relatively adaptable feed grain.

Each of these chemical constituents is confined to a specific compartment or structure in the grain, and each has a unique set of processing and nutritional characteristics. Unfortunately, a notable shortage of detailed studies relating to groat morphology and microchemistry exists in the literature. Cereal grains have long been notoriously difficult to examine with the microscope, but recent technological advances have led to numerous publications on the morphological intricacies of wheat, barley, and rice kernels. Oat morphology has been largely ignored for many years, and until recently it was necessary to consult either dated or vague descriptions of groat anatomy and composition. Bonnett (1961) examined oat vegetative tissues, while minimizing data on grain morphology. Renewed interest in groat structure has been shown by Angold (1979), Bechtel and Pomeranz (1981), Youngs et al (1982), and Saigo et al (1983).

The remainder of the material in this chapter will outline the morphological characteristics of the groat and emphasize the basic chemical nature of each morphological entity. An important feature of this crop is that it displays an outstanding degree of genetic diversity. Thus, in addition to variation in structure and composition of materials *within* the groat, there is often pronounced variation in concentration and structural organization between groats of different cultivars. In several ways the oat is unique, and some of this variation is noted. This chapter is intended merely as an overview of the basic structural and chemical properties of the domestic oat kernel (*Avena sativa*), with occasional reference to the common wild oat (*A. fatua*). More detailed discussions relating to specific grain constituents are provided by several authors in the companion chapters.

II. MICROSCOPIC TECHNIQUES

Morphological analysis is essentially microscopic analysis. Beyond noting basic kernel characteristics such as size, shape, and color, high-quality microscopic technique is essential for defining precisely the many subtle structural and chemical interactions among grain constituents. Several grain compounds are stored in structures only a fraction of a micrometer in diameter, and certainly most are beyond the resolving limit of the human eye.

A. Electron Microscopy

Of the two basic types of microscopes, electron and light, the former is perhaps the most widely used at present. Scanning electron microscopy (SEM) is particularly useful and offers the distinct advantages of high resolution, ease of specimen preparation, and a surface view of structures in detail that is quite unapproachable with other instruments. In cereal studies, it is admirably suited to examining seed surfaces, starch or other granules, and spatial relationships between components. Scanning electron micrographs of mature groats and endosperm protein bodies were obtained using standard preparative methods. In summary, samples were mounted on double-sided adhesive tape on aluminum stubs and vacuum-coated with approximately 400 Å of gold-palladium. They were examined under an accelerating voltage of 25 kV in a Stereoscan Mark II-A SEM (Cambridge Scientific Instruments Ltd.). Transmission electron microscopy (TEM) was not used in this study, but excellent samples of its application to oat endosperm analysis have been furnished recently by Bechtel and Pomeranz (1981) and Saigo et al (1983).

B. Light Microscopy

Unfortunately, no cytochemical procedures have been devised to identify or localize many cereal components selectively in either the SEM or TEM. Therefore, unless the morphological characteristics of a particular component (such as starch and some protein bodies) are known from other methods, their chemical identity, and hence their role in physiological, nutritional, or industrial situations, can only be estimated. In contrast, various simple modifications of the light microscope are admirably suited to routine detection of morphological *and* chemical characteristics of cereal components. Most grain reserves are packaged in structures of sufficient size to be resolved with the light microscope, and the expanding list of reagents, dyes, and other techniques (e.g., immunofluorescence) that selectively label particular components is impressive. The common modifications of the light microscope, namely bright-field, polarizing, and fluorescence microscopy, are of particular value for cereal grain analysis, and all three modifications may be combined in one instrument to provide a very powerful device for qualitative chemical analysis.

With the exception of the scanning electron micrographs and four of the plates (nos. 5, 6, 14, and 15), all illustrations in this chapter are fluorescence photomicrographs of representative sections of mature groat tissues that were prepared by a range of methods suitable for demonstrating specific grain constituents. Briefly, sections were cut either by hand using a clean razor blade

Plates 1–7: (1) Low magnification view, using FC I (see description of fluorescence filter combinations in Table II) of the outer regions of the groat. Note intense autofluorescence in the cell walls of the bran components, especially the aleurone layer, pericarp, and trichomes. There is little or no natural fluorescence in the endosperm cell walls. Bar = 100 μm; al = aleurone layer, end = starchy endosperm, p = pericarp, and t = trichomes. (2) Higher magnification of similar material. Note characteristic blue autofluorescence in aleurone cell walls and bright autofluorescent deposits in aleurone cytoplasm. Bar = 100 μm. (3) Similar to (1) but stained with Congo red to show mixed linkage β-glucan. Note that β-glucans are highly localized in the aleurone cell walls (arrows) as well as the adjacent starchy endosperm (*). Phenolic compounds are blue. Bar = 100 μm. (4) Section of portion of aleurone layer and adjacent starchy endosperm after staining with ANS (8-anilino-1-naphthalene sulfonic acid) to show protein. Note the marked difference in staining intensity of protein bodies in starchy endosperm compared with those in aleurone layer. Bar = 100 μm. (5 and 6) Same section as viewed by polarizing microscopy (5) or bright-field microscopy after Alizarine Red staining (6). Both methods reveal high concentrations of phytin crystals (arrows) in the aleurone cells. Bar = 100 μm. (7) Similar to (5) and (6) but photographed using fluorescence microscopy and FC III after acriflavine HCl staining. Phytin crystals are fluorescent red (arrows). Bar = 65 μm.

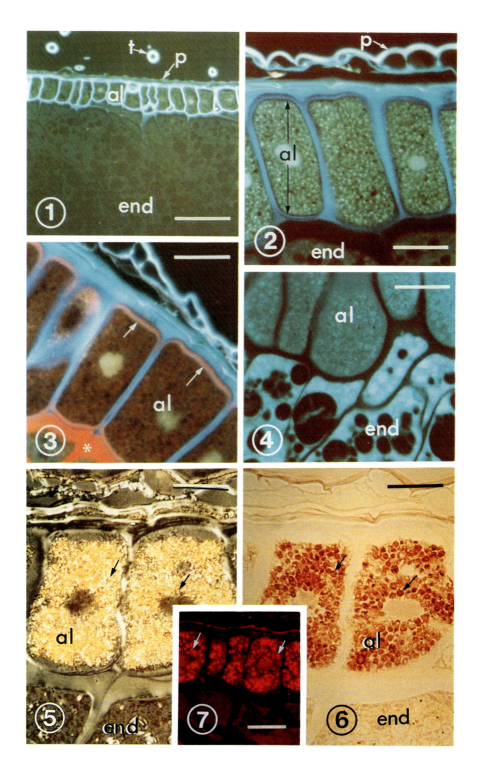

Plates 8–11: (8) Hand-cut section of groat bran after treatment with cyanogen bromide/*p*-aminobenzoic acid showing intense fluorescent yellow staining (FC II) of the niacin reserves (arrows) in the aleurone layer. Bar = 100 μm; al = aleurone layer and end = starchy endosperm. (9) Similar to (8) but stained with *p*-dimethylaminocinnamaldehyde to show reserves of fluorescent red (FC II) aromatic amines (arrows) in the aleurone cells. Bar = 100 μm. (10) Periodate/Schiffs-stained section (FC III) showing periodate-sensitive residues in starch, aleurone cell walls, and in the middle lamella of the endosperm cell walls. Bar = 100 μm; s = starch. (11) Calcofluor-stained section (FC I) showing the high concentration of mixed-linkage β-glucans in the endosperm cell walls (arrows). Bar = 100 μm.

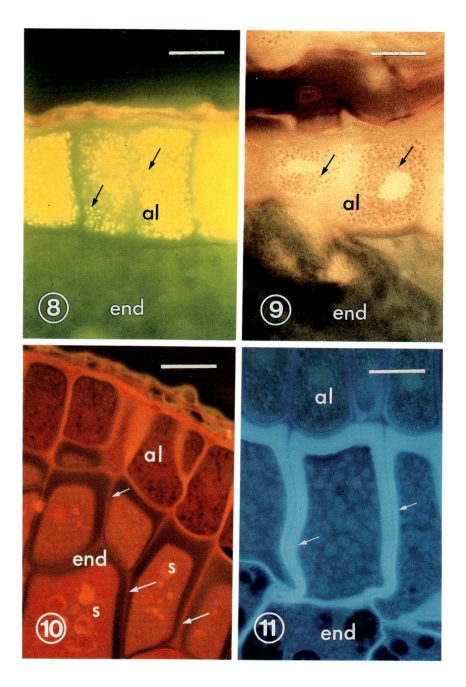

Plates 12–18: (12) Low magnification view of the endosperm and aleurone layer of high-protein oat variety (Hinoat) after Calcofluor staining to show high concentrations of β-glucans in the subaleurone region of the grain (arrows). Bar = 100 μm; al = aleurone layer and end = starchy endosperm. (13) Similar to (12) but of a lower-protein variety (Elgin), showing the absence of specialized β-glucan accumulation in the subaleurone region (arrows). Bar = 100 μm. (14) Section of high-protein groat (Hinoat) stained with Acid Fuchsine and viewed using bright-field optics to show extensive development of protein bodies (arrows) in the subaleurone cells. Bar = 65 μm. (15) Similar to (14) but of a lower protein variety (Elgin), showing smaller size and lower concentration of subaleurone protein bodies. Starch grains (s) are unstained. Bar = 65 μm. (16) Hand-cut section of groat stained with diphenylboronate (FC II) to show distribution of fluorescent yellow flavonoids (*) in the starchy endosperm and aleurone layer. The germ (g) is generally devoid of stainable material except where the aleurone layer is adjacent to it. Bar = 100 μm. (17) Section of low-lipid groat (Exeter) endosperm stained with Nile blue (FC II) to show droplets of storage lipid (arrows). Bar = 65 μm. (18) Similar to (17) but of a high-lipid variety (CI4492), showing high concentrations of lipid droplets (arrows) in the endosperm. Bar = 65 μm.

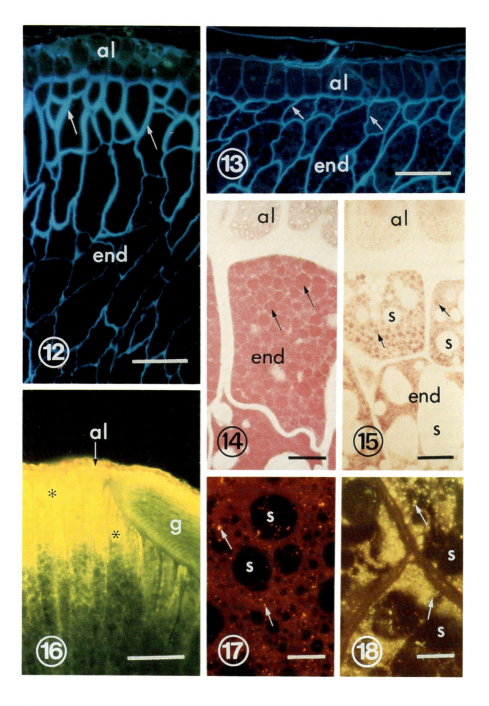

Plates 19–23: (19) Section of groat stained with the periodate/acriflavine HCl (PAA) reaction (FC II), showing the typical accumulation of starch (s) in the endosperm. Bar = 100 μm; al = aleurone layer and end = starchy endosperm. (20) Periodate/Schiffs-stained (FC III) central endosperm tissue showing the morphology of compound (s) and simple (*) starch granules. Bar = 100 μm. (21) View of part of germ, showing blue autofluorescence (arrows) in cell walls of the scutellar epithelium (se) and scutellar parenchyma (sp) cells. Bar = 100 μm. (22) Section of scutellar parenchyma tissue starch with Nile blue (FC II), showing envelope of fluorescent yellow lipid droplets (arrows) that encircle each protein body. Bar = 100 μm. (23) Section of germ and adjacent starchy endosperm after staining with both Calcofluor (blue, for β-glucans) and PAA (yellow, for starch). Starch and β-glucans are absent from germ tissues. Bar = 100 μm.

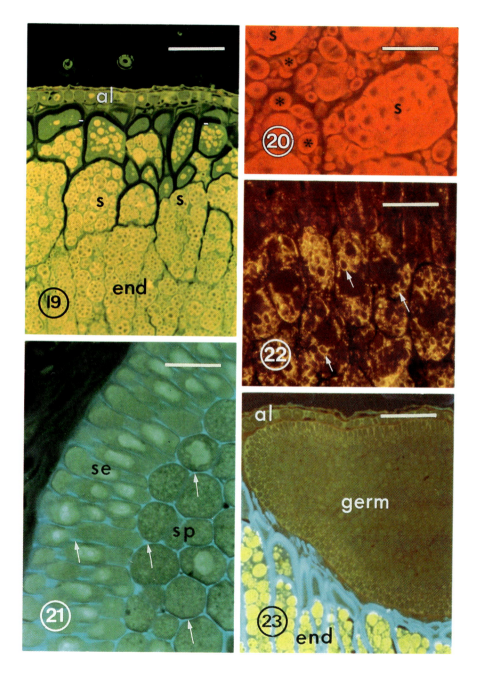

(for routine, low-resolution examination and to minimize extraction or mobilization of compounds) or after glutaraldehyde fixation and subsequent embedding in glycol methacrylate plastic (GMA; Feder and O'Brien, 1968) or urea-glutaraldehyde-glycol methacrylate (modified GMA; Pease, 1973). Embedded sections were cut 0.1–1.0 μm thick, using glass knives, on an ultramicrotome and affixed to glass slides. GMA permits routine high-resolution examination, and the plastic medium is compatible with a wide variety of cytochemical procedures, including digestion of sections with specific enzymes (e.g., β-glucanase; Wood et al, 1983) that are used to confirm the identity of components. The modified GMA embedding technique minimizes extraction of storage lipids from grain tissues and is useful in conjunction with lipid-specific staining techniques such as Nile blue A (Hargin et al, 1980).

Once sections have been mounted on glass slides, they may be examined without further treatment, using polarizing microscopy to detect anisotropic substances (such as phytin crystals, starch, and cell walls) or fluorescence microscopy to detect the few naturally fluorescent (autofluorescent) groat compounds (such as lignin and phenolic acids). However, most constituents are not detectable by these methods and must be stained with specific bright-field stains (diachromes) or fluorescent stains (fluorochromes). Because the fluorescence microscope offers distinct advantages over bright-field techniques in sensitivity and chemical specificity (Fulcher and Wong, 1980, 1982), most of the micrographs were obtained after appropriate staining procedures using a Zeiss Universal microscope equipped with a III RS epi-illuminating condenser, an HBO 200 W mercury arc illuminator, and a limited range of specific fluorescence filter combinations. Table I summarizes the bright-field and fluorescence staining methods used to demonstrate groat components, and Table

TABLE I
Microscopic Methods for Cereal Analysis

Component	Method	Reference
Protein	8-Anilino-1-naphthalene sulfonic acid (ANS)	Gates and Oparka (1982)
	Acid Fuchsine	Fulcher and Wong (1980)
Starch and periodate-sensitive cell walls	Periodate/Schiffs (PAS)	O'Brien and McCully (1981)
	Periodate/acriflavine HCl (PAA)	Fulcher and Wood (1983)
Storage lipids	Nile blue A	Hargin et al (1980)
Mixed-linkage β-glucans	Calcofluor	Wood and Fulcher (1978)
	Congo red	Fulcher and Wood (1983)
Phenolic acids	Autofluorescence	Fulcher et al (1972a)
		Fincher (1976)
Flavonoids	Boric acid esters	Fulcher and Wong (1980)
Phytin	Polarizing optics	Fulcher (1972)
	Alizarin Red S	Fulcher (1972)
	Acriflavine HCl	Yiu et al (1982)
Niacin	Cyanogen bromide/ p-aminobenzoic acid	Fulcher et al (1981)
Aromatic amines	p-Dimethylaminobenzaldehyde or p-dimethylaminocinn- amaldehyde	Fulcher et al (1981)

II lists the spectral characteristics of fluorescence exciter and barrier filter combinations (FC I, II, and III) that are suitable for use with fluorescence staining methods. Complete details for each method are available elsewhere (Wood and Fulcher, 1978; Fulcher and Wong, 1980; Hargin et al, 1980; O'Brien and McCully, 1981; Fulcher, 1982; Yiu et al, 1982; Fulcher and Wood, 1983). All microscopic techniques must be applied in conditions that least disrupt the normal organization of the component under investigation. All cereal components are affected to some degree by the methods used during preparation for any form of microscopy; thus, interpretation of constituent identity and distribution is to some extent a function of preparation methods.

Most of the methods illustrated in this chapter can be applied with little modification to millstreams, solutions, baked products, and other processed materials. Indeed, the methods have been developed not only for their specificity and sensitivity, but largely for their potential for rapid scanning of large numbers of breeding lines and industrial samples for important characteristics. In the past few years, several North American laboratories have become increasingly interested in oat chemistry and processing, and in the area of process development the microscope has frequently proven to be invaluable. In addition to simple particle analysis (i.e., evaluating size and shape), which relates directly to mill characteristics, sieving procedures, and so forth, the microscope is useful in analyzing a wide range of additional properties through the application of specific markers or stains for individual chemical constituents (see Table I). These procedures often provide rapid compositional evaluation of the products resulting from a particular process, and an experienced microscopist with a knowledge of groat morphology should be able to *estimate* the influence of a particular procedure on the various groat components. Although it does not readily permit accurate quantitative assessments, at the very least the microscope is capable of providing a good estimate of chemical and physical effects at various stages of processing. For this reason, the microscope has become increasingly popular as an essential instrument in process development laboratories.

III. DEVELOPMENT OF THE OAT KERNEL

Cereal morphologists are well aware that a cereal kernel is the product of a long developmental process that begins with the union of genetic material (fertilization) within the ovary of the flower and ends several weeks later with the formation of a large, relatively dry, mature kernel, or caryopsis, surrounded by several layers of protective tissue. As a member of the grass family, oats share

TABLE II
Spectral Characteristics of Typical
Fluorescence Filter Combinations (FC) for Cereal Analysis

Combination	Maximum Transmission (nm)	
	Exciter Filter	Barrier Filter
FC I	365	>418
FC II	450–490	>520
FC III	546	>590

many developmental and morphological features with other common cereals and, as in other cereals, the initial event is one of "double fertilization." At anthesis, pollen grains are deposited on the feathery receptor structure (stigma) on the upper part of the ovary, where one or more of the grains germinate to produce pollen tubes, which penetrate the ovary. Several hours after anthesis, one pollen tube releases two male haploid nuclei into the ovary, where they unite with a haploid egg nucleus and two associated haploid polar nuclei. The product of the former union becomes the diploid embryo, and the latter union gives rise to the triploid endosperm. During the subsequent six weeks, these two new genetic products slowly divide, cellularize, and differentiate into the embryo and endosperm.

At maturity, the embryo is differentiated into 1) vegetative tissues, including shoot and root apices, primary leaves, adventitious roots, and protective structures (the coleoptile and coleorhiza); and 2) the scutellum, which contains a variety of reserves in parenchyma and epithelial tissues and which probably absorbs and translocates nutrients from the endosperm to the germinating embryo during germination. The function and morphology of the scutellum of several grasses, including oats, have been considered in some detail by Smart and O'Brien (1979a, 1979b, 1979c). In combination, the embryonic axis and scutellum comprise only 2.8–3.7% (by weight) of the mature, ungerminated, dehulled oat grain (Youngs, 1972; Youngs and Peterson, 1973).

The other product of fertilization, the triploid endosperm, differentiates during development into two distinct tissues, the starchy endosperm and the aleurone layer. At maturity, the starchy endosperm is the largest grain fraction, contributing between 55.8 and 68.3% of the weight in common oat cultivars (Youngs, 1972; Youngs and Peterson, 1973). It is the primary source of nutrients for use by the embryo during germination, containing all of the starch and much of the protein and lipid reserves in the kernel. Surrounding the starchy endosperm (and part of the embryo) is a single layer of aleurone cells, which secrete a variety of hydrolytic enzymes during germination to digest and mobilize the starchy endosperm reserves. Postfertilization events relating to oat kernel development (division, cellularization, differentiation of tissues, and synthesis of reserves) have received only occasional attention from microscopists (Bonnett, 1961; Peterson and Saigo, 1978; Saigo et al, 1983). However, the overall pattern of cellular differentiation in oats is similar to that of wheat, which has been described in considerable detail (Evers, 1970; Mares et al, 1975; Morrison, 1975; Morrison et al, 1975, 1978; Morrison and O'Brien, 1976; Briarty et al, 1979). In addition, the extensive literature describing major metabolic events associated with oat kernel filling, including the contribution of tissues surrounding the kernel, has been reviewed admirably by Youngs et al (1982).

To this point, only those events relating to embryo and endosperm development have been considered. It is equally important to recognize, however, that these events take place within several layers of maternal (ovary) tissue, which become progressively reduced in thickness during endosperm and embryo development. At maturity, these maternal tissues (the nucellus, seed coat, and pericarp) are highly compressed and provide the primary protective structures around the kernel. The caryopsis, or groat, consisting of embryo, endosperm, and outer layers, is in turn contained within two floral bracts (the lemma and palea), which at maturity are dehydrated, fibrous structures commonly termed the *hulls*. In most domestic oat cultivars, the hulls may contribute up to 30% of the dry weight of the complete oat kernel and usually remain on the groat after threshing. In hull-less varieties, the hulls are less tightly appressed to the groat and are usually removed by routine threshing procedures. A detailed description of the outer tissues of the wild oat kernel has been published recently by Morrison and Dushnicky (1982).

On the basis of this very simple description of oat development, it is not difficult to appreciate that the different genetic origins of the major components of the kernel give rise to morphologically and chemically different entities, all highly compressed and all intimately associated with one another. And because the developing groat is dependent upon vegetative structures of the parent plant for most of its nutrients, it is equally important to recognize that the final product is influenced strongly by the environment (e.g., soil moisture and nitrogen levels, day length, and temperature). It is also a feature of cereal grains that endosperm cell maturation occurs not uniformly throughout the kernel, but sequentially from the center of the kernel toward the periphery (Evers, 1970). Therefore, there is usually a gradation in structure and composition from the inner to the outer endosperm, and although they may be subtle, the differences can often be recognized microscopically. In some instances (e.g., niacin, aromatic amines), reserves may be restricted entirely to one tissue. Such gradation and compartmentalization influence markedly the composition of products derived from different regions of the groat.

IV. MICROCHEMICAL ORGANIZATION OF THE MATURE GROAT

The remainder of this chapter emphasizes the microchemical properties of the groat. However, because the hulls contribute significantly to total kernel weight (as much as 30%) and are an integral part of the harvested grain of all but a few commercial hull-less types, their chemistry and structure should be considered at least briefly. Essentially the hulls are leaflike structures that tightly enclose the groat and provide protection during development. During early development, they contain several different tissues, including photosynthetic and vascular tissues for nutrient transport, and they contribute significantly to groat nutrition (Jennings and Shibles, 1968). At maturity, however, the hulls are dry, brittle, and devoid of significant metabolic activity. Instead, as a result of gradual degeneration of cellular functions, they persist primarily as compressed layers of cell wall material containing a range of carbohydrate and lignin polymers.

Whistler (1950) reported 30% hemicellulose in oat hulls, whereas Rasper (1979) found approximately 50% hemicellulose associated with almost 40% crude cellulose and 10% crude lignin. The contribution of oat hulls to the total dietary fiber content of whole oat preparations is therefore considerable. They are not easily digested by nonruminants, and most oat breeding programs attempt to increase the nutritional quality of oats by reducing hull percentages. Pomeranz et al (1973) analyzed several oat species and found an average of 4.2% protein in the hulls. The amounts are minor, however, and would contribute only 1–2% of the total kernel protein.

In addition to cellulose, hemicellulose, lignin, and protein, recent studies have demonstrated that water extracts of wild oat hulls contain short-chain fatty acids and several phenolic compounds, including ferulic, caffeic, and *p*-coumaric acid, as well as vanillin and protochatechualdehyde (Chen et al, 1982). These authors were unable to demonstrate that these compounds made any significant contribution to seed dormancy in wild oats, although they did indicate that simple hull removal from a particular strain of dormant wild oats dramatically

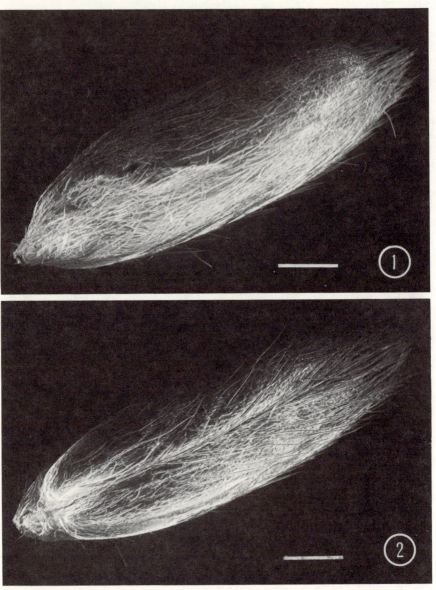

Figs. 1 and 2. Scanning electron micrographs of kernels (groats) of a common domestic oat (var. Sentinel) showing the dorsal or upper surface (Fig. 1) and the ventral or crease side (Fig. 2) after hull removal. The germ area is visible at the lower left of Fig. 1. Note the extensive trichome development. Bar = 1.25 mm.

increased germination percentage in the population. Black (1959) had earlier shown a marked negative effect of hulls on oat germination, and although the mechanism of the influence is poorly understood, it is at least apparent that the hulls may play a significant role in germination processes.

After the hulls are removed, the remaining groat is not unlike other common cereals in general morphology. However, the groat is usually longer and more slender than wheat or barley kernels, and most domestic oat cultivars are also more extensively covered by hairs, or trichomes, which often obscure other surface details. The trichomes are hollow, single-celled projections of the pericarp, which are clearly resolved with the SEM (Figs. 1 and 2). On the dorsal surface, a slight oval indentation indicates the location of the embryo at the proximate end of the groat (Fig. 1), and a crease is visible on the ventral surface (Fig. 2). The groat is attached at its embryo end to the parent plant and, during development, nutrients are transported into the grain by vascular tissues located at the deepest point in the crease. The conducting tissue is situated near the central axis of the groat, and although probably not functional at maturity, it remains in this location as a thin strand of fibrous tissue. The vascular bundle in the groat has not received particular attention from cereal anatomists, but it is superficially similar to equivalent structures in the wheat grain, which have been described in detail by Zee and O'Brien (1970a, 1970b, 1971).

Although the groat is particularly hairy in comparison with other cereal grains, breeders have been rewarded in their efforts to manipulate groat "hairiness." Several breeding lines have been developed in which hairiness is pronounced only at the distal end of the groat (Fig. 3), whereas other lines occasionally demonstrate excessive trichome development (Fig. 4). The wide variation in groat hairiness among cultivars is but one example of the genetic diversity in the oat crop.

For discussion purposes, it is useful to consider that the groat contains three morphological and chemically distinct components. Each of these—the bran, germ, and starchy endosperm—may in fact be composed of several different tissues; but their collective structural characteristics are so distinctly different from each other that, with mechanical disruption, the groat tends to fragment through natural cleavage regions into the three fractions. In practice, its high lipid content and soft texture make the groat difficult to separate cleanly into the three fractions by commercial methods. Nonetheless, because the several different tissues of each fraction tend to remain aggregated (acting in products as a single constituent), and because most other cereals can be milled readily into similar fractions, equivalent terminology is useful in discussing all cereals. The terms *bran, germ,* and *starchy endosperm* are traditional description of *commercial* fractions, however, and do not accurately reflect the genetic, chemical, or functional characteristics of each fraction. Figure 5 is a diagram of the anatomy of a groat, showing the relationships in longitudinal (left) and cross sections (lower right). Approximations of the three major industrial fractions are shown within the figure as (A) bran, (B) starchy endosperm, and (C) part of the germ adjacent to the starchy endosperm. All subsequent micrographs are enlargements of tissues in cross sections taken from groats roughly at level (C) in the longitudinal diagram.

A. Bran

Although the starchy endosperm contributes in excess of 80% of the dry weight of the groat, the properties of the bran most influence the quality characteristics of oats. Indeed, those characteristics, which are so important to a germinating grain (in terms of metabolic activity; large reserves of phytin, vitamins, protein, and lipid; and tough fibrous cell walls), also "make or break" oat products.

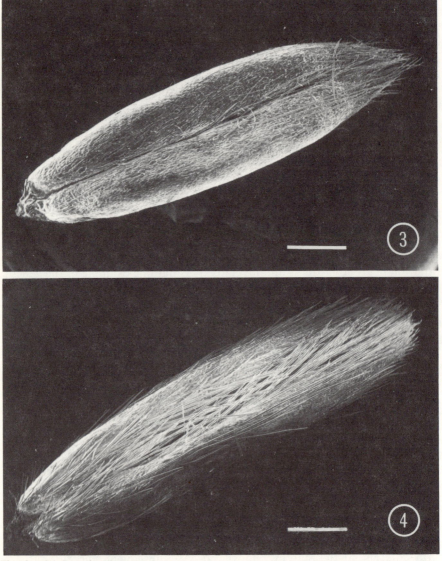

Figs. 3 and 4. Scanning electron micrographs of groats showing examples of the wide range in groat trichome development, from minor (Fig. 3) to excessive (Fig. 4). Material supplied by V. D. Burrows. Bar = 1.25 mm.

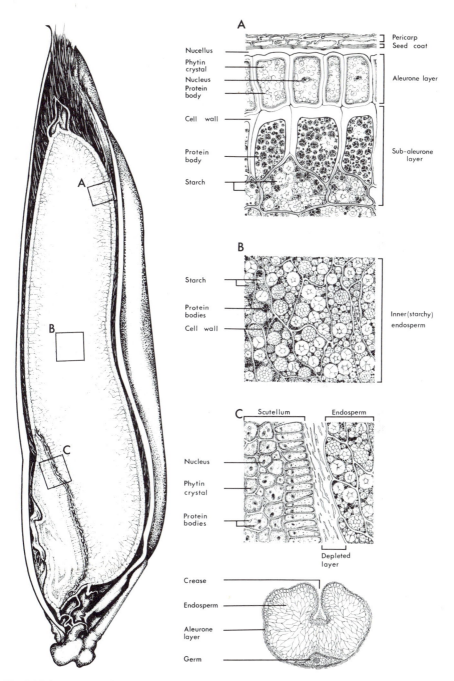

Fig. 5. Major structural features of the oat kernel. On the left is a kernel (with hulls on) that has been split longitudinally to reveal the approximate size and location of the major tissues. At the lower right is a view of a cross section of the groat. (A), (B), and (C) are higher magnifications of portions of the bran, starchy endosperm, and germ.

Bran is essentially the envelope of the groat, composed of the outermost pericarp, immediately under which is found a double-layered cuticle or testa. Next is the nucellus, a thin, compressed layer. All three structures are remnants of the ovary. The developing and expanding grain caused them to become highly compressed and, occasionally, indistinguishable from one another. Although these tissues initially contained metabolically active cells, compression results in a loss of cytoplasm, with only the structural components (cell walls) remaining. Thus, the layers are primarily carbohydrate and fibrous, probably with an abundance of lignin or related phenolic compounds, which make the layers rather tough and difficult to digest. The chemical composition of the structures, including the waxy material in the testa, is largely unknown, although Kolattukudy (1977) has proposed that cutin, the major component in plant cuticles, is a polymer of aliphatic fatty acids containing small amounts of ferulic and coumaric acid esters.

Perhaps the most important component of the bran is the aleurone layer, which is located immediately adjacent to the nucellus and, with the outer layers, surrounds both the starchy endosperm and all but a small portion of the germ. The layer is usually one cell thick, each cuboidal cell filled with distinctive cytoplasmic material. The aleurone layer is by far the thickest part of the bran fraction, and although it is developmentally and genetically a part of the starchy endosperm, its tenacious adherence to the outer layers during processing has resulted in its traditional recognition as a major component of bran. The oat aleurone layer is approximately 50–150 μm thick.

One of the most interesting features of the aleurone layer is the apparently high concentration of phenolic compounds in the cell walls. Phenolic compounds are often autofluorescent, and low-magnification fluorescence microscopy highlights the distribution of suspected phenolics (Plate 1). Bright blue autofluorescence is characteristic of all bran layers, including the trichomes, but most of it is found in the aleurone layer. High magnifications show clearly that the blue fluorescence is associated primarily with the cell walls (Plate 2). The precise chemistry of the fluorescing compounds in oats is not known, but both wheat (Fulcher et al, 1972a) and barley (Fincher, 1976) contain significant concentrations of ferulic acid esterified to the carbohydrate polymers in the wall. The oat aleurone autofluorescence is probably also the result of concentrations of ferulic acid. Its microscopic appearance is indistinguishable from that of wheat and barley; it undergoes a characteristic spectral shift in alkaline media (Fulcher et al, 1972a); and it has been identified as a constituent of oatmeal containing mostly bran (Durkee and Thivierge, 1977). Although the total ferulic acid content of cereal grains is not particularly high, the compound is obviously highly localized.

The function of ferulic acid in the grain is not clear, but its location suggests that it may be related to protection of the grain from microorganisms, or perhaps to structural properties of the aleurone cell wall. Certainly, ferulic acid inhibits the growth of many seedborne microorganisms (Fulcher, 1972), and phenolic molecules similar to ferulic acid combine readily with proteins to form a less digestible complex. Treatment of bran with alkali (e.g., potassium hydroxide) releases much of the fluorescent substance, which accounts for the common experience that alkaline processing often leads to excessive, and quite undesirable, protein-phenol interactions.

An additional surprising feature of the aleurone cell wall is the suspected presence of mixed linkage β-(1→3)(1→4)-D-glucans. Using Congo red staining, which has a considerable affinity for β-glucans and related carbohydrates (Wood and Fulcher, 1978), highly localized deposits of stained fluorescent material are demonstrated along the inner region of the aleurone cell wall (Plate 3).

Exposure of GMA-embedded sections to a specific mixed-linkage β-glucanase obtained from *Bacillus subtilis* (Wood et al, 1983) completely removes the Congo red staining of the inner wall layer. The concentrations of the β-glucans in the oat aleurone layer are relatively minor in comparison with those in the underlying starchy endosperm, but their presence may contribute significantly to the water-binding capacity of the bran and hence to its efficacy as dietary fiber. Similar β-glucans have also been detected recently in wheat and barley aleurone cells (Bacic and Stone, 1981). The inner aleurone wall of wheat may also contain protein (Fulcher, 1972; Bacic and Stone, 1981), but this has not yet been confirmed for oats. At the very least, the groat aleurone cell wall is a tough, complex structure that is relatively resistant to penetration by microorganisms. Recent studies in this laboratory have confirmed the very old observation (Girard, 1884) that the aleurone layer is also quite resistant to mammalian digestive enzymes.

Inside its rather impermeable cell wall, each aleurone cell is packed with numerous individual protein bodies (aleurone grains), and there is some indication that the aleurone proteins are chemically somewhat different from those found in the adjacent starchy endosperm. For example, application of fluorescent protein stains such as anilino-naphthalene sulfonic acid (ANS) (Fulcher and Wong, 1980; Gates and Oparka, 1982) invariably results in different stain intensities in the two tissues (Plate 4); very little stain is taken up by the protein matrix of the aleurone grains in comparison with that of the starchy endosperm. Such differences in stain intensity may simply indicate differences in protein concentration in the two tissues, but they may also reflect differences in protein composition. Published comparisons of amino acid composition of oat bran and starchy endosperm (Pomeranz et al, 1973) and isolated aleurone and endosperm protein bodies (Donahowe and Peterson, 1983) have not demonstrated significant differences between the two tissues, although in wheat the aleurone layer may contain two to three times the basic amino acid content of the starchy endosperm (Fulcher et al, 1972b).

Recent evidence from this laboratory, however, suggests that substantial differences in amino acid profiles may indeed exist among the major groat tissues. Amino acid analysis of hand-dissected fractions of domestic oats showed that the lysine content of bran protein (which is primarily derived from the aleurone layer) was approximately 25% higher than that of the endosperm protein, and similar increases were observed for threonine, serine, and alanine. In contrast, significant reductions were noted for bran levels of glutamic acid, isoleucine, leucine, and phenylalanine. Complete data for the cultivar Scott (17.7% protein) are shown in Table III; high (21%) and low (14%) protein cultivars have also been analyzed and show similar amino acid profiles, but these data are not included. In any case, the results in Table III indicate that significant differences exist in amino acid composition between the bran and starchy endosperm, although a comparison of SDS-polyacrylamide gel electrophoresis patterns of isolated oat endosperm and aleurone proteins has shown only minor differences (Donahowe and Peterson, 1983).

Regardless of possible differences in protein composition between aleurone and starchy endosperm, however, the fact remains that oat bran, which includes the aleurone layer, is a major contributor to the total protein complement of the groat. In high-protein cultivars, such as Goodland, the bran may contain up to 32.5% protein and contribute more than 40% of the dry weight of the grain (Youngs, 1972). Lower-protein cultivars are only slightly less impressive in this regard, and in seven different cultivars that were hand-dissected and analyzed by Youngs (1972), it was apparent that the bran typically contains roughly double the protein concentration of the starchy endosperm. Overall, bran contains approximately half of the total groat protein.

Within each protein body are regions that are unstained by any protein-specific dye, and the composition of the nonprotein regions of the aleurone grain has intrigued and perplexed cereal chemists for over 100 years. Recently, the electron microscope has been used to demonstrate that two separate and chemically distinct types of structures are actually embedded in the protein matrix of oat (Bechtel and Pomeranz, 1981), wheat (Fulcher et al, 1981), and barley (Jacobsen et al, 1971) aleurone grains. One of these structures, the globoid, has been identified as the primary site of phytin accumulation in wheat and barley (Jacobsen et al, 1971; Fulcher et al, 1981) and in oat embryos (Buttrose, 1978). Phytin is a relatively insoluble salt (calcium/magnesium/potassium) of myoinositol hexaphosphate (Lolas et al, 1976), and, although the oat aleurone globoids have not been shown to contain phytin, they are virtually indistinguishable microscopically from wheat and barley phytin deposits (Bechtel and Pomeranz, 1981). Chemical analyses have demonstrated that oat bran contains three to five times more potassium, calcium, magnesium, and

TABLE III
Percentage of Amino Acid Distributed in Groat Tissues[a]

	Whole Groats	Starchy Endosperm	Bran	Germ
Aspartic acid	8.68	8.58	8.77	10.73
Threonine	3.42	3.60	3.91	4.79
Serine	4.80	6.04	6.60	5.81
Glutamic acid	19.98	20.80	16.12	13.01
Proline	6.29	5.98	5.90	4.74
Glycine	9.21	8.26	10.56	11.17
Alanine	6.94	6.23	8.37	9.74
Cystine	0.58	1.17	2.08	0.17
Valine	6.65	6.12	6.33	6.78
Methionine	1.91	2.19	1.78	2.05
Isoleucine	4.46	4.40	3.68	3.76
Leucine	8.22	8.15	7.27	7.18
Tyrosine	2.63	2.81	2.40	1.95
Phenylalanine	4.51	4.42	3.90	3.43
Histidine	2.17	2.11	2.10	2.35
Lysine	3.81	3.42	4.29	6.55
Arginine	5.73	5.73	5.89	6.59

[a] Cultivar Scott: percentage of protein = 17.7 (N × 6.25). Data are given as percentages of amino acids analyzed, with each analysis conducted in triplicate.

phosphorus than the endosperm (Peterson et al, 1975). Phytin globoids are optically anisotropic, or crystalline, and are therefore microscopically detectable using polarizing optics as shown in Plate 5. They can also be detected with conventional bright-field optics after staining with Alizarin Red S (Plate 6), which has been used to localize phytin reserved in wheat (Fulcher, 1972). A new staining technique using the fluorochrome acriflavine HCl has been developed for more sensitive detection of phytin globoids in wheat and rapeseed (Fulcher, 1982; Yiu et al, 1982), and its application to sections of groat tissues emphasizes the highly localized nature of phytin distribution (Plate 7). The technique is far more sensitive and specific for phytin crystals than conventional polarizing and bright-field staining methods. There may be several phytin globoids or crystals in each aleurone grain, and often regions of the groat vary considerably in the number of crystals per cell (usually more in the ventral than in the dorsal aleurone tissues).

The significance of the high aleurone-phytate levels to grain physiology can only be estimated. Obviously, the complex is an excellent source of phosphorus and cations for the germinating kernel, and occasional suggestions have been made that it may act as an energy source or dormancy initiator (Lolas et al, 1976). Recent interest in phytin has shifted to the nutritional impact of the complex, and it is becoming increasingly apparent that dietary phytin of cereal origin adversely affects absorption, by chelation, of trace elements such as zinc (see review by Erdman, 1981). Because oat phytin is apparently sequestered in a fashion similar to that of other cereals, the dietary effects of oat phytin might well be similar to those exerted by other grains.

The other inclusion in each aleurone grain is not only one of the most fascinating groat components, but one that has had almost as many labels attached to it as there are cereal microscopists. In the past few years, the structure has been termed a *protein-carbohydrate* body (Jacobsen et al, 1971; Bechtel and Pomeranz, 1981); but there is little solid evidence for such a label and absolutely no hint of its biological function in the grain. The structure can be stained selectively using the periodate/Schiffs (PAS) reaction to demonstrate periodate-sensitive carbohydrates (Fulcher et al, 1981; Table I), but until recently little additional information has been available. It has been known for decades, however, that wheat bran is one of the richest natural sources of the B vitamin niacin (Kodicek, 1940). Oats are not far behind, and if one accepts the premise that most groat constituents are compartmentalized, then significant concentrations of niacin should occur in specific structures. This is the case, and fluorescent staining using cyanogen bromide and *p*-aminobenzoic acid (Table I) shows clearly that impressive reserves of niacin are found in the periodate-sensitive carbohydrate body (Plate 8). Thus, the "protein-carbohydrate" body is in fact a "niacin" body, and the nutritional properties of the niacin reserves no doubt are influenced strongly by its location, associated molecules, and morphology.

No information appears to be in the literature regarding the chemistry of the niacin complex in oats. However, in the 1940s, a group in the United Kingdom directed by Kodicek began to isolate and characterize the complex in wheat. A large portion of wheat bran niacin is not digestible by mammalian systems, and Kodicek's original intent was to determine the chemical basis for its nutritional unavailability (e.g., Kodicek and Wilson, 1960). Many years later, we are still puzzled by this nutritional peculiarity, but Kodicek's studies (Mason et al, 1973; Mason and Kodicek, 1973a, 1973b) have shown that niacin is intimately associated with a range of other molecular species. In addition to the carbohydrate (primarily glucose) polymer to which niacin is "bound," various phenolic acids, an aromatic amine, several simple sugars such as xylose and arabinose, and perhaps protein have also been found associated with niacin (Kodicek and Wilson, 1960). Several of these constituents are relatively easy to detect microscopically in oats as well. The carbohydrate component in wheat has been illustrated (Fulcher et al, 1981) and is similar in oats. There is also an autofluorescent component that is not unlike ferulic acid (Plate 2). The other major component, an aromatic amine, is also detectable microscopically by reaction with *p*-dimethylaminocinnamaldehyde (Table I) to yield a deep-red reaction product (Plate 9). The precise chemistry of the aromatic amine in the niacin body is unknown, but Mason and Kodicek (1973b) found *o*-aminophenol in the complex after hydrolysis, and this has been confirmed for oats (Fulcher et al, 1981). If the compound is in fact *o*-aminophenol, then there is a very powerful antioxidant at the core of each aleurone grain. Such a molecule might be expected to influence several physiological functions in the groat and may also contribute to a range of previously unexplained industrial problems (such as occasional "browning" reactions in cereal products). Based on microscopic comparison, the groat appears to contain higher concentrations of the amine than any other cereal. As in other cereals, the niacin/aromatic amine complex is confined entirely to the aleurone layer.

Obviously, aleurone grains, or protein bodies, are very complex, containing a protein matrix in which discrete deposits of phytin, niacin, phenolics, and carbohydrates are embedded, each in specific regions of the bodies. In turn, each aleurone grain is surrounded by lipid droplets (similar to those in the germ) and a cytoplasmic network of endoplasmic reticulum, mitochondria, and plastids (Bechtel and Pomeranz, 1981). A prominent nucleus is located at the center of each aleurone cell. All of these components are densely packed within the thick, fibrous cell wall, and it is little wonder that workers in the last century found that their own digestive systems had little impact on cereal bran (Girard, 1884).

Any discussion of bran microchemistry would be incomplete without at least brief mention of the contribution of the aleurone layer to germination. Since the last century, it has been well known that the aleurone layer is a major source of hydrolytic enzymes that degrade and permit translocation of reserve materials from the endosperm to the developing embryo of the germinating grain. Some of these enzymes—e.g., lipase (Matlashewski et al, 1982; Urquhart et al, 1983)—are present in the bran at maturity, whereas others—e.g., α-amylase and maltase—are synthesized primarily during germination (Simpson and Naylor, 1962; Naylor, 1966). Synthesis of hydrolytic enzymes by the aleurone layer during germination appears to be at least partially controlled by embryo-derived hormones such as gibberellin (Simpson and Naylor, 1962; Simpson, 1965; Naylor, 1966) and perhaps abscisic acid (Quail and Carter, 1969) and ethylene (Meheriuk and Spencer, 1964; Adkins and Ross, 1981). However, although Schander (1934) demonstrated rather elegantly that an intimate and subtle metabolic relationship exists between the embryo and aleurone layer during germination, the extent to which degradation of the starchy endosperm by aleurone-derived enzymes is dependent upon hormones produced by the embryo is not at all clear. For example, embryo-derived gibberellins are important factors in the control of aleurone function in wild oats (Simpson, 1965; Taylor and Simpson, 1980). However, Upadhyaya et al (1982) have demonstrated that response to hormones and germination inhibitors among genetically pure lines of wild oats is variable, and Simpson (1978) has suggested that the aleurone layer in domesticated oats and nondormant wild oats may even be somewhat autonomous with respect to hormonal supply.

In any case, it is unlikely that control of aleurone activity can be attributed to a *single* metabolic event such as hormone stimulation, and the extensive review of Simpson (1978) emphasizes that germination mechanisms must be viewd as a function of a wide range of factors, including metabolic, structural, and environmental influences. Recent evidence relating to wheat, for example, suggests that response to hormones is a function of temperature-dependent membrane transitions in the aleurone layer, which may, in part, be controlled by the degree of dehydration of the kernel during maturation (Armstrong et al, 1982; Norman et al, 1982).

B. Starchy Endosperm

Depending on the cultivar, the starchy endosperm of domestic oats may contribute anywhere from 55% to almost 70% of the weight of the mature groat (Youngs, 1972). The tissue contains major reserves of starch, protein, lipid, and β-glucans (gums), all of which are hydrolyzed during germination by enzymes derived from the aleurone layer (and possibly the scutellum) to provide nutrients for the growing embryo. Unlike the surrounding aleurone layer and adjacent germ (Fig. 5), the mature starchy endosperm of oats and other cereals is considered to be relatively inactive metabolically. Minor amounts of various enzymes may exist in the endosperm at maturity (especially if the grain is harvested early), but the potential impact of endosperm enzymes on nutritional and processing characteristics does not appear to be well documented.

Compositionally and structurally, the endosperm is perhaps the simplest of all groat tissues. It is composed of only one cell type, each an individual storehouse of reserves. Over the decades, oat breeders have successfully increased the total size of the starchy endosperm and often have been able to alter the ratio of the major components of the groat, but the general architecture of the endosperm is not unlike that of the wild oat (cf. Fig. 5 and Simpson, 1978, Plate 1). Each cell sequesters significant reserves of starch, protein, lipid, and β-glucans, and it is primarily the concentrations of these components that vary from cell to cell.

Each endosperm cell is surrounded by a cell wall with distinctive physical and chemical properties quite different from those in other grain tissues. Until recently, identification and localization of specific cereal cell wall carbohydrates have depended primarily upon cell wall isolation, selective extraction, and chemical analysis, followed by electron microscopy of chemically defined fractions. In studies of barley, for example (Fincher, 1975), this approach has yielded considerable information regarding the spatial arrangement of particular carbohydrates within the wall, but it is not particularly amenable to rapid and specific visualization of carbohydrates for screening purposes. In contrast, however, several simple fluorescence microscopic techniques have been devised for routine detection of endosperm cell wall constituents (for more complete technical details, see Fulcher and Wong, 1980, and Fulcher and Wood, 1983), and these have been applied with some success to all major cereals.

One of the methods, the fluorescent PAS method (Table I), is a particularly useful procedure for detecting periodate-sensitive carbohydrates in endosperm tissue. The PAS and related periodate/acriflavine (PAA) methods have been used for several years as reliable indicators of vicinal hydroxyl groups in carbohydrates (O'Brien and McCully, 1981). When applied to oat sections, the highest concentration of periodate-sensitive material in the endosperm wall occurs in the middle lamella (Plate 10). The chemical composition of reactive material in the wall has not been defined further, but the staining results at least indicate that different wall components are not distributed uniformly throughout the wall. Strongly PAS-positive materials constitute a relatively minor part of a typical oat endosperm cell wall.

Although the middle lamella portion of the endosperm cell wall shows marked PAS positivity, the remainder of the structure is only lightly stained by the technique, and until fairly recently other microscopic staining techniques with demonstrable chemical specificity for endosperm wall components have not been available. However, an extensive series of studies on the interaction of fluorescent dyes with endosperm cell walls (primarily in oats) has shown that Calcofluor and Congo red can be used as sensitive and specific markers for mixed-linkage β-glucans (Wood and Fulcher, 1978; Fulcher and Wong, 1980; Wood, 1980, 1981, 1982; Fulcher and Wood, 1983; Wood et al, 1983). Application of either Congo red (Plate 3) or Calcofluor (Plate 11) to oat endosperm sections shows clearly that the majority of the cell wall is composed of this $(1\rightarrow3)(1\rightarrow4)$-$\beta$-D-glucan, a high-viscosity carbohydrate that has been isolated from oat flour by conventional methods (Wood et al, 1977, 1978). This suggests that the endosperm cell wall is the primary reserve of mixed-linkage β-glucan, and this view is confirmed by the observation that material stainable with Calcofluor or Congo red can be completely removed from tissue sections by pretreatment with a specific mixed-linkage β-glucan hydrolase (Wood et al, 1983). However, in some oat cultivars, considerable variation in gum content may be found from one region of the endosperm to another. In the mid or inner endosperm, the cell walls are usually relatively thin, whereas walls of cells immediately adjacent to the aleurone layer (the subaleurone layer) may be up to four to five times thicker (Plate 12). This appears to be particularly true of high-protein cultivars, and comparisons of equivalent tissues in high-protein (Plate 12) and low-protein (Plate 13) breeding lines further emphasize the considerable compositional variation between different cultivars. Chemical analyses have also indicated that there is a genetic component to β-glucan variation in oats (Wood et al, 1978).

Mixed-linkage β-glucans appear to be important constituents in most cereal endosperm cell walls. In barley, these β-glucans may comprise 75% of the endosperm cell wall (Fincher, 1975). Although they have not been quantitated on a cell wall basis, microscopic (Plate 11; Fulcher and Wood, 1983) and extraction studies (Wood et al, 1978) indicated that oat endosperm β-glucan levels are also relatively high. Prentice et al (1980) have subsequently found β-glucan levels of 6.6 and 4.8% in two domestic oat varieties. Wheat contains significantly lower concentrations of the polymer, but even in wheat, over 90% of endosperm cell wall glucose occurs as β-glucan (Bacic and Stone, 1980).

The physiological role of β-glucans in oats has been largely ignored and has been considered only superficially in relation to other cereals. However, in barley they have been implicated as osmoregulators because of their marked hygroscopicity and as additional energy sources (supplementing starch) for the growing embryo (see Bamforth, 1982, for discussion). Aastrup (1979) has observed that changes in environmental moisture levels during grain development affect barley β-glucan levels, and this observation may support the osmoregulatory hypothesis. It is not yet known if oat β-glucan levels are similarly affected by moisture during development.

Some recent interest has also centered on the potential regulatory role of β-glucan in germination processes. Based on the observations of investigators at the turn of the century (e.g., Brown and Escombe, 1898; Mann and Harlan, 1915) that endosperm cell wall degradation is a prerequisite for diffusion of hydrolytic enzymes during germination, β-glucans (and β-glucan hydrolases) have received increasing attention as potentially important determinants of germination rate, at least in barley (Munck, 1981). This possibility has not yet been investigated in oats, but barleys with improved malting (germination) characteristics apparently have been identified in breeding programs by application of the Calcofluor staining technique to detect thin-walled varieties (S. Aastrup, *personal communication*).

Although the bran and germ contain the highest levels of basic amino acids (Table III), in comparison with wheat or barley, oat endosperm proteins are also relatively rich in basic amino acids, including lysine, and may contain up to double the concentration of this essential amino acid (Table III; Zarkadas et al, 1982). Oat endosperm protein is very similar in amino acid balance to rice endosperm, which is generally considered superior to other cereals in nutritional quality. As with several other oat components, protein concentration varies considerably, and breeders have produced several oat cultivars (e.g., Goodland, Dal, and Hinoat) with typically higher protein concentrations than traditional cultivars. Because protein quality and content are of some importance in a primary end use of the crop (livestock nutrition), development of high-protein varieties has been a major goal of oat breeders for several years. Groat protein levels, although primarily under genetic control, also may vary in response to environmental conditions (Youngs and Gilchrist, 1976). As discussed earlier, the contribution of the bran to total groat protein is typically much higher than in other cereals, and the starchy endosperm provides correspondingly less ($\sim$40–50%) (Youngs et al, 1982) than, for example, the starchy endosperm of wheat, which usually provides 70–75% of the total grain protein. The oat endosperm values may, in fact, represent significant underestimates because of technical difficulties associated with obtaining clean separation of bran and endosperm for analysis; regardless of their accuracy, however, the endosperm remains an important protein source. In any case, oats are normally consumed in the human diet as whole-grain products, whereas wheat typically has the bran removed.

Chemically, oat endosperm proteins are also quite distinct from those of most other cereals. In wheat, the salt-soluble globulins are relatively minor endosperm components; but in oats they appear to be the major fraction, providing over 55% of the endosperm protein (Peterson and Smith, 1976). Albumins, prolamins (which predominate in most other cereals), and glutelins are also present. Youngs et al (1982) have reviewed the rather extensive literature describing oat endosperm proteins; not surprisingly, the different protein fractions exhibit significant differences in amino acid composition and in sequence of development. During maturation, globulin production increases significantly relative to other protein fractions (Peterson and Smith, 1976).

At maturity, wheat and barley endosperm proteins appear microscopically as a relatively homogeneous matrix in which the starch granules are embedded. In contrast, oat endosperm proteins are packaged in roughly spherical protein bodies. Saigo et al (1983) have recently provided strong electron microscopic evidence that storage protein is synthesized on membrane-bound ribosomes and is subsequently transported within the cisternae of the endoplasmic reticulum to vacuoles, where it condenses to form protein bodies. Although it has not yet been determined whether all oat protein fractions are synthesized and accumulated by the same mechanism, there are indications that there is a degree of compartmentalization within the protein body. Saigo et al (1983) and Bechtel and Pomeranz (1981) have described at least two structurally distinct components within the bodies, and Saigo et al (1983) have suggested that one of the inclusions may represent either lipid or prolamin, a minor oat storage protein. Earlier studies also noted arginine-rich areas in the oat endosperm (Fulcher et al 1972b).

Typically, protein body size varies considerably, ranging from approximately 0.2 to 6.0 μm in diameter. The larger bodies are invariably concentrated in the outer endosperm cells (i.e., in the subaleurone region), and it is here that differences in overall groat protein content are most dramatically illustrated. In high-protein types such as Hinoat ($\sim$23% protein), the subaleurone cells appear to "specialize" in protein production, so that this region of the groat is composed mostly of large protein bodies, with very few starch grains (Plate 14). In groats from lower-protein cultivars such as Elgin ($\sim$15% protein), the protein bodies are still relatively large in the subaleurone cells, but the protein concentration in this region is diluted with large starch grains (Plate 15). Endosperm protein bodies are not difficult to prepare in the laboratory, and their considerable size range can be shown readily using the SEM. Using this approach, the large protein bodies apparently are aggregates of much smaller bodies (Figs. 6 and 7), although it is not clear whether there are any compositional differences between the various subunits of the aggregates, as might be expected from the results of Saigo et al (1983). Flavonoid residues have also been detected in the protein-rich subaleurone tissues (Plate 16; Fulcher and Wong, 1980), but their precise distribution in relation to protein body morphology has not been determined.

A comprehensive analysis of lipid composition in 12 different oat cultivars was published by Sahasrabudhe (1979). In confirming yet another example of wide chemical variation in different cultivars, the author noted that total lipid content in the groat may range from 2 to 12%. These figures agree closely with those (3–11.5%) reported by Brown and Craddock (1972), who determined that the average concentration in most oat varieties was 5–9% lipid. Most importantly, the neutral lipid fraction (primarily triglyceride) is the most variable, whereas the levels of the other lipid fractions remain relatively constant (Sahasrabudhe, 1979). Palmitic, oleic, and linoleic acids are the major components in the oat neutral lipid fractions. Both the total lipid and fatty acid composition appear to be highly heritable (Youngs et al, 1982), and extensive studies by Youngs and

Forsberg (1979) and Gullord (1980) have indicated no significant correlation between lipid and protein content in the groat.

In most cereals, the bran and germ contain the highest concentrations of lipid. In wheat, for example, the aleurone layer may contain approximately 8–20%

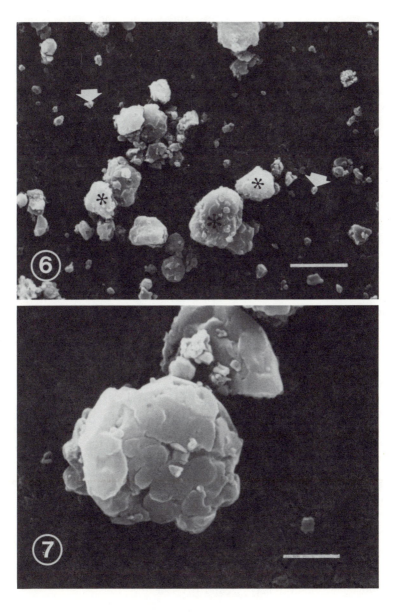

Figs. 6 and 7. Scanning electron micrographs of endosperm protein bodies. Fig. 6, large (*) and small (arrows) protein bodies isolated from Hinoat groats. Fig. 7, view similar to Fig. 6 but at higher magnification to show the compound nature of the large protein bodies. Bars both equal 100 μm.

lipid (60–75% of which is triglyceride), whereas the endosperm contains only 0.8–1.0% lipid, with typically lower proportions of neutral components (Hargin et al, 1980). In the oat groat, the aleurone layer and germ are also rich in lipids, but, as shown in a limited number of analyses, the endosperm may also contain as much as 6–8% total lipid (Youngs et al, 1979). Therefore, the endosperm is primarily responsible for synthesizing and storing the relatively high lipid concentrations that are characteristic of oats but not of other cereals.

Microscopic detection and localization of cereal endosperm lipids have been generally unsuccessful, although several authors have described a variety of inclusions that were thought to contain lipids on the basis of electron microscopic observations (see Hargin et al, 1980, for discussion). Recently, however, a simple fluorescence staining technique based on the interaction between Nile blue A and neutral lipids has allowed sensitive and specific identification of lipid droplets (spherosomes) in wheat endosperm (Hargin et al, 1980). The method induces intense yellow fluorescence in droplets of neutral lipid and, as expected, reveals high concentrations of lipid droplets in the oat endosperm. The droplets are hexane-extractable and, not surprisingly, the concentration of droplets found in low-lipid cultivars (such as Exeter, Plate 17) is far lower than that of high-lipid types (such as CI4492, Plate 18). The droplets are distributed rather uniformly throughout the cytoplasm between the protein bodies in any given cell; but, as with the proteins and β-glucans described previously, the highest concentrations of lipid droplets in the tissue invariably are found in the subaleurone cells. The detailed chemistry and the changes that occur during the development of oat lipids have been reviewed by Youngs et al (1982).

As with most cereals, the largest single component in the groat is starch (Plate 19), and, according to Youngs et al (1982), several of the physical characteristics of the compound are not unlike those of wheat starch. As expected, starch content is highly variable among different cultivars, ranging from 43 to 64% of the groat (Youngs, 1974; Paton, 1977), and apparently showing a negative correlation with protein content (MacArthur and D'Appolonia, 1979). Although it is also chemically somewhat similar to wheat and barley starch, oat starch may exhibit superior retrogradation and pasting properties (Paton, 1977, 1981). In addition, oat starch is quite unlike that of wheat or barley in that much of the polymer occurs as a compound or aggregate grain composed of several multifaceted individual granules (Plate 20; Bechtel and Pomeranz, 1981), which are similar to those found in rice (Bechtel and Pomeranz, 1978). However, unlike rice, a proportion of the starch also occurs in single, unaggregated granules (Plate 20). The aggregates range in size from 20 to 150 μm in diameter, whereas individual granules are typically only 4–10 μm across.

Lipids, protein, and β-glucans increase in concentration toward the outer regions of the starchy endosperm (i.e., the subaleurone region). As one might expect, starch concentration shows the reverse pattern; it is greatest in the midendosperm and lowest in the subaleurone region (Plate 19). Microscopic detection of starch is a relatively simple matter using appropriate periodate-based fluorescence methods (PAS and PAA, Table I).

C. Germ

Like the aleurone layer, the germ is a viable structure capable of considerable metabolic activity, and it is from the germ that a mature plant arises. The germ actually consists of several structurally and functionally distinct tissues, but most of these, such as the coleoptile, coleorhiza, and rudimentary leaves and roots of the embryonic axis, represent only a tiny fraction of the groat and are of little consequence to the processor.

Approximately at its midpoint, the embryonic axis is attached to a relatively large structure, the *scutellum*, which in turn is constructed of two distinct tissues—the *parenchyma* and the *epithelium*. The scutellar parenchyma probably represents over 80% of the germ weight. It is constructed of a mass of roughly spherical cells that apparently function as nutrient storehouses. The surface of the scutellum is differentiated into elongated cells (the epithelium), which act as absorbing tissue for assimilating nutrients as they are released by hydrolysis from the starchy endosperm during germination. As germination proceeds, the scutellar parenchyma develops vascular tissue for the transport of these nutrients from the absorptive surface of the scutellum to the embryonic axis (Swift and O'Brien, 1971). The size of the germ, relative to the endosperm and bran, is shown in Fig. 5.

In many respects, the composition of the scutellar parenchyma parallels that of the aleurone layer. Its cell walls are much thinner, but they contain high concentrations of autofluorescent material (Plate 21). Smart and O'Brien (1979a, 1979b, 1979c) have demonstrated that the autofluorescent substance in wheat, barley, and probably in oats is the ubiquitous ferulic acid. Like the aleurone layer, each parenchyma cell also contains numerous protein bodies that contain typical crystalline phytin globoids, as described by Buttrose (1978), and that are indistinguishable from those found in the aleurone layer (Plates 5–7). Each protein body is surrounded by an envelope of neutral lipid droplets (Plate 22), and little, if any, detectable starch or β-glucan exists in the scutellum at maturity (Plate 23). One notable difference between the aleurone layer and scutellum is that the protein bodies of the latter are devoid of niacin deposits, as indicated by microchemical methods. Thus, any niacin that might appear in chemical analyses of germ fractions would probably be derived from the aleurone layer, which is adjacent to the outer surface of the germ (Plate 23).

As shown in Table III, the amino acid composition of the germ is strikingly different from that of the starchy endosperm. Lysine content, for example, is approximately 90% higher in the germ than in the starchy endosperm, whereas the glutamic acid content is 37% lower. Most other amino acids also show notable increases or decreases. During development, the germ and starchy endosperm undergo cell division and expand as two independent structures, both enclosed by the surrounding bran layers. As the two structures near maturity, considerable pressure is exerted at the boundary between them. This pressure apparently results in damage to the starchy endosperm adjacent to the germ. The contents of the endosperm cells are degraded at this point, leaving only a few compressed cell walls (Plate 23). Not surprisingly, the region is commonly referred to as the *depleted layer* (Fig. 5), or as the *intermediate layer* (Palmer, 1982), and the chemical and structural discontinuity between the germ and the endosperm forms a natural cleavage along which the two tend to separate when subjected to certain processing conditions.

Because of its several chemical and structural similarities to the aleurone layer, the scutellum (i.e., most of the germ) might be expected to act similarly as a dietary component or during processing. In addition, the suggestion of Mann and Harlan (1915) that the scutellum secretes hydrolytic enzymes into the starchy endosperm during the early stages of germination has been revived in studies on both barley (MacGregor and Matsuo, 1982; Palmer, 1982) and oats (Fulcher and Wood, 1983). Although actual secretion of enzymes by the scutellum has not yet been demonstrated satisfactorily, there is some indication that the scutellum shares some of the hydrolytic functions of the aleurone layer.

V. SUMMARY

The groat is a discrete organism capable of numerous and diverse metabolic activities. It contains a wide range of chemical constituents, including proteins, starch, lipid, vitamins, phenolic compounds, and enzymes, and each of these chemical components is found in specific structures in specific locations in the groat. The concentrations and distribution of each constituent may vary in different oat varieties.

A selection of microscopic methods has been used to illustrate the major constituents in oats. The bran consists of several distinct tissues and contains protein bodies, neutral lipids, ferulic acid, and significant concentrations of niacin, phytin, and aromatic amines. The starchy endosperm is the primary source of starch, protein, and β-glucans. The germ is chemically similar to the bran in several respects but does not appear to contain niacin or aromatic amines.

ACKNOWLEDGMENTS

S. I. Wong provided excellent technical assistance, and his drawing, shown in Fig. 5, is gratefully acknowledged. Thanks are also due to V. D. Burrows, R. Oughton, J. Lacapra, and S. Itz for continuing discussion and for provision of experimental materials or technical assistance. This chapter is Contribution No. 795A from the Plant Research Centre, Agriculture Canada.

LITERATURE CITED

AASTRUP, S. 1979. The effects of rain on β-glucan content in barley grains. Carlsberg Res. Commun. 44:381-393.

ADKINS, S. W., and ROSS, J. D. 1981. Studies in wild oat seed dormancy. I. The role of ethylene in dormancy breakage and germination of wild oat seeds (*Avena fatua* L.). Plant Physiol. 67:358-362.

ANGOLD, R. E. 1979. Cereals and bakery products. Pages 75-138 in: Food Microscopy. J. G. Vaughan, ed. Academic Press, New York.

ARMSTRONG, C., BLACK, M., CHAPMAN, J. M., NORMAN, H. A., and ANGOLD, R. 1982. The induction of sensitivity to gibberellin in aleurone tissue of developing wheat grains. I. The effect of dehydration. Planta 154:573-577.

BACIC, A., and STONE, B. A. 1980. A $(1\rightarrow3)$ and $(1\rightarrow4)$-linked β-D-glucan in the endosperm cell walls of wheat. Carbohydr. Res. 82:372-377.

BACIC, A., and STONE, B. A. 1981. Chemistry

and organization of aleurone cell wall components from wheat and barley. Aust. J. Plant Physiol. 8:475-485.

BAMFORTH, C. W. 1982. Barley β-glucans. Brew. Dig. June, pp. 22-27.

BECHTEL, D. B., and POMERANZ, Y. 1978. Ultrastructure of the mature ungerminated rice (*Oryza sativa*) caryopsis. The starchy endosperm. Am. J. Bot. 65:684-692.

BECHTEL, D. B., and POMERANZ, Y. 1981. Ultrastructure and cytochemistry of mature oat (*Avena sativa* L.) endosperm. The aleurone layer and starchy endosperm. Cereal Chem. 58:61-69.

BLACK, M. 1959. Dormancy studies in seed of *Avena fatua*. I. The possible role of germination inhibitors. Can. J. Bot. 37:393-402.

BONNETT, O. T. 1961. The oat plant: Its histology and development. Ill. Agric. Exp. Stn. Bull. 672. 112 pp.

BRIARTY, L. G., HUGHES, C. E., and

EVERS, A. D. 1979. The developing endosperm of wheat—a stereological analysis. Ann. Bot. 44:641-658.

BROWN, C. M., and CRADDOCK, J. C. 1972. Oil content and groat weight of entries in the world oat collection. Crop Sci. 12:514-515.

BROWN, H. T., and ESCOMBE, F. 1898. On the depletion of the endosperm of *Hordeum vulgare* during germination. Proc. R. Soc. London 63:3-25.

BUTTROSE, M. S. 1978. Manganese and iron in globoid crystals of protein bodies from *Avena sativa* and *Casuarina*. Aust. J. Plant Physiol. 5:631-640.

CHEN, F. S., MacTAGGART, J. M., and ELOFSON, R. M. 1982. Chemical constituents in wild oat (*Avena fatua*) hulls and their effects on seed germination. Can. J. Plant Sci. 62:155-161.

DONAHOWE, E. T., and PETERSON, D. M. 1983. Isolation and characterization of oat aleurone and starchy endosperm protein bodies. Plant Physiol. 71:519-523.

DURKEE, A. B., and THIVIERGE, P. A. 1977. Ferulic acid and other phenolics in oat seeds (*Avena sativa* cv. Hinoat). J. Food Sci. 42:551-552.

ERDMAN, J. W., Jr. 1981. Bioavailability of trace minerals from cereals and legumes. Cereal Chem. 58:21-26.

EVERS, A. D. 1970. Development of the endosperm of wheat. Ann. Bot. 34:547-555.

FEDER, N., and O'BRIEN, T. P. 1968. Plant microtechnique: Some principles and new methods. Am. J. Bot. 55:123-142.

FINCHER, G. B. 1975. Morphology and chemical composition of barley endosperm cell walls. J. Inst. Brew. London 81:116-122.

FINCHER, G. B. 1976. Ferulic acid in barley cell walls: A fluorescence study. J. Inst. Brew. London 82:347-349.

FULCHER, R. G. 1972. Observations on the aleurone layer with emphasis on wheat. Ph.D. thesis, Monash University, Melbourne, Australia.

FULCHER, R. G. 1982. Fluorescence microscopy of cereals. Food Microstructure 1:167-175.

FULCHER, R. G., O'BRIEN, T. P., and LEE, J. W. 1972a. Studies on the aleurone layer. I. Conventional and fluorescence microscopy of the cell wall with emphasis on phenol-carbohydrate complexes in wheat. Aust. J. Biol. Sci. 25:23-34.

FULCHER, R. G., O'BRIEN, T. P., and SIMMONDS, D. H. 1972b. Localization of arginine-rich proteins in mature seeds of some members of the Gramineae. Aust. J. Biol. Sci. 25:487-497.

FULCHER, R. G., O'BRIEN, T. P., and

WONG, S. I. 1981. Microchemical detection of niacin, aromatic amine, and phytin reserves in cereal bran. Cereal Chem. 58:130-135.

FULCHER, R. G., and WONG, S. I. 1980. Inside cereals, a fluorescence microchemical view. Pages 1-26 in: Cereals for Food and Beverages: Recent Progress in Chemistry and Technology. G. E. Inglett and L. Munck, eds. Academic Press, New York.

FULCHER, R. G., and WONG, S. I. 1982. Fluorescence microscopy of cereal grains. Can. J. Bot. 60:325-329.

FULCHER, R. G., and WOOD, P. J. 1983. Identification of cereal carbohydrates by fluorescence microscopy. Pages 111-147 in: New Frontiers in Food Microstructure. D. B. Bechtel, ed. Am. Assoc. Cereal Chem., St. Paul, MN.

GATES, P. J., and OPARKA, K. J. 1982. The use of the fluorescent probe 9-anilino-1-naphthalene sulphonic acid (ANS) as a histochemical stain in plant tissues. Plant Cell Environ. 5:251-256.

GIRARD, A. 1884. Memoire sur le composition chimique et la valeur alimentaire des diverse parties du grain de froment. Ann. Chim. Phys. 3:289-355.

GULLORD, M. 1980. Oil and protein content and its relation to other characters in oats. Acta Agric. Scand. 30:216-218.

HARGIN, K. D., MORRISON, W. R., and FULCHER, R. G. 1980. Triglyceride deposits in the starchy endosperm of wheat. Cereal Chem. 57:320-325.

JACOBSEN, J. V., KNOX, R. B., and PYLIOTIS, N. A. 1971. The structure and composition of aleurone grains in the barley aleurone layer. Planta 101:189-209.

JENNINGS, V. M., and SHIBLES, R. M. 1968. Genotypic differences in photosynthetic contributions of plant parts to grain yield in oats. Crop Sci. 8:173-175.

KODICEK, E. 1940. Estimations of nicotinic acid in animal tissues, blood, and certain foodstuffs. II. Applications. Biochemistry 34:724-726.

KODICEK, E., and WILSON, P. W. 1960. The isolation of niacytin, the bound form of nicotinic acid. Biochem. J. 76:27P-28P.

KOLATTUKUDY, P. E. 1977. Lipid polymers and associated phenols, their chemistry, biosynthesis and role in pathogenesis. Recent Adv. Phytochem. 11:185-246.

LOLAS, G. M., PALAMIDIS, N., and MARKAKIS, P. 1976. The phytic acid-total phosphorus relationship in barley, oats, soybeans, and wheat. Cereal Chem. 53:867-871.

MacARTHUR, L. A., and D'APPOLONIA, B. L. 1979. Comparison of oat and wheat carbohydrates. II. Starch. Cereal Chem. 56:458-461.

MacGREGOR, A. W., and MATSUO, R. R. 1982. Starch degradation in endosperms of barley and wheat kernels during initial stages of germination. Cereal Chem. 59:210-216.

MANN, A., and HARLAN, H. V. 1915. Morphology of the barley grain with reference to its enzyme-secreting areas. U.S. Dep. Agric. Bull. 183:1-32.

MARES, D. J., NORSTOG, K., and STONE, B. A. 1975. Early stages in the development of wheat endosperm. I. The change from free nuclear to cellular endosperm. Aust. J. Bot. 23:311-326.

MASON, J. B., GIBSON, N., and KODICEK, E. 1973. The chemical nature of the bound nicotinic acid of wheat bran: Studies of nicotinic acid-containing macromolecules. Br. J. Nutr. 30:297-302.

MASON, J. B., and KODICEK, E. 1973a. The chemical nature of the bound nicotinic acid of wheat bran: Studies of partial hydrolysis products. Cereal Chem. 50:637-646.

MASON, J. B., and KODICEK, E. 1973b. The identification of o-aminophenol and o-aminophenyl glucose in wheat bran. Cereal Chem. 50:646-654.

MATLASHEWSKI, G. J., URQUHART, A. A., SAHASRABUDHE, M. R., and ALTOSAAR, I. 1982. Lipase activity in oat flour suspensions and soluble extracts. Cereal Chem. 59:418-422.

MEHERIUK, M., and SPENCER, M. 1964. Ethylene production during germination of oat seeds and *Penicillium digitatum* spores. Can. J. Bot. 42:337-340.

MORRISON, I. N. 1975. Ultrastructure of the cuticular membranes of the developing wheat grain. Can. J. Bot. 53:2077-2087.

MORRISON, I. N., and DUSHNICKY, L. 1982. Structure of the covering layers of the wild oat (*Avena fatua*) caryopsis. Weed Sci. 30:352-359.

MORRISON, I. N., KUO, J., and O'BRIEN, T. P. 1975. Histochemistry and fine structure of developing wheat aleurone cells. Planta 123:105-116.

MORRISON, I. N., and O'BRIEN, T. P. 1976. Cytokinesis in the developing wheat grain; division with and without a phragmoplast. Planta 130:57-67.

MORRISON, I. N., O'BRIEN, T. P., and KUO, J. 1978. Initial cellularization and differentiation of the aleurone cells in the ventral region of the developing grain. Planta 140:19-30.

MUNCK, L. 1981. Barley for food, feed, and industry. Pages 427-459 in: Cereals: A Renewable Resource. Theory and Practice. Y.

Pomeranz and L. Munck, eds. Am. Assoc. Cereal Chem., St. Paul, MN.

NAYLOR, J. M. 1966. Dormancy studies in seed of *Avena fatua*. V. On the response of aleurone cells to gibberellic acid. Can. J. Bot. 44:19-32.

NORMAN, H. A., BLACK, M., and CHAPMAN, J. M. 1982. The induction of sensitivity to gibberellin in aleurone tissue of developing wheat grains. II. Evidence for temperature-dependent membrane transitions. Planta 154:578-586.

O'BRIEN, T. P., and McCULLY, M. E. 1981. The Study of Plant Structure: Principles and Selected Methods. Termarcarphi Pty. Ltd., Melbourne, Aust.

PALMER, G. H. 1982. A reassessment of the pattern of endosperm hydrolysis (modification) in germinated barley. J. Inst. Brew. London 88:145-153.

PATON, D. 1977. Oat starch. I. Extraction, purification and pasting properties. Staerke 29:149-154.

PATON, D. 1979. Oat starch: Some recent developments. Staerke 31:184-189.

PATON, D. 1981. Behaviour of Hinoat oat starch in sucrose, salt, and acid solutions. Cereal Chem. 58:35-39.

PEASE, D. C. 1973. Glycol methacrylate copolymerized with glutaraldehyde and urea as an embedment retaining lipids. J. Ultrastruct. Res. 45:124-137.

PETERSON, D. M., and SAIGO, R. H. 1978. Oat seed globulin—extraction, characterization, synthesis, and location. Plant Physiol. 61:(Suppl.) 78.

PETERSON, D. M., SENTURIA, J., YOUNGS, V. L., and SCHRADER, L. E. 1975. Elemental composition of oat groats. J. Agric. Food Chem. 23:9-13.

PETERSON, D. M., and SMITH, D. 1976. Changes in nitrogen and carbohydrate fractions in developing oat groats. Crop Sci. 16:67-71.

POMERANZ, Y., YOUNGS, V. L., and ROBBINS, G. S. 1973. Protein content and amino acid composition of oat species and tissues. Cereal Chem. 50:702-707.

PRENTICE, N., BABLER, S., and FABER, S. 1980. Enzymic analysis of beta-D-glucans in cereal grains. Cereal Chem. 57:198-202.

QUAIL, P. H., and CARTER, O. G. 1969. Dormancy in seeds of *Avena ludoviciana* and *Avena fatua*. Aust. J. Agric. Res. 20:1-11.

RASPER, V. F. 1979. Chemical and physical characteristics of dietary cereal fiber. Pages 93-115 in: Dietary Fibers: Chemistry and Nutrition. G. E. Inglett and S. I. Falkehag, eds. Academic Press, New York.

SAHASRABUDHE, M. R. 1979. Lipid composition of oats (*Avena sativa* L.). Am. Oil Chem. Soc. 56:80-84.

SAIGO, R. H., PETERSON, D. M., and HOLY, J. 1983. Development of protein bodies in oat starchy endosperm. Can. J. Bot. 61:1206-1215.

SCHANDER, H. 1934. Keimungsphysiologische Studien iiber die Bedeutung der Alevronschicht bei *Oryza* vnd anderen Gramineen. Z. Bot. 27:433-515.

SIMPSON, G. M. 1965. Dormancy studies in seed of *Avena fatua*. IV. The role of gibberellin in embryo dormancy. Can. J. Bot. 43:793-816.

SIMPSON, G. M. 1978. Metabolic regulation of dormancy in seeds—a case history of the wild oat (*Avena fatua*). Pages 167-220 in: Dormancy and Developmental arrest. M. E. Clutter, ed. Academic Press, New York.

SIMPSON, G. M., and NAYLOR, J. M. 1962. Dormancy studies in seed of *Avena fatua*. III. A relationship between maltese, amylases, and gibberellin. Can. J. Bot. 40:1660-1673.

SMART, M. G., and O'BRIEN, T. P. 1979a. Observations on the scutellum. I. Overall development during germination in four grasses. Aust. J. Bot. 27:391-401.

SMART, M. G., and O'BRIEN, T. P. 1979b. Observations on the scutellum. II. Histochemistry and autofluorescence of the cell wall in mature grain and during germination of wheat, barley, oats, and ryegrass. Aust. J. Bot. 27:403-411.

SMART, M. G., and O'BRIEN, T. P. 1979c. Observations on the scutellum. III. Ferulic acid as a component of the cell wall in wheat and barley. Aust. J. Plant Physiol. 6:485-491.

SWIFT, J. G., and O'BRIEN, T. P. 1971. Vascular differentiation in the wheat embryo. Aust. J. Bot. 19:63-71.

TAYLOR, J. S., and SIMPSON, G. M. 1980. Endogenous hormones in after-ripening wild oat (*Avena fatua*) seed. Can. J. Bot. 58:1016-1024.

UPADHYAYA, M. K., NAYLOR, J. M., and SIMPSON, G. M. 1982. The physiological basis of seed dormancy in *Avena fatua* L. I. Action of the respiratory inhibitors sodium azide and salicylhydrozamic acid. Physiol. Plant. 54:419-424.

URQUHART, A. A., ALTOSAAR, I., MATLASHEWSKI, G. J., and SAHASRABUDHE, M. R. 1983. Localization of lipase activity in oat grains and milled oat fractions. Cereal Chem. 60:181-183.

WHISTLER, R. L. 1950. Xylan. Pages 269-290 in: Advances in Carbohydrate Chemistry, Vol. 5. C. S. Hudson and S. M. Cantor, eds. Academic Press, New York.

WOOD, P. J. 1980. The interaction of direct dyes with water soluble substituted celluloses and cereal β-glucans. Ind. Eng. Chem. Prod. Res. Dev. 19:19-22.

WOOD, P. J. 1981. The use of dye/polysaccharide interactions in β-D-glucanase assay. Carbohydr. Res. 94:C19-C23.

WOOD, P. J. 1982. Factors affecting precipitation and spectral changes associated with complex-formation between dyes and β-D-glucan. Carbohydr. Res. 102:283-293.

WOOD, P. J., and FULCHER, R. G. 1978. Interaction of some dyes with cereal β-D-glucans. Cereal Chem. 55:952-966.

WOOD, P. J., FULCHER, R. G., and STONE, B. A. 1983. Studies on the specificity of interaction of cereal cell wall components with Congo red and calcofluor. Specific detection and histochemistry of (1→4)(1→3)-β-D-glucan. J. Cereal Sci. 1:95-110.

WOOD, P. J., PATON, D., and SIDDIQUI, I. R. 1977. Determination of β-glucan in oats and barley. Cereal Chem. 54:524-533.

WOOD, P. J., SIDDIQUI, I. R., and PATON, D. 1978. Extraction of high-viscosity gums from oats. Cereal Chem. 55:1038-1049.

YIU, S. H., POON, H., FULCHER, R. G., and ALTOSAAR, I. 1982. The microscopic structure and chemistry of rapeseed and its products. Food Microstructure 1:135-143.

YOUNGS, V. L. 1972. Protein distribution in the oat kernel. Cereal Chem. 49:407-411.

YOUNGS, V. L. 1974. Extraction of a high protein layer from oat groat bran and flour. J. Food Sci. 39:1045-1046.

YOUNGS, V. L., and FORSBERG, R. A. 1979. Protein-oil relationships in oats. Crop Sci. 19:798-802.

YOUNGS, V. L., and GILCHRIST, K. D. 1976. Note on protein distribution within oat kernels of single cultivars that differ in protein concentration. Cereal Chem. 53:947-949.

YOUNGS, V. L., and PETERSON, D. M. 1973. Protein distribution in the oat (*Avena sterilis* L.) kernel. Crop Sci. 13:365-367.

YOUNGS, V. L., PETERSON, D. M., and BROWN, C. M. 1982. Oats. Pages 49-105 in: Advances in Cereal Science and Technology, Vol. 5. Y. Pomeranz, ed. Am. Assoc. Cereal Chem., St. Paul, MN.

YOUNGS, V. L., PÜSKÜLCÜ, M., and SMITH, R. R. 1977. Oat lipids. I. Composition and distribution of lipid components in two oat cultivars. Cereal Chem. 54:803-812.

ZARKADAS, C. G., HULAN, H. W., and PROUDFOOT, F. G. 1982. A comparison of the amino acid composition of two commercial oat groats. Cereal Chem. 59:323-327.

ZEE, S. Y., and O'BRIEN, T. P. 1970a. A

special type of tracheary element associated with xylem discontinuity in the floral axis of wheat. Aust. J. Biol. Sci. 23:783-791.

ZEE, S. Y., and O'BRIEN, T. P. 1970b. Studies on the ontogeny of the pigment stand in the caryopsis of wheat. Aust. J. Biol. Sci. 23:1153-1171.

ZEE, S. Y., and O'BRIEN, T. P. 1971. Vascular transfer cells in the wheat spikelet. Aust. J. Biol. Sci. 24:35-49.

CHAPTER 4

SUGARS AND NONSTARCHY POLYSACCHARIDES IN OATS

LINDA A. MacARTHUR-GRANT
Department of Cereal Science and Food Technology
North Dakota State University
Fargo, North Dakota

I. INTRODUCTION

In contrast to other cereal grains, little information is available in the literature that pertains specifically to the sugars and nonstarchy polysaccharides of oat flour (endosperm) and oat bran. In addition, there is much variability in reported data because of the sources of the material used in these investigations. For this reason, a definition of terms appears to be mandatory. In this review, the term *whole oats* is the oat kernel plus hulls, whereas *oat groat* is the dehulled oat kernel. *Oat flour* refers to the milled groat, which includes both endosperm and bran, and *oat bran* pertains to the outer covering of the oat groat. This covering can be separated, but not entirely, from the endosperm by sieving.

II. SUGARS

A. Occurrence and Types of Sugars in Oats

Total sugar content of whole oats and oat groats has been reported to be 1.4 and 0.9–1.3%, respectively, whereas oat bran contains considerably more total sugars (2.6–3.4%) (Matz, 1969; MacArthur and D'Appolonia, 1979). A definite varietal influence appears to be associated with total sugar content in oat flour and bran. Cultivars having higher protein content generally, but not always, have a higher total sugar content. In any case, the amount is usually less than in hard red spring (HRS) wheat flour and bran, probably because of greater contamination with glucofructans, which are found in significant amounts in wheat with only trace amounts detectable in oats.

The free sugar concentration in oats is small in comparison with that in other cereal grains (Table I) (McLeod and Preece, 1954). Sugars most frequently reported in various oat fractions, in decreasing amounts, include sucrose, raffinose, glucose, fructose, maltose, glucodifructose, and fructosans. In addition, a tetrasaccharide, stachyose, and a pentasaccharide, verbascose, have been reported in oat endosperm and bran (MacArthur and D'Appolonia, 1979). Values for seven individual free sugars extracted from oat bran and endosperm (MacArthur and D'Appolonia, 1979) are shown in Table II.

Sucrose was the predominant sugar, with amounts being higher in the oat samples than in the HRS wheat samples. Raffinose was the next predominant sugar, with values for oat flour and bran being less than those for wheat. Maltose was found in very small amounts in the oat samples compared with those in the HRS samples, and fructose and glucose values were similar in all flours and brans examined. Of the remaining free sugars, stachyose was of particular interest. It has been found in most grass seeds that contain raffinose, and its detection is, in most cases, facilitated by the absence of fructosans (French, 1954). Since the fructosan content of mature oats is very low, a significant amount of stachyose was detected in the oat flour and bran samples, whereas only a trace of stachyose was found in the bran of the HRS wheat sample. Stachyose has been reported previously in wheat bran (Saunders and Walker, 1969). Verbascose, a pentasaccharide that has also been observed as a free sugar of oats, has been reported in wild oats (*Avena fatua*) (French, 1954). The

TABLE I
Free Sugars and Oligosaccharides of Different Cereal Grains[a]

Sugar or Oligosaccharide	Barley	Wheat	Rye	Oats	Maize
Fructosans[b]	780	1,030	3,940	89	0
Raffinose	450	331	419	192	186
Glucodifructose	250	406	750	38	0
Maltose	90	Trace	Trace	Trace	Trace
Sucrose	908	836	1,857	639	783
Glucose	107	92	77	52	50
Fructose	26	57	98	91	55
Total	2,611	2,752	7,141	1,101	1,074

[a] Source: MacLeod and Preece (1954); used by permission. Results expressed as mg/100 g of grain.
[b] Soluble in 80% ethanol.

TABLE II
Analysis of Seven Individual Free Sugars in Oat and Wheat Flour and Bran[a]

Sample Source	Sucrose	Raffinose	Maltose	Stachyose	Verbascose	Fructose	Glucose
Oat flour							
Dal	0.63	0.26	0.03	0.07	⋯	0.03	0.07
Froker	0.45	0.16	0.01	0.08	⋯	0.05	0.06
Cayuse	0.40	0.20	0.02	0.08	⋯	0.02	0.06
Wheat flour							
Waldron	0.16	0.05	0.07	⋯	⋯	0.02	0.05
Oat bran							
Dal	2.28	0.48	0.05	0.20	0.03	0.04	0.10
Froker	2.66	0.30	0.01	0.24	0.03	0.07	0.08
Cayuse	1.70	0.29	0.02	0.20	0.05	0.03	0.07
Wheat bran							
Waldron	1.75	1.30	0.03	0.04	⋯	0.05	0.09

[a] Source: MacArthur and D'Appolonia (1979); used by permission. Results expressed as percentages.

simultaneous occurrence of raffinose, stachyose, and verbascose in oats is not uncommon, but the three have not commonly been reported together in the Gramineae (Preece, 1957). The 1:1:3 ratio of fructose, glucose, and galactose obtained by MacArthur and D'Appolonia (1979) from an unknown peak associated with the oat brans suggested trace amounts of verbascose.

B. Changes in Sugar Content of Oats During Maturation

When the changes in individual sugars during maturation are compared, the greatest difference occurs in raffinose. Colin and Belval (1934) investigated the common cereal grains with respect to their contents of fructosans and raffinose. They found that both oligosaccharides had completely disappeared from the embryo by the time the grain reached maturity. According to Preece (1957), the amount of fructosan present in oats is a function of the degree of grain ripeness. Therefore, a fully ripe sample would have virtually no fructosan present. Schulbach (1953) also reported the presence of fructosans in unripe oat kernels but did not elaborate on its duration during the maturation process, or on possible differences noted among cultivars. According to Peterson and Smith (1976), total nonstructural carbohydrates increased with maturation, the increases ranging from 56.4–61.2% initially to 60.4–71.3% at maturity. Concentrations of fructosans were initially high (12.0–19.0%), but they declined to trace levels by maturity.

The significance of the moderate to high levels of fructosans is not fully understood, but they appear to be a temporary storage form for sugars during early kernel development. It is not known whether the monosaccharides released upon hydrolysis are respired or converted to starch. Cataldo et al (1975) found that the difference in protein concentration between high- and low-protein cultivars could be largely attributed to a greater relocation of storage carbohydrates from plant tissue into the developing groats. The lower levels of fructosans observed in the high-protein cultivars may be coincidental or may indicate some difference in carbohydrate metabolism that affects protein concentration. Thus, rates of carbohydrate accumulation in the groat may be an important factor in the determination of protein concentration at maturity. In contrast, Peterson et al (1975) determined the quantity of storage carbohydrates in different parts of the oat plant at early anthesis and at maturity and found no relationship between percentage of total nonstructural carbohydrates in the mature panicle and percentage of total nitrogen in the oat groat.

C. Changes in Sugar Content of Oats During Processing and Storage

Very little information is reported in the literature concerning changes in sugar content of oats during processing and storage, particularly for longer storage times.

Minokova et al (1974) measured the changes in sugar content in oat kernels during storage at different moisture levels and reported that the total amount of free sugars did not change during 26 days of storage. Changes in the quantities of individual sugars during processing of whole oat groats into oatmeal or flour have been investigated by Karchik et al (1976a, 1976b). Using descending chromatography, these workers found that hydrothermal processing of certain oat products decreased the total reducing sugars, whereas maltose increased to values greater than those of the control because of the enzymatic hydrolysis of the starch. They also reported that the processing necessary to make oat flour led to a considerable increase in glucose, maltose, and fructose.

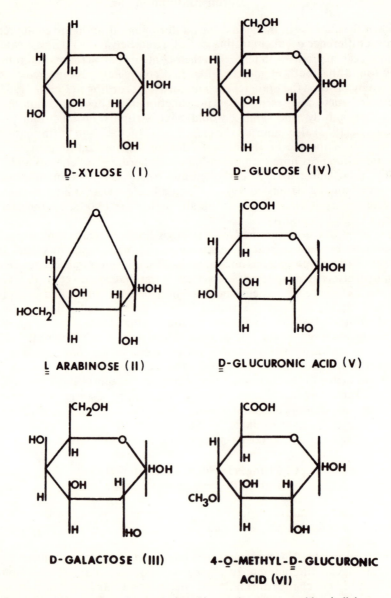

Fig. 1. Pentose and hexose sugars frequently found in cereal pentosans and hemicelluloses.

III. HEMICELLULOSES AND PENTOSANS

A. Occurrence and Composition

Studies have been conducted on hemicelluloses and pentosan-type material derived from nearly every part of the oat plant. Upon acid hydrolysis, the hemicelluloses and pentosans yield derivatives of the pentosans and hexoses. The monomeric units most frequently found in cereal pentosans and hemicelluloses are the pentose sugars D-xylose (I) and L-arabinose (II) (Fig. 1). In addition, certain hexose sugars and their derivatives have been reported, including D-galactose (III), D-glucose (IV), D-glucuronic acid (V), and 4-0-methyl-D-glucuronic acid (VI). In general, the hemicelluloses and pentosans that may be derived from cereals and grasses are characterized by the presence of L-arabinofuranose residues linked as single-unit side chains to a backbone of D-xylopyranose residues. Aspinall and Wilkie (1956) conducted structural studies on a pure hemicellulose from oat straw. They found that the xylan molecule was composed of 40–45 D-xylopyranose residues, with the main chain carrying two side chains linked through positions 2 and 3 of the D-xylose residues and terminated by L-arabinofuranose and 4-0-methyl-D-glucuronic acid residues, respectively (Fig. 2). They concluded that this xylan was of a similar molecular size to that of wheat-straw xylan (Fig. 3); however, it was not possible to isolate a xylan from the oat straw devoid of arabinose residues, as was the case for the wheat straw.

Hemicelluloses isolated by Preece and Hobkirk (1954) from oat husk using a 4% aqueous sodium hydroxide solution in the cold appeared to be quite similar in character to those isolated from various cereal straws with respect to their

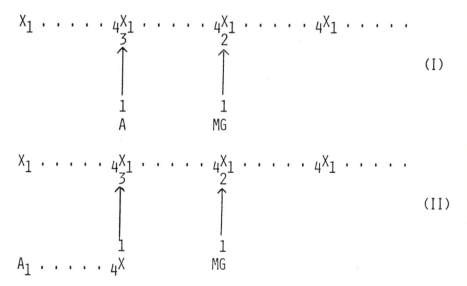

Fig. 2. Diagrammatic structures (I and II) of oat straw hemicellulose according to Aspinall and Wilkie (1956). A = L-arabinofuranose, X = D-xylopyranose, and MG = 4-0-methyl-D-glucuronic acid. Subscripts refer to carbon atoms at which adjacent sugars are joined.

relatively low specific viscosity, inclusion of uronic acid residues, and abundance of pentosan units. In contrast, endospermic hemicelluloses are essentially free of uronic acid residues, contain moderate to large proportions of glucosan, and have high specific viscosities. A hemicellulosic glucan was isolated from the young leaves of the oat plant by Frazer and Wilkie (1971). The pure hemicellulose was a β-glucan with D-glucopyranose residues linked glycosidically β-1,3 and β-1,4. Structural analyses and specific rotation indicated that the glucan was similar to those isolated from various cereal endosperms. A structurally similar glucan (lichenin) was isolated from oat endosperm by Peat et al (1957) and was shown to have D-glucopyranose residues linked β-1,3 and β-1,4 that were degraded by "cellulase" and "laminarase" (Parrish et al, 1960). Total hemicellulose (with a galactose-glucose-arabinose-xylose ratio of 2.8:11.4:10.0:23.5) was extracted from the holocellulose of the leaves of young oat plants by treatments with 5.0 and 24.0% potassium hydroxide solution by Reid and Wilkie (1969c). Fractionation of total hemicellulose (Fig. 4, Table III) resulted in the isolation of a pure acidic galactoarabinoxylan.

The galactoarabinoxylan was similar to other land-plant xylans in that it had a backbone of 1,4-linked D-xylopyranose residues. It possessed terminal nonreducing arabinofuranose residues, but it was unusual in having galactose residues that were both terminal and nonterminal. According to Morikawa and Senda (1978), oat coleoptile cell walls are reported to be composed of 15–40% β-cellulose, 30–70% hemicellulose, 3–8% pectic substances, and 6–12% protein. Several hemicellulosic polysaccharides were isolated from oat coleoptile cell walls (Labavitch and Ray, 1978). Included in these was a glucuronoarabinoxylan that has been reported to be similar to acidic polysaccharides characterized from a number of monocot tissues (Wada and Ray, 1978). The linkages consist of a β-1,4 linked xylan backbone with substituents, primarily arabinose, on the 2 and/or 3 positions of about half of the backbone residues, which resembles those in the arabinoxylans of cultured monocot cells reported by McNeil et al (1975). The presence of a noncellulosic glucan was also reported by Wada and Ray (1978) as a prominent structural oat coleoptile cell wall component, which was shown to be a β-linked polymer containing both 1,4- and 1,3-linked glucosyl residues in a ratio of 2:1. In addition, a small amount of xyloglucan polysaccharides similar to those described for cell walls of dicots were detected.

Morikawa and Senda (1978) found that noncellulosic polysaccharides in the cell wall tissue of oat seed coat, as well as cellulose microfibrils, had an oriented

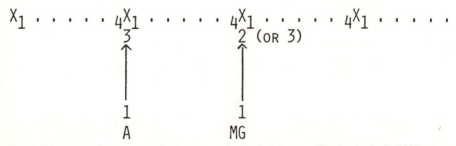

Fig. 3. Diagrammatic structure of wheat straw hemicellulose according to Aspinall (1959). A = L-arabinofuranose, X = D-xylopyranose, and MG = 4-0-methyl-D-glucuronic acid. Subscripts refer to carbon atoms at which adjacent sugars are joined.

structure. These authors studied the changes that took place in this orientation mode during extension growth as well as upon mechanical extension of the walls. Anderson and Greenwood (1955) used a graded extraction procedure to investigate the polysaccharides present in groats. They found that all of the polysaccharide fractions obtained were contaminated with protein, which they did not attempt to remove. Complete separation of the polysaccharide material into products of varying solubility thus was difficult. The differentiation,

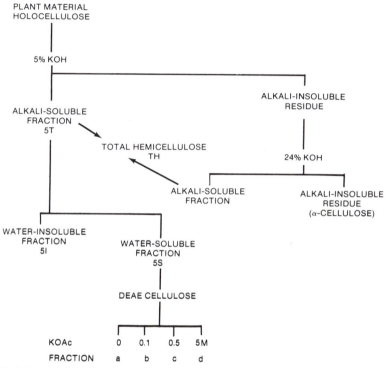

Fig. 4. Isolation and fractionation of oat leaf total hemicellulose. (Reprinted, with permission, from Reid and Wilkie, 1969c)

TABLE III
Analytical Data of Isolated and Fractionated Oat Leaf Total Hemicellulose (TH)[a]

| | | Isolated Fractions[b] | | | KOAc Fractions from Diethylaminoethyl Cellulose Fractionation | | | |
Sugar	TH	5T	5S	5I	a	b	c	d
Galactose	2.8	3	2	0	2	4	2.8	···
Arabinose	10	10	10	10	10	10	10	···
Xylose	23.5	20	15	55	26	14	18.5	···
Glucose	11.4	10	5	25	10	2	0	···
Percentage of TH	100	···	64	13	3.1	3.5	15	3.8

[a] Source: Reid and Wilkie (1969b); used by permission.
[b] 5T = alkali-soluble, 5S = water-soluble, 5I = water-insoluble.

therefore, between the hemicellulosic and water-soluble material was not clear. However, water-soluble pentosans and glucosans other than starch appeared to be present. Oat hulls have been reported to contain the highest concentration of pentosans, an average of 29.5%, along with 16.7% lignin (Whistler, 1950), whereas whole oats and oat groats contained 14 and 4%, respectively (Matz, 1969). Oat pentosans are characterized by insolubility in water, solubility in 2–4% alkaline aqueous solution, negative optical rotation (levorotatory), and absence of reducing groups (Whistler and Smart, 1953).

In addition to these investigations, studies related to the changes in polysaccharides of the maturing oat plant have received considerable attention. These studies have been concerned primarily with the relationship between plant maturity and hemicellulosic composition, with particular reference to the nonendospermic tissues of the oat plant. Reid and Wilkie (1969a, 1969b) reported that, in any one part of the oat plant, increasing maturity brings an increase in the percentage of xylose and a decrease in the percentage of arabinose and glucose residues in each total hemicellulose fraction. In the total hemicelluloses of the oat leaf, the percentage of galactose also decreases.

Buchala and Wilkie (1973) further examined the changes in total hemicellulose composition of the leaf and stem tissues of field-grown oat plants and found that, in addition to an increase in xylose residues and a decrease in arabinose and glucose residues, the ratio of 4-0-methyl-D-glucuronosyl to D-glucuronosyl residues increases with maturity. The results of this study are shown in Table IV.

TABLE IV
Composition of Total Hemicelluloses of Maturing Oat Plants
in Molar Percentages (Reducing Sugars = 100%)[a]

Tissue	Days from Sowing to Harvest	Rhamnose	Arabinose	Xylose	Galactose	Glucuronic Acid	4-0-Methyl-glucuronic Acid	Glucose
Stem	81	0.8	11.7	61.2	0.9	9.1	5.6	10.6
	106	0.4	11.2	62.8	0.9	9.4	4.4	10.6
	137	0.5	10.2	67.8	1.1	7.4	8.2	8.2
	162	0.4	10.8	70.7	1.2	7.6	3.3	6.1
Top leaf	106	0.5	16.5	58.7	1.9	11.4	5.0	5.9
	137	1.0	16.2	58.7	3.7	13.9	4.9	1.5
	162	0.7	15.1	62.0	3.0	8.6	7.5	3.0
Middle two leaves	106	1.0	18.0	54.1	5.3	10.5	4.6	6.3
	137	0.1	16.3	55.3	5.7	10.6	3.7	8.4
	162	0.3	14.3	58.4	3.4	10.5	6.1	6.9
Leaf and bottom leaf	56	0.8	19.5	34.5	3.6	16.5	3.8	20.9
	81	1.0	18.9	41.9	3.0	11.8	2.8	20.4
	106	1.3	17.2	48.0	3.3	14.5	2.9	12.6
	137	0.9	16.4	56.8	5.9	7.3	4.6	8.1

[a] Source: Buchala and Wilkie (1973); used by permission. Values determined by gas-liquid chromatography or derived acetates of neutral sugars and by borotritide determination of acid sugars. Values not corrected for xylose residues remaining glycosidically linked to uronic acid residues after hydrolysis.

B. Water-Soluble Nonstarchy Polysaccharides

In attempting to elucidate the composition of the water-soluble (WSNP) and water-insoluble nonstarchy polysaccharides (WINP) of oat flour and bran, various techniques of selective fractionation have been utilized. Figure 5 shows the isolation and purification procedure used by MacArthur and D'Appolonia (1980). The extraction and purification techniques used by these workers to prepare crude oat WSNP are the same as those commonly used for wheat flour WSNP. Once the amylase-treated WSNP was obtained (Table V), two different

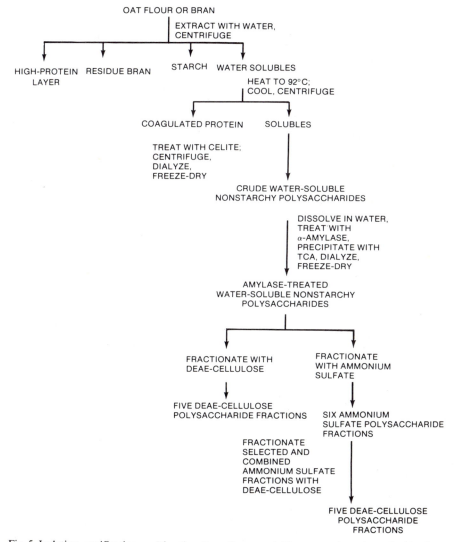

Fig. 5. Isolation, purification, and fractionation of water-soluble nonstarchy polysaccharides from oat flour or oat bran. (Reprinted, with permission, from MacArthur and D'Appolonia, 1980)

fractionation methods were utilized. One method used a stepwise elution from a column of diethylaminoethyl (DEAE) cellulose and produced five distinct fractions. The other method, using different concentrations of ammonium sulfate to precipitate the polysaccharide material, produced six fractions. Preece and Hobkirk (1953) also used a graded ammonium sulfate fractionation procedure to investigate the WSNP derived from the whole oat kernel. Table VI shows the results of these two studies. The major units identified from the gums were glucosan, xylan, araban, and galactan. In addition, separate pentose and hexose oligosaccharides were detected in the mother liquor fraction. Preece and Hobkirk (1953) found β-glucan apparently uncontaminated by pentosan in the

TABLE V
Percentage Yield of Crude and Amylase-Treated Water-Soluble Nonstarchy Polysaccharides[a]

Sample	Crude Material from Flour or Bran (db)	Amylase-Treated Material		
		From Crude Pentosans	From Flour or Bran (db)	From Flour or Bran (protein free)
Wheat flour	1.3	64.3	0.8	0.66
Oat flour	2.1	65.5	1.4	1.38
Oat bran	4.2	65.4	2.8	2.76

[a] Source: MacArthur and D'Appolonia (1980); used by permission.

TABLE VI
Approximate Percentage Composition of Principal Gum Fractions Obtained by Ammonium Sulfate Precipitation

Fraction	Ammonium Sulfate Concentration	Gum	I[a]	II[b]	
				Flour	Bran
1	20	Glucosan	100	95	100
		Xylan	⋯	2	Trace
		Araban	⋯	3	Trace
2	30	Glucosan	93	100	95
		Xylan	1	Trace	2
		Araban	6	Trace	3
3	40	Glucosan	88	100	97
		Xylan	8	Trace	1
		Araban	4	Trace	2
4	50	Glucosan	⋯	90	85
		Xylan	⋯	5	7
		Araban	⋯	5	8
5	60	Glucosan	⋯	80	70
		Xylan	⋯	10	15
		Araban	⋯	10	14
6	Mother liquor	Glucosan	15	30	34
		Xylan	14	9	13
		Araban	40	31	27
		Galactan	31	30	26

[a] Data from Preece and Hobkirk (1953). Analyzed from whole oat kernel.
[b] Data from MacArthur and D'Appolonia (1980). Analyzed from oat endosperm and bran.

10% ammonium sulfate fraction, whereas MacArthur and D'Appolonia (1980) found the 20, 30, and 40% fractions to be relatively pure β-glucans with very slight pentosan contamination, particularly for the oat bran. The probability of separating hexosan from pentosan is increased by using ammonium sulfate fractionation.

Preece and Hobkirk (1953) reported the amounts of pentosan material in oats to be minimal but found similarities to wheat, rye, and barley upon fractionation with ammonium sulfate. They pointed out that the xylan-araban ratio in the 40% fraction from oats is essentially the same as corresponding fractions collected from wheat, rye, and barley.

MacArthur and D'Appolonia (1980) reported similar results utilizing DEAE-cellulose column chromatography. A further attempt was made by MacArthur and D'Appolonia (1980) to obtain an essentially pure arabinoxylan from oat flour or bran. They selected and combined the ammonium sulfate fractions highest in pentosan content and subjected this material to DEAE-cellulose chromatography. The results (Table VII) indicate less glucose for both the oat flour and oat bran fractions than was originally obtained in the ammonium sulfate fractions.

Fraction 2 of the oat flour was found to be an essentially pure β-glucan. The oat bran contained the most arabinose and xylose in the first three of the five fractions, with fraction 4 containing only a trace of glucose.

C. Water-Insoluble Nonstarchy Polysaccharides

MacArthur and D'Appolonia (1980) also examined the water-insoluble nonstarchy polysaccharides (WINP) of oat flour and bran. A schematic diagram (Fig. 6) shows the isolation and purification procedure utilized. The designated *high-protein layer* corresponds to the fraction referred to as *tailings, squeegee,* or *amylodextrin* by investigators working with wheat flour (D'Appolonia and MacArthur, 1975). Youngs (1974) has described a procedure for the isolation of this high-protein fraction. The yields of crude WINP from the oat flour and oat bran high-protein layer fractions were reported as 11.2 and 4.8%, respectively.

TABLE VII
**Component Sugars in Hydrolyzed Diethylaminoethyl Cellulose Fractions
from Combined Ammonium Sulfate Fractions[a]**

| Sample | Fraction | Percentage of | | | |
		Arabinose	Xylose	Galactose	Glucose
Oat flour	F1	27.6	31.6	⋯	40.8
	F2	Trace	Trace	⋯	100.0
	F3	24.4	8.6	27.1	39.9
	F4	25.0	2.3	47.4	25.3
	F5	15.5	11.2	⋯	73.3
Oat bran	F1	28.2	33.5	⋯	38.3
	F2	18.8	21.0	⋯	60.2
	F3	27.9	19.3	⋯	52.8
	F4	⋯	⋯	⋯	⋯
	F5	8.7	6.3	⋯	85.0

[a] Source: MacArthur and D'Appolonia (1980); used by permission.

Following purification with α-amylase, recovery of WINP from the oat bran high-protein layer was significantly higher than from the oat flour fraction. These authors have previously reported that pentosans associated with the "sludge" fraction (WINP) of HRS wheat flour had a higher arabinoxylan content than did the corresponding WSNP. Table VIII shows the composition of the polysaccharide after DEAE-cellulose column chromatography fractionation. These data and the paper chromatogram (Fig. 7) of the hydrolyzed DEAE-cellulose WINP fractions show that essentially pure arabinoxylans were obtained in DEAE-cellulose fractions 1 and 2 of the WINP of the oat bran.

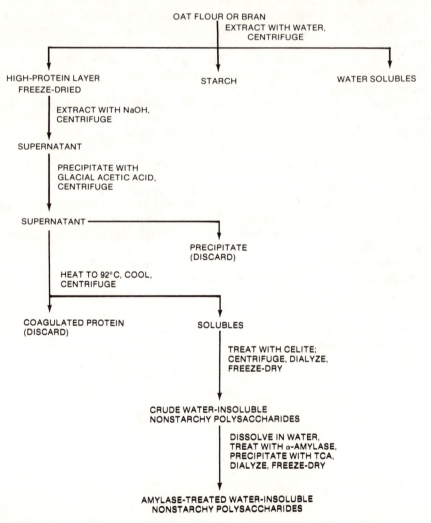

Fig. 6. Isolation, purification, and fractionation of water-insoluble nonstarchy polysaccharides from oat flour or oat bran. (Reprinted, with permission, from MacArthur and D'Appolonia, 1980)

TABLE VIII
Component Sugars in Water-Insoluble Nonstarchy Polysaccharides as Found
by Diethylaminoethyl Fractionation[a]

		Percentage of			
Sample	Fraction	Arabinose	Xylose	Galactose	Glucose
Oat bran					
(high-protein layer)	Unf.	18.9	13.9	Slight trace	67.2
	F1	49.9	50.1	...	...
	F2	53.4	46.6	...	...
	F3	45.7	37.6	Slight trace	16.7
	F4	33.8	29.0	Trace	37.2
	F5	10.7	7.9	...	81.4
Oat flour					
(high-protein layer)	Unf.	23.1	13.8	Trace	63.1
	F1	34.5	32.5	...	33.0
	F2	29.0	19.9	...	51.1
	F3	27.2	19.7	Trace	53.1
	F4	26.2	13.7	9.0	51.1
	F5	11.6	10.0	...	78.8

[a] Source: MacArthur and D'Appolonia (1980); used by permission.

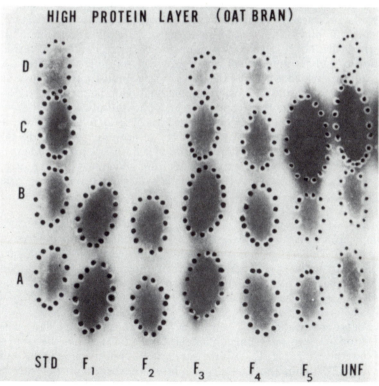

Fig. 7. Paper chromatograms of component sugars in hydrolyzed, unfractionated and fractionated, water-insoluble nonstarchy polysaccharides extracted from high-protein layer. A = arabinose, B = xylose, C = glucose, and D = galactose; F = fraction. (Reprinted, with permission, from MacArthur and D'Appolonia, 1980)

IV. UTILIZATION OF OAT HULLS AND PENTOSANS

Commercial utilization of oat hull and other pentosans lies in the manufacture of furfural and related furan compounds. Ten years ago, oat hulls satisfied approximately 22% of the annual furfural demand; but the demand for furan chemicals has far outstripped the available supply of oat hulls, which have been crowded into relative insignificance by the use of corncobs, bagasse, and rice hulls (Shukla, 1975).

Furfural is a member of the furan (C_4H_4O) family of chemicals. Commercial procedures of furfural production from pentosans involve a sequence of hydrolysis to pentoses followed by a cyclodehydration of monomeric pentose to furfural, all in a single stage (Shukla, 1975).

In general, the furans are more reactive than their benzene analogues, undergoing electrophilic substitutions easily and participating in a variety of additive reactions. In addition to its reactivity, which makes it an important intermediate for the manufacture of various chemicals (Fig. 8), furfural has excellent solvent properties. The list of industrial uses of furfural and other oat hull products is long and varied (Shukla, 1975). Furfural is used in the following ways: 1) selective solvent extraction of crude petroleum; 2) separation of butadiene from contaminating hydrocarbons; 3) as an intermediate in the nylon industry; 4) production of formaldehyde furfural resins required for the manufacture of pipes, tanks, and reaction vessels; 5) production of tetra-hydrofurfuryl alcohol, a solvent for dyes, resins, lacquers, paints, and varnish; 6) production of polytetramethylene ether glycol by cationic polymerization of tetrahydrofuran, which is used by the polyurethane industry for the production of elastomers and thermoplastics; 7) commercial manufacture of D-xylose; 8) manufacture of phenolic resin glues and phenolic and plywood adhesives; 9) production of hydrogen peroxide explosives from ground hull; 10) production of antiskid tread composition; 11) as a filter aid in breweries; 12) production of construction board material; 13) production of cellulose pulp; plus many other uses. Oat hulls also exhibit cariostatic properties, which may lead to such new uses as sweeteners or chewing gum to provide protection from dental caries.

V. SUMMARY

In comparison with other cereal grains, the occurrence of total and free sugars in oats is small. Those sugars most frequently reported in various oat fractions (i.e., groat, bran, hull, etc.) include sucrose, raffinose, glucose, fructose, maltose, stachyose, verbascose, glucodifructose, and fructosans. During maturation, raffinose undergoes the greatest change, whereas during processing, total reducing sugars decline and glucose, maltose, and fructose increase.

Whole oats contain about 14% pentosans (mainly araban and xylan), with the concentration of pentosans being higher in the hull (29%) than in the groat (4%). Most of the crude fiber (11–12%) of the oat kernel is concentrated in the hull and consists of 16.7% lignin and 29.4% α-cellulose. Another major polysaccharide present in oat groat is β-D-glucan, present at a level of 12–15%. This polysaccharide is discussed at length in Chapter 6. The most spectacular commercial utilization of oat hull and pentosans is in the production of furfural and related furan compounds. Other potential industrial applications include the use of hulls as an abrasive and the use of hulls or straw as a filler in certain plastic resin products or building materials.

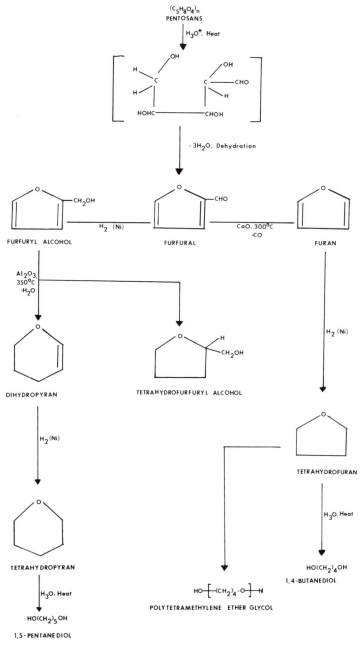

Fig. 8. Production schematics of various industrial chemicals from oat hull pentosans. (Reprinted, with permission, from Shukla, 1975)

LITERATURE CITED

ANDERSON, D. M. W., and GREENWOOD, C. T. 1955. An investigation of the polysaccharide content of oats, *Avena sativa* L. J. Sci. Food Agric. 6:587-592.

ASPINALL, G. O. 1959. Structural chemistry of the hemicelluloses. Adv. Carbohydr. Chem. 14:429-468.

ASPINALL, G. O., and WILKIE, K. C. B. 1956. The constitution of an oat straw xylan. J. Chem. Soc. p. 1072-1076.

BUCHALA, A. J., and WILKIE, K. C. B. 1973. Uronic acid residues in the total hemicelluloses of oats. Phytochemistry 12:655-659.

CATALDO, D. A., SCHRADER, L. E., PETERSON, D. M., and SMITH, D. 1975. Factors affecting seed protein concentration in oats. I. Metabolism and distribution of N and carbohydrate in two cultivars that differ in groat protein concentrations. Crop Sci. 15:19-23.

COLIN, H., and BELVAL, H. 1934. La raffinose dans les cereales. C. R. Hebd. Seances Acad. Sci. 196:1825-1831.

D'APPOLONIA, B. L., and MacARTHUR, L. A. 1975. Comparison of starch, pentosans and sugars of some conventional height and semidwarf hard red spring wheat flours. Cereal Chem. 52:230-239.

FRAZER, C. G., and WILKIE, K. C. B. 1971. A hemicellulosic glucan from oat leaf. Phytochemistry 10:199-204.

FRENCH, D. 1954. Unit-cell data for maltose hydrate and some acyl saccharide derivatives. Acta Crystallogr. 7:136-143.

KARCHIK, S. N., MELNIKOV, E. M., and SHABLOVSKAYA, I. S. 1976a. Change in the content of sugars in oat products, buckwheat and rice groats during heat processing. Vopr. Pitan. 4:82-85.

KARCHIK, S. N., MELNIKOV, E. M., and KOROLEW, A. I. 1976b. Effect of steaming on the carbohydrate composition of oatmeal. Prikl. Biokhim. Mikrobiol. 12:93-95.

LABAVITCH, J. M., and RAY, P. M. 1978. Structure of hemicellulosic polysaccharides of *Avena sativa* coleoptile cell walls. Phytochemistry 17:933-937.

MacARTHUR, L. A., and D'APPOLONIA, B. L. 1979. Comparison of oat and wheat carbohydrates. I. Sugars. Cereal Chem. 56:455-457.

MacARTHUR, L. A., and D'APPOLONIA, B. L. 1980. Comparison of nonstarchy polysaccharides in oats and wheat. Cereal Chem. 57:39-45.

MacLEOD, A. M., and PREECE, I. A. 1954. Studies on the free sugars of the barley grain. V. Comparison of sugars and fructosans with those of other cereals. J. Inst. Brew. London 60:46-55.

MATZ, S. A. 1969. Cereal Science. AVI Publishing Co., Westport, CT.

McNEIL, M., ALBERSHEIM, P., TAIZ, L., and JONES, R. L. 1975. The structure of plant cell walls. VII. Barley aleurone cells. Plant Physiol. 53:64-68.

MINOKOVA, Y. A., POPOV, M. P., and STARODUBTSEVA, A. I. 1974. Change in the sugar level in oats during storage. Izv. Vyssh. Uchebn. Zaved. Pishch. Tekhnol. 1:52-57.

MORIKAWA, H., and SENDA, M. 1978. Infrared analysis of oat coleoptile cell walls and oriented structure of matrix polysaccharides in the walls. Plant Cell Physiol. 19:327-336.

PARRISH, F. W., PERLIN, A. S., and REESE, E. T. 1960. Selective enzymolysis of poly-β-D-glucans and the structure of the polymers. Can. J. Chem. 38:2094-2104.

PEAT, S., WHELAN, W. J., and ROBERTS, J. G. 1957. The structure of lichenin. J. Chem. Soc. p. 3916-3924.

PETERSON, D. M., SCHRADER, L. E., CATALDO, D. A., YOUNGS, V. L., and SMITH, D. 1975. Assimilation and remobilization of nitrogen and carbohydrates in oats, especially as related to groat protein concentration. Can. J. Plant Sci. 55:19-28.

PETERSON, D. M., and SMITH, D. 1976. Changes in nitrogen and carbohydrate fractions in developing oat groats. Crop Sci. 16:67-71.

PREECE, I. A. 1957. Cereal Carbohydrates. R. Inst. Chem. Monogr. W. Heffer and Sons, Cambridge, England.

PREECE, I. A., and HOBKIRK, R. 1953. Non-starchy polysaccharides of cereal grains. III. Higher molecular gums of common cereals. J. Inst. Brew. 59:385-392.

PREECE, I. A., and HOBKIRK, R. 1954. Non-starchy polysaccharides of cereal grains. V. Some hemicellulose fractions. J. Inst. Brew. 60:490-496.

REID, J. S. G., and WILKIE, K. C. B. 1969a. Polysaccharides of the oat plant in relationship to plant growth. Phytochemistry 8:2045-2051.

REID, J. S. G., and WILKIE, K. C. B. 1969b. An acidic galactoarabinoxylan and other pure hemicelluloses in oat leaf. Phytochemistry 8:2053-2058.

REID, J. S. G., and WILKIE, K. C. B. 1969c. Total hemicelluloses from oat plants at different stages of growth. Phytochemistry

8:2059-2065.

SAUNDERS, R. M., and WALKER, H. G., Jr. 1969. The sugars of wheat bran. Cereal Chem. 46:85-92.

SCHLUBACH, H. H. 1953. Uber den kohlen-hydratstoff-wechsel der getreidearten. Experientia 9:230-234.

SHUKLA, T. P. 1975. Chemistry of oats: Protein foods and other industrial products. Crit. Rev. Food Sci. Nutr. October. pp. 383-431.

WADA, S., and RAY, P. M. 1978. Matrix polysaccharides of oat coleoptile cell walls. Phytochemistry 17:923-931.

WHISTLER, R. L. 1950. Advances in Carbohydrate Chemistry, Vol. 5. Academic Press, New York.

WHISTLER, R. L., and SMART, C. L. 1953. Polysaccharide Chemistry. Academic Press, New York.

YOUNGS, V. L. 1974. Extraction of a high-protein layer from oat groat bran and flour. J. Food Sci. 39:1045-1046.

CHAPTER 5

OAT STARCH: PHYSICAL, CHEMICAL, AND STRUCTURAL PROPERTIES

DAVID PATON
Food Research Centre
Research Branch
Agriculture Canada
Ottawa, Ontario, Canada

I. INTRODUCTION

Unlike other world cereal crops such as wheat, corn, barley, and rice, oats have never been considered a subject of major scientific or industrial study. The reasons for this are quite evident, as previous chapters in this monograph clearly indicate, yet oats are still considered a major cereal crop in North America and in parts of western and central Europe.

Most of the detailed examinations of the morphology and the physical, chemical, and functional properties of cereal starches have been notably devoid of the inclusion of oat starch by way of comparison. MacMasters et al (1947), in a study of the potential uses of oats and other small grains for starch production, concluded that apart from a smaller granule size, oat starch had an amylose content similar to that of wheat and corn. Clendenning and Wright (1945) analyzed oat starch by an acid digestion method and showed this starch to have a much higher lipid content (1.2%) than other cereal starches. A study of the role of cereal starches in breadmaking by Prentice et al (1954) indicated that oat starch had a crumb-softening effect on composite wheat breads but that no more than 30% could be incorporated without substantial reduction in loaf volume. The same study showed that rye and cassava starches tended to increase the firmness of the loaf crumb. More recent studies of a similar nature have been conducted by Hoseney et al (1971), who concluded that granule size was not the major criterion for baking performance, since small wheat starch granules were found to function normally in composite or wheat flour replacement systems.

The first substantial information on the physical and chemical properties of oat starch was published in the mid-1950s by the research group at the University of Edinburgh, Scotland, under the leadership of C. T. Greenwood. As part of a voluminous study on starches, oat starch was fractionated into its amylose and amylopectin components, and important physical parameters of these fractions were examined. Details of this work are discussed below. Apart from Greenwood's work, no other studies on oat starch of any major significance were conducted until Paton (1977) showed that the pasting properties of an oat starch derived from a high-protein cultivar were atypical of most other cereal starches. This work and that of Wu et al (1973) on wet fractionation of the oat kernel heralded a renewed interest in the chemistry and utilization of oats as a source of potentially useful food ingredients.

II. MORPHOLOGICAL CHARACTERISTICS

Figures 1A and 1B represent sections through a groat of a low-protein (14–16%, dry basis [db]) and a high-protein (19–22%) cultivar of oats. The view shown is in the area of the aleurone-subaleurone interface. The starch granules tend to exist as clusters of individual granules and are located in each cell with neighboring discrete protein bodies (dark spots on the micrographs). The higher-protein cultivar trends to be almost devoid of starch in the location adjacent to the aleurone layers, whereas starch granules and protein bodies are more evenly distributed in the lower-protein cultivar. The floury endosperm of the high-protein cultivar also contains a lesser proportion of starch granules than the lower-protein cultivar (not shown). Such cellular organization is in contrast to that of wheat or barley, in which starch granules are embedded in a near-homogeneous protein matrix.

 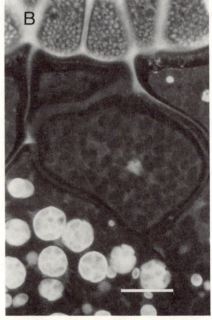

Fig. 1. Photomicrographs of sections through low-protein (A) and high-protein (B) oat groat, showing location of starch granules and protein bodies. Bars: A = 45 μm, B = 50 μm.

Figures 2A and 2B are photomicrographs of oat starch and wheat starch, respectively. Oat starch is highly aggregated and may exist in the size range of 30–60 μm in diameter. Individual starch granules are irregular in shape, often of a polyhedral configuration, and exhibit weak birefringence (Fig. 2A). The average granule diameter is 7–10 μm, within a range of 3–12 μm. The micrograph of wheat starch, at the same magnification, is presented for comparison. Oat starch is more akin to rice starch in both size and shape.

III. ISOLATION OF OAT STARCH

Oat starch has been isolated in the laboratory by several different methods. Unlike wheat starch, oat starch cannot be freely washed from the grain by selective hydration and agglomeration of the protein component. However, as Youngs (1974) demonstrated, an aqueous slurry of oat flour may be centrifuged to produce layered sediments, the heaviest of which contains prime starch. Repeated washing of this layer produces starch with a protein content in the range of 0.44–0.6% (db) and a total extractable lipid content in the range of 0.67–1.11% (db). The latter confirms the values, found in earlier studies by Clendenning and Wright (1945) and by Acker and Becker (1971, 1972), of 1.3% protein and 0.91–1.07% (db) lipid. Starch isolation by sedimentation of a flour slurry is not practical on an industrial scale, since removing the hydrated bran and storage protein layers would be extremely difficult. MacMasters et al (1947) used a variant of the corn-steeping process to wet-mill oats. The resulting starch yields were in the range of 65–85% (by weight) of the dehulled groat, depending on the groat protein content. However, the protein content in the starch was quite high, in the range of 0.6–1.1% ($N \times 5.7$).

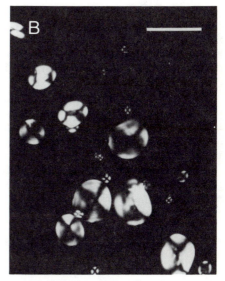

Fig. 2. Photomicrographs of oat starch (A) and wheat starch (B) under polarized light. Bars: A = 40 μm, B = 45 μm.

Wu et al (1973) and Paton (1977) extracted starch from dehulled oats by solubilizing the protein in alkaline solutions. In the former, $0.1 N$ NaOH was the agent used, whereas Paton (1977) made use of a 20% (w/w) Na_2CO_3 solution to adjust the pH of the slurry to 10.0. Sodium carbonate was selected to give maximum protein dissolution while ensuring that the starch was prohibited from chemical swelling. Starch yields were found in the 43–61% range in 23 cultivars having a groat protein content of 14–24%. Starch isolated by this method analyzed < 0.4% protein and < 0.3% total extractable lipid. The low lipid content probably results from the use of an alkaline reagent, which tends to remove saponified lipid with the extracted protein. Paton (1977) reported total phosphorus values in the 0.06–0.08% range, which is considerably higher than that found for wheat (0.058%), corn (0.022%), and waxy maize (0.018%) but of the same order as that for potato starch (0.070%). Recently acquired unpublished data indicate that the higher phosphorus content of oat starches may be caused by phospholipid, which is imbedded in the starch and is difficult to extract by methods that do not destroy the granule structure.

Adkins and Greenwood (1966) used the technique of soaking oat groats in water (6–8° C) containing mercuric chloride and carefully macerating the softened grain. The macerate was screened to remove bran and the protein removed by a combination of sedimentation in water and shaking with toluene. Such a procedure, although not practical, ensured that the starch was recovered in high yield and was minimally degraded or chemically modified. This starch was exhaustively extracted with hot 80% MeOH and had a residual protein content of 0.24%.

Several processes have been devised at the industrial level to fractionate the oat groat. Hohner and Hyldon (1977) used essentially the same procedure as described by Paton (1977) and produced an oat starch containing 0.51% protein and 0.40% total extractable lipid. Oughton (1980), in a patent assigned to Dupont of Canada, adopted the rather unconventional approach of simultaneously defatting and sequentially fractionating oat groats by means of selected organic solvents in a system utilizing a plurality of hydrocyclones. These workers were primarily interested in recovering oat protein fractions of high purity; as a result, the residual starch fraction contained up to 7.0% protein.

Recently, Burrows et al (1984) disclosed a process whereby whole or dehulled oats are steeped in an aqueous medium at 50° C for 28 hr. Under these conditions, the grain does not germinate but the endospermic portion is liquefied and can be separated from the bran or hulls by squeezing or by grinding and sieving. This low-bran endosperm may then be subjected to further wet processing to yield a starch of high purity.

From a technical standpoint, therefore, some form of wet extraction of oat groats is probably the preferred route to take to obtain starch of acceptable purity, although in terms of economic feasibility and practicality such approaches may be uncertain.

IV. FUNCTIONAL BEHAVIOR OF OAT STARCH

Before examining the physical and chemical behavior of oat starch, it would be more in context to describe its functional nature, since most of the recent studies on oat starch have been conducted to explain specific functional performance.

Perhaps the most important functional characteristic of any starch is its behavior in water under conditions of varying heat treatments, namely its viscosity (i.e., consistency) or pasting characteristic. Pasting behavior of starches is in general so well known that a full description of the procedure and an explanation of the events occurring throughout will not be given here. The early work of MacMasters et al (1947) showed that oat starch, isolated by a variant of the corn-steeping process, had pasting behavior very similar to that of corn starch. Since wheat, corn, and rye starches, in general, behave quite similarly in their cooking or pasting behavior, it is not surprising that oat starch has received almost no attention from this point of view. It was not until Paton (1977) published the results of a study on numerous oat starches that some rather unique paste behavior was identified (Fig. 3, Table I). Starch paste characteristics were determined using a novel recording viscometer known as the

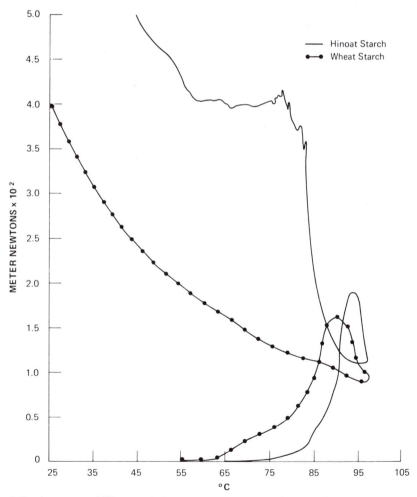

Fig. 3. Pasting curves of Hinoat and wheat starch at 9% (db) solids. (Reprinted, with permission, from Paton, 1979)

Ottawa Starch Viscometer (Voisey et al, 1977); the test sample was brought from ambient to 95°C by means of heat transferred from a near-boiling water bath, and cooling was performed by draining the boiling water and replacing it with running tap water at < 15°C. Pasting was exceedingly rapid, the total time taken to complete the heating, cooking, and cooling cycle being about 12 min.

Perhaps the most important feature is a more intense mixing action in the viscometer bowl. Recently, Doublier (1981) has shown as part of a fundamental rheology study that the flow behavior of wheat starch pastes is markedly influenced by the intensity of mixing during the preparation of the paste. Thus, the manner in which the consistency of starch pastes is developed and recorded with the Ottawa Starch Viscometer is more representative of the way in which starch is cooked in an industrial setting. By analyzing each pasting curve in terms of ratios of key reference points on the curve, distinct starch behavioral differences were identified. In Table I, the ratio P/H is a measure of the susceptibility of swollen starch granules to shear as a result of stirring at a fixed temperature (95°C). The higher the value in this column, the greater the susceptibility. Most of the oat cultivars examined gave starches that had high shear susceptibility and behaved in this respect like the waxy starches. However, several were more like the behavior experienced with normal cereal starches, such as wheat and corn.

TABLE I
Pasting Properties of Starches[a]

Starch	Initial Swelling (°C)	Peak Temp. (°C)	Peak Torque (P; cm/g)	Hold Torque (H; cm/g)	P/H	T_{30}[b] (cm/g)	Torque at 25°C (T; cm/g)	Temp. at T_{30} (°C)	Increase in T_{30} (%) Relative to H
Corn	72.0	89.7	144	116	1.24	160	304	67.7	37.9
Wheat	68.5	93.1	158	112	1.41	188	326	65.8	67.8
Rice	79.1	91.3	130	88	1.47	156	252	68.0	77.3
Potato	60.4	78.7	400	208	1.92	344	1,000	71.0	65.4
Arrowroot	65.1	75.8	320	148	2.16	232	490	73.6	56.7
Waxy maize	72.6	79.9	198	112	1.76	128	172	77	14.3
Commercial oat	64.8	93.8	132	88	1.50	176	304	67.0	100.0
Hinoat									
OTT/70	65.0	93.8	162	91	1.78	385	432	72.5	323.0
OTT/71	66.2	93.6	158	93	1.70	372	418	72.3	300.0
OTT/73	68.0	91.5	164	94	1.75	375	416	71.7	298.9
REG/72	67.5	93.1	166	93	1.79	368	440	72.6	295.7
Glen/73	70.0	93.1	144	86	1.67	224	344	70.6	160.5
Smith 0N/73	67.0	94.1	156	112	1.39	240	312	67.0	114.3
Smith 500 N/73	67.0	93.0	158	112	1.41	224	324	67.5	100.0
Rodney									
REG/67	62.0	91.0	148	105	1.41	240	340	70.5	128.6
Glen Coop/73	64.0	89.2	128	80	1.60	188	316	67.8	135.0
REG/73	69.2	92.2	152	88	1.72	464	560	72.2	427.3
Glen/73	72.0	90.8	126	76	1.65	360	464	72.1	373.7
OTT/67	68.0	91.0	164	88	1.86	440	520	72.6	400.0
Garry	62.4	90.6	162	94	1.72	232	336	71.6	146.8
Harmon	62.0	91.1	150	96	1.56	184	304	67.1	91.7
Stormont	62.0	90.1	162	112	1.44	236	348	67.8	110.7

[a] Source: Paton (1979); used by permission.
[b] T_{30}-torque developed after first 30 sec of cooling.

The designation T_{30}, which represents the torque value attained by each paste after the first 30 sec of the cooling cycle, is taken as a measure of the initial development of setback. Most oat starches develop unusually high values of torque within this period, and this must be regarded as atypical behavior. The temperature at which this rise in consistency occurs is 3–5° C higher than that found for other cereal starches. The torque at 25° C represents that of the final setback. Interestingly, 75–80% of this value is achieved within the first 30 sec of cooling for the majority of oat starches. The starches examined in Table I were isolated from random samples of oat cultivars without consideration of year grown, area of location, or cultural practices. The possibility of these factors having some influence on the behavior of the starch pastes should not be ruled out. Nor should protein content of the oat cultivar be neglected: the cultivars Garry, Harmon, and Stormont have protein contents in the range of 12–15% of the groat weight, the Rodney cultivars are in the range of 14–17%, and the Hinoat samples are in excess of 17%. Again, no firm conclusions should be drawn about the relationship between starch behavior and protein content without the undertaking of a wide and critical study of location, agricultural practices, and fertilizer use. The last variable is illustrated by the samples identified as Smith 0 N and 500 N, where nitrogen fertilizer at 0 and 500 kg/ha had been applied to test plots grown with the cultivar Hinoat. Both conditions had the effect of producing starches that possessed paste properties quite unlike those of other Hinoat samples but very like those of other cereal starches.

Paton (1979) published further studies on oat starch showing the response of torque to starch concentration during the cooling cycle. Torque was measured for an oat starch cultivar (Hinoat) and a commercial corn starch when the paste temperatures had reached 75 and 25° C, respectively. In contrast to the previous measurements (Paton 1977), torque was monitored directly as a function of temperature, using a thermocouple cemented to the measuring blade of the paddle and connected to an X-Y recorder. Time was therefore not a recorded variable. The results are shown in Fig. 4. The paste torque/temperature response for oat starch is markedly higher than for corn starch under the same conditions.

Workers have long recognized that $7M$ urea is a potent solvent for starch even at room temperature. Since this dissolution is time dependent, and owing to the rapidity of paste development with the Ottawa Starch Viscometer, Paton (1979) was able to perform pasting curves in $7M$ urea for corn and oat starches. Again, the most distinct differences were observed during the cooling cycle (Fig. 5). The cooling curves for corn pastes in water and $7M$ urea did not differ significantly, whereas those for oat starch did, the $7M$-urea curve shifting to lower temperatures. Cooking wheat and oat starches under neutral (pH 6.8) and acidic (pH 3.0) conditions also indicated that the high setback for oat starch was destroyed under acidic conditions. Paton tentatively concluded that $7M$ urea and mild acid produced the effect of interfering with the ability of the swollen starch macromolecules to trap water and that the abnormally high viscosity thus developed was probably associated with a corresponding retardation in the extent of retrogradation of oat starch gels. In determining routine amylograms of oat starch in the absence and presence of sodium carboxymethylcellulose, MacArthur and D'Appolonia (1979) noted that this compound reduced the development of high setback viscosity. Their observations resemble those of Paton working with $7M$ urea and mild acid and may represent a similar interference during the cooling of starch pastes. Further evidence of such interference can be gathered from work recently published by Paton (1981) on the behavior of oat starch in the presence of salts, sucrose, and acid solutions.

Table II illustrates data for the measurement of gel strength of different starches at room temperature 1 hr after pasting and after gel storage at 2°C for 24 hr. Gel strength was determined according to the procedure of Voisey and Emmons (1966), using a cheese curd tension knife equipped with an electronic strain gauge. The values given in Table II represent the maximum force (in grams) required to cut the gel. Again, some starches toughen substantially, whereas others show only small increases in gel strength. Paton (1977) described the latter gels as being more translucent, less firm, and more elastic and adhesive than corn starch. Thus, the oat starches coded Hinoat OTT/73, Hinoat REG/72,

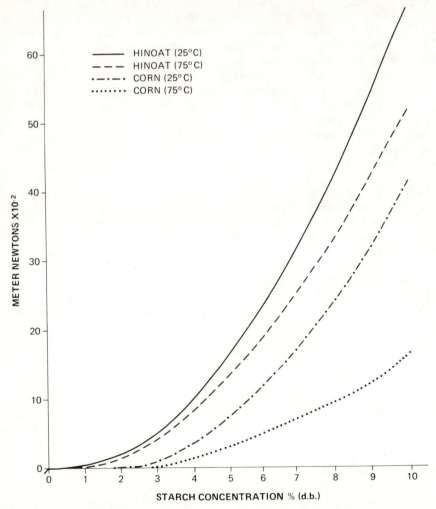

Fig. 4. Effect of concentration on the cooling curves of Hinoat and corn starch pastes. (Reprinted, with permission, from Paton, 1979)

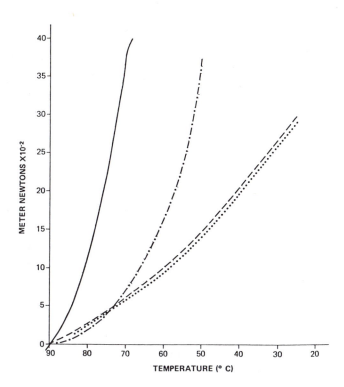

Fig. 5. Cooling curves of Hinoat and corn starch pastes cooked in water and 7*M* urea solution showing Hinoat-water (solid line), Hinoat-urea (dotted-dashed line), corn-water (dashed line), and corn-urea (dotted line). (Reprinted, with permission, from Paton, 1979)

TABLE II
Characteristics of Starch Gels (9% solids db)[a]

Starch	Gel Description[b]	Gel Strength[c]		
		A (g)	B (g)	B/A
Hinoat				
OTT/73	T, WS, sl e, t	324	365	1.13
REG/72	T, WS, e, t	390	394	1.01
Glen/73	o, MS, c	210	527	2.51
Rodney				
REG/73	T, WS, e, t	356	390	1.09
Glen Coop/73	o, MS, c	184	596	3.24
Glen/73	T, WS, sl e, t	287	327	1.14
Commercial oat	vo, FS, c	179	645	3.60
Corn	vo, FS, c	177	622	3.51
Wheat	vo, FS, c	...	...	...
Harmon	vo, FS, c	176	636	3.61
Garry	o, MS, c	179	545	3.05
Stormont	o, MS, c	183	609	3.33

[a] Source: Paton (1977); used by permission.
[b] o = Opaque, vo = very opaque, T = translucent, WS = weak set, MS = medium set, FS = firm set, sl e = slight elastic, e = elastic, t = tacky, and c = cleanly cuttable.
[c] A = at room temperature, B = following storage for 24 hr at 2°C.

Rodney REG/73, and Glen Coop Glen/73 are of this type. The starches from the oat cultivars Harmon, Garry, and Stormont exhibit gel behavior similar to that of corn, wheat, and a commercial oat starch. We may infer from Table II that protein content bears no apparent relationship to starch paste gel strength. Hinoat Glen/73 has a B/A ratio more typical of Harmon, Garry, and Stormont, whereas two of the lower-protein cultivars (Rodney REG/73 and Glen Coop Glen/73) have paste gels similar to those of the higher-protein cultivars. Errors in starch extraction and paste measurement could not account for such deviations in behavior, and workers have suggested that starch composition may play a major role in determining gel behavior.

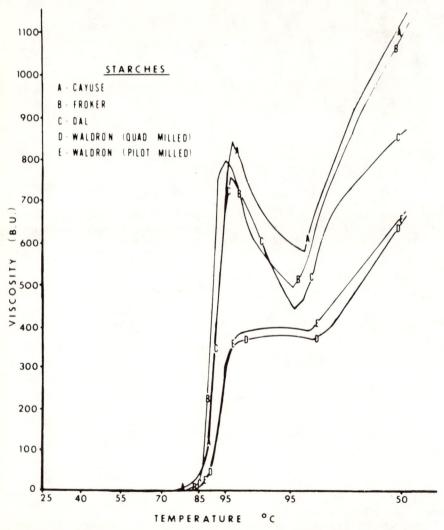

STARCHES

A - CAYUSE
B - FROKER
C - DAL
D - WALDRON (QUAD MILLED)
E - WALDRON (PILOT MILLED)

Fig. 6. Amylograph curves of oat and wheat starch. (Reprinted, with permission, from MacArthur and D'Appolonia, 1979)

MacArthur and D'Appolonia (1979), in a comparative study of oat and wheat starches, displayed amylograms of three oat starches and a wheat starch (Fig. 6). A close examination of Fig. 6 confirms, in part, the previous findings of Paton (1977). Oat pastes upon cooling developed higher torque values than those of wheat. MacArthur and D'Appolonia (1979) did not describe their oat starch gels, stating that higher setback torques for oat starches are generally associated with a greater tendency toward starch paste retrogradation but presenting no direct evidence. These conclusions are in marked contrast to those of Paton (1977). Figures 7 and 8 show that $0.1N$ Na_2SO_4, NaCl, and $CaCl_2$ also decreased the magnitude of the high setback viscosity of oat starch, whereas wheat was comparatively unaffected. In separate experiments (Paton, 1981; Table III), increasing the effective sucrose concentration in mixtures of oat starch and water reduced the shear sensitivity of the swollen granules, as indicated by the breakdown ratio (P/ H), and also reduced the ability of this starch to develop

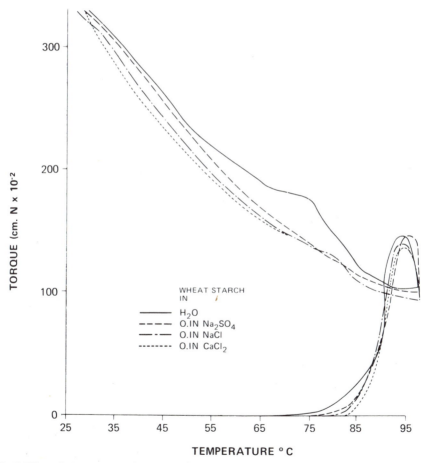

Fig. 7. Effect of salts on the pasting curves of wheat starch. (Reprinted, with permission, from Paton, 1981)

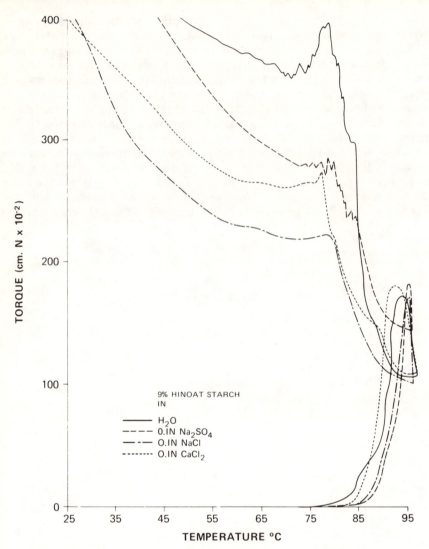

Fig. 8. Effect of salts on the pasting curve of Hinoat starch. (Reprinted, with permission, from Paton, 1981)

high setback viscosity (ratios C_1/H and C_2/H). In contrast, sucrose-wheat starch pastes actually increased in torque value upon cooling, with a corresponding small increase in the setback ratios. Incorporating citric acid at the 0.5% level (based on the total weight of slurry) into starch-sucrose pastes did not alter the magnitude of the peak torque as long as sucrose was present. However, the time taken to reach the peak was progressively extended (Figs. 9 and 10). Thus, sucrose prevented the total acid breakdown of wheat and oat starches but did little to protect the oat starch from an acid reduction of high setback viscosity. No consistent relationship could be found between paste composition and gel strength.

TABLE III
Behavior of Wheat and Hinoat Oat Starch in Presence of Sucrose[a]

Starch	Torque		Breakdown Ratio (P/H)	Torque		Setback Ratios	
	Peak (P)	Hold (H)[b]		C_1 (80° C)	C_2 (68° C)	C_1/H	C_2/H
Wheat							
H_2O	148	104	1.42	148	184	1.42	1.77
Added sucrose, %							
10	166	112	1.48	184	210	1.64	1.87
20	178	120	1.48	200	226	1.67	1.88
30	192	128	1.50	204	240	1.60	1.88
40	184	136[c]	1.35	226	256	1.65	1.88
Hinoat Oat							
H_2O	176	108	1.63	396	362	3.67	3.35
Added sucrose, %							
10	182	104	1.75	384	350	3.69	3.36
20	192	116	1.66	352	334	3.04	2.88
30	212	176	1.20	258	294	1.47	1.67
40	214	184[c]	1.15	208	244	1.13	1.33

[a] Source: Paton (1981); used by permission.
[b] Holding time at 97° C, 3 min except where noted.
[c] Holding time at 97° C, 15 min.

It is therefore important to consider these functional characteristics in terms of the physical, chemical, and structural properties of oat starches.

V. PHYSICAL, CHEMICAL, AND STRUCTURAL PROPERTIES

The literature before the 1970s is almost devoid of information on the physicochemical characteristics of oat starch. Only a few pockets of specific information were produced and directed toward an examination of the subcomponent fractions. Granular starches, in themselves, possess some characteristic physical properties, the most common of which are their uptake of iodine, known as iodine affinity or iodine-binding capacity; their swelling power at different temperatures in water; and the solubility of the starch under similar conditions. Matz (1969) reported that oat starch exhibited a higher water absorption at room temperature than other cereal starches. Table IV, which is taken from the work of MacArthur and D'Appolonia (1979), shows no significant difference in water absorption between the starches of three oat

TABLE IV
Physicochemical Properties of Oat and Wheat Starches[a]

Starch Source	Absolute Density at 30°C (g/ml)	Water-Binding Capacity[b] (%)	Intrinsic Viscosity [η]	Amylose (%)	Protein[b] (%)	Total Lipid[b] (%)
Dal	1.46814	85.0	1.59	27.9	0.60	1.11
Froker	1.45569	87.0	1.42	25.5	0.59	0.67
Cayuse	1.45890	86.0	1.45	25.9	0.44	0.81
Waldron[c]	1.46806	86.0	1.80	25.6	0.22	0.48

[a] Source: MacArthur and D'Appolonia (1979); used by permission.
[b] Results expressed on a dry basis.
[c] Sample obtained from Brabender Quadrumat Jr. flour mill.

cultivars (Dal, Froker, and Cayuse) and a wheat starch isolated from the cultivar Waldron. Tables IV and V contrast data from the work of MacArthur and D'Appolonia (1979) and Paton (1979). It is apparent from Paton's studies (Table V) that starch isolated from a high-protein oat cultivar (Hinoat OTT/73 in Table I), here designated as Oat 1, possessed different properties from starch of a

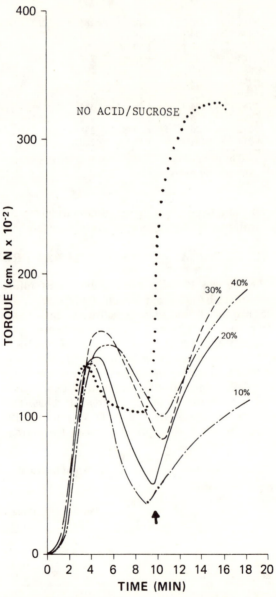

Fig. 9. Effect of sucrose concentration on pasting curve of wheat starch in the presence of citric acid. Arrow = start of cooling period. (Reprinted, with permission, from Paton, 1981)

lower-protein cultivar (Rodney REG/67 in Table I) designated as Oat 2. The groat protein contents of the cultivars were 22 and 16%, respectively. Oat 1 had a higher swelling power, higher intrinsic viscosity, and lower solubility and apparent amylose content than did Oat 2.

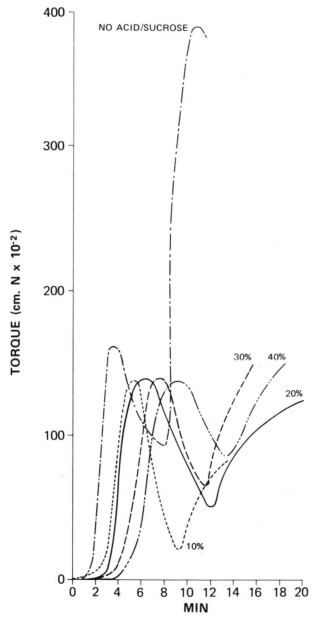

Fig. 10. Effect of sucrose concentration on pasting curve of Hinoat starch in the presence of citric acid. (Reprinted, with permission, from Paton, 1981)

In contrast, the work of MacArthur and D'Appolonia (1979) indicated that the three oat cultivars (Dal, Froker, and Cayuse) had higher amylose content than the cultivars examined by Paton, although both sets of data showed similar values for starch intrinsic viscosity. MacArthur and D'Appolonia (1979) (Table VI) also showed intrinsic viscosity data for amylose and amylopectin components fractionated from the starch of their three oat cultivars. Oat amylose tended to have higher values (2.46–2.99 dl/g) than did wheat starch (2.33 dl/g), implying that oat amylose is more linear and would be expected to exhibit a higher hydrodynamic volume in the test solvent. Oat amylopectin, on the other hand, was shown to have slightly lower intrinsic viscosity values than wheat amylopectin, implying a higher degree of branching in this polymeric fraction. No indication was given by these authors of the yields of their fractions nor of the purity.

Several oat cultivars grown in West Germany have also been examined in our laboratory. The data (Table VII) are contrasted with an oat starch (Hinoat,

TABLE V
Physical Characteristics of Cereal Starches[a]

Starch	Iodine Affinity at 30°C (%)	Amylose (%)	Swelling Power at 95°C	Solubility at 95°C	$[\eta]$ (dl/g)
Wheat	4.82	25.3	19.8	47.6	1.54
Corn	4.42	23.2	19.2	21.7	1.58
Oat 1	3.19	16.8	23.6	16.8	1.62
Oat 2	3.54	18.6	19.4	26.0	1.50

[a] Source: Paton (1979); used by permission.

TABLE VI
Intrinsic Viscosity of Oat and Wheat Starch Fractions[a]

Starch Source	Amylose $[\eta]$	Amylopectin $[\eta]$
Dal	2.46	2.07
Froker	2.55	1.73
Cayuse	2.99	1.70
Waldron[b]	2.33	1.80

[a] Source: MacArthur and D'Appolonia (1979); used by permission. Average of two determinations.
[b] Sample obtained from Brabender Quadrumat Jr. flour mill.

TABLE VII
Comparative Physical Data for Oat, Wheat, and Corn Starches

Starch	Iodine Affinity	Amylose (%)	Swelling Power (95°C)	Solubility (95°C)
German oats				
Erbgraf	3.52	18.52	18.13	12.9
Selma	3.33	17.52	24.54	14.8
Pitol	3.48	18.31	23.96	16.5
Bottus	3.56	18.73	23.11	14.4
Etich	3.70	19.47	20.08	16.4
Hinoat	3.26	17.15	23.60	16.8
Wheat	4.82	25.37	19.80	47.2
Corn	4.42	23.26	19.20	21.7

OTT/73), wheat, and corn. All of these oat starches were similar to each other but differed from corn and wheat. Although not shown here, the pasting curves of starches isolated from these German cultivars were more like those presented by Paton (1977). However, none was as pronounced in the rapid cooling response to rising consistency as was observed for several Canadian cultivars.

What are the influences of climatic and other agronomic conditions on the properties of oat starches? MacArthur and D'Appolonia (1979) suggested that a relationship may exist between groat protein content and starch paste behavior, although none was evident from the data of Table IV. A similar suggestion arises from the results of Paton (1977, 1979), although tempered somewhat by a recognition of the possible influences of agronomic factors. The data shown in Table VII are from oat cultivars that had groat protein contents in the range of 9–12%. Such values are considerably lower than are usually found for typical North American cultivars, and certainly substantially less than some of those examined by Paton (17–23%). The similar yet not identical paste behavior of the German oat starches to some of Paton's oat starches does not bear out any definitive correlation between protein content and starch paste consistency.

The only early comprehensive study on the fractionation and characterization of subcomponents of oat starch was that of Arbuckle and Greenwood (1958). An iodine affinity value of 5.13 (corresponding to an amylose content of 27.0%) was found for the oat variety Milford. These workers examined the digestibility of oat and wheat amyloses obtained under different leaching conditions and determined that the presumed linear amylose could not be quantitatively converted to maltose by the action of a purified β-amylase. There was the distinct possibility that, in both oat and wheat amyloses, minor branch points existed that prevented total conversion but that these branch points might be randomly distributed.

Later work by Banks and Greenwood (1967) (Table VIII) on an unspecified oat cultivar showed that a pure amylose could only be produced by repeated crystallization of the first-obtained amylose-thymol complex. Oat amylose gave only a 78% conversion to maltose under the action of β-amylase but could be totally converted by the concurrent action of β-amylase and a debranching enzyme. These results also demonstrate the existence of branch points in amylose. Further, these workers found that oat amylose had the highest limiting viscosity value of all of the cereal starch amyloses examined. However, the light-scattering studies gave rise to molecular weights for amyloses that were not

TABLE VIII
Iodine Affinities of Amylose and Amylopectin Fractions
of Various Cereal Starches[a]

Source of Starch	Iodine-Binding Capacity (%)	
	Amylose	Amylopectin
Barley	19.6	0.6
Oats	19.5	0.5
Rye	19.7	0.4
Wheat I	19.7	0.5
Wheat II	19.3	0.6

[a] Source: Banks and Greenwood (1967); used by permission.

in the same order as the corresponding limiting viscosity numbers, a fact that again supports branching in amylose. This study also indicated no significant difference in the physical properties of cereal starch amylopectins, except that oat amylopectin had a slightly lower average chain length ($\overline{CL}$).

Banks and Greenwood (1967) were the first to demonstrate the existence of fractions intermediate in behavior to amylose and amylopectin (Table IX). An amylopectin subfraction termed *anomalous amylopectin* was identified for all cereal starches and was present to the extent of 4.5% in oat. It was characterized as being essentially similar to amylopectin except that its average chain length was somewhat larger—in other words, the molecule was less branched. An *anomalous amylose* fraction was also found in amounts equal to anomalous amylopectin, but this fraction was fairly highly branched since conversion into maltose by the combined action of β-amylase and a debranching enzyme was less than complete. The significance of the existence of these starch fractions in terms of starch functionality was never addressed in this study.

Paton (1979), using a gel-permeation chromatographic technique, showed that wheat, corn, and oat starch could be fractionated as illustrated in Fig. 11. The starches were different in elution profile, with both the oat and wheat starches exhibiting a fraction in addition to the excluded amylopectin and the eluted amylose. This intermediate fraction may be the same one referred to by Banks and Greenwood (1967). However, as can be seen clearly from the curves,

TABLE IX
Properties of Amylopectin and Intermediate Material Obtained
from Some Cereal Starches[a]

Starch	Polysaccharide	Amount (%)	IBC[b] (%)	$\overline{CL}$ [c]	$[\beta]$[d] (%)	$[\beta + Z]$[e] (%)	$[\eta]$ (ml/g)
Barley	Amylopectin	~65	0.6	20	57	56	160
	Thymol-amylopectin	4	2.6	···	···	···	···
	Anomalous amylopectin	4	2.6	27	62	63	165
	Anomalous amylose	0	···	···	···	···	···
Oats	Amylopectin	~65	0.5	18	54	55	145
	Thymol-amylopectin	9	4.3	···	···	···	···
	Anomalous amylopectin	4.5	0.8	25	60	60	145
	Anomalous amylose	4.5	8.5	···	61	93	170
Rye	Amylopectin	~65	0.4	20	58	57	150
	Thymol-amylopectin	4	2.6	···	···	···	···
	Anomalous amylopectin	4	2.4	27	57	60	170
	Anomalous amylose	0	···	···	···	···	···
Wheat I	Amylopectin	~65	0.5	21	56	56	150
	Thymol-amylopectin	9	5.7	···	···	···	···
	Anomalous amylopectin	5	1.2	27	62	64	135
	Anomalous amylose	4	11.8	···	56	94	180
Wheat II	Amylopectin	~65	0.6	20	54	55	150
	Thymol-amylopectin	6	5.0	···	···	···	···
	Anomalous amylopectin	4	0.2	25	55	55	155
	Anomalous amylose	2	12.5	···	66	97	185

[a] Source: Banks and Greenwood (1967); used by permission.
[b] Iodine-binding capacity.
[c] Average chain length.
[d] Amount of fraction converted into maltose by action of β-amylase.
[e] Amount of fraction converted into maltose by joint action of β-amylase and a debranching Z enzyme.

all starch fractions are not eluted in finely discrete peaks; rather, they occur as broad bands. Since the support gel used in this fractionation has a separation power in the range of 300,000 to 20×10^6 daltons, the peak assigned to intermediate material probably consists of several species that would include Banks and Greenwood's anomalous amylose and anomalous amylopectin.

Figure 12 illustrates the change in the ratio of the absorbances (615/540 nm) of iodine-treated fractions over the elution profile for wheat and oat starch. These two wavelengths respectively represent the λ_{max} for amylose and amylopectin complexes. By determining the absorption of each tube at these two wavelengths against tube number, an elution profile is developed in which each eluted fraction may be considered in terms of being amylose-like or amylopectin-like. Clearly, the nature of the starch species in the oat starch (Hinoat cultivar) is quite different from that of wheat starch and tends to favor a greater or altered type of branching. Quantitative analysis of the chromatograms shown in Fig. 11 shows that oat starch contains 56% amylopectin, 26% intermediate material, and 18% amylose. The corresponding data for wheat are 56, 16, and 28%.

Attempts in our laboratory to fractionate and purify Hinoat starch subcomponents quantitatively by thymol or butanol treatment have not been successful. Based upon iodine-complexing data, the purest amylopectin still retained up to 12% of linear material. The purest amylose, after repeated

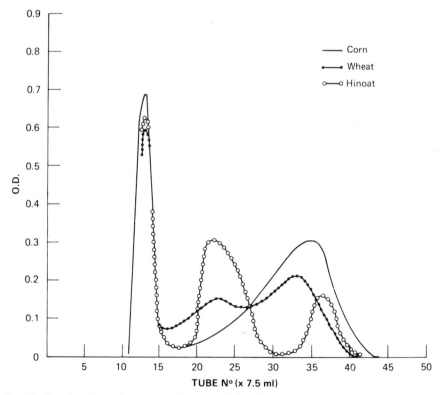

Fig. 11. Fractionation of cereal starches by gel filtration chromatography. (Reprinted, with permission, from Paton, 1979)

recrystallization from 1-butanol, could not bind iodine to an extent greater than 15.8%. This is to be contrasted with the data of Banks and Greenwood (1967), who found values in the range of 18.6–19.0%. Neither was it possible to obtain pure intermediate fractions from any of the mother liquors. Thus, it has not been possible to isolate and characterize the individual oat starch fractions as is the usual practice with other starches.

Recent studies (*unpublished data*) from our laboratory examined the structure of Hinoat oat starch by means of selected degradation by β-amylase and pullulanase, followed by gel-permeation chromatography. As expected, the action of β-amylase produces a typical bimodal distribution of limit dextrin and maltose. The extent of conversion of oat starch was found to be 66%, somewhat higher than that previously reported (Paton, 1979). This level results from improved dissolution of the granule and the use of a longer digestion period. The average chain length of the products of β-amylase action was 35 for oat and 24 for corn starch. Pullulanase, in hydrolyzing α-1,6 linkages, produces a trimodal distribution of DP species with a degree of polymerization of 60, 45, and 15, respectively. Again, the average chain length of debranched products is much higher for oat than for corn (55 and 26, respectively) (Table X). The products of pullulanase debranching have been chromatographed on Sephacryl S-300 (Fig. 13). The unusual oat starch distribution in and immediately after the void volume could arise from some long, linear main chains of amylopectin, from intermediate material previously demonstrated in this starch, or both. The

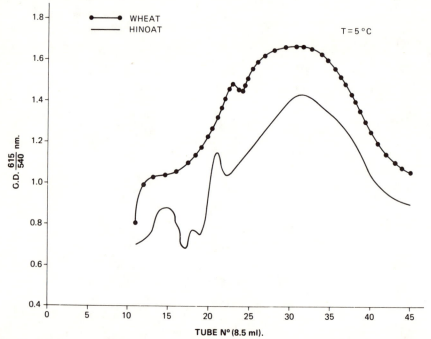

Fig. 12. Fractionation of cereal starches showing change in optical density 615/540 nm ratios. (Reprinted, with permission, from Paton, 1979)

pullulanase-digested material can be 97% converted to maltose upon further action of β-amylase, indicating that all fractions are substantially linear. Our conclusions at this point are that Hinoat oat starch contains an intermediate species that has a predominantly linear structure but is lightly branched, the side chains themselves being rather long and linear.

The pasting behavior of oat starches does not seem obviously dependent on amylose content, swelling power, or solubility. From a rheological standpoint, the presence of approximately 25% of the starch as an intermediate species might be expected to influence the behavior of cooked starch in dispersion and its

TABLE X
Quantitative Analysis of Starches Digested by Pullulanase

Starch	Carbohydrate weight (%)			
	DP > 60	DP $= 45$	DP $= 15$	$\overline{CL}$ [a]
Corn	30	12	57	27
Waxy corn	5	20	74	26
Oat (Hinoat)	39	11	50	55

[a]Average chain length of debranched starch.

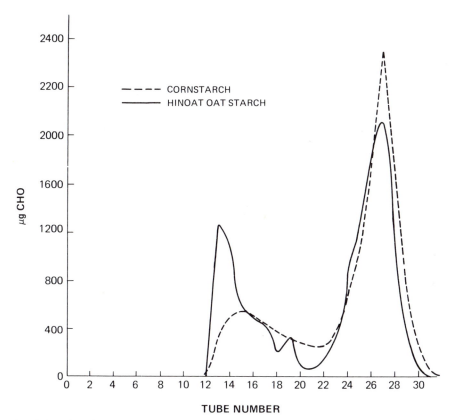

Fig. 13. Fractionation on Sephacryl S-300 of pullulanase-debranched oat and corn starches.

subsequent behavior on cooling. The existence of a lightly branched, open structure should be conducive to the trapping of water as the temperature of the hot starch paste is lowered. Such a structure would also be susceptible to ionic effects and would further be influenced by the presence of other hydrophilic substances, such as sugars. Paton (1977) previously reported that several oat starch pastes exhibited a stable gel characteristic upon refrigerated storage. Some samples could be frozen and thawed without the induction of major syneresis.

Some recent studies of oat starch by differential scanning calorimetry have revealed the presence of a rather sizable amylose-lipid complex (Paton, 1985, *unpublished results*). Figure 14 clearly shows this endotherm and the δH value of 3.57 J/g for the peak at 102.3°C. The reversibility of this complex is also indicated by the dotted line, which represents a rerun of the first scan following a cooling period from 140 to 40°C. Applying the methodology of Kugimiya and Donovan (1981) using synthetic lysolecithin, it may be shown that a fully lysolecithin-saturated oat starch displays an endotherm in the 102–105°C region with a δH of 5.95 J/g. We may thus calculate that in oat starch, 60% of the natural amylose is complexed by starch lipid during the process of heating the starch in an aqueous environment. Apparently, the lipid content of the oat starches described by Paton (1977) using petroleum distillate as a solvent severely underestimates the true lipid content of these starches. That this lipid is strongly bound within the starch granule has also been suggested and indeed shown by Morrison et al (1984). These researchers determined the amylose content of six oat starches of Canadian origin colorimetrically before and after removing any internally bound lipids by refluxing the starch in a 3:1 mixture (v/v) of *n*-propanol/water at 100°C in screw-capped vials at a solvent:starch ratio of 20:1

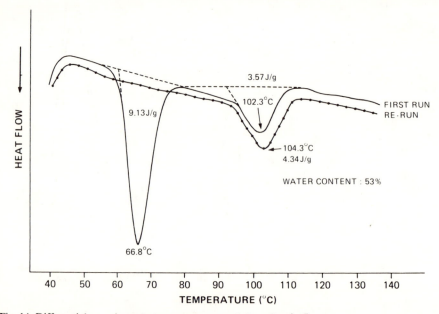

Fig. 14. Differential scanning calorimetry of oat starch (var. Sentinel).

TABLE XI
Amylose and Lipid Contents of Oat Starches[a]

Variety	Amylose (%)			Lipids (mg/100 g)			
	Total	Apparent	ΔAM[b]	FFA[c]	lysoPL[d]	Total	As FAME[e]
Sentinel	25.2	17.3	7.9	411	977	1,388	991
Lodi	26.9	18.3	8.6	462	1,061	1,523	1,091
Hinoat	27.0	18.3	8.8	382	998	1,380	971
CI 4492	28.2	20.2	8.0	455	1,035	1,490	1,069
Terra	28.3	20.1	8.2	421	1,055	1,476	1,044
Exeter	29.4	20.7	8.6	268	1,081	1,349	898

[a] Source: Morrison et al (1984); used by permission. Analysis performed on dry weight basis.
[b] Difference between total and apparent amylose contents.
[c] Free fatty acids.
[d] Lysophospholipid.
[e] Fatty acid methyl esters.

(Table XI). The presence of 1.35–1.52% total lipid by the reflux method differs markedly from that reported by Paton (1977), and this lipid causes a large underestimation of true amylose content of the starch. The apparent amylose contents of Morrison et al (1984) are within the range reported by Paton (1977). Morrison et al (1984) also reported that the values for the difference between true and apparent amylose content was highest for oat starches.

It is even more surprising that a starch that has now been shown to have a normal cereal amylose content should exhibit pasting behavior like that shown in Fig. 3. The question arises as to the role, nature, and function of the internal starch lipids in contributing to paste characteristics. If lipids are removed by the procedure of Morrison et al (1984), oat starch shows no evidence of the formation of an amylose-lipid complex (Fig. 15). Further, if pasting curves are determined for a native and completely defatted oat starch, the cooling curve for the defatted starch is more typical of wheat starch (Fig. 16). We have observed that this cooked paste is short in texture, quite opaque, and sets up into a firm gel, all in marked contrast to the native oat starch.

In the relatively few studies conducted to date, considerable variability has been observed in the physicochemical and functional behavior of oat starches. Because many factors influence structure and functional behavior, it has not yet been possible to determine definitively the reasons for unusual behavior of starch pastes.

VI. COMMERCIAL INTERESTS AND OPPORTUNITIES

It is generally accepted that less than 15% of the total oat production ever leaves the farm, a figure that has remained relatively constant over the past 15–20 years. Of the small percentage entering the food industry, most oats are still used in very traditional products—for example, oatmeal, muffins, and cookies. Oats, as a potential source of new and perhaps novel ingredients, appear to have escaped the same vigorous investigation given to other cereals such as wheat, barley, rye, and even triticale. Kessler and Hicks (1949) and Waggle (1968) obtained patents on the use of oat flour in adhesive preparations, which in all likelihood were a function of the properties of the starch component. It seems timely, therefore, in light of the most recent information on the pasting behavior of oat starch, to reexamine applications based upon adhesive properties. Industrial glues and wallpaper pastes might be worthy of consideration. However, since such applications are usually highly specific and frequently confidential, evaluations of this kind should and must be conducted by industrial ingredient suppliers. Paton (1974) produced limited information on acetylation, hydroxyethylation, and oxidation of oat flours and found some treatments that were as effective as for wheat flours. Such substituted wheat and corn starch ethers and esters find use as wet-end additives in the production of brown paper and cardboard products. The pharmaceutical industries are large users of refined starches for tableting and coating agents. Rice starch was traditionally used for this purpose because of its small granule size, but rising costs and unavailability of stable supplies are making this use prohibitive. Oat starch should be considered a potential replacement for rice starch in these applications. Another area where oat starch may make inroads is in photocopy papers.

A further constraint on oat starch production lies in the economics of the process itself. Although data have been published on physicochemical properties of gum, protein, and bran from oats, no single commercial use of an oat component provides the necessary driving force for utilization of oat fractions.

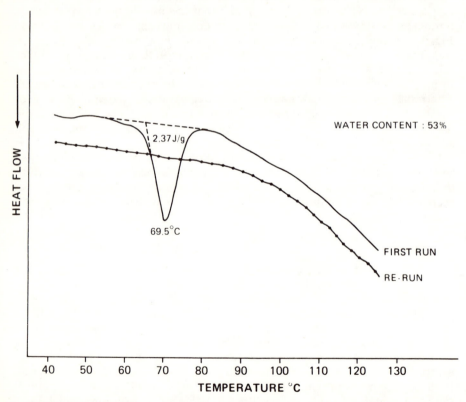

Fig. 15. Differential scanning calorimetry of oat starch (var. Sentinel) defatted by the procedure of Morrison et al (1984).

Although dry fractionation processes are preferred, aqueous processes should not be ruled out until the added value based on specific uses can be determined. Nonaqueous fractionation processes may prove hazardous and, as already discussed, do not yield a starch of high purity. Theoretically, there are few current starch applications (except those based upon waxy corn) that could not be filled by oat starch. However, much more cooperation is needed among research groups to further this end, and a greater effort and commitment are necessary on the part of cereal processing companies.

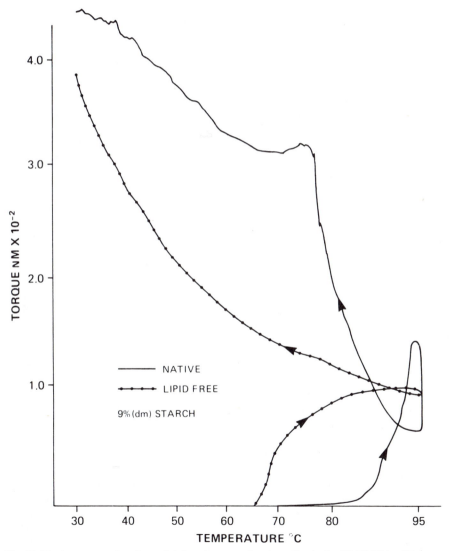

Fig. 16. Pasting curves of native and defatted oat starches (var. Sentinel) of 9.0% (db) solids.

VII. SUMMARY

Oat starch may be isolated from groat flour by any wet extraction method that does not severely damage the functionality of the starch granules. Such extractive procedures also aid in lowering the residual protein and free lipid content to levels comparable to those of other isolated starches. The granules are generally irregular in shape, weakly birefringent, and have an average granule diameter of the same order of magnitude as rice starch. The physical and functional properties of oat starch vary over a wide range for a spectrum of oat cultivars and may be influenced by climate, agronomic conditions, and perhaps even genetic control. Amylose contents have been found in the 16–27% range. Pasting properties and characteristics such as swelling power and solubility also appear variable, with some cultivars exhibiting major departures from accepted behavior. The response of some oat starch pastes to temperature during the cooling cycle is perhaps the most pronounced of these abnormal characteristics; contrary to accepted belief, this rapid development of cold viscosity does not appear related to the tendency of the starch paste to retrograde. Oat starch gels are more elastic, adhesive, and translucent and show greater stability under storage conditions than corresponding pastes made from unmodified wheat or corn starches.

Over the same range of starch concentration, oat starch pastes exhibit higher viscosities during the cooling cycle than wheat or corn starches, although oat starch granules when cooked tend to be more sensitive to shear. The cooling curves are markedly influenced by the presence of salts, sucrose, gums, and mild acids. Much more research is needed to describe this behavior and also to determine the role (if any) of phosphorus, since oat starches contain almost the same levels of phosphate as potato starch. Although free lipids have been extracted from oat starch granules and shown not to influence the pasting behavior, this aspect should undoubtedly be given greater attention in light of the latest data for internally bound lipids, especially phospholipids, on the pasting behavior of oat starch.

The least understood area of oat starch chemistry is that relating to structural studies. Thus far, oat starch isolated from a high-protein variety has been shown to have an amylose of slightly lower molecular weight than corn or wheat amylose and has been found to contain a lightly branched, substantially linear, intermediate material. Gel chromatographic examination of starch solutions, debranched by the action of pullulanase, would seem to indicate that this oat starch is fundamentally different from wheat and corn that has been similarly treated. All of these studies have been done on whole starches, but greater emphasis needs to be placed on the physicochemical and structural properties of the various subfractions of this starch. Limited attempts to fractionate starch components quantitatively from this high-protein cultivar have not proved successful.

One major impediment to the development of technologies for the isolation of oat starch is the need to establish the agronomic stability of any exhibited, potentially beneficial characteristic. Much more research must be done to establish the influence of geographic, climatic, and soil conditions on starch behavior; to survey the World Oat Collection to determine the frequency of occurrence of such characteristics; and to establish any relationship between genetic control of starch composition and starch functional behavior. Finally, industry must rekindle its efforts to evaluate existing and new technologies at the pilot plant level and conduct end-use evaluations for oat starch. Such efforts, in concert with present research on the chemistry of oats and with the anticipated future assistance of the plant geneticist, are necessary if potential opportunities for food and nonfood uses of oat starch are to be fully exploited.

ACKNOWLEDGMENT

This chapter is Contribution No. 480 of the Food Research Centre, Agriculture Canada.

LITERATURE CITED

ACKER, L., and BECKER, G. 1971. Recent studies on lipids of cereal starches. II. Lipids of various types of starch and their binding to amylose. Staerke 23:419-424.

ACKER, L., and BECKER, G. 1972. The lipids of cereal starches. Gordian 72:275-278.

ADKINS, G. K., and GREENWOOD, C. T. 1966. The isolation of cereal starches in the laboratory. Staerke 18:213-218.

ARBUCKLE, A. W., and GREENWOOD, C. T. 1958. Physiochemical studies on starches. XIII. The fractionation of oat and wheat starches. J. Chem. Soc. 2626-2629.

BANKS, W., and GREENWOOD, C. T. 1967. Fractionation of laboratory-isolated cereal starches using dimethylsulphoxide. Staerke 19:394-398.

BURROWS, V. D., FULCHER, R. G., and PATON, D. 1984. Processing aqueous treated cereals. U.S. patent 4,435,429.

CLENDENNING, K. A., and WRIGHT, D. E. 1945. Polarimetric determination of starch in cereal products. III. Composition and specific rotatory power of starches in relation to source and type. Can. J. Res. 23B:131-138.

DOUBLIER, J. L. 1981. Rheological studies on starch; flow behaviour of wheat starch pastes. Staerke 33:415-420.

HOHNER, G. A., and HYLDON, R. G. 1977. Oat groat fractionation process. U.S. patent 4,028,468.

HOSENEY, R. C., FINNEY, K. F., POMERANZ, Y., and SHOGREN, M. D. 1971. Functional (breadmaking) and biochemical properties of wheat flour components. VIII. Starch. Cereal Chem. 48:191-201.

KESSLER, C. C., and HICKS, W. L. 1949. Adhesive from cereal flour. U.S. patent 2,466,172.

KUGIMIYA, M., and DONOVAN, J. W. 1981. Calorimetric determination of the amylose content of starches based on formation and melting of the amylose-lysolecithin complex. J. Food Sci. 46:765-770.

MacARTHUR, L. A., and D'APPOLONIA, B. L. 1979. Composition of oat and wheat carbohydrates. II. Starch. Cereal Chem. 56:458-461.

MacMASTERS, M. M., WOLF, M. J., and SECKINGER, H. L. 1947. The possible use of oat and other cereal grains for starch production. Am. Miller Process. 75(1):82-83.

MATZ, S. A. 1969. Oats. Pages 78-96 in: Cereal Science. S. A. Matz, ed. AVI Publishing Co., Westport, CT.

MORRISON, W. R., MILLIGAN, T. P., and AZUDIN, M. N. 1984. A relationship between amylose and lipid contents of starches from diploid cereals. J. Cereal Sci. 2:257-271.

OUGHTON, R. W. 1980. Process for the treatment of comminuted proteinaceous material. Canadian patent 1,080,700.

PATON, D. 1974. The effects of chemical modification on the pasting characteristics of a high-protein oat flour (Hinoat). Cereal Chem. 51:641-647.

PATON, D. 1977. Oat starch. I. Extraction, purification and pasting properties. Staerke 29:149-153.

PATON, D. 1979. Oat starch: Some recent developments. Staerke 31:184-187.

PATON, D. 1981. Behavior of Hinoat oat starch in sucrose, salt, and acid solutions. Cereal Chem. 58:35-39.

PRENTICE, N., CUENDET, L. S., and GEDDES, W. F. 1954. Studies on bread staling. V. Effect of flour fractions and various starches on the firming of bread crumb. Cereal Chem. 31:188-206.

VOISEY, P. W., and EMMONS, D. B. 1966. Modification of the curd firmness test for cottage cheese. J. Dairy Sci. 49:93-96.

VOISEY, P. W., PATON, D., and TIMBERS, G. E. 1977. The Ottawa Starch Viscometer—A new instrument for research and quality control applications. Cereal Chem. 54:534-557.

WAGGLE, D. H. 1968. Dextrin adhesive composition. U.S. patent 3,565,651.

WU, Y. V., CLUSKEY, J. E., WALL, J. S., and

INGLETT, G. E. 1973. Oat protein concentrates from a wet-milling process: Composition and properties. Cereal Chem. 50:481-488.

YOUNGS, V. L. 1974. Extraction of a high protein layer from oat groats, bran and flour. J. Food Sci. 39:1045-1046.

OAT β-GLUCAN: STRUCTURE, LOCATION, AND PROPERTIES

PETER J. WOOD
Food Research Centre
Central Experimental Farm
Agriculture Canada
Ottawa, Ontario, Canada

I. INTRODUCTION

In 1942, a nonstarchy glucan was isolated from oats and shown to be similar to lichenin from the lichen Iceland moss (Morris, 1942). A similar polysaccharide in barley, now known to be $(1\rightarrow3)(1\rightarrow4)$-linked β-D-glucan, was recognized as having a significant role in malt and beer production. Consequently, commercial interest has stimulated research into barley endospermic β-D-glucan, but there has been no equivalent stimulus for similar research on oats, and very little work has been published beyond the early structural studies. This situation may well change over the next few years as a consequence of new interest in oat gum (which is composed mainly of the β-D-glucan) as a potentially valuable industrial hydrocolloid and as an important, physiologically active dietary component of possible therapeutic value.

In this chapter, information available on oat β-D-glucan—hereinafter referred to simply as oat β-glucan—is discussed with respect to location, structure, extraction, analysis, and properties. Dye interaction and β-glucanases are also discussed. Where data on barley β-glucan are relevant, such as in analytical methodology, these also are reviewed.

II. OCCURRENCE

The mixed-linkage β-glucan of oat and barley endosperm belongs to a family of unbranched polysaccharides composed of $(1\rightarrow4)$- and $(1\rightarrow3)$-linked β-D-glucopyranosyl units in varying proportions. This polysaccharide has been identified in a variety of tissues of the main cereals of commerce (Preece and Hobkirk, 1953; Buchala and Wilkie, 1970; Woolard et al, 1976; Nevins et al,

1978; Anderson et al, 1978) and is also present in ryegrass (Smith and Stone, 1973), bamboo (Wilkie and Woo, 1976), and a variety of other grasses (Stinard and Nevins, 1980). It has been suggested to be transiently present in mung bean (Buchala and Franz, 1974) and is present in a number of lichens (Peat et al, 1957; Takeda et al, 1972; Nishikawa et al, 1974).

This review is mainly concerned with the endospermic oat β-glucan. The nonendospermic β-glucans of oats have been studied by Wilkie and Buchala and co-workers as part of an extensive survey of the hemicelluloses of the Gramineae, and this subject has recently been reviewed (Wilkie, 1979). In leaves and stem, glucose-containing polysaccharides generally comprised 5–25% of the total hemicellulose. Exact estimates of the various polysaccharide fractions, in particular of the β-glucan, were difficult to obtain, as was reliable comparative structural information, because of the complex fractionation procedures required (Reid and Wilkie, 1969a; Fraser and Wilkie, 1971). In all tissues, there was a trend toward a decrease in the proportion of β-glucan in the total hemicellulose fraction, and in the proportion of (1→3)-linkages, with increasing maturity (Reid and Wilkie, 1969b; Buchala and Wilkie, 1971). Although considerable chemical information is available on the nonendospermic β-glucans of oats, there is little detailed information on physical properties. A β-glucan isolated from oat leaf contained (1→3)- and (1→4)-β-linked D-glucopyranosyl units in the ratio of 1.00:1.65. This material had a high degree of polymerization (DP) and was insoluble in water (Fraser and Wilkie, 1971). Difficulties in fractionation were indicative of a high degree of polydispersity and polymolecularity, and possibly of aggregation phenomena (Reid and Wilkie, 1969a; Fraser and Wilkie, 1971; Wilkie, 1979). A glucan isolated from oat coleoptiles contained (1→4)- and (1→3)-β-linked D-glucopyranosyl units in the ratio of 2:1 and had a molecular weight of 240,000–260,000 and an intrinsic viscosity of 3.88 dl/g (Wada and Ray, 1978; Labavitch and Ray, 1978).

The presence of glucose in hydrolysates from plant polysaccharide preparations was frequently, but often incorrectly, attributed to degraded cellulose, starch, or possibly glucomannans and galactoglucomannans. It was largely because of the efforts of Wilkie and Buchala and their co-workers (e.g., Buchala and Wilkie, 1970) that the widespread occurrence of the mixed-linkage β-glucan in cereals and grasses was recognized.

III. STRUCTURE

The structural similarity between oat β-glucan and lichenin, recognized by Morris (1942), was later established by Acker et al (1955a, 1955b) and Peat et al (1957).

Acid hydrolysis of the purified polymer showed that it contained only glucose, and partial hydrolysis and acetolysis gave cellobiose, laminaribiose, cellotriose, and oligosaccharides containing both 3-O-linked glucose and 4-O-linked glucose. The identification of mixed linkage oligosaccharides and the lack of solely (1→3)-linked oligomers other than laminaribiose indicated that the glucan was not a mixture of (1→4)- and (1→3)-linked polysaccharides, and that the (1→3)-linkages occurred mostly singly.

The classical techniques of periodate oxidation and methylation analysis were used to reveal the overall average structure of oat and barley β-glucan and of lichenin. Hydrolysis of a completely methylated β-glucan yields 2,3,4,6-tetra-O-methyl-D-glucose from the nonreducing end; 2,3,6-tri-O-methyl-D-glucose from 4-linked units; and 2,4,6-tri-O-methyl-D-glucose from 3-linked units. Branch points may be detected from di-O-methyl-glucose (or monomethyl ether if doubly branched) with a corresponding increase in the yield of the tetra-O-methyl-D-glucose. The proportion of (1→3)- and (1→4)-linked units may also be deduced from periodate consumption data, since each 4-O-substituted glucose will consume one mole of periodate, whereas 3-O-substituted units, lacking the necessary vicinal glycol unit, will be resistant. (In practice, the polysaccharide is remarkably resistant to oxidation, requiring days for complete reaction.) Measurement of the consumption of periodate by the polysaccharide suggested a ratio of β-(1→4)- to (1→3)-linkages of 3.2:1 (Peat et al, 1957), though somewhat lower values for (1→4)-linked residues were found by Acker et al (1955a) and Goldstein et al (1965). Acker et al (1955b) also used methylation analysis to demonstrate β-(1→4)-linkages and β-(1→3)-linkages and possibly a small (< 2%) amount of branching. Smith and Montgomery (1959) described methylation analysis of intact and partially degraded glucans. For oat β-glucan of DP 1,000 (undegraded), the ratio of (1→4)- to (1→3)-linked glucopyranosyl units was 1.77:1.00; for DP 120, 2.24:1.00; and for DP 20, 2.54:1.00. Traces of di-O-methyl and equivalent tetra-O-methyl-D-glucose suggested some branching. Unfortunately, details of these studies were not published.

Periodate consumption data and methylation analysis can only provide an average description of the polysaccharide structure, and additional techniques are required to elucidate linkage sequence.

A powerful technique for detecting runs of (1→3)-linked units is based on Smith degradation, in which a periodate-oxidized polysaccharide is reduced with sodium borohydride and subjected to partial acid hydrolysis. The oxidized and reduced units are sensitive to acid and readily hydrolyzed, but the intact (1→3)-linked units are relatively stable to hydrolytic conditions. Runs of β-(1→3)-linked glucopyranosyl units therefore yield laminaribiose and higher laminarisaccharides terminated at the reducing end by erythritol. Sequences of up to four (1→3)-linked glucopyranosyl units were reported for oat β-glucan by Goldstein et al (1965). Similar sequences have been reported for barley β-glucan (Igarashi and Sakurai, 1966; Fleming and Kawakami, 1977).

Some caution is required in interpretation of periodate oxidation and Smith degradation data for polysaccharides, such as oat β-glucan, that are very resistant to oxidation. Lichenin, for example, showed an initial second-order rate constant for oxidation that was one-hundredth of the value for α-linked amylose (Aalmo et al, 1978). Periodate-resistant residues may be created by formation of interresidue hemiacetal bonds (Ishak and Painter, 1971), and, if care is not taken to ensure complete oxidation, some of the intact glucose or glucose oligomers obtained on subsequent hydrolysis may arise from unoxidized (1→4)-linked glucopyranosyl units. In structural studies of oat β-glucan, therefore, it is necessary to use long oxidation times, but this runs the risk of overoxidation (Perlin and Suzuki, 1962) and of chain scission (Scott and Tigwell, 1973). More recently, Woodward et al (1983) have shown that incomplete hydrolysis during Smith degradation may have led to misidentification of (1→3)-linked oligomers in previous studies.

Evaluation of the oligosaccharides released by a "cellulase" from *Strepto-*

myces and a "laminaranase" from *Rhizopus arrhizus* (Parrish et al, 1960) confirmed and extended the suggestions of Peat et al (1957) and showed that the polymer was mainly composed of two structural sequences, one a tetrameric unit in which single $(1\rightarrow3)$-β-linkages alternated with two $(1\rightarrow4)$-β-linkages and the other a pentameric unit in which one $(1\rightarrow3)$-β-linkage alternated with three $(1\rightarrow4)$-β-linkages. Thus, the polymer was essentially a chain of cellotriose and cellotetraose units joined by single $(1\rightarrow3)$-β-linkages. There was no evidence to suggest the occurrence of isolated $(1\rightarrow4)$-linkages or of contiguous $(1\rightarrow3)$-linkages. Application of enzyme techniques to the Iceland moss lichenin (Perlin and Suzuki, 1962) revealed differences in sequence, lichenin apparently containing a higher proportion of the tetrameric unit and therefore a more regular structure than cereal β-glucan.

Preece et al (1960), using crude cereal enzyme preparations, tentatively identified laminaritriose and laminaritetraose, suggesting that contiguous $(1\rightarrow3)$-β-linked units exist in both oat and barley β-glucan. Only minor differences in oligosaccharide production were detected between the two cereal β-glucans. Clarke and Stone (1966), using a $(1\rightarrow4)$-β-D-glucanase preparation from *Aspergillus niger*, concluded that lichenin possessed a linkage sequence somewhat different from oat and barley β-glucan and obtained evidence for a small amount of contiguous $(1\rightarrow3)$-linkages.

In recent years, a number of workers used specific enzymes to identify mixed-linkage β-glucans and to estimate the proportions of $(1\rightarrow4)$- and $(1\rightarrow3)$-linked units (Nevins et al, 1978). A $(1\rightarrow3)$ $(1\rightarrow4)$-β-D-glucan 4-glucanohydrolase from *Bacillus subtilis* (Rickes et al, 1960; Huber and Nevins, 1977; Anderson et al, 1978) only attacks the β-$(1\rightarrow4)$-linkage of a 3-substituted glucose in the mixed-linkage β-glucan, as indicated schematically in Fig. 1. Separation and identification of the oligosaccharide degradation products can therefore yield structural information. Runs of $(1\rightarrow3)$- and $(1\rightarrow4)$-linked glucose units will not be hydrolyzed and, in the absence of significant amounts of such residues, the percentage of $(1\rightarrow3)$-linkages can be approximately calculated from the relative amounts of 3-O-β-cellobiosyl-D-glucose and 3-O-β-cellotriosyl-D-glucose formed, assuming that the subunits are linked by a β-$(1\rightarrow4)$-glucosidic bond. Bio-gel P-2 chromatography and more recent high-performance liquid chromatographic methods (Wood, 1985) allow separation and estimation. The ratio of these two oligosaccharides has been calculated for enzyme treatments of oat β-glucan (ammonium sulfate precipitated), oat gum and the original oat grain, and for barley β-glucan and intact grain. There was consistently a higher proportion of tetrasaccharide produced from oats than from barley (P. J. Wood and J. Weisz, *unpublished results*). This is not reflected in any considerable difference in $(1\rightarrow4)$- to $(1\rightarrow3)$-linkage ratios. A range in mole ratio of tri:tetra of

```
-- GLC(1→3)GLC(1↓4)GLC(1→4)GLC(1→4)GLC(1→3)GLC(1↓4)GLC(1→4)GLC(1→3)GLC(1↓4) --

          ENZYME HYDROLYSES BONDS INDICATED, ↓, TO GIVE:
                  GLC(1→4)GLC(1→4)GLC(1→3)GLC
                  GLC(1→4)GLC(1→3)GLC
```

Fig. 1. Hydrolysis of oat β-glucan by $(1\rightarrow3)(1\rightarrow4)$-$\beta$-D 4-glucanohydrolase (EC.3.2.1.73) from *Bacillus subtilis*. GLC = glucose.

1.72–2.64 leads to a decrease of less than 1% in the proportion of (1→4)-linkages (from 70.3 to 69.4%). Indeed, the full range possible is only from 75 to 66%, assuming no longer runs of (1→4)- or (1→3)-linkages are present. Analysis of the proportion of linkages is thus not a good indicator of the relative proportions of the two major repeating units. Furthermore, the distribution of these along the polymer chain or between molecules is unknown. The enzyme digestion produces, in addition to trisaccharide and tetrasaccharide, pentasaccharide and higher oligomers, including 2–5% of an insoluble residue that yields glucose on hydrolysis. Similar material isolated from barley has been shown to contain up to nine adjacent (1→4)-linkages (Woodward et al, 1983).

The importance of linkage analysis lies in the effect that the sequence of these has on polysaccharide conformation, or shape, in solution, since this has a profound effect on physical properties (Rees, 1977). For example, the greater the length of uninterrupted β-(1→4)- or β-(1→3)-linked chains, the greater the similarity of those portions of the molecule to cellulose or curdlan, both highly associated, water-insoluble polymers. High-resolution nuclear magnetic resonance (NMR) techniques provide a powerful tool for investigating these aspects; in a study of lichenin from Iceland moss (Gagnaire et al, 1975), three different anomeric doublets were observed, each showing a coupling constant characteristic of the β-configuration. These presumably arose from the anomeric protons of 1) a 4-O-substituted glucose linked to a 3-position, 2) a 4-O-substituted glucose linked to a 4-position, and 3) a 3-O-substituted glucose linked to a 4-position. It was noteworthy that no signals attributable to a 3-O-substituted glucose linked to a 3-position (i.e., laminaran-like) were detected. On the basis of other evidence, it was suggested that lichenin might possess a ribbonlike (i.e., highly extended) conformation, with hydrogen bonding along the chain.

The ^{13}C NMR spectra of a viscous oat gum extract (containing ∼ 80% β-glucan) and of an ammonium sulfate purified oat β-glucan (98–100% β-glucan) are shown in Fig. 2. The spectra, obtained at 62.8 MHz, were similar to those obtained for a purified oat β-glucan by Dais and Perlin (1982) at 100 MHz. There were clearly 17 resonances, and, with decreased line broadening, an extra resonance could be detected at ∼ 79 ppm. The assignments shown in Table I are those given by Dais and Perlin (1982). The main point to note is that the resolved signals from 3-O-linked units (86.73, 76.11, 72.06, and 68.20 ppm) are essentially single peaks, supporting the view that these units occur in isolation. Signals from the 4-O-linked units (e.g., ∼ 79.7 ppm) show slightly different shifts, depending upon their positions in the pentameric and tetrameric repeat units of the polysaccharide. These results were obtained with both the purified and the crude β-glucan, indicating that purification had not preferentially isolated a nonrepresentative fraction (P. J. Wood and Preston, *unpublished results*).

Most of the evidence thus suggests that oat-β-glucan has the structure shown in Fig. 3, similar to that described for lichenin by Gagnaire et al (1975). Oat and barley β-glucans, however, contain a significant proportion of the pentameric repeat unit containing three consecutive (1→4)-linkages. Indeed, the evidence from enzyme studies suggests more of this sequence in oat β-glucan than in barley. There is tentative evidence to suggest that longer runs of consecutive (1→4)-linkages and contiguous (1→3)-linkages might exist as minor features and constituents. These may have missed rigorous identification because of

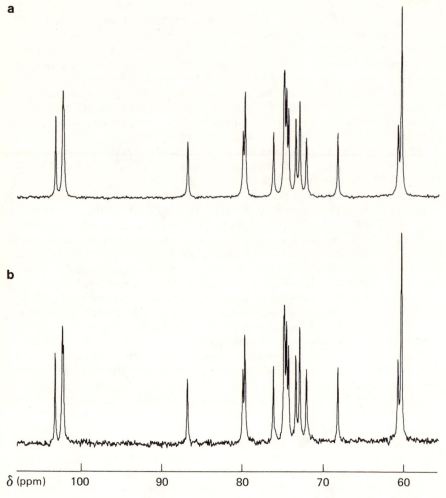

Fig. 2. ^{13}C-Nuclear magnetic resonance spectra at 62.8 M Hz of (a) purified oat β-glucan ($\sim$ 2%) and (b) high-viscosity oat gum extract ($\sim$ 1%) in dimethyl sulfoxide (d6) at 90°C with 112,000 and 141,000 scans, respectively. Continuous decoupling was used with a pulse width of 45° and total delay between pulses of 0.6 sec. (Courtesy C. M. Preston)

inadequate sensitivity of techniques or because purification has led to the selection of a particular subfraction of β-glucan (Wilkie, 1979). The molecule is unbranched. The possibility that oat β-glucan may exist linked to protein in the endosperm cell wall, as has been reported for barley (Forrest and Wainwright, 1977), has not been investigated.

Although oat β-glucan soluble in water or dilute alkali does not appear to contain significant amounts of contiguous (1→3)-linkages, there is histochemical evidence to suggest minor amounts of (1→3)-β-D-glucan associated with the endosperm cell walls (Fulcher et al, 1977; Wood and Fulcher, 1984; R. G. Fulcher, *unpublished observations*).

TABLE I
**^{13}C Nuclear Magnetic Resonance Chemical Shifts at 62.8 MHz
and Resonance Assignments for Oat β-Glucan in DMSO$_{d6}$ at 90°C**

Chemical Shift (δ) (ppm)	Assignment
103.21	C-1[a] linked to 3G[b]
102.32, 102.21	C-1 linked to 4G
86.73	C-3 of 3G
79.84, 79.61	C-4 of 4G
76.11	C-5 of 3G
74.82, 74.74	C-3 of 4G
74.49, 74.23	C-5 of 4G
73.36, 72.89	C-2 of 4G
72.06	C-2 of 3G
68.20	C-4 of 3G
60.68	C-6 of 3G
60.27	C-6 of 4G

[a] Carbon atom giving resonance.
[b] 3G = 3-O-substituted glucose residue; 4G = 4-O-substituted glucose residue.

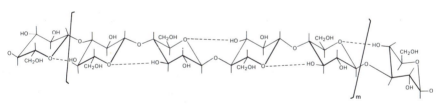

Fig. 3. Main structural units of oat β-D-glucan. Although the tetrameric and pentameric units are repeated, their distribution along the polysaccharide chain is not known. Broken lines represent possible intramolecular hydrogen bonds. (Adapted from Gagnaire et al, 1975)

IV. EXTRACTION AND PURIFICATION

The quantitative isolation of β-glucan from cereals is not easy because of difficulties in complete extraction of the β-glucan free from contaminating starch, pentosan, and protein. Plant polysaccharides are notoriously difficult to isolate readily in pure form, and indeed attempts to do so may be misleading (Wilkie, 1979).

In most early studies, oat β-glucan was extracted with water at about room temperature. It was recognized that endogenous enzymes in the seed resulted in a degraded product if an enzyme deactivation step, usually refluxing 75–80% ethanol, was not carried out. Until the studies of Wood et al (1977, 1978), little had been published on extraction of oat β-glucan, but much information is available on barley and is briefly reviewed here. Although parallels probably do exist, it should not be assumed that the barley methodology can be applied uncritically to oats.

A variety of different extraction methods have been used in studies of barley β-glucan in attempts to maximize yield and viscosity of the extract and to obtain the most nativelike or undegraded material while avoiding contamination with starch and other components. Temperatures below the gelatinization temperature of starch, and avoidance of strong alkali, minimized starch solubilization. However, with mild conditions, extraction was incomplete, and it was recognized that yields were improved with enzyme-active flour or alkali extraction (Preece and Hobkirk, 1954).

Acid extraction has been used to optimize yields of barley β-glucan, particularly where viscosity was to be used as a measure of β-glucan and its influence on malting potential (e.g., Greenberg and Whitmore, 1974), and high yields of barley β-glucan (5.5%) were obtained with hot 4.4% trichloroacetic acid (Luchsinger et al, 1958). Bathgate and Dalgliesh (1974) used successive treatments with 90% dimethylsulfoxide, hot water, and NaOH (4%) to obtain yields of about 4.5% (of pearled flour). In studies of barley endosperm cell walls, exhaustive extraction has been done with water (40–65°C) and $1M$ NaOH containing sodium borohydride (1%, w/v) (Fincher, 1975; Forrest and Wainwright, 1977). Other workers (Mares and Stone, 1973; Anderson and Stone, 1978) have applied classical polysaccharide extraction procedures (to wheat and ryegrass endosperm cell walls) and shown the value of $8M$ urea as a β-glucan extractant. Evidence that the β-glucan in barley endosperm cell walls contained peptide linkages (Forrest and Wainwright, 1977) led to the use of hydrazine as an extractant, since this disrupts the peptide bonds but does not affect the glycosidic linkages (Martin and Bamforth, 1981).

No similar body of information exists on the extraction of oat β-glucan. Wood et al (1977) described an alkaline extraction of oat gum, outlined schematically in Fig. 4, which was essentially adopted from a wet-milling process for separation of oat starch and protein (Hohner and Hyldon, 1977; Paton, 1977). Use of mild alkali (pH 10) and carbonate rather than sodium hydroxide minimized starch gelatinization and solubilization, even at elevated temperatures (Wood et al, 1978). With alkaline extraction, prior enzyme deactivation of flour was not necessary to optimize yields—indeed, deactivated flour showed slightly reduced yield—but the extraction conditions did not inactivate enzymes; therefore, the mixture was kept cold during pH adjustment, and centrifugation and addition of 2-propanol (isopropyl alcohol, IPA) were done as rapidly as possible. Care was needed in isolation of the IPA precipitate, or oat gum. The initial precipitate retained considerable water and, if the initial flour was not enzyme inactive, should not be left at room temperature, since some viscosity loss was noted under these circumstances. The precipitate produced a coherent pellet on centrifuging and was therefore disrupted and dewatered in a high-speed homogenizer in 100% IPA before centrifuging and drying in air with gentle warming. The crude isolated gum was then analyzed for starch, total glucan, and other components. Routinely, three successive extractions were used; but this did not represent the total gum content, as further extractions continued to release gum (Fig. 5). Histochemical examination of extracted flour has since confirmed the presence of residual β-glucan amounting to about one-third of the total (Wood and Fulcher, 1978; Wood, 1985) (Fig. 6).

Factors affecting yields and viscosities of extracts were studied (Wood et al, 1977, 1978; *unpublished results*). Because cultivar (and possibly environment) appears to influence β-glucan content and viscosity (Wood et al, 1978), comparative studies should use the same batch of flour, prepared from dehulled oats (groats). Since some seed damage, leading to enzyme activity, may occur during dehulling, groats (and flour) should be stored frozen.

Decreasing particle size increased extraction efficiency. Defatting (hexane extraction) did not significantly affect yields of gum or oat β-glucan, whereas enzyme deactivation produced a slight but significant decrease in yield (Table II). Increasing the temperature of extraction increased yield (Table III) similar to effects noted for aqueous extraction of barley (Fleming and Kawakami, 1977; Prentice et al, 1980). Extended extraction periods at elevated temperature may lead to losses, presumably because of alkaline degradation (Madacsi et al, 1983). Use of carbonate/bicarbonate buffer for extraction rather than adjusting pH by addition of sodium carbonate was of no advantage. At low ionic strength the pH dropped during extraction, whereas at high ionic strength evolution of carbon dioxide during pH adjustment led to excessive foaming.

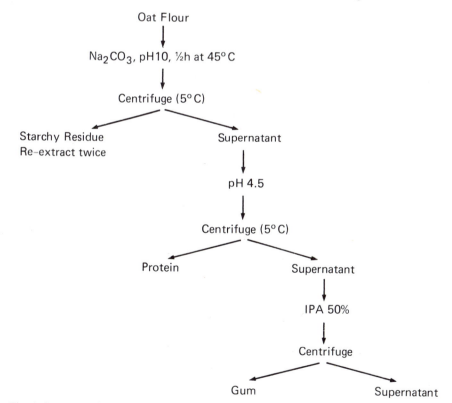

Fig. 4. Summary of oat gum extraction procedure. Gum extracts obtained after reextraction of starchy residue are referred to as extracts 1, 2, and 3; if dye precipitation method is used, Calcofluor is added to the first supernatant. (Reprinted, with permission, from Wood et al, 1977)

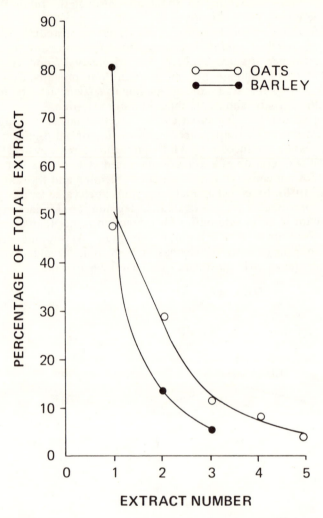

Fig. 5. Distribution of gum (expressed as percentage of total extracted) between successive extractions of oats (o) and barley (●). (Reprinted, with permission, from Wood et al, 1977)

Fig. 6. A, Transverse section of barley endosperm showing intense Calcofluor fluorescence (arrows) in starchy endosperm cell walls; a = aleurone layer and se = starchy endosperm. B, Transverse section of oat endosperm, also showing Calcofluor staining throughout the endosperm cell walls. Note greater thickness of walls in subaleurone region (sa). Scale as in A. C, Portion of the subaleurone region of oat endosperm, showing lamellar organization of thickened subaleurone cell walls (arrows). D, Portion of coarse oat flour particle before alkaline extraction. E, as in D, and to same scale, but after first alkaline extraction. Most of the inner endosperm cell walls have been removed during treatment, leaving only those materials (arrows) adhering directly to the aleurone layer. F, as in D, and to same scale, but after third alkaline extraction. Thin layer of Calcofluor-positive material (arrows) adheres to the aleurone layer. (Reprinted, with permission, from Wood and Fulcher, 1978)

Oat gums typically showed glucan, pentosan, ash, and protein values as indicated in Table IV. No attempt was made to analyze for other possible components, such as lipid, phenolics, or phytate (Acker et al, 1955a).

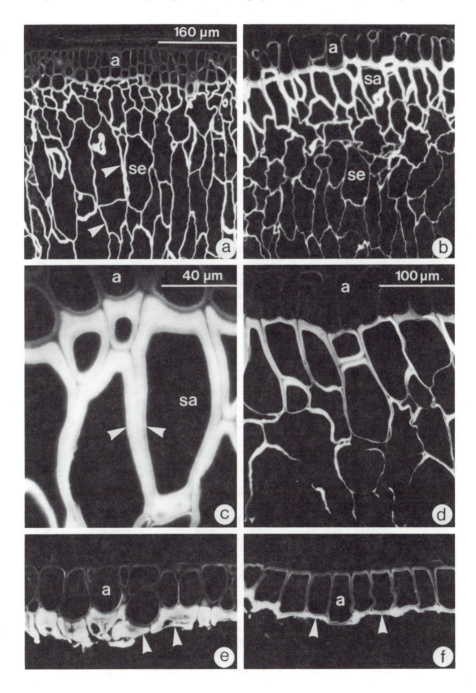

A variety of methods have been used to purify crude extracts of oat or barley gum to give a pure β-glucan fraction, but the most successful and generally applicable was that developed by Preece and co-workers (Preece and Mackenzie, 1952; Preece and Hobkirk, 1953), who showed that almost pure, pentosan-free glucan could be precipitated by 20–30% ammonium sulfate. Application of this procedure to high-viscosity gum isolated as described by Wood et al (1978) gave 75–86% yields of precipitate containing ~ 98% glucan on a dry weight basis.

The procedure of precipitation by 50% IPA (Wood et al, 1977) afforded considerable purification from pentosan, with a ratio of 20:1 of glucose to pentose in the precipitated gum, compared with 9:1 in the original crude solution. In a further graded precipitation of the gum, 25% IPA gave an almost pentosan-free fraction in 63% yield (P. J. Wood and J. Weisz, *unpublished results*).

Direct, or cellulose-substantive, dyes such as Calcofluor (C.I. 40622; CI Fluorescent Brightener 28) and Congo red (C.I. 22120) form complexes with cereal β-glucans (Wood and Fulcher, 1978; Wood 1980a, 1980b, 1982). Addition of Calcofluor or Congo red to an aqueous solution of oat gum yielded a precipitate that after isolation and acid hydrolysis showed only glucose by paper chromatography (Fig. 7). A clean separation from pentose- and galactose-

TABLE II
Effect of Defatting and Enzyme Deactivation on Yield of β-Glucan Extracted from Oats[a]

Treatment	Extract	Glucan Yield (Percentage [db] of Flour)	Starch Content (Percentage [db] of Glucan)	β-Glucan Yield (Percentage [db] of Flour)
Untreated flour	1	2.36	0.52	2.35
	2	1.10	0.69	1.09
	3	0.46	1.96	0.45
Total		3.92	0.72	3.89
Defatted flour	1	2.44	4.23	2.32
	2	1.20	5.09	1.13
	3	0.39	11.40	0.34
Total		4.02	5.18	3.78
Enzyme deactivated flour	1	2.20	2.97	2.10
	2	0.98	5.04	0.93
	3	0.53	10.49	0.47
Total		3.71	4.99	3.50

[a] Source: Wood et al (1978); used by permission. Cultivar was Hinoat, 1975; data represent averages of duplicate extractions.

TABLE III
Effect of Temperature on Yield of β-Glucan Extracted from Oats[a]

Temperature (°C)	β-Glucan Yield (Percentage [db] of Flour)
5	3.41
25	3.66
45	3.89
63	4.11
80	4.59

[a] Source: Wood et al (1978); used by permission. Cultivar was Hinoat, 1975; data represent averages of duplicate extractions.

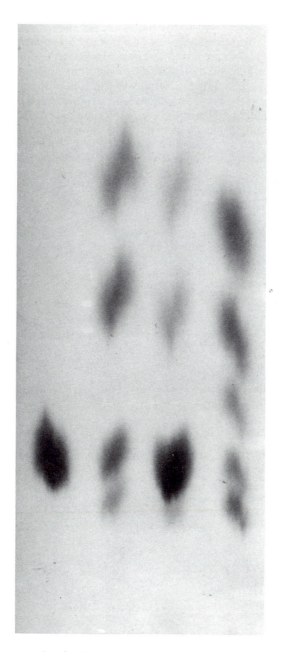

Fig. 7. Paper chromatography of acid hydrolysates of (from left to right) material precipitated by Calcofluor from oat gum, material not precipitated by Calcofluor, original oat gum, and standard monosaccharides. Identity of monosaccharides from bottom to top: galactose, glucose, mannose, arabinose, and xylose.

containing fractions (subsequently isolated from the supernatant by addition of IPA) was obtained. Similarly, addition of Calcofluor to the crude supernatant from a carbonate buffer extract of oat flour (as shown in Fig. 4) yielded a dye-glucan precipitate. Dispersion in hot water and reprecipitation by IPA (three volumes) gave a product analyzed as shown in Table V. The β-glucan yields obtained from precipitation by IPA and by Calcofluor were in good agreement, but the Calcofluor-precipitated material was less contaminated by ash, protein, and pentosan. The ratio between glucose and pentose was 110:1 in Calcofluor extract 1, compared with 9:1 in the crude alkaline extract. This was a simple procedure for isolating β-glucan from oats, but the initially precipitated complex contained a considerable amount of bound dye ($\sim$ 50% by weight). Although this amount was reduced (to $\sim$ 5%) by dispersion in water and reprecipitation with IPA, and was further reduced by repeating this process, final traces of dye were difficult to remove. Treatment with diethylaminoethyl cellulose reduced the dye content to $<$ 0.1%.

TABLE IV
Percentage Analysis of Gum from Enzyme-Deactivated Flour from Hinoat Oats[a]

Extract	Gum Yield[b]	Glucan[c,d]	Arab.[c,e]	Xyl.[c,e]	Ash[c]	Protein[c,f]	Total[c]
1	1.66	66	1.4	1.6	2.4	19.9	91.3
2	1.18	78	1.3	1.6	1.1	12.2	94.2
3	0.82	81	1.1	1.3	1.0	6.9	91.3

[a] Combined samples from six extractions.
[b] Percentage (db) of flour.
[c] Percentage (db) of gum extract.
[d] Anhydro-glucose determined by cysteine/sulfuric acid.
[e] Arab. = anhydro-arabinose; xyl. = anhydro-xylose. Determined by gas-liquid chromatography.
[f] N $\times$ 6.25.

TABLE V
Percentage Comparison of Yields and Analysis of β-Glucan Fractions Isolated from Oat Flour by Calcofluor (Calc 1, 2, 3) and Isopropyl Alcohol (IPA 1, 2, 3)

Sample	Gum Yield[a]	β-Glucan[b]	Dye[c]	Ash[c]	Protein[c,d]	Arab.[c,e]	Xyl.[c,e]
Calc 1	1.96	1.73	7.2	1.1	6.4	0.3	0.4
Calc 2	1.19	1.02	3.0	0.6	8.1	0.7	0.6
Calc 3	0.78	0.70	5.0	0.7	8.1	0.9	0.9
Calcofluor total	3.93	3.45	...	...	...	...	...
IPA 1	2.30	1.64	...	4.8	19.4	1.6	1.6
IPA 2	1.66	1.34	...	2.8	10.2	1.4	1.7
IPA 3	0.81	0.60	...	5.9	8.8	2.1	1.7
Isopropyl alcohol total	4.77	3.58	...	...	...	...	...

[a] Precipitate isolated (less dye content for Calcofluor precipitates) as percentage (db) of ethanol-extracted flour.
[b] Percentage of ethanol-extracted flour (db) as determined by cysteine/sulfuric acid assay.
[c] Percentage of isolated precipitate (db).
[d] N $\times$ 6.25, corrected for N content of dye.
[e] Arab. = anhydro-arabinose; xyl. = anhydro-xylose. Determined by gas-liquid chromatography.

V. ANALYSIS

No method presently exists to quantitatively extract cereal *β*-glucan uncontaminated by starch. Both polysaccharides yield glucose on hydrolysis, and starch is present in considerable excess; this poses problems for the analyst. Rapid and simple methods were required for malting and feed barley, particularly in breeding programs. Viscosity has been used for this purpose (Greenberg and Whitmore, 1974; Bendelow, 1975; Aastrup, 1979) but has not been used with oats. In general, the relationship between *β*-glucan concentration and viscosity is variable, and this method cannot provide reliable estimates of *β*-glucan content (Anderson et al, 1978; Wood et al, 1978), although it may be useful for comparing extracts (Aastrup, 1979).

Extracts of oat or barley grain therefore normally contain both α-glucan (starch) and *β*-glucan. Total glucan can be determined by acid hydrolysis and specific measurement of released glucose, either chromatographically or by an enzyme technique such as the glucose oxidase-peroxidase assay (e.g., Lloyd and Whelan, 1969). Starch can be determined by specific and quantitative hydrolysis to glucose using α-amylase and glucoamylase; the difference between total glucan and starch is *β*-glucan. Such a method was described by Fleming et al (1974) for barley. A similar method was described by Wood et al (1977) for oats, the main modifications being the use of carbonate (pH 10) for extraction (thus avoiding an enzyme deactivation step) and the use of a specific colorimetric method (cysteine-sulfuric acid) for total glucose (thus eliminating the need for acid hydrolysis and neutralization). A comparison of estimates of glucan content in seven samples of oat gum by the colorimetric method and by glc analysis of a hydrolysate ($1 M$ H_2SO_4, 4 hr, 100° C) showed an average value 5% higher by the colorimetric procedure. This possibly reflects the generally known, but frequently ignored, losses that occur during acid hydrolysis. The estimated recovery of *β*-glucan (that is, release of anhydro-glucose) typically changes with time, as shown in Fig. 8 for a sample of ammonium-sulfate-purified oat *β*-glucan. A maximum release of anhydro-glucose of ~ 92% was obtained after 45 min to 1

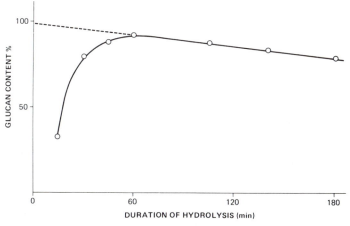

Fig. 8. Rate of hydrolysis of oat *β*-glucan by $0.5 M$ trifluoracetic acid at 120° C.

hr of hydrolysis with $0.5M$ trifluoroacetic acid, and extrapolation of the subsequent degradation portion of the curve to zero time gave values of 98–100% recovery of glucan (P. J. Wood and J. Weisz, *unpublished results*).

Considerable care is required in the choice of glucoamylase for starch analysis, since commercial sources of bacterial and fungal glucoamylase may be contaminated by β-glucanase systems capable of releasing glucose from cereal β-glucan. The preparation used by Fleming et al (1974) and subsequently by Wood et al (1977) was Agidex, produced by British Drug Houses from *Aspergillus*.

The method of Wood et al (1977), which was designed to determine β-glucan in crude oat gum isolated by a potential commercial process (Hohner and Hyldon, 1977), routinely involved a precipitation step using IPA as described previously. A more rapid analysis, which did not require isolation of gum extracts, was also described; since starch contamination was low, assay for this could be omitted. For this procedure, glucan yields of $3.80 \pm 0.05\%$ (dry weight basis) were found, compared with $3.40 \pm 0.09\%$ for the IPA precipitation method.

In recent years, methods using specific β-glucanases have been reported. Anderson et al (1978) described the use of the specific $(1\rightarrow3)(1\rightarrow4)$-$\beta$-D-glucan 4-glucanohydrolase from *B. subtilis*, already discussed. The enzyme, freed from α-amylase and other polysaccharide-degrading enzymes, was used to hydrolyze β-glucan in ethanol-extracted, enzyme-inactive flour. The oligosaccharide reaction products were extracted into ethanol, acid hydrolyzed, and estimated as glucose by the glucose-oxidase procedure. This is not a rapid procedure but has the advantage of being directly applicable to flour, so that problems with extraction are avoided. The enzyme apparently released all β-glucan from barley, since a repeat incubation of the residue produced no further products. A recent refinement of this method uses β-glucosidase to hydrolyze the oligosaccharide reaction products (McCleary and Glennie-Holmes, 1985).

Two methods for cereal β-glucans have recently been reported that depend on enzymatic conversion of β-glucan to glucose (Prentice et al, 1980; Martin and Bamforth, 1981). The requirements are the converse of those for starch measurement in that the β-glucanase must not be contaminated by glucoamylase and conversion to glucose must be consistent and quantitative. Both procedures appear to meet these requirements, although, in the original technique of Martin and Bamforth, loss of glucose by adsorption onto enzyme necessitated a calibration using a standard β-glucan. Subsequent reports suggest that this is not necessary but that the choice of conditions for the removal of glucoamylase is critical because of variation between enzyme batches (Bourne et al, 1982; Prentice, 1982).

Martin and Bamforth (1981) used hydrazine as an extractant, which apparently achieves maximum release of β-glucan. Prentice et al (1980) used a carbonate extraction procedure similar to that of Wood et al (1977, 1978) but with differences in the grinding technique and with a greater ratio of liquids to solids and extended extraction time. These factors may explain the higher starch solubilization noted for barley at 45°C compared with that reported by Wood et al (1977) for oats. Complete extraction was apparently achieved at 80°C, although there is a danger of loss through alkaline degradation under these conditions (Madacsi et al, 1983). Since a direct β-glucan measurement was used rather than a difference method, the large contamination with starch could be tolerated. A recent method describes a convenient analysis using perchloric acid as extractant before determination of β-glucan (and starch) enzymically (Ahluwalia and Ellis, 1984).

A nonenzymatic specific method for β-glucan analysis has recently been reported, based on precipitation by dyes as described earlier (Wood and Weisz, 1984). For analysis, the complex of dye and β-glucan was precipitated directly from carbonate buffer extracts, or solutions, by Calcofluor and hydrolyzed with $0.5M$ trifluoroacetic acid. Released glucose was determined by high-performance liquid chromatography using aqueous elution from a Bio-Rad HPX-85 heavy metal carbohydrate column. Elution profiles of the hydrolysate (Fig. 9) essentially only showed the presence of glucose. Values for β-glucan

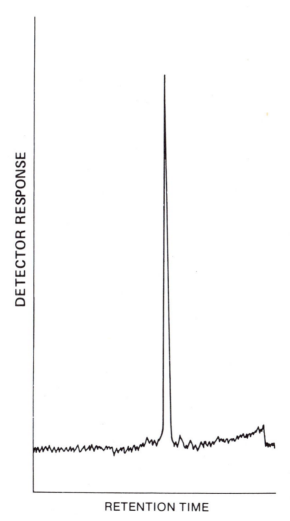

Fig. 9. High-performance liquid chromatographic trace of hydrolysate from oat β-glucan precipitated by Calcofluor (monitored by refractive index).

TABLE VI
β-Glucan Content of Oat Grain Estimated by Various Methods

β-glucan Content (%)	Reference	Comments
1.4	Preece and MacKenzie (1952)	Possibly contaminated by starch
5.7	Acker et al (1955a)	Coarse or bran fraction
1.3	Peat et al (1957)	Structural study
0.9	Parrish et al (1960)	Structural study
2.8–4.6	Wood et al (1977, 1978)	Dehulled oat
2.5	Anderson et al (1978)	Whole oat
4.8–6.6	Prentice et al (1980)	Whole oat
2.7–5.4	McCleary and Glennie-Holmes (1985)	Whole oat
3.8–5.5	P. J. Wood and J. Weisz (*unpublished*)	Dehulled oat

obtained by this method were not significantly different from those obtained from the difference between values for total glucan and starch. A specific precipitation of cereal β-glucan by copper was recently reported (Madacsi et al, 1983) and applied to measurement of β-glucan in sorghum. Carbonate at pH 10 was used for extraction in both the dye method and the copper method, and yields thus depended on temperature and duration of extraction. Recent results (Wood, 1985) suggest that about two-thirds of the total β-glucan analyzed was extracted at 45°C and more than 90% at 80°C.

A method using adsorption of Calcofluor to β-glucan has recently been described (Jensen and Aastrup, 1981; Jørgensen, 1983), in which the fluorescence intensity of a suspension of material (from beer, wort, malt, and barley) was measured. An advantage was that for barley and malt no extraction was required, since measurement was on a stained suspension of flour, but the method required correlation with an independent analysis.

In the early literature, most quoted values for barley β-glucan content in the seed were in a range of about 1–2.5% (Fleming et al, 1974; Bourne et al, 1976), although Luchsinger et al (1958) reported 5.5% and Bathgate and Dalgliesh (1974) reported more than 4%. Recently, more exhaustive extractions and the application of enzyme methods have led to higher reported values: 6.2–7.2% (Fleming and Kawakami, 1977), 3.6–6.4% (Anderson et al, 1978), 4.5–8.2% (Prentice et al, 1980), and 5.4–8.6% (Martin and Bamforth, 1981). Values reported for oat β-glucan content are shown in Table VI, in which the last five are from analytical studies. Since oat hulls are a significant proportion by weight of the seed, it is important to know on what basis percentages are being reported. This is not always clear in the literature, but in the absence of contrary statements, whole seed values are assumed.

Differences between cultivars have been noted, but studies required to distinguish genetic and environmental effects have not been carried out.

VI. PROPERTIES

Very few data on the physical properties of oat β-glucan have been published. Aqueous solutions of purified oat β-glucan show specific optical rotations of about −8°C to −12°C, values similar to those of barley β-glucan (Preece and Hobkirk, 1953; Parrish et al, 1960; Clarke and Stone, 1966). More negative values indicate pentosan contamination, and more positive values suggest starch contamination.

Oat gum, as prepared by Wood et al (1977, 1978), dissolved in hot water and, more readily, in dilute alkali to give highly viscous solutions that slowly lost viscosity on standing. In alkali or with enzyme-deactivated samples, this loss cannot be attributed to enzyme action, and therefore it probably arises from changes in molecular organization. After prolonged periods, some precipitation may occur or, in concentrated (1%, w/v) solution, gelation may take place. With gum samples from undeactivated flour in water, the loss of viscosity was more rapid, suggesting contamination by β-glucanase. Because of these problems, dry dimethylsulfoxide (DMSO) was chosen as a suitable solvent for viscosity measurements since these solutions showed unchanged viscosity over eight months of storage at 4–5° C.

Difficulty was encountered in obtaining reproducible viscosities at 0.5% (w/v) using capillary viscometry (Wood et al, 1978). This possibly resulted in part from the extreme concentration dependence of viscosity at that concentration (Fig. 10) and the high shear sensitivity. Nevertheless, extremely high viscosities, in excess

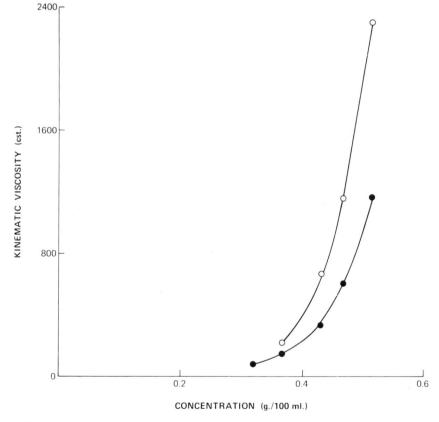

Fig. 10. Effect of oat gum concentration (extract 2 from enzyme-deactivated Hinoat flour) on viscosity: o = water and ● = dimethyl sulfoxide. (Reprinted, with permission, from Wood et al, 1978)

of 1,000 cSt at 20° C, were recorded for some samples at 0.5% (w/v). Such values are considerably higher than previously quoted values for oat β-glucan, which ranged from $\sim$ 2–3 cSt (Preece and MacKenzie, 1952; Preece and Hobkirk, 1953: quoted as specific viscosity) to 260 cSt (Clarke and Stone, 1966: quoted as relative viscosity). (Specific and relative viscosity were converted to centistokes by assuming water density = 1 g/ml and η H_2O = 1.) In view of the different conditions employed, such comparisons should be treated with caution.

For routine comparison of samples, a concentration of 0.2% (w/v) in DMSO was chosen. In the extraction procedure described by Wood et al (1978) (Fig. 4), the viscosity of extract 1 was always lower than that of extracts 2 and 3, regardless of cultivar or whether enzyme-deactivated flour was used (Tables VII and VIII). This probably reflects a slower solubilization of the more viscous material and higher protein contamination in the first extract. Extracts 2 and 3

TABLE VII
Analysis and Viscosity of Gum Extracts from Oat Cultivars Hinoat and Rodney[a]

Sample Extract	Gum Yield[b]	Starch Content[c]	β-Glucan Yield[b]	Viscosity Kinematic[d] (cst)	Viscosity Limiting[e] $[\eta]_{c \to o}$ (dl/g)
Hinoat					
1	2.22	1	1.55	11.4	10.3
2	1.33	2	1.16	43.1	16.9
3	0.69	4	0.53	41.9	17.8
Total	4.24	2	3.24	...	...
Rodney					
1	2.14	1	1.70	5.7	5.8
2	0.86	2	0.71	21.9	13.9
3	0.47	4	0.35	32.1	16.2
Total	3.47	2	2.76	...	...

[a] Source: Wood et al (1978); used by permission. Data represent averages of duplicate extractions and analyses.
[b] Percentage (db) of flour from dehulled grain.
[c] Percentage (db) of gum extract.
[d] 0.2% (w/v) in dimethyl sulfoxide.
[e] In dimethyl sulfoxide.

TABLE VIII
Comparison of Viscosities of Gum from Untreated and Enzyme-Deactivated Flour from Hinoat Oats[a]

Sample Extract	Kinematic Viscosity[b] (cst) Mean	Kinematic Viscosity[b] (cst) SD
Untreated		
1	9.1	0.6
2	37.8	1.6
3	41.5	3.2
Deactivated		
1	23.1	1.7
2	49.2	6.3
3	50.8	4.0

[a] Source: Wood et al (1978); used by permission. Data represent averages of six extractions.
[b] 0.2% in dimethyl sulfoxide.

normally had quite similar viscosities. All extracts from enzyme-deactivated flour showed higher viscosity than the equivalent sample from untreated flour, demonstrating that even with care in temperature control and rapid processing, enzymatic degradation occurred during isolation of the gum. As would be expected if enzyme were responsible, the greatest difference was noted between the first extracts.

Measurements of intrinsic viscosity, or limiting viscosity number $[\eta]_{c \to 0}$, in DMSO gave values ranging from 5.8 dl/g for extract 1, cultivar Rodney, to 17–18 dl/g for extracts 2 and 3 from Hinoat. A similar value (17.5 dl/g) was found with 7M urea as solvent. These values are high for neutral polysaccharides, for which values < 10 dl/g are normally quoted (Banks and Greenwood, 1975; Mitchell, 1979); the latter values, however, are for aqueous systems.

A value of 3.88 dl/g was quoted by Wada and Ray (1978) for an oat coleoptile mixed-linkage $β$-glucan. Forrest and Wainwright (1977) reported values of 15–18 dl/g for a high-molecular-weight $β$-glucan isolated from barley endosperm cell walls. Intrinsic viscosity depends on molecular size and shape according to the equation

$$[\eta]_{c \to 0} = KM^a$$

where M is molecular weight and K and a are constants for the particular polymer type.

The high values of $[\eta]_{c \to 0}$ for oat gum suggest a high-molecular-weight, extended molecule; the data of Gagnaire et al (1975) suggest an extended ribbonlike conformation.

High-molecular-weight, extended polysaccharides are shear sensitive. The viscosity measurements described thus far, because of limitations on sample quantity, were carried out in capillary viscometers with efflux times such that shear was low to zero. More recent investigations (P. J. Wood, *unpublished results*) have studied shear effects at 25°C on Hinoat oat gum, using Haake cylinder and cup (NV) attachments on the Food Research Institute Universal Food Rheometer (Voisey and Randall, 1977).

Figure 11 shows the effect of shear rate on apparent viscosity for a high-viscosity Hinoat oat gum in water, 40% sucrose, and 1M NaCl and, for comparison, a sample of medium-viscosity hydroxyethyl cellulose. The sample of oat gum was more viscous at all concentrations and showed a greater shear sensitivity. No hysteresis effects were noted during a cycle of increasing and decreasing shear rate.

The validity of the power law equation

$$\eta_{rel} = KS^{n-1},$$

where η_{rel} is the relative viscosity (ratio of sample to solvent viscosity) and S is shear rate, was studied by plotting log η_{rel} against log S (Fig. 12). Correlation coefficients ≥ 0.99 showed that the equation was valid over the range of shear rates studied. The flow behavior index n and the consistency index K were calculated from slope and intercept (Table IX). The small values that were observed for n, decreasing with increasing concentration, indicated considerable departure from Newtonian behavior—i.e., high shear sensitivity. This behavior

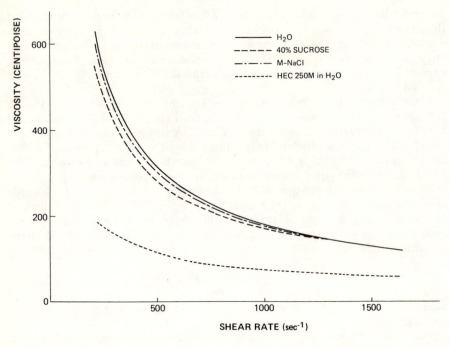

Fig. 11. Effect of shear rate on apparent viscosity of solutions of high-viscosity (extract 3) oat gum (1%, w/v) in water, 40% sucrose, $1M$ NaCl, and hydroxyethyl cellulose (HEC) (1%, w/v) in water.

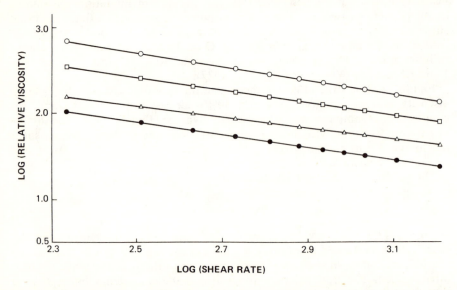

Fig. 12. Relationship between logarithms of shear rate and of relative viscosity for high-viscosity (extract 3) oat gum: o = 1% in water, □ = 0.73% in water, △ = 0.55% in water, and ● = 1% in 40% sucrose.

TABLE IX
Flow Behavior Index (n) and Consistency Index (K) for Oat Gum,
Hydroxyethylcellulose (HEC 250M), Guar Gum, and Locust Bean Gum

Sample	Concentration (g/L)	n H₂O	n 40% Sucrose	K H₂O	K 40% Sucrose
Oat gum	7.3	0.25	0.33	1.9×10^4	1.8×10^3
	10	0.19	0.26	5.1×10^4	5.1×10^3
HEC 250M	10	0.41	0.44	4.8×10^3	9.0×10^2
	20	0.26	0.30	3.5×10^4	7.4×10^2
Guar gum[a]	7.5	0.32	0.34	4.5×10^3	1.1×10^3
	10	0.28	0.30	1.04×10^4	2.4×10^3
Locust bean gum[a]	10	0.30	0.30	1.5×10^4	3.4×10^3

[a] Data from Elfak et al (1979).

is typical of high-molecular-weight polysaccharides, the intermolecular interactions and molecular orientation of which may be greatly affected by flow. These results and the high values observed for $[\eta]_{c \to 0}$ suggest that oat gum, or β-glucan, is a high-molecular-weight, extended molecule. Unlike guar and locust bean gum, n appeared to be affected somewhat by high sucrose concentrations. The consistency index was also affected by solvent. It is possible that added sucrose reduced the effective hydrodynamic volume of the polysaccharide, but this would require dilute solution studies for confirmation.

The behavior in $1M$ NaCl showed that oat gum does not behave as a polyelectrolyte (Fig. 11), although, at this salt concentration, viscosity loss and precipitation were observed in some samples.

Data on the molecular weight of oat β-glucan are also sparse. Acker et al (1955b) reported values for oat β-glucan of 27,500–63,000, and Smith and Montgomery (1959) reported a value of about 160,000, whereas Podrazky (1964) reported 26,800. Gel filtration studies in this laboratory (P. J. Wood, *unpublished results*) using Sephacryl S-400 (Pharmacia) or (for high-performance liquid chromatography) TSK-60 (Bio-Rad) point to molecular weights much greater than these values, indeed greater than the high-molecular-weight standard, Blue Dextran (i.e., $> 2 \times 10^6$).

VII. INTERACTION OF OAT β-GLUCAN WITH DYES

Calcofluor, a commercial fluorescent whitening agent, is a useful microscopic stain for viewing plant and fungal cell walls (Darken, 1961; Hughes and McCully, 1975). Barley endosperm cell walls, which contain about 75% β-glucan (Fincher, 1975) but little if any cellulose, the usual substrate for the dye, stained intensely with Calcofluor. It is now clear that this staining of oat and barley endosperm cell walls (Fig. 6 and Chapter 3) is caused by a specific, or at least selective, affinity of β-glucans for such dyes as Calcofluor and Congo red (Wood and Fulcher, 1978; Wood 1980a, 1980b; Wood and Fulcher, 1983; Wood et al, 1983). Studies of complex formation between dye and polysaccharide in solution, and of staining behavior, have provided the foundation for a variety of new approaches to the isolation, identification, analysis, and localization of cereal β-glucans and β-glucanases.

Complex formation may be detected in solution by increase in viscosity, formation of precipitate, bathochromic (red) shifts in the absorption spectra of the dyes, and increase in fluorescence intensity of the dyes. The greatest interactions between polysaccharide and dye were observed for cereal $(1\rightarrow3)(1\rightarrow4)$-$\beta$-D-glucans and other polysaccharides containing contiguous $(1\rightarrow4)$-linked β-D-glucopyranosyl units (Wood, 1980a).

Factors affecting the precipitation of oat β-glucan by Calcofluor have been studied (Wood and Fulcher, 1978; Wood, 1980a, 1982). In water, quantitative precipitation of oat β-glucan was obtained at high concentrations (0.4–0.9%, w/v) but not at low concentration (0.09%, w/v) (Fig. 13). Salt inhibited precipitation of 0.8% (w/v) solutions of glucan relative to dilute buffers (pH 6.0 and 8.0) and water (Fig. 14). A comparison of precipitation of 0.6% and 0.3% (w/v) solutions of oat β-glucan in phosphate buffer at pH 7.0 at ionic strengths 0.05 and 0.2 showed that the more concentrated β-D-glucan solution required more Calcofluor to produce the same degree of precipitation and that increase in ionic strength inhibited the precipitation (Wood, 1982). Despite this, low concentrations of oat β-glucan ($\sim$ 0.1%, w/v) could be quantitatively precipitated by Calcofluor in buffers of ionic strength 0.2, at both pH 7.0 and 10.0 (Fig. 13). It was thus possible to isolate oat β-glucan directly from carbonate buffer extracts of oat flour, as already discussed.

At concentrations at which precipitation did not readily occur, oat β-glucan induced bathochromic shifts of 20–30 nm in the absorption maximum of Congo red and 14–17 nm in that of Calcofluor (Wood, 1980a, 1980b, 1982). Fluorescence intensity increases of up to 7-fold were observed for Calcofluor and 12-fold to 13-fold for Congo red (which is, however, only weakly fluorescent).

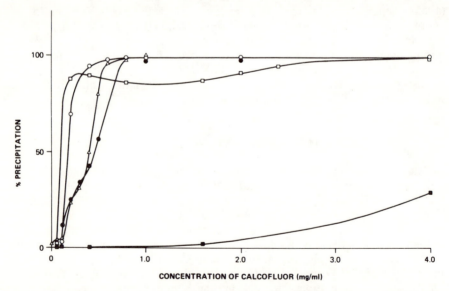

Fig. 13. Effect of Calcofluor concentration on precipitation of different concentrations of oat β-glucan in water and buffers of ionic strength 0.2: o = 0.9% in water; □ = 0.43% in water; ■ = 0.09% in water; Δ = 0.08% in phosphate buffer, pH 7.0, I 0.2; and • = 0.12% in carbonate buffer, pH 10.0, I 0.2.

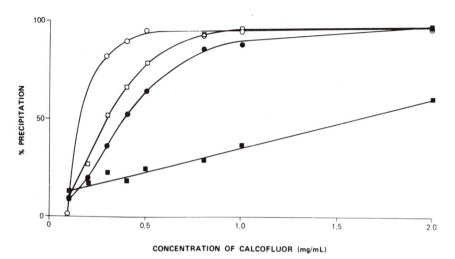

Fig. 14. Effect of Calcofluor concentration on precipitation of oat β-glucan (~ 0.8%, w/v) by Calcofluor in different media: o = water; □ = phosphate buffer, pH 8.0, I 0.05; ● = phosphate buffer, pH 6.0, I 0.05; and ■ = 1*M* NaCl.

Electrolyte increased the wavelength shifts and fluorescence intensity increases relative to water. These spectral changes were useful indicators of complex formation in solution and were observed with as little as 0.5–1.0 μg of oat β-glucan per milliliter. Although there was a linear relationship between oat β-glucan concentration and absorbancy of Congo red (10 μg/ml) and of Calcofluor (10 μg/ml) in, for example, ionic strength 0.2 carbonate buffer, the narrow range of absorbancy and β-glucan concentration (0–5 μg/ml) made this approach most useful for qualitative estimation of β-glucan. The complex of dye and β-glucan tended to precipitate in the presence of electrolyte, particularly with Congo red (10 μg/ml) in 0.5*M* NaCl, in which precipitation was observed with as little as 2.5 μg/ml of oat β-glucan. This phenomenon, along with a variability in results according to the mode of mixing of glucan and dye, hindered the development of reliable quantitative data and determination of binding constants.

VIII. SPECIFICITY OF DYE INTERACTIONS AND HISTOCHEMISTRY OF OAT β-GLUCAN

Precipitation and changes in spectra were used to study the specificity of the interaction of dyes with cereal β-glucan (Wood and Fulcher, 1978; Wood, 1980a; Wood et al, 1983). Major interaction (precipitation and large spectral shifts) was shown only by polysaccharides containing contiguous (1→4)-linked β-D-glucopyranosyl units—in particular, (1→3)(1→4)-β-D-glucans. In contrast, (1→3)-linked β-D-glucans such as curdlan required alkali for dissolution, and addition of 1*M* NaCl was then necessary for major wavelength shifts to be observed with both dyes. Starch, dissolved in alkali containing 1*M* NaCl, showed considerable interaction with Congo red but not with Calcofluor (Wood, 1982). Some other polysaccharides, such as glucomannan, induced smaller changes in the spectra of the dyes (Wood, 1980a); but most other polysaccharides, including the endospermic pentosans of cereals, showed no interaction with the dyes (Wood et al, 1983). Thus, complex formation with Congo red and Calcofluor may be used specifically to detect and estimate cereal $(1{\rightarrow}3)(1{\rightarrow}4)$-$\beta$-D-glucan in solution. The ability of oat β-D-glucan to bind the dyes was progressively lost during incubation with the specific $(1{\rightarrow}3)(1{\rightarrow}4)$-$\beta$-D-glucanohydrolase from *B. subtilis* (Wood, 1981). The specificity of the dye-binding, and in particular the fluorescent staining of cereal β-glucan in oat, barley, and wheat grain by Congo red and Calcofluor, can thus be confirmed by use of the *B. subtilis* enzyme (Wood et al, 1983). The combination of these specific reagents thus confirmed the presence of $(1{\rightarrow}3)(1{\rightarrow}4)$-$\beta$-D-glucan in oat endosperm cell walls, particularly in the subaleurone layer, and demonstrated its presence in oat aleurone cell walls (Fig. 8 of Chapter 3; Fig. 6).

The aleurone cells and thick cell-walled subaleurone endosperm are most resistant to attrition during milling and therefore become part of the coarse or bran fraction. It is easy to see microscopically, particularly in a variety such as Hinoat, that a considerable proportion of the subaleurone region is composed of β-glucan (Fig. 8 of Chapter 3; Fig. 6), and it is therefore not surprising that coarse or bran fractions of oat flour have a high β-glucan, or gum, content (Acker et al, 1955a; Hohner and Hyldon, 1977; P. J. Wood, *unpublished results*). Commercial oat bran contains 7–8% β-glucan; regular rolled oats contain about half this amount.

Wood and Fulcher (1978) used Calcofluor staining to monitor the process of alkaline extraction of β-glucan from oat flour (Fig. 6) and demonstrated that the thick subaleurone cell walls were the most resistant to extraction. Treatment with the β-glucanase from *B. subtilis* confirmed that unextracted mixed-linkage β-glucan was responsible for the residual staining of particles.

Specific fluorescence microscopy of cereal β-glucan using Congo red and Calcofluor staining is thus a powerful tool for studying varieties and processes (Fulcher and Wood, 1983; Fulcher et al, 1984).

Extensive studies have been carried out on isolated endosperm cell walls from barley and other cereals (Fincher, 1975; Forrest and Wainwright, 1977; Bacic and Stone, 1980). Selvendran and Du Pont (1980) recently described a new method for isolation of oat cell walls. Analysis of fragments of the endosperm cell wall revealed low values for nonstarchy, noncellulosic glucan, relative to pentosan, which suggest that a significant amount of β-glucan was lost during isolation. Use of Calcofluor or Congo red during isolation of cell walls and fractions therefrom should help in assessing such material losses.

IX. β-GLUCANASES

There is very little published information on β-glucanase systems in oats, unlike barley, for which considerable literature is available (Luchsinger, 1966; Bourne and Pierce, 1972; Thompson and Laberge, 1977). Preece and co-workers reported a high level of endo-β-glucanase for ungerminated oats relative to barley (Preece et al, 1954; Preece, 1957), and the necessity to inactivate this enzyme before aqueous extractions of gum has been commented on.

In barley, three main types of endo-β-D-glucanase activity have been reported, namely endo-$(1{\rightarrow}4)$-β-D-glucanase, endo-$(1{\rightarrow}3)$-β-D-glucanase, and endo-

$(1\rightarrow3)(1\rightarrow4)$-β-D-glucanase.[1] Measurements of these activities are normally made viscometrically using O-(carboxymethyl) cellulose or O-(hydroxyethyl) cellulose, O-(carboxymethyl) pachyman (α-$(1\rightarrow3)$-β-linked glucan), and barley (or oat) β-glucan (Manners and Marshall, 1969). Viscosities of high-molecular-weight polysaccharides are rapidly reduced by random chain cleavage, and this degradation effect is expected of endoenzyme activity, whereas exoenzyme action—removing units successively from the chain end—has little immediate effect on viscosity.

Using essentially the methods of Manners and Wilson (1974) and Manners and Marshall (1969), the three endo-β-glucanase activities were measured in flours from ungerminated oats, using oat β-glucan as substrate for endo-$(1\rightarrow3)(1\rightarrow4)$-β-glucanase measurements (Wood et al, 1978). Table X shows values found for two samples of the cultivar Hinoat and one sample of Rodney. All three types of activity were shown by each flour, and in each case numerical values for activity were in the order $(1\rightarrow3) > (1\rightarrow3)(1\rightarrow4) > (1\rightarrow4)$. Although these values depend to a large extent on the substrate properties, endo-$(1\rightarrow3)$-β-glucanase activity seemed significantly greater. A similarly high activity is noted for this enzyme in germinating seeds, which is somewhat puzzling in view of the apparent lack of appreciable amounts of substrate.

The crude enzymes used to obtain the results of Table X were isolated by acetate buffer extraction and ammonium sulfate precipitation. Use of alkaline extraction (as for gum preparation) followed by neutralization, centrifugation, dialysis of the supernatant, and lyophilization gave an extract that continued to show all three activities. The $(1\rightarrow4)$-β- and $(1\rightarrow3)(1\rightarrow4)$-β-glucanase activities were 7–8% of the acetate-extracted activities, but the $(1\rightarrow3)$-β-glucanase showed 40% activity. This demonstrated that β-glucanase might be carried through the gum extraction procedure. Further evidence of this was discussed earlier in this chapter. The initial effect of these enzymes on gum preparations is to decrease reproducibility of viscosity determinations and decrease overall viscosity. Changes in yield of gum may not be apparent until extensive degradation has

---

[1] $(1\rightarrow4)$-β-D-glucan 4-glucanohydrolase (EC.3.2.1.4), $(1\rightarrow3)$-β-D-glucan 3-glucanohydrolase (EC.3.2.1.39), and $(1\rightarrow3)(1\rightarrow4)$-β-D-glucan 4-glucanohydrolase (EC.3.2.1.73), respectively.

TABLE X
Endo β-Glucanase Activities of Oat Flour

Sample	Substrate[a]	Linkage	Activity[b]
Hinoat (1975)	HEC	β-$(1\rightarrow4)$	2.9
	CMP	β-$(1\rightarrow3)$	46.2
	OG	β-$(1\rightarrow4)(1\rightarrow3)$	9.5
Hinoat (1973)	HEC	β-$(1\rightarrow4)$	1.0
	CMP	β-$(1\rightarrow3)$	49.3
	OG	β-$(1\rightarrow4)(1\rightarrow3)$	3.3
Rodney (1967)	HEC	β-$(1\rightarrow4)$	1.1
	CMP	β-$(1\rightarrow3)$	12.7
	OG	β-$(1\rightarrow4)(1\rightarrow3)$	3.5

[a] HEC = O-(hydroxyethyl) cellulose, 250M; CMP = O-(carboxymethyl) pachyman; and OG = oat gum.
[b] Expressed as rate of change of reciprocal specific viscosity units per 100 g of flour per minute.

taken place. Enzyme-degraded gum is more rapidly solubilized, leading to an increased yield in extract 1 and, depending upon the extent of degradation, perhaps an overall increase in yield.

An enzyme that solubilizes barley β-glucan has been described in germinating barley (Bamforth et al, 1979). The enzyme appears to be a carboxypeptidase that releases endospermic cell wall β-glucan from a peptide-β-glucan macromolecule. Such an enzyme might also be present in oats.

X. SUMMARY

Recent studies on the preparation and properties of oat gum, the major component of which is oat β-glucan, arose out of a surge of interest in novel food proteins in the last decade. The development of high-protein cultivars made oats an attractive possibility as a protein resource, but wet-milling processes designed to separate starch and protein encountered problems with viscosity buildup as β-glucan was solubilized. Such problems may be turned to advantage if economic processes can be designed to isolate the gum, since the aqueous solution properties of this material make it ideally suited as an industrial hydrocolloid. No such commercial process exists at present, although a patent describes alkaline extraction of a coarse fraction of oat flour to yield 7–17% oat gum, depending on the cultivar (Hohner and Hyldon, 1977). The key to this process, and possibly to any future commercially successful process, is to exploit the concentration of β-glucan in the thick subaleurone cell walls in certain oat cultivars. Such fractions are also of particular value because of their high protein content (Chapter 3). Processes that achieve maximum concentration of thick subaleurone cells, in particular the cell walls, before contact with aqueous systems should be sought to maximize yield and ease of oat gum extraction. Once within an aqueous environment, enzyme activity must be minimized, or else enzyme deactivation must be achieved by means that do not damage the commercial value of such coproducts as starch and protein.

Considerable rheological and applications testing is required to establish potential markets for oat gum. Data presented in this chapter, however, have shown that oat gum achieves high viscosity at low concentration, is extremely pseudoplastic at 0.5% and greater concentration, and is essentially stable to sugar (40%) and salt. It therefore seems to compare favorably with high-viscosity neutral polysaccharides such as some of the substituted celluloses, guar gum, and locust bean gum. Indeed, low-shear, low-concentration viscosities seem as great or greater than those normally reported for most commercial gums (see Whistler and BeMiller, 1973), although differences in methods of measurement make valid comparisons difficult. The prime area of interest for oat gum should therefore be as a thickening agent in typical food applications, such as ice cream, sauces, and salad dressings. The range of potential applications, too wide to discuss here, are well covered in, for example, texts by Whistler and BeMiller (1973) and Glicksman (1982).

The recent recognition that oat bran, possibly by virtue of its gum content, exerts potentially beneficial physiological activity when consumed as part of the human diet (Chapter 11) provides further motivation for study of oat β-glucan. Large quantities of oat gum are required to allow development of palatable products that can be used in extensive clinical trials. The economics of production of oat gum and its utilization in food products for a specific medical market, such as for diabetics and patients with hypercholesterolemia, will differ from those involved in general commerce. There would, no doubt, be problems with labeling, regulations, and analysis.

Studies of the metabolic fate of oat *β*-glucan and monitoring of dietary intake will require improved methods of analysis of this dietary fiber. Recognition that a portion of oat *β*-glucan is water soluble and that this solubility may be increased both by endogenous and added (impure *α*-amylase) enzymes should help avoid analytical errors. Exploitation of dye interactions seems to offer potential for rapid and specific methods for detection and analysis.

In conclusion, more complete analytical data on cultivar variations in amounts, distribution, and extractability of the *β*-glucan are required. Our knowledge of structure, physical properties, and organization within the cell wall has improved little in the last 20 years. In view of its potential as a new industrial hydrocolloid and as a nutritionally valuable dietary component, further studies of oat gum are warranted.

ACKNOWLEDGMENT

This chapter is Contribution No. 489 of the Food Research Centre, Agriculture Canada.

LITERATURE CITED

AALMO, K. M., ISHAK, M. F., and PAINTER, T. J. 1978. Possibilities for selective Smith-degradation. Carbohydr. Res. 63:C3-7.

AASTRUP, S. 1979. The relationship between the viscosity of an acid flour extract of barley and the *β*-glucan content. Carlsberg Res. Commun. 44:289-304.

ACKER, L., DIEMAIR, W., and SAMHAMMER, E. 1955a. The lichenin of oats. I. Properties, preparation and composition of the muciparous polysaccharides. Z. Lebensm. Unters. Forsch. 100:180-188.

ACKER, L., DIEMAIR, W., and SAMHAMMER, E. 1955b. The lichenin of oats. II. Determination of molecular weight and further studies on constitution. Z. Lebensm. Unters. Forsch. 102:225-231.

AHLUWALIA, B., and ELLIS, E. E. 1984. A rapid and simple method for the determination of starch and *β*-glucan in barley and malt. J. Inst. Brew. London 90:254-259.

ANDERSON, M. A., COOK, J. A., and STONE, B. A. 1978. Enzymatic determination of 1,3:1,4-*β*-glucans in barley grain and other cereals. J. Inst. Brew. London 84:233-239.

ANDERSON, R. L., and STONE, B. A. 1978. Studies on *Lolium multiflorum* endosperm in tissue culture. II. Structural studies on the cell walls. Aust. J. Biol. Sci. 31:573-586.

BACIC, A., and STONE, B. A. 1980. A (1→3)- and (1→4)-linked *β*-D-glucan in the endosperm cell-walls of wheat. Carbohydr. Res. 82:372-377.

BAMFORTH, C. W., MARTIN, H. L., and WAINWRIGHT, T. 1979. A role for carboxypeptidase in the solubilisation of barley *β*-glucan. J. Inst. Brew. London 85:334-338.

BANKS, W., and GREENWOOD, C. T. 1975. Starch and Its Components. Edinburgh University Press, Edinburgh, Scotland.

BATHGATE, G. N., and DALGLIESH, C. E. 1974. The diversity of barley and malt *β*-glucans. Proc. Am. Soc. Brew. Chem. 33:32-36.

BENDELOW, V. M. 1975. Determination of non-starch polysaccharides in barley breeding programmes. J. Inst. Brew. London 81:127-130.

BOURNE, D. T., JONES, M., and PIERCE, J. S. 1976. *β*-Glucan and *β*-glucanases in malting and brewing. Tech. Q. Master Brew. Assoc. Am. 13:3-7.

BOURNE, D. T., and PIERCE, J. S. 1972. *β*-Glucan and *β*-glucanase—review. Tech. Q. Master Brew. Assoc. Am. 9:151-157.

BOURNE, D. T., POWLESLAND, T., and WHEELER, R. E. 1982. The relationship between total *β*-glucan of malt and malt quality. J. Inst. Brew. London 88:371-375.

BUCHALA, A. J., and FRANZ, G. 1974. A hemicellulosic *β*-glucan from the hypocotyls of *Phaseolus aureus*. Phytochemistry 13:1887-1889.

BUCHALA, A. J., and WILKIE, K. C. B. 1970. Non-endospermic hemicellulosic β-glucans from cereals. Naturwissenschaften 57:496.

BUCHALA, A. J., and WILKIE, K. C. B. 1971. The ratio of β(1→3) to β(1→4) glucosidic linkages in non-endospermic hemicellulosic β-glucans from oat plant (*Avena sativa*) tissues at different stages of maturity. Phytochemistry 10:2287-2291.

CLARKE, A. E., and STONE, B. A. 1966. Enzymic hydrolysis of barley and other β-glucans by a β-(1→4)-glucan hydrolase. Biochem. J. 99:582-588.

DAIS, P., and PERLIN, A. S. 1982. High field ^{13}C NMR spectroscopy of β-D-glucans, amylopectin and glycogen. Carbohydr. Res. 100:103-116.

DARKEN, M. A. 1961. Applications of fluorescent brighteners in biological techniques. Science 133:1704-1705.

ELFAK, A. M., PASS, G., and PHILLIPS, G. O. 1979. The effect of shear rate on the viscosity of solutions of guar gum and locust gum. J. Sci. Food Agric. 30:439-444.

FINCHER, A. B. 1975. Morphology and chemical composition of barley endosperm cell walls. J. Inst. Brew. London 81:116-122.

FLEMING, M., and KAWAKAMI, K. 1977. Studies of the fine structure of β-D-glucans of barleys extracted at different temperatures. Carbohydr. Res. 57:15-23.

FLEMING, M., MANNERS, D. J., JACKSON, R. M., and COOKE, S. C. 1974. The estimation of β-glucan in barley. J. Inst. Brew. London 80:399-404.

FORREST, I. S., and WAINWRIGHT, T. 1977. The mode of binding of β-glucans and pentosans in barley endosperm cell walls. J. Inst. Brew. London 83:279-286.

FRASER, C. G., and WILKIE, K. C. B. 1971. β-Glucans from oat leaf tissues at different stages of maturity. Phytochemistry 10:1539-1542.

FULCHER, R. G., SETTERFIELD, G., McCULLY, M. E., and WOOD, P. J. 1977. Observations on the aleurone layer. II. Fluorescence microscopy of the aleurone-sub-aleurone junction with emphasis on possible β-1,3-glucan deposits in barley. Aust. J. Plant Physiol. 4:917-928.

FULCHER, R. G., and WOOD, P. J. 1983. Identification of cereal carbohydrates by fluorescence microscopy. Pages 111-147 in: New Frontiers in Food Microstructure. D. B. Bechtel, ed. Am. Assoc. Cereal Chem., St. Paul, MN.

FULCHER, R. G., WOOD, P. J., and YIU, S. H. 1984. Insights into food carbohydrates through fluorescence microscopy. Food

Technol. 38:101-106.

GAGNAIRE, D., MARCHESSAULT, R. H., and VINCENDON, M. 1975. Nuclear magnetic resonance of lichenin. Tetrahedron Lett. 45:3953-3956.

GLICKSMAN, M. 1982. Food Hydrocolloids. CRC Press, Boca Raton, FL.

GOLDSTEIN, I. J., HAY, G. W., LEWIS, B. A., and SMITH, F. 1965. Controlled degradation of polysaccharides by periodate oxidation, reduction and hydrolysis. Pages 361-370 in: Methods in Carbohydrate Chemistry, Vol. 5. R. L. Whistler, ed. Academic Press, New York.

GREENBERG, D. C., and WHITMORE, E. T. 1974. A rapid method for estimating the viscosity of barley extracts. J. Inst. Brew. London 80:31-33.

HOHNER, G. A., and HYLDON, R. G. 1977. Oat groat fractionation process. U.S. patent 4,028,468.

HUBER, D. J., and NEVINS, D. J. 1977. Preparation and properties of a β-D-glucanase for the specific hydrolysis of β-D-glucans. Plant Physiol. 80:300-304.

HUGHES, J., and McCULLY, M. E. 1975. The use of an optical brightener in the study of plant structure. Stain Technol. 50:319-329.

IGARASHI, O., and SAKURAI, Y. 1966. Studies on the non-starchy polysaccharides of the endosperm of naked barley. Agric. Biol. Chem. 30:642-645.

ISHAK, M. F., and PAINTER, T. 1971. Formation of inter-residue hemiacetals during the oxidation of polysaccharides by periodate ion. Acta Chem. Scand. 25:3875-3877.

JENSEN, S. A., and AASTRUP, S. 1981. A fluorometric method for measuring 1,3:1,4-β-glucan in beer, wort, malt and barley by use of Calcofluor. Carlsberg Res. Commun. 46:87-95.

JØRGENSEN, K. G. 1983. An improved method for determining β-glucan in wort and beer by use of Calcofluor. Carlsberg Res. Commun. 48:505-516.

LABAVITCH, J. M., and RAY, P. M. 1978. Structure of hemicellulosic polysaccharides of *Avena sativa* coleoptile cell walls. Phytochemistry 17:933-937.

LLOYD, J. B., and WHELAN, W. J. 1969. An improved method for enzymic determination of glucose in the presence of maltose. Anal. Biochem. 30:467-470.

LUCHSINGER, W. W. 1966. Carbohydrases of barley and malt. Cereal Sci. Today 11:69-75, 82-83.

LUCHSINGER, W. W., ENGLISH, H., and KNEEN, E. 1958. Influence of malting and brewing on barley gums. Proc. Am. Soc. Brew. Chem. pp. 40-46.

MADACSI, J. P., PARRISH, F. W., and ROBERTS, E. J. 1983. Nonenzymic method for determination of beta-glucan in the presence of starch. J. Am. Soc. Brew. Chem. 41:161-162.

MANNERS, D. J., and MARSHALL, J. J. 1969. Studies on carbohydrate metabolising enzymes. XXII. The β-glucanase system of malted barley. J. Inst. Brew. London 75:550-561.

MANNERS, D. J., and WILSON, G. 1974. Purification and properties of an endo-$(1\rightarrow3)$-β-D-glucanase from malted barley. Carbohydr. Res. 37:9-22.

MARES, D. J., and STONE, B. A. 1973. Studies on wheat endosperm. I. Chemical composition and ultrastructure of the cell walls. Aust. J. Biol. Sci. 26:793-812.

MARTIN, H. L., and BAMFORTH, C. W. 1981. An enzymic method for the measurement of total and water soluble β-glucan in barley. J. Inst. Brew. London 87:88-91.

McCLEARY, B. V., and GLENNIE-HOLMES, M. 1985. Enzymic quantification of $(1\rightarrow3)(1\rightarrow4)$-β-D-glucan in barley and malt. J. Inst. Brew. London 91:285-295.

MITCHELL, J. R. 1979. Rheology of polysaccharide solutions and gels. Pages 51-72 in: Polysaccharides and Foods. J. M. V. Blanshard and J. R. Mitchell, eds. Butterworths Press, Woburn, MA.

MORRIS, D. L. 1942. Lichenin and araban in oats (*Avena sativa*). J. Biol. Chem. 142:881-891.

NEVINS, D. J., YAMAMOTO, R., and HUBER, D. J. 1978. Cell wall β-D-glucans of five grass species. Phytochemistry 17:1503-1505.

NISHIKAWA, Y., OHKI, K., TAKAHASHI, K., KURONO, G., FUKUOKA, F., and EMORI, M. 1974. Studies on the water soluble constituents of lichens. II. Antitumour polysaccharides of *Lasalia, Usnea* and *Cladonia* species. Chem. Pharm. Bull. 22:2692-2702.

PARRISH, F. W., PERLIN, A. S., and REESE, E. T. 1960. Selective enzymolysis of poly-β-D-glucans, and the structure of the polymers. Can. J. Chem. 38:2094-2104.

PATON, D. 1977. Oat starch. I. Extraction, purification and pasting properties. Staerke 29:149-153.

PEAT, S., WHELAN, W. J., and ROBERTS, J. G. 1957. The structure of lichenin. J. Chem. Soc. 3916-3924.

PERLIN, A. S., and SUZUKI, S. 1962. The structure of lichenin: Selective enzymolysis studies. Can. J. Chem. 40:50-56.

PODRAZKY, V. 1964. Some characteristics of cereal gums. Chem. Ind. London 712-713.

PREECE, I. A. 1957. Malting relationships of barley polysaccharides. Wallerstein Lab. Commun. 20:147-161.

PREECE, I. A., AITKEN, R. A., and DICK, J. A. 1954. Non-starchy polysaccharides of cereal grains. VI. Preliminary study of the enzymolysis of barley glucosan. J. Inst. Brew. London 60:497-507.

PREECE, I. A., GARG, N. K., and HOGGAN, J. 1960. Enzymic degradation of cereal hemicelluloses. III. Oligo-saccharide production from β-glucan. J. Inst. Brew. London 66:331-337.

PREECE, I. A., and HOBKIRK, R. 1953. Non-starchy polysaccharides of cereal grains. III. Higher molecular gums of common cereals. J. Inst. Brew. London 59:385-392.

PREECE, I. A., and HOBKIRK, R. 1954. Non-starchy polysaccharides of cereal grains. V. Some hemicellulose fractions. J. Inst. Brew. London 60:490-496.

PREECE, I. A., and MacKENZIE, K. G. 1952. Non-starchy polysaccharides of cereal grains. II. Distribution of water-soluble gum-like materials in cereals. J. Inst. Brew. London 58:457-464.

PRENTICE, N. 1982. Purification of beta-glucanase for beta-D-glucan assays. Cereal Chem. 59:231-232.

PRENTICE, N., BABLER, S., and FABER, S. 1980. Enzymic analysis of beta-D-glucans in cereal grains. Cereal Chem. 57:198-202.

REES, D. A. 1977. Polysaccharide Shapes. Chapman and Hall, London.

REID, J. S. G., and WILKIE, K. C. B. 1969a. Polysaccharides of the oat plant in relationship to plant growth. Phytochemistry 8:2045-2051.

REID, J. S. G., and WILKIE, K. C. B. 1969b. Total hemicelluloses from oat plants at different stages of growth. Phytochemistry 8:2059-2065.

RICKES, E. L., HAM, E. A., MOSCATELLI, E. A., and OTT, W. H. 1960. The isolation and biological properties of a β-glucanase from *B. subtilis*. Arch. Biochem. Biophys. 69:371-375.

SCOTT, J. E., and TIGWELL, M. J. 1973. Periodate induced viscosity decreases in aqueous solutions of acetal and ether-linked polymers. Carbohydr. Res. 28:53-59.

SELVENDRAN, R. R., and Du PONT, M. S. 1980. An alternative method for the isolation and analysis of cell wall material from cereals. Cereal Chem. 57:278-283.

SMITH, F., and MONTGOMERY, R. 1959. The Chemistry of Plant Gums and Mucilages. Rheinhold Publishing Corp., New York.

SMITH, M. M., and STONE, B. A. 1973. Chemical composition of the cell walls of *Lolium multiflorum* endosperm. Phytochemistry 12:1361-1367.

STINARD, P. S., and NEVINS, D. J. 1980. Distribution of non-cellulosic β-D-glucans in grasses and other monocots. Phytochemistry 19:1467-1468.

TAKEDA, T., FUNATSU, M., SHIBATU, S., and FUMIKO, F. 1972. Polysaccharides of lichens and fungi. V. Antitumour active polysaccharides of lichens of *Evernia, Acroscyphys* and *Alectoria* spp. Chem. Pharm. Bull. 20:2445-2449.

THOMPSON, R. G., and LABERGE, D. W. 1977. Barley endosperm cell walls: A review of cell wall polysaccharides and cell wall degrading enzymes. Tech. Q. Master Brew. Assoc. Am. 14:238-243.

VOISEY, P. W., and RANDALL, C. J. 1977. A versatile food rheometer. J. Texture Stud. 8:339-358.

WADA, S., and RAY, P. M. 1978. Matrix polysaccharides of oat coleoptile cell walls. Phytochemistry 17:923-931.

WHISTLER, R. L., and BEMILLER, J. N., eds. 1973. Industrial Gums, 2d ed. Academic Press, New York.

WILKIE, K. C. B. 1979. The hemicelluloses of grasses and cereals. Pages 215-264 in: Advances in Carbohydrate Chemistry and Biochemistry, Vol. 36. R. Tipson and D. Horton, eds. Academic Press, New York.

WILKIE, K. C. B., and WOO, S. L. 1976. Non-cellulosic β-D-glucans from bamboo, and interpretive problems in the study of all hemicelluloses. Carbohydr. Res. 49:399-409.

WOOD, P. J. 1980a. Specificity in the interaction of direct dyes with polysaccharides. Carbohydr. Res. 85:271-287.

WOOD, P. J. 1980b. The interaction of direct dyes with water soluble substituted celluloses and cereal β-glucans. Ind. Eng. Chem. Prod. Res. Dev. 19:19-23.

WOOD, P. J. 1981. The use of dye-poly-saccharide interactions in β-D-glucanase assay. Carbohydr. Res. 94:C19-23.

WOOD, P. J. 1982. Factors affecting precipitation and spectral changes associated with complex formation between dyes and β-D-glucans. Carbohydr. Res. 102:283-293.

WOOD, P. J. 1985. Dye-polysaccharide interactions—recent research and applications. Pages 267-278 in: New Approaches to Research on Cereal Carbohydrates. R. D. Hill and L. Munck, eds. Elsevier, Amsterdam.

WOOD, P. J., and FULCHER, R. G. 1978. Interaction of some dyes with cereal β-glucans. Cereal Chem. 55:952-966.

WOOD, P. J., and FULCHER, R. G. 1983. Dye interactions: A basis for specific detection and histochemistry of polysaccharides. J. Histochem. Cytochem. 31:823-826.

WOOD, P. J., and FULCHER, R. G. 1984. Specific interactions of aniline blue with (1→3)-β-D-glucan. Carbohydr. Polym. 4:49-72.

WOOD, P. J., FULCHER, R. G., and STONE, B. A. 1983. Studies on the specificity of interaction of cereal cell wall components with Congo Red and Calcofluor. Specific detection and histochemistry of (1→3)(1→4)-β-D-glucan. J. Cereal Sci. 1:95-110.

WOOD, P. J., PATON, D., and SIDDIQUI, I. R. 1977. Determination of β-glucan in oats and barley. Cereal Chem. 54:524-533.

WOOD, P. J., SIDDIQUI, I. R., and PATON, D. 1978. Extraction of high viscosity gums from oats. Cereal Chem. 55:1038-1049.

WOOD, P. J., and WEISZ, J. 1984. Use of Calcofluor in analysis of oat beta-D-glucan. Cereal Chem. 61:73-75.

WOODWARD, J. R., FINCHER, G. B., and STONE, B. A. 1983. Water soluble (1→3)(1→4)-β-D-glucans from barley (*Hordeum vulgare*) endosperm. II. Fine structure. Carbohydr. Polym. 3:207-225.

WOOLARD, G. R., RATHBONE, E. B., and NOVELLIE, L. 1976. A hemicellulosic β-D-glucan from the endosperm of sorghum grain. Carbohydr. Res. 51:249-252.

CHAPTER 7

OAT STORAGE PROTEINS

DAVID M. PETERSON
A. CHRIS BRINEGAR
U.S. Department of Agriculture
Agricultural Research Service and Department of Agronomy
University of Wisconsin
Madison, Wisconsin

I. INTRODUCTION

The superior nutritional value of oat protein has long been recognized. Although oat protein quality does not approach that from animal sources, it surpasses that of the other cereals, as shown by feeding tests and by comparison of its amino acid balance with the FAO standard (FAO/WHO, 1973). At the same time, protein concentration in the oat groat (dehulled kernel) is considerably higher (the range of cultivated varieties in the United States is about 15–20%) than that of other cereals. Despite this nutritional advantage, the consumption of oat products by humans has accounted for only 5–10% of U.S. production, the majority of the crop being used for animal feed.

Over the last 10–15 years, oat breeders and the oat milling industry have shown considerable interest in the possibility of developing new cultivars of oats that contain higher percentages of protein, an improved amino acid balance, or both, as compared with currently grown cultivars. In part, this interest resulted from a public awareness that children in certain underdeveloped areas of the world suffered from a shortage of protein even though their diets were sufficient in calories, a condition known as kwashiorkor. It was believed that development and introduction of high-protein grains would contribute to the alleviation of this condition at a reasonable cost. At the same time, oat breeders discovered that certain lines of the wild hexaploid oat *Avena sterilis* L., obtained from collections in the Mediterranean region, contained considerably higher percentages of protein than did cultivated oats, *A. sativa* L. It was suggested that high-protein genes from *A. sterilis* could be transferred into adapted cultivars, increasing their protein levels. Thus, a perceived need to increase protein and the possibility of doing so coincided.

From the current perspective, is there a need to increase oat protein or to improve its amino acid balance? For the most part, North Americans, Europeans, and the populations in other developed countries consume a varied diet containing adequate protein, and oats and oat products are a minor component. Other factors, such as high cholesterol and saturated fats and inadequate fiber in American diets, suggest that benefits would result from increased use of oats as a source of dietary protein. In some underdeveloped areas, dietary protein may come primarily from one or a few sources, usually of plant origin, and the quantities of essential amino acids can be marginal. A grain with high-quality protein, such as oats, could be beneficial to people in these circumstances. Overcoming cultural and dietary habits would be a major obstacle in introducing any new foods. Also, oats are not adapted to growth in tropical areas, where much of the world's malnourished population lives. There are areas in temperate regions where cultivation of high-protein oats could be introduced or increased to serve as a useful dietary component for malnourished people.

As animal feed, high-protein oats can be beneficial if they provide protein more cheaply than alternative sources such as soybean meal. Livestock farmers usually grow their own oats, but they must purchase a protein concentrate such as soybean meal. The economics of replacing all or part of the protein concentrate with oats may be favorable in many circumstances, and the value of high-protein oats has been demonstrated in diets for swine (Wahlstrom and Libal, 1975; Anderson et al, 1978; Wahlstrom and Libal, 1979) and poultry (Maruyama et al, 1975; Sibbald, 1979). For nonruminants, the amino acid balance of protein sources is important. Several studies have shown that supplementation of oats with lysine and methionine improved their nutritional value (e.g., Maruyama et al, 1975). Therefore, improving the amino acid balance of oat protein genetically might be of value.

This chapter does not attempt to review results of breeders' efforts to increase oat protein. Breeders have succeeded in releasing several high-protein cultivars, such as Dal, Otee, and Hinoat, which usually contain up to 20% groat protein. Exotic germ plasm, mutation breeding, and traditional methods have been used (Frey, 1977).

One must bear in mind the bioenergetic costs of increasing the protein in oats. Protein is produced at about twice the bioenergetic cost to the plant as carbohydrate (Penning de Vries et al, 1975). Thus, increased protein percentage requires more energy, and yield is reduced unless increased inputs of energy and N are achieved (Bhatia and Rabson, 1976). This theoretical relationship may explain the often noted negative correlation between yield and protein concentration. Furthermore, synthesis of the essential amino acids requires more energy than synthesis of glutamine, glutamic acid, and some other nonessential ones (Mitra et al, 1979). If a genetic improvement of amino acid balance could be achieved, it too might depress yield. Thus, breeding for increased protein or improved amino acid balance may inadvertently decrease yield, which would not be acceptable in most circumstances.

It would seem reasonable to expect that such a highly nutritious and economical protein source would be used more for human consumption, both in the developed and underdeveloped countries. In the former, this use would require considerable research into the properties of oat protein and the development of novel applications in food products. Our knowledge of oat proteins is substantially less than that of wheat, corn, barley, or rice. In underdeveloped countries, the problems may be more agronomic and cultural.

An important reason to study the basic structure and synthesis of oat storage proteins is the future possibility of genetic engineering. Techniques are becoming available whereby genes can be transferred from one plant species into another and expressed in the new host. One might envision oat protein genes that could be transferred into wheat to improve its protein quality; or, conversely, genes for wheat gluten that could be transferred into oats, allowing the baking of bread using flour solely from oats.

The objective of this chapter is to bring together and interpret data on the characteristics of oat storage proteins. It considers their chemical and physical properties and introduces recent results of studies on their mode of synthesis and cellular location. Then it discusses changes in these proteins that occur with seed maturation and germination and the effects of the plant's growth environment on protein composition. Finally, it presents the limited data available in the literature on functionality and food applications of oat protein. Throughout, the reader will note the paucity of information available on oat storage proteins as compared with those of the other major cereals. This is unfortunate because oat proteins have unique features that could be developed in novel ways.

II. CLASSIFICATION AND SOLUBILITY FRACTIONATION

The classic technique of Osborne (1910), and variations of it, have often been used to extract and classify oat storage proteins (Michael et al, 1961; Ewart, 1968; Wu et al, 1972; Völker, 1975; Peterson, 1976; Peterson and Smith, 1976; Wieser et al, 1980). Under this scheme, ground, defatted seeds are successively extracted with water, salt solution, alcoholic solution, and dilute alkali or acid to yield albumins, globulins, prolamins, and glutelins, respectively. Most cereals have a high percentage of prolamins, the alcohol-soluble fraction, but oats and rice are exceptions. All workers agree that the prolamin (avenin) content of oats is low, with estimates ranging from about 4 to 14% of total protein. The water-soluble fraction, albumins, which is considered to be primarily enzymes, is also a minor component (9–20% of total protein).

There has been disagreement on the proportion of protein that is globulin. Accurate determination of its quantity requires complete and exclusive extraction, and the Osborne method does not meet this requirement. The value of 80% salt-soluble globulins reported in a frequently cited review (Brohult and Sandegren, 1954) may be too high according to more recent reports. Peterson and Smith (1976) reported that globulins were the major protein component of oats, accounting for 46 and 50% of total N (56 and 54% of recovered protein) in the cultivars Goodland and Orbit, respectively (Table I). In another study, Peterson (1976) found about 52% globulins among several cultivars, environments, and fertility levels. In both of these studies, the glutelins accounted for 21–27% of total protein. By contrast, several German workers have reported that glutelins are the major fraction and that the concentration of globulins is only 12–19% (Michael et al, 1961; Völker, 1975; Wieser et al, 1980). It is unlikely that German cultivars vary so distinctly from those grown in the United States. We attribute the low globulin percentages obtained to less rigorous extraction conditions. For example, Michael et al (1961) used dilute K_2SO_4 (concentration not specified) rather than the frequently used $1 M$ NaCl solution. Peterson (1978) showed that at equal ionic strength the salts of divalent ions extracted less globulin than those of monovalent ions. Wieser et al (1980) used only 0.4M NaCl, which has been shown to extract less than half of the globulin fraction (Peterson, 1978). We have observed that globulin will precipitate from a 1M NaCl solution at 4°C, so extraction at room temperature may be required for more complete recovery. Any unextracted globulin will be attributed to the glutelin fraction (Robert et al, 1983a). We have found that electrophoretic profiles of alkaline extracts include bands similar in mobility to those of the globulin fraction, even following exhaustive extraction with 1M NaCl at pH 8, which suggests that the glutelin fraction contains residual globulin (see Section III).

III. CHARACTERIZATION

A. Amino Acid Composition

The amino acid composition of oat groats has been reported by numerous workers (Ewart, 1967; Hischke et al, 1968; Eggum, 1969; Tkachuk and Irvine, 1969; Robbins et al, 1971). The most comprehensive report is that of Robbins et al (1971), who surveyed 289 samples representing oat cultivars grown commercially in the United States and Canada during the period 1900–1970, promising experimental lines, and important sources of germ plasm in oat-breeding programs (Table II). Typical of seed storage proteins, the combined glutamine-glutamic acid fraction, reported as glutamic acid, is relatively high. Lysine, considered to be nutritionally limiting in oat protein, averaged 4.2 g/100 g of amino acids recovered, higher than in other cereals but still below the recommended FAO reference standard of 5.5 g/100 g (FAO/WHO, 1973). Threonine, another essential amino acid, averaged 3.3 g/100 g as compared with 4.0 g/100 g for the FAO reference standard. Variability among samples for these nutritionally limiting amino acids was low. These results were a disappointment to breeders, who had hoped to find more genetic diversity for these essential amino acids.

Recently, Zarkadas et al (1982) reported high methionine levels, 3.76 and 3.23% of total amino acids, for Oxford and Sentinel, two Canadian cultivars. These values, considerably higher than all previous reports, are of interest because methionine has also been considered nutritionally limiting in oats

TABLE I
Composition of Oat Storage Protein as Determined by Solubility Fractionation[a]

Cultivar	Albumins	Globulins	Prolamins	Glutelins	Protein Recovery (%)
Goodland[b]	11	56	9	23	83
Orbit[b]	10	54	13	23	93
AuSable[c]	19	52	8	21	79
Rodney[c]	16	53	9	22	82
Clintland 64[c]	14	52	7	27	86

[a] Given as percentage of total recovered.
[b] Data from Peterson and Smith (1976).
[c] Data from Peterson (1976).

(Bressani et al, 1963; Kies et al, 1972). These workers analyzed methionine as methionine sulfone from performic acid oxidized samples (Moore, 1963), but Tkachuk and Irvine (1969) reported identical recovery of methionine from standard 6N HCl 25-hr hydrolysis, as compared with methionine sulfone from performic acid oxidized samples. We were unable to confirm the high methionine values for Sentinel and Oxford by analysis of independently obtained samples in our laboratory (*unpublished data*).

These workers (Zarkadas et al, 1982) also determined tryptophan in Oxford and Sentinel oats. Their levels of 1.19 and 1.35% were somewhat lower than those reported by Tkachuk and Irvine (1969) but higher than the value reported by Eggum (1969).

The relationship between protein concentration and protein quality has received some attention. Hischke et al (1968) reported a negative correlation of −0.86 between lysine (as percentage of protein) and protein concentration in a study involving seven cultivars. Robbins et al (1971) also found a negative correlation between lysine and protein percentage, but it was much less (−0.18) than the previous report. However, threonine and methionine were also negatively correlated with percentage of protein (−0.47 and −0.39, respectively) (Robbins et al, 1971). These results suggest that plant breeders developing higher-protein cultivars than those presently available should be mindful of potential adverse effects on protein quality. However, three points are noteworthy: 1) the negative correlation of lysine with protein is much less in oats than in wheat; 2) although lysine as percentage of protein tends to decline as protein percentage increases, lysine as percentage of groat dry weight increases; 3) it has not been demonstrated conclusively that the small differences in protein

TABLE II
Protein Concentration and Amino Acid Composition of 289 Oat Groat Samples[a]

Protein or Amino Acid	Maximum	Minimum	Mean	SD	CV
Protein	24.4	12.4	17.1	2.010	11.7
Lysine	5.2	3.2	4.2	0.324	7.6
Histidine	3.1	1.2	2.2	0.316	14.1
Ammonia	3.0	2.5	2.7	0.079	2.9
Arginine	7.8	6.2	6.9	0.231	3.3
Aspartic acid	9.9	8.3	8.9	0.278	3.1
Threonine	3.5	3.0	3.3	0.098	2.9
Serine	4.8	3.8	4.2	0.187	4.4
Glutamic acid	26.9	21.9	23.9	0.938	3.9
Proline	5.8	3.8	4.7	0.416	8.7
Half-cystine	2.6	0.6	1.6	0.442	26.8
Glycine	5.5	4.4	4.9	0.210	4.3
Alanine	5.5	4.2	5.0	0.187	3.7
Valine	5.7	4.9	5.3	0.106	2.0
Methionine	3.3	1.0	2.5	0.347	13.9
Isoleucine	4.1	3.4	3.9	0.088	2.2
Leucine	7.8	4.8	7.4	0.205	2.8
Tyrosine	4.4	2.3	3.1	0.232	7.4
Phenylalanine	5.7	4.9	5.3	0.138	2.6

[a] Source: Robbins et al (1971); ©American Chemical Society, used by permission. Proteins given as N × 6.25 (db), %; amino acids given as g/100 g of amino acids and ammonia recovered.

quality among oat cultivars are nutritionally significant to the extent that they can be measured by feeding tests (Hischke et al, 1968).

The solubility fractions have distinctive amino acid profiles (Waldschmidt-Leitz and Zwisler, 1963; Draper, 1973). The prolamin fraction is distinguished by

TABLE III
Amino Acid Composition of Protein Solubility Fractions, Cultivar Condor[a]

Amino Acid	Albumin	Globulin	Prolamin	Glutelin	Whole Meal
Lysine	7.5	5.6	3.0	4.9	4.4
Histidine	2.6	2.9	1.6	3.0	1.8
Arginine	4.8	9.8	4.4	9.2	7.0
Aspartic acid	11.2	8.9	3.1	9.7	8.9
Threonine	5.1	3.6	2.1	4.3	3.4
Serine	6.0	4.9	2.7	4.9	5.4
Glutamic acid	12.5	20.3	34.7	17.0	21.8
Proline	5.6	5.5	8.4	5.3	6.9
Half-cystine	1.3	1.3	3.8	0.8	3.4
Glycine	6.1	5.4	2.2	4.9	5.5
Alanine	7.3	5.7	4.1	4.9	5.3
Valine	6.0	4.9	5.4	5.3	4.7
Methionine	2.2	1.8	3.4	1.7	1.5
Isoleucine	4.4	4.3	3.4	4.8	3.7
Leucine	7.8	6.9	9.8	7.8	7.0
Tyrosine	2.8	2.4	1.6	4.8	4.1
Phenylalanine	6.6	5.9	6.6	6.6	5.1

[a] Recalculated from Draper (1973). Data given as g/100 g of amino acids recovered.

TABLE IV
Protein Concentration and Amino Acid Composition
of Hand-Separated Oat Groat Tissues, Cultivar Orbit[a]

Protein or Amino Acid	Whole Groats	Embryonic Axis	Scutellum	Bran	Starchy Endosperm
Protein	13.8	44.3	32.4	18.8	9.6
Lysine	4.5	8.2	6.9	4.1	3.7
Histidine	2.4	3.9	3.6	2.2	2.2
Ammonia	2.7	1.9	1.8	2.5	2.9
Arginine	6.8	8.3	9.0	6.8	6.6
Aspartic acid	8.7	10.2	9.7	8.6	8.5
Threonine	3.4	5.0	4.7	3.4	3.3
Serine	4.6	4.8	5.0	4.8	4.6
Glutamic acid	21.7	14.2	14.9	21.1	23.6
Proline	5.5	3.3	3.6	6.2	4.6
Half-cystine	2.1	0.5	1.0	2.4	2.2
Glycine	5.2	6.3	6.2	5.4	4.7
Alanine	5.0	7.2	6.9	5.1	4.5
Valine	5.5	6.0	6.2	5.5	5.5
Methionine	2.2	2.2	2.1	2.1	2.4
Isoleucine	3.9	3.9	3.8	3.8	4.2
Leucine	7.6	7.1	7.1	7.4	7.8
Tyrosine	3.0	2.9	3.0	3.5	3.3
Phenylalanine	5.2	4.2	4.4	5.1	5.6

[a] Source: Pomeranz et al (1973). Proteins given as N × 6.25 (db), %; amino acids given as g/100 g of amino acids and ammonia recovered.

its high levels of glutamine-glutamic acid and proline and its low level of lysine as compared with the other fractions (Table III) (Draper, 1973). Thus, the relatively scarce oat prolamin is typical of other cereal prolamins and is considered a nutritionally poor protein. The albumin fraction is closer to crude vegetative plant protein in amino acid balance, showing higher levels of lysine, asparagine-aspartic acid, and alanine, and this fraction is lowest in glutamine-glutamic acid. The compositions of globulin and glutelin tend to be intermediate between those of albumin and prolamin, and similar to each other. Wieser et al (1981) determined the amide content of the various solubility fractions of oats. They calculated that 87% of the aspartate and glutamate residues of the prolamin fraction were present as asparagine and glutamine, whereas in the globulin fraction, only 61% were amidated.

The amino acid composition of hand-separated oat groat tissues was reported by Pomeranz et al (1973). Amino acid profiles of the embryonic tissues, embryonic axis and scutellum, resembled the albumin fraction with higher lysine and glutamine-glutamic acid concentrations (Table IV). The bran, which included the aleurone layer, and the starchy endosperm were nearly alike in amino acid composition, except that bran was higher in proline. Their amino acid profiles were characteristic of globulin. Since these two tissues comprise more than 95% of the total groat weight (Youngs, 1972), it is obvious that the embryo has very little influence on the overall amino acid balance.

Pomeranz et al (1973) analyzed samples of commercial oat products for amino acid composition (Table V). Amino acid composition of oat flakes was similar to that of the groats from which they were produced. Light oats, which have a higher percentage of hull, were more characteristic of the hull amino acid profile

TABLE V
Protein Concentration and Amino Acid Composition of Commercial Oat Products[a]

Protein or Amino Acid	Heavy Oats	Light Oats	Groats	Hulls	Flakes
Protein	13.4	9.6	18.9	5.7	17.6
Lysine	4.2	5.2	3.9	4.9	4.1
Histidine	2.4	2.7	2.3	2.4	2.4
Ammonia	3.3	3.8	3.2	3.6	3.2
Arginine	6.4	6.3	6.2	6.8	6.0
Aspartic acid	9.2	11.1	9.0	10.5	9.0
Threonine	3.3	4.1	3.1	4.1	3.1
Serine	4.0	4.5	3.9	4.6	4.0
Glutamic acid	21.6	20.0	22.4	20.3	22.7
Proline	5.7	3.1	6.2	2.4	6.1
Half-cystine	1.7	0.4	2.0	0.5	1.7
Glycine	5.1	6.0	5.0	6.1	5.0
Alanine	5.1	5.5	5.0	5.4	5.0
Valine	5.8	6.2	5.7	6.4	5.8
Methionine	2.3	1.5	2.5	1.5	2.4
Isoleucine	4.2	4.5	4.3	4.5	4.3
Leucine	7.5	7.6	7.4	7.9	7.5
Tyrosine	2.6	2.4	2.5	2.9	2.1
Phenylalanine	5.4	5.3	5.5	5.3	5.6

[a] Source: Pomeranz et al (1973). Proteins given as N × 6.25 (db), %; amino acids given as g/100 g of amino acids and ammonia recovered.

(high lysine, asparagine-aspartic acid, and threonine and low proline, cystine, and methionine) than were heavy oats.

B. Isolation and Physical Properties

As mentioned in the previous section, the traditional Osborne method of separating proteins does not necessarily result in the quantitative extraction of homogeneous polypeptides as detected by sodium dodecyl sulfate-polyacrylamide gel electrophoresis (SDS-PAGE). An initial 50mM Tris-HCl extraction of defatted oat flour at pH 8 yields numerous polypeptides ranging in molecular weight from 5,000 to 70,000, many of the polypeptides representing enzymes or their subunits (Fig. 1, lane B).

When the subsequent NaCl extract was dialyzed against cold distilled water, the supernatant protein had an SDS-PAGE pattern (Fig. 1, lane C) very similar to that of the Tris extract. The precipitated fraction (globulin) was composed of only two size classes of polypeptides (Fig. 1, lane D). The amount of globulin precipitated by this step is very dependent on the extent of NaCl removal by dialysis and the quality of the water used, but typical recovery of globulin is 40–50 mg/g of flour. At higher loadings, minor polypeptides near 55,000–60,000 mol wt were visible in this globulin preparation and were more apparent if the initial Tris extraction was omitted. The possibility that these minor components are residual precursor forms of the lower-molecular-weight globulin polypeptides, and therefore copurify with them, has been considered and is discussed in a later section.

After removal of the salt-extractable proteins from the residue, avenin (the oat prolamin) can be extracted with 52% (v/v) ethanol and then precipitated by adding two volumes of cold 1.5% (w/v) NaCl (Kim et al, 1978). SDS-PAGE of avenin shows several polypeptides ranging from 22,000 to 33,000 mol wt, with the major species at 30,000 (Fig. 1, lane E). There was no obvious similarity to the globulin fraction. However, when a final extraction of the residue is made with a buffer of 1% (w/v) SDS, 50mM 2-mercaptoethanol, and 50mM Tris-HCl, pH 8, to recover the remaining protein, the SDS-PAGE pattern was predominantly a mixture of globulin and avenin polypeptides with a few other uncharacterized polypeptides (Fig. 1, lane F). Extraction with 0.1N NaOH instead of the SDS buffer (to obtain the Osborne "glutelin" fraction) yielded essentially the same pattern (not shown), although resolution was diminished because of alkaline degradation of the protein.

Fig. 1. Sodium dodecyl sulfate-polyacrylamide gel electrophoresis (SDS-PAGE) of oat (cultivar Froker) proteins fractionated by solubility. Samples were prepared in 2% SDS, 0.1M dithiothreitol, 1mM EDTA, 20% sucrose, 50mM Tris-HCl, pH 6.8; heated briefly in a boiling water bath; and then electrophoresed on 15% running and 4% stacking gels, as described by Laemmli (1970). Staining was with Coomassie blue R-250. A, Standard proteins (phosphorylase B, bovine serum albumin, ovalbumin, carbonic anhydrase, soybean trypsin inhibitor, and lysozyme). B, Tris extract of defatted oat flour (50mM Tris-HCl, pH 8.0, room temperature, 45 min, 25 ml/g of flour). C and D, Supernatant and precipitate, respectively, of a 1M NaCl extract of the residue (in Tris buffer as above) after dialysis of the extract against H$_2$O at 4°C. E, Ethanol (52%, v/v) extract of residue precipitated by dilution with two volumes of 1.5% NaCl (same extraction conditions as above). F, Residual protein extracted with 1% SDS, 50mM 2-mercaptoethanol, 50mM Tris-HCl, pH 8.0 (same extraction conditions as above, except at 2.5 ml/g of flour). (Reprinted, with permission, from Brinegar, 1983)

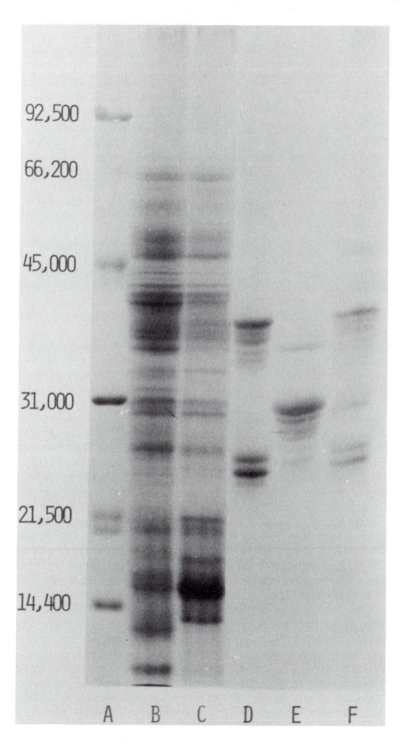

Since it is difficult to assign specific polypeptides as albumin storage proteins due to simultaneous extraction of enzymes, we are hesitant to classify any specific polypeptides of oats as distinct albumin *storage* proteins. The same holds for the glutelin fraction, as the majority of polypeptides appear to be identical to globulin and avenin polypeptides. These may have escaped solubilization in earlier fractions because of extensive disulfide cross-linking or an association with membranes or other cellular components. In our judgment, the only biochemically defined classes of oat storage proteins found in any significant quantity are the globulin and avenin proteins. Since the globulin polypeptides occur in the water-soluble and residue (glutelin) fractions in addition to the salt-extracted fraction, we must conclude that the globulin, as a percentage of the total seed protein, is actually higher than the 55% observed for the salt-extractable globulin alone and is possibly as high as 75%. These observations have been confirmed by recent studies (Robert et al, 1983a; Colyer and Luthe, 1984).

Peterson (1978) characterized the globulin fraction by ultracentrifugation and SDS-PAGE. The native protein had a sedimentation coefficient ($S_{20,w}$) of 12.1, with a minor component observed at 18.3 (possibly a dimeric form). Sedimentation equilibrium analysis yielded a native molecular weight of 322,000. Continuous SDS-PAGE (Weber and Osborn, 1969) of the globulin could resolve only two major polypeptide bands at approximately 22,000 and 32,000 mol wt. A consideration of the weight ratio of the two polypeptide bands, as determined from their amino acid composition and their molecular weights, suggested that the native globulin was composed of six each of these polypeptides. Some minor bands in the region of 55,000 mol wt were also observed; however, the globulin was isolated by precipitation of a $1.0M$ NaCl extract by cold water dialysis, which results in a higher proportion of these minor polypeptides than the previously described extraction procedure. Kim et al (1979a) have reported molecular weights (by SDS-PAGE) of polypeptides from their globulin preparation in the vicinities of 13,000–22,000, 38,000, and 62,000, with a significant percentage of the total protein stain in the region of 62,000. The absence of a disulfide reducing agent in their samples could explain the increased proportion of high-molecular-weight species.

More recent characterization of oat globulin reported large and small polypeptide size ranges (molecular weights) of 32,500–37,500 and 22,000–24,000 (Brinegar and Peterson, 1982a) (Fig. 2, lanes 1 and 2); 33,000–40,000 and 20,000–25,000 (Robert et al, 1983b); 36,000-40,000 and 20,000–23,000 (Rossi and Luthe, 1983); and 35,000–40,000 and 20,000–25,000 (Walburg and Larkins, 1983). These polypeptide groups will be referred to as α and β (Brinegar and Peterson, 1982a), respectively. These same studies also confirmed that the α and β polypeptides are linked by disulfide bonds. The unreduced $\alpha\beta$ species shown in Fig. 2 (lane 3) are approximately 53,000–58,000 mol wt, but other estimates range from 50,000 to 70,000. Therefore, the native molecular structure is now thought to be a hexamer of the disulfide-linked $\alpha\beta$ species—i.e., $(\alpha\beta)_6$.

In addition to the size differences between the globulin α and β polypeptides, their charge properties are also dissimilar, with the β polypeptides being substantially more basic. We have taken advantage of their charge characteristics to develop a method of separating the α and β polypeptide groups by ion exchange chromatography (DEAE Sepharose CL-6B in $6M$ urea) of the reduced and S-carboxyamidomethylated globulin (Brinegar and Peterson, 1982a) (Fig. 3). Walburg and Larkins (1983) have obtained similar results using a Sephadex ion exchange technique. We have found DEAE cellulose to give less successful results. Partial separation has been obtained using gel filtration (Sephacryl S-300) in $6M$ guanidine-HCl, pH 5 (*data not shown*), but rechromatography was necessary to purify each polypeptide group.

Isoelectric focusing (IEF) in $6M$ urea (Fig. 4) not only shows the isoelectric point (pI) ranges of the α and β polypeptides (5.9–7.2 and 8.7–9.2, respectively), but also further demonstrates the heterogeneity within each group. We have carefully studied the globulin IEF pattern under various conditions of running time, loading method and position, pH gradients, vertical vs. horizontal apparatus, gel composition, and sample preparation and have concluded that

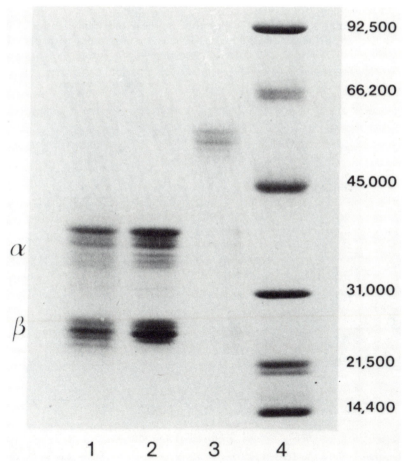

Fig. 2. Sodium dodecyl sulfate-polyacrylamide gel electrophoresis (linear 5–20% gradient) of oat globulin, with lane 1 reduced; lane 2, reduced and carboxyamidomethylated; lane 3, unreduced; and lane 4, standard proteins. (Reprinted, with permission, from Brinegar and Peterson, 1982a)

loading the sample at or near the basic (cathodic) end of the gel causes some chemical modification or deterioration of the globulin (especially the basic polypeptides), resulting in anomalous patterns (Brinegar and Peterson, 1982a). Patterns are very reproducible when run as described in Fig. 4, and our finding that disulfide reduction of the sample before IEF gives the same result as a previously reduced and iodoacetamide alkylated globulin has made sample preparation simpler, especially when running several samples at once (e.g., for cultivar or species comparison). The globulin fractions (reduced and alkylated with 4-vinylpyridine) of several cultivars have been compared using IEF by Robert et al (1983b), who found essentially the same type of heterogeneous pattern and pI ranges as we have described, although they reported the β group as slightly more basic (pI = 9.0–10.0). There were minor differences among cultivars, but all shared a similar pattern. Two-dimensional electrophoresis (IEF followed by SDS-PAGE) detected even more heterogeneity in the α and β groups (Brinegar and Peterson, 1983a; Walburg and Larkins, 1983), with as many as 30 α and 15 β polypeptides resolved.

Amino acid analyses of the whole globulin and the α and β polypeptide groups (Peterson, 1978; Brinegar and Peterson, 1982a) are consistent with their IEF

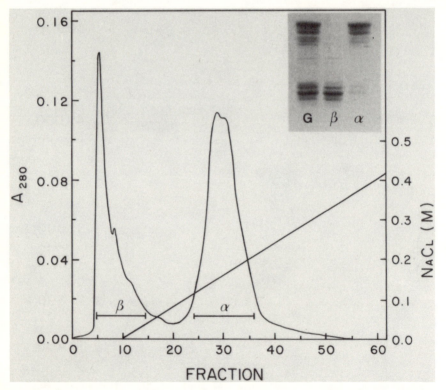

Fig. 3. Separation of the α and β polypeptides by DEAE Sepharose CL-6B chromatography ($6M$ urea, 20mM Tris-HCl, pH 8.2) of reduced and carboxyamidomethylated oat globulin. Inset: Sodium dodecyl sulfate-polyacrylamide gel electrophoresis of separated α and β polypeptides compared with reduced, alkylated globulin (G). (Reprinted, with permission, from Brinegar and Peterson, 1982a)

mobilities. The basic amino acid content (lysine, histidine, and arginine) of the β polypeptide groups is 27% higher (on a molar basis) than the α group and most likely has a higher ratio of amide to acidic amino acid (Table VI). As would be expected, the amino acid content of the whole globulin is intermediate or nearly

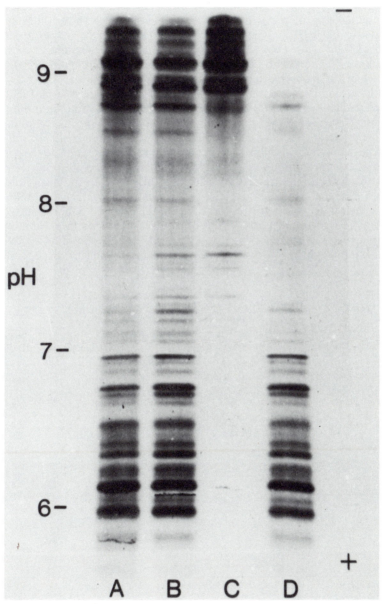

Fig. 4. Isoelectric focusing (in 6M urea) of oat globulin. A, Reduced with dithiothreitol; B, reduced and carboxyamidomethylated; C, isolated β polypeptides; D, isolated α polypeptides. Samples were loaded near the acidic end of the gel and run by the method of Brinegar and Peterson (1982a).

equal to those of the α and β groups. Of particular interest is the similarity in amino acid composition between the oat globulin and soybean glycinin (11S protein). We have also noted analogies between the oat globulin and legume 11S storage proteins regarding their polypeptide and native molecular weights, interchain disulfide bonding, native structure, and acidic and basic nature of the two polypeptide groups (Peterson, 1978; Brinegar and Peterson, 1982a). N-terminal sequencing of the β polypeptide group has also shown substantial homology to a basic (B) polypeptide of glycinin (Walburg and Larkins, 1983). Antibodies raised against oat 12S globulin reacted against the α and β subunits of pea legumin (Robert et al, 1985b) and rice glutelin (Robert et al, 1985c). The significance of these observations is discussed in a later section.

Recently, globulin fractions with sedimentation coefficients of 7S and 3S have been identified (Burgess et al, 1983). These fractions are less abundant than the 12S globulin described above. It has been suggested that 3S and 7S globulins are similar to the vicilin fractions of legumes (Adeli and Altosaar, 1984); indeed, immunological similarities have been demonstrated (Robert et al, 1985a).

Avenin, the oat prolamin, was extensively studied by Kim et al (1978). A crude preparation of avenin (extracted as described earlier) from the hexaploid *A. nuda* was separated from nonprotein contaminants by chromatography on Sephadex G-100, using an acetic acid-urea buffer (Fig. 5). Separation of the components (based on charge) by starch gel electrophoresis in an aluminum lactate-urea (pH 3.2) buffer system (Landry et al, 1965) resolved at least eight different bands, which were assigned to the α, β, and γ groups (fast,

TABLE VI
Amino Acid Composition of Oat Globulin and Its α and β Polypeptides Compared with Soybean Glycinin[a]

Amino Acid	α (Acidic)	β (Basic)	Globulin	Soybean Glycinin[b]
Aspartic acid	8.2	11.4	9.2	11.8
Threonine[c]	3.8	4.7	4.1	4.2
Serine[c]	6.9	6.8	7.0	6.6
Glutamic acid	22.1	14.2	19.1	18.8
Proline	5.0	5.0	4.9	6.3
Glycine	8.4	6.0	7.5	7.8
Alanine	5.7	6.9	6.0	6.7
Valine	6.0	6.8	6.4	5.6
Cysteine[d]	0.9	1.1	1.1	1.1
Methionine	0.4	1.5	0.9	1.0
Isoleucine	4.4	5.6	4.8	4.6
Leucine	7.5	7.2	7.4	7.2
Tyrosine	3.2	3.7	3.5	2.5
Phenylalanine	5.6	4.6	5.2	3.9
Lysine	2.3	3.9	2.9	4.1
Histidine	2.0	2.5	2.2	1.8
Arginine	6.4	7.2	6.6	5.9
Tryptophan[e]	1.2	0.7	1.0	...

[a] Source: Brinegar and Peterson (1982a); used by permission. Data presented as mole %.
[b] Data from Kitamura and Shibasaki (1975).
[c] Extrapolated from 24- and 72-hr hydrolysis values.
[d] Determined as carboxymethylcysteine.
[e] Determined spectrophotometrically.

intermediate, and slow mobilities, respectively) (Fig. 6). Further fractionation of the avenin components by ion exchange on SP-Sephadex C-50 resulted in the isolation of the γ1 and γ4 species, which were found to have nearly identical molecular weights (22,000), very similar amino acid composition, and threonine as the N-terminal amino acid. Even the nonhomogeneous ion exchange fractions were found to be extremely similar to the purified γ components, suggesting a very high degree of homology. A later report (Kim et al, 1979b) revised the avenin molecular weights to 20,000–34,000, which compare well with the 22,000–33,000 values determined in our laboratory (Fig. 1, lane E). Comprehensive electrophoretic studies of avenin (Kim and Mossé, 1979; Kim et al, 1979b) have shown that electrophoretic comparison can be useful in determining the phylogenetic relationships of *Avena* species, but Robert et al (1983c) reported considerable interspecific heterogeneity by SDS-PAGE and IEF.

The most prominent feature of the avenin amino acid composition (Wu et al, 1972; Draper, 1973; Kim et al, 1979a; Wieser et al, 1980) is the extremely high content of glutamine-glutamic acid (~ 35 mole %) and the fairly high levels of leucine and proline (11 and 10 mole %, respectively). Such values are typical of other cereal prolamins (Wieser et al, 1980). An interesting characteristic, however, that sets avenin apart from the other prolamins is its obviously different N-terminal sequence. In a comparison of Gramineae prolamin N-terminal sequences, Bietz (1982) found that avenin shared no homology with the prolamins in other members of the subfamily Festucoideae (wheat, rye, barley)

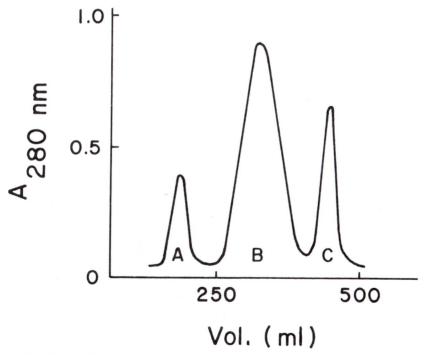

Fig. 5. Purification of avenin (peak B) by Sephadex G-100 chromatography in acetic acid-urea. (Reprinted, with permission, from Kim et al, 1978)

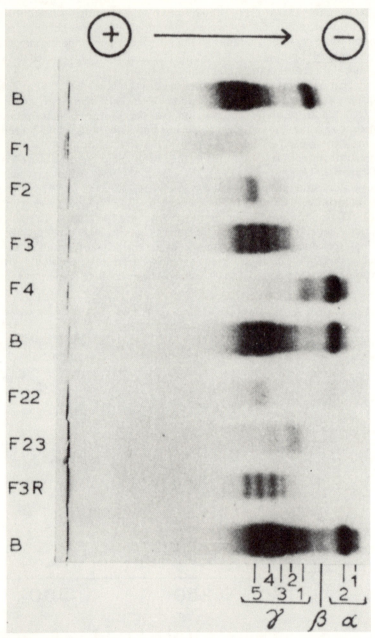

Fig. 6. Starch gel electrophoresis of purified avenin (B) and its components fractionated by SP-Sephadex C-50 ion exchange chromatography; α = fast, β = intermediate, and γ = slow mobilities. (Reprinted, with permission, from Kim et al, 1978)

or the subfamily Panicoideae (maize, *Tripsacum*, sorghum, pearl millet). An electrophoretic comparison of several cereal prolamins (urea-lactate, pH 3.6) showed that the avenin polypeptides generally migrated faster than those of other species (Laurière and Mossé, 1982).

C. Proteolytic Enzymes and Inhibitors

There is little published information available on proteolytic enzymes in oat groats. Donhowe and Peterson (1983) found activity against a casein substrate from a preparation of aleurone layer protein bodies, but a protein body preparation from starchy endosperm had no activity. The pH optimum was between 5 and 6. No attempt was made to purify or characterize the enzyme. Sutcliffe and Baset (1973) measured a protease activity in endosperm extracts of germinating oats. The activity, measured at pH 8 against casein, increased rapidly over the first two days of germination and less rapidly thereafter (Fig. 7).

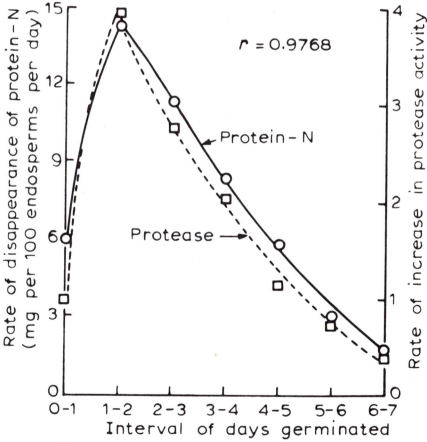

Fig. 7. Rate of disappearance of protein-N from the endosperm and rate of increase of protease activity in endosperm of germinating oats. (Reprinted, with permission, from Sutcliffe and Baset, 1973)

Trypsin inhibitor activity was found in water, saline, and sulfosalicylic acid extracts of oat seeds (Lorenc-Kubis, 1969). Trypsin inhibitor of oat flour was completely destroyed by pepsin and was thermolabile, suggesting that it poses no serious problem for the consumption of oats (Laporte and Trémolières, 1962).

IV. SYNTHESIS AND CELLULAR LOCALIZATION

A. Synthesis

Variations in the percentage of the Osborne fractions during groat development have been studied for several cultivars (Ohm and Peterson, 1975; Peterson and Smith, 1976). Most obvious was the fairly linear increase in the salt-soluble globulin fraction soon after anthesis, which tapered off at 20–25 days to approximately 55% of the total protein. A linear increase was also observed for the prolamin fraction but at a much lower rate, attaining only about 10% at maturity. Electrophoretic analysis (SDS-PAGE) of total protein extracts from

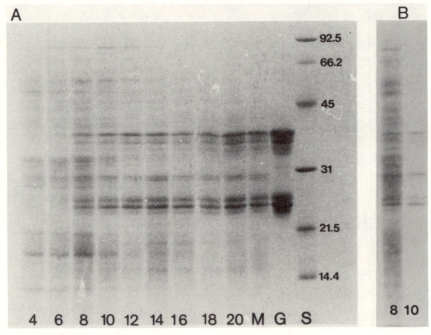

Fig. 8. Time course of protein accumulation in developing groats analyzed by sodium dodecyl sulfate-polyacrylamide gel electrophoresis (SDS-PAGE). A, Primary groats were ground and extracted with 2% SDS, 0.1 M dithiothreitol, 1mM EDTA, 20% sucrose, 50mM Tris-HCl, pH 6.8, for 10 min in a boiling water bath, then electrophoresed as described in Fig. 1 and stained with Coomassie blue R-250. Numbers refer to the age of the groats (days after anthesis) at the time of extraction; G = purified globulin, M = total protein extracted from mature groats, and S = standard proteins. B, In vivo labeling of protein in 8- and 10-day primary groats by the uptake of ^{3}H-leucine through excised panicles. After 6 hr, the protein was extracted and electrophoresed as above and the tritium-labeled polypeptides were visualized by fluorography. (Reprinted, with permission, from Brinegar, 1983)

groats ranging in age from four days to maturity (Fig. 8A) confirms those measurements (Brinegar, 1983). The mature globulin α and β polypeptides are barely visible in groats younger than six days, but by eight days they comprise about half of the protein; after 10 days, most of the protein is globulin. Bands near 30,000 mol wt, which represent the major avenin polypeptides, are visible from the earliest stages of development. Polypeptides near 17,000–20,000, 26,000, 33,000, and 52,000 mol wt are present early but diminish in proportion by 10–12 days. This disappearance may be a dilution effect rather than an actual degradation of these polypeptides or their conversion to other species. Robert et al (1983b) have also observed the early synthesis, then subsequent decrease in proportion, of 20,000 and 33,000–35,000 mol wt polypeptides. They also showed by SDS-PAGE without a reducing agent that these polypeptides formed disulfide-linked species similar to those found in mature seeds, only with slightly lower molecular weights. This observation suggested that the early polypeptides were actually related to, but not necessarily precursors of, the mature globulin polypeptides, raising the possibility that two families of globulin genes are expressed at different stages of development. The disappearance of the 52,000 mol wt species and the concomitant increase in a component near 58,000–60,000 mol wt over the period of 4–12 days after anthesis (Fig. 8A) may also reflect this change in globulin gene expression, since data show that the α and β polypeptides are derived from a higher-molecular-weight precursor (see below). In vivo labeling experiments (uptake of ^{3}H-leucine through excised panicles) with developing seeds have shown a rather abrupt decrease in the synthesis of several polypeptides between 8–10 days of age, whereas the mature globulin α and β polypeptides and avenin polypeptides rapidly increased in proportion (Fig. 8B).

The oat globulin is thought to be synthesized primarily on polyribosomes attached to the endoplasmic reticulum (ER). These membrane-bound polyribosomes were found to be twice as active as free polyribosomes in the incorporation of labeled amino acids, owing to their higher population of the larger polyribosome size classes (Luthe and Peterson, 1977). Analysis of a total polyribosome fraction from developing (about 12-day-old) seeds by sucrose density gradient centrifugation (Fig. 9) showed that messenger RNA (mRNA) species that could accommodate over 12 ribosomes were present (Brinegar, 1983).

Several laboratories have recently characterized the in vitro translation products of membrane-bound polyribosomes (Matlashewski et al, 1982) and polyadenylated mRNA isolated from polyribosomes of developing seeds (Brinegar and Peterson, 1982b; Rossi and Luthe, 1983; Walburg and Larkins, 1983). In all cases, much of the radioactive label was incorporated into polypeptide species near 60,000 mol wt, which were immunoprecipitable with antibodies raised against the globulin, indicating that the globulin was initially synthesized as a family of precursors containing both the α and β sequences. Additional evidence for this was given by Walburg and Larkins (1983), who showed that antiserum raised against globulin α polypeptides alone were capable of immunoprecipitating the 60,000 mol wt precursors. Luthe and Peterson (1977) had previously concluded from their in vitro translations of oat polyribosomes that the oat globulin was synthesized as the individual α and β polypeptides, although they did report a small percentage of label incorporated into 55,000 mol wt polypeptides. In view of the recent data, their wheat germ cell-free translation system may have been producing prematurely terminated polypeptides—i.e., not completely translating the entire mRNA sequence.

Although Rossi and Luthe (1983) and Walburg and Larkins (1983) did not detect any notable size differences between the globulin precursors synthesized in vitro and the species labeled in vivo, Brinegar and Peterson (1982b) found the in vitro products to be approximately 2,000 mol wt larger (Fig. 10), which suggested the presence of a signal sequence. Such extra sequences are commonly linked to the N-terminus of secretory proteins during synthesis and facilitate their transport through membranes (Blobel and Dobberstein, 1975). During or after transport, the signal sequence is proteolytically removed. In the case of the oat globulin, this transport and signal sequence cleavage most likely occurs across the rough endoplasmic reticulum (RER) membrane. Support for this idea comes from electron micrographs (Saigo et al, 1983), which show a high population of RER that contained aggregated protein in the lumen. Continuity of the RER lumen with vacuoles may provide a path for newly synthesized proteins, which migrate to the vacuole and coalesce into protein bodies.

At some point after synthesis, the globulin precursors are cleaved to form the individual α and β polypeptides. The enzyme responsible for this posttranslational proteolysis has not been isolated, but short-term radioactive labeling of globulin in developing seeds has shown that only precursor polypeptides are present in ER preparations, whereas both the precursor and the α and β polypeptides are labeled in fractions containing protein bodies (Adeli and Altosaar, 1983; Adeli et al, 1984; A. C. Brinegar and D. M. Peterson, *unpublished data*). Therefore, the precursor-cleavage enzyme is probably associated with developing protein bodies. It is not known when the disulfide bonds are formed between the α and β sequences in relation to precursor cleavage and assembly into the holoprotein, but the presence of a disulfide-isomerase in the ER of developing wheat endosperm (Roden et al, 1982) that

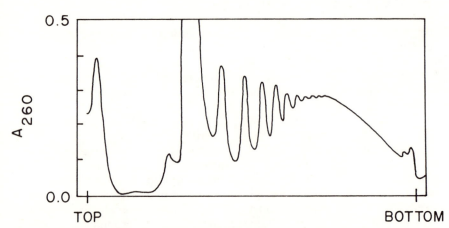

Fig. 9. Sucrose gradient analysis of total polyribosomes from 12-day-old primary groats isolated by the method of Brinegar and Peterson (1982b). The polyribosomes were fractionated by sucrose gradient centrifugation according to Luthe and Peterson (1977), except that centrifugation was for 2 hr using linear 10–40% sucrose gradients.

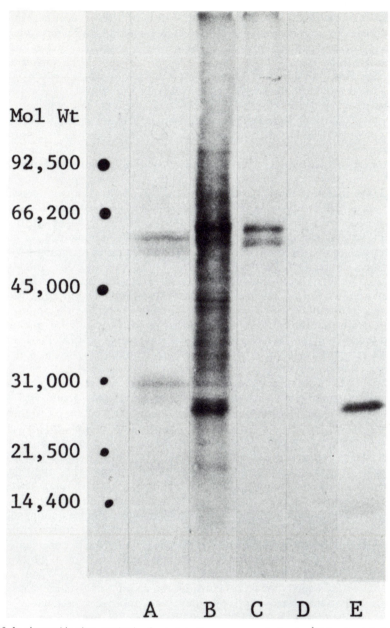

Mol Wt

92,500 ●

66,200 ●

45,000 ●

31,000 ●

21,500 ●

14,400 ●

A   B   C   D   E

Fig. 10. In vivo and in vitro synthesis of oat globulin precursors. A, In vivo ^{3}H-leucine-labeled seed extract showing globulin precursors (58,000–62,000 mol wt) and avenin (30,000 mol wt); B, total ^{35}S-methionine-labeled in vitro translation products of polyadenylated RNA from total polyribosomes; C, immunoprecipitated globulin precursors (60,000–64,000) from in vitro translation; D, in vitro translation products immunoprecipitated with preimmune (control) antibodies; E, in vitro translation products without added RNA. Samples were analyzed by sodium dodecyl sulfate-polyacrylamide gel electrophoresis (linear 7.5–20% gradient) and fluorography. (Reprinted, with permission, from Brinegar and Peterson, 1982b)

catalyzes intramolecular disulfide bonding during or after transport across the membrane (Scheele and Jacoby, 1982) might also have an analogous role in the oat ER by directing disulfide bonding of α and β sequences in the globulin precursor. Minor glycosylation may be another uncharacterized posttranslational modification, since Peterson (1978) detected approximately 6 mol of neutral sugar per mole of native globulin. The degree to which these various modifications contribute to the observed size and charge heterogeneity is unclear. Most of the globulin heterogeneity may be caused by a multigene family coding for several slightly different proteins.

Immunoprecipitations of the in vitro translation products using antiavenin antibodies also gave evidence for a precursor form of avenin (Brinegar, 1983) (Fig. 11). Some of the immunoprecipitated products were slightly larger (by 1,000–2,000 mol wt) than the corresponding purified avenin polypeptides, suggesting the presence of a signal sequence. A short signal sequence is known to occur in zein, the prolamin of maize (Larkins, 1981). Since the avenin polypeptides synthesized in vitro were in relatively low abundance, and other polypeptides were detected in the immunoprecipitate (including globulin precursors), more work is needed to isolate and characterize the avenin synthesized in vitro to prove its precursor nature conclusively.

As mentioned earlier, the oat globulin has several physical and chemical properties in common with the legume 11S storage proteins (legumins). The demonstration that the oat globulin is synthesized as 60,000 mol wt precursors extends the analogy, since the legumins of soybean, broad bean, and pea are also initially synthesized as 60,000 mol wt precursors containing both the acidic and basic sequences (for references and further discussion, see Brinegar and Peterson, 1982a, 1982b). The question of whether the oat globulin precursor is similar with regard to signal sequence, arrangement of acidic and basic sequences, and sequence homology must await confirmation by the sequencing of cloned globulin genes. This type of storage protein biosynthesis and structure in plants as diverse as oat and seed legumes may be a demonstration of an efficient and evolutionarily advantageous method of protein packaging passed on from a common ancestor and conserved despite many other physiological divergences. Aside from being of taxonomical interest, it is tempting to speculate about the potential for gene transfer between oats and the seed legumes. The demonstrated similarities between oat globulin and 11S legume globulins suggest that further investigation into gene homology between their precursor sequences and even regulatory genes might be fruitful. One could further speculate that transfer of genes for storage globulins between legumes and oats might be more easily accomplished than, say, between legumes and corn. When vector technology is developed, one might consider the transfer of glycinin genes from soybeans to oats to increase lysine content and improve functionality. Such transfers are beyond the realm of possibility now, but could conceivably be attempted within the next decade.

B. Localization

Whole-kernel oats are lower in protein concentration than oat groats (Youngs and Senturia, 1976; Weaver et al, 1981) because the hull (lemma and palea of the floret), which is removed in producing groats, is quite low in protein (Youngs,

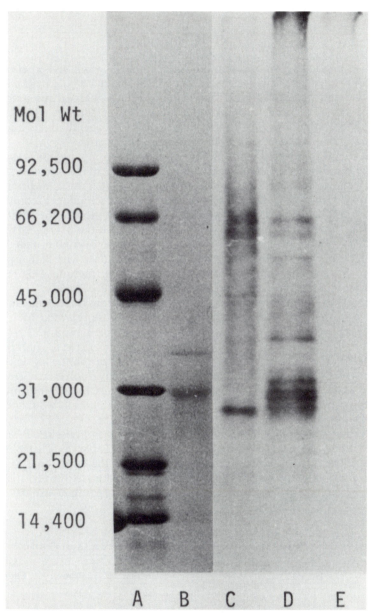

Fig. 11. In vitro synthesis of avenin. Protein-stained sodium dodecyl sulfate-polyacrylamide gel electrophoresis (15%): A, standard proteins; B, avenin. Fluorograms: C, total ^{35}S-methionine-labeled in vitro translation products of polyadenylated RNA from total polyribosomes; D, immunoprecipitated translation products using antiavenin antibodies; E, immunoprecipitated translation products using preimmune (control) antibodies. (Reprinted, with permission, from Brinegar, 1983)

1972). Based on 105 samples, Youngs and Senturia (1976) demonstrated that groat protein percentage could be predicted from whole-kernel protein percentage by the equation:

$$\hat{P}_G = 0.81 + 1.27\ P_O$$

where $\hat{P}_G$ = predicted groat protein percentage and P_O = whole-kernel protein percentage. The correlation coefficient between measured whole-oat and groat protein concentrations was 0.93; the 95% confidence interval for predicting individual values of groat protein fell within ±1.9 percentage points of the regression line. After dehulling, further milling of oat groats into flakes and flour did not significantly alter the protein percentage (Weaver et al, 1981).

Within the groat, protein is not uniformly distributed. Youngs (1972) showed that the embryonic axis was highest in protein percentage and the starchy endosperm lowest (Table VII). Values for the scutellum and bran were intermediate. However, because the embryonic axis and scutellum accounted for only about 3% of the total groat weight, their contribution to the total groat protein was quite small. Youngs (1972) found that, on the average, the bran contributed 49% and the starchy endosperm 45% of the total groat protein. There was considerable variation among cultivars, however, those with higher protein tending to have a greater contribution from the bran fraction. This resulted primarily from a thicker bran layer in the higher-protein cultivars.

Draper (1973) found considerably lower values for embryo and endosperm N percentage in the European cultivar Condor than those reported by Youngs (1972). He reported 3.10% N (19.4% protein) for the embryo and 1.92% N (12.0% protein) for the endosperm. However, the embryo was reported to contain 9.1% of the total dry matter (whole-kernel basis), so the embryo's contribution to total N (protein) was 17.7%. Although these discrepancies may result from differences between Condor and the U.S. cultivars studied by Youngs, Draper indicates that his embryo fraction contained some starchy material, which would tend to increase its weight and decrease its N percentage.

To determine the quality of the protein in the various groat tissues, Pomeranz et al (1973) analyzed them for amino acid composition (Table IV). They observed a similarity between the amino acid composition of the embryonic axis and scutellum and between the bran and starchy endosperm. However, the germ tissues were higher in the basic amino acids and asparagine-aspartic acid,

TABLE VII
Protein Concentration of Oat Groats and Hand-Separated Oat Groat Fractions[a]

Cultivar	Groats	Embryonic Axis	Scutellum	Bran	Starchy Endosperm
Orbit	13.8	44.3	32.4	18.8	9.6
Lodi	14.6	36.5	26.2	19.6	10.7
Garland	14.8	40.5	28.9	18.5	10.9
Froker	15.5	26.3	28.0	20.7	9.7
Portal	16.5	35.3	29.1	23.0	10.3
Dal	20.8	40.9	24.2	26.5	13.5
Goodland	22.5	40.7	32.4	32.5	17.0

[a] Source: Youngs (1972); used by permission. Concentrations given as N × 6.25 (db), %.

threonine, glycine, alanine, and valine, whereas the endosperm tissues were higher in glutamine-glutamic acid, proline, cystine, tyrosine, and phenylalanine. Similar differences were noted by Draper (1973) between an embryo and an endosperm fraction, except that arginine was not higher in the embryo. Draper also found that methionine, isoleucine, and leucine were higher in the embryo than in the starchy endosperm, the latter being lower in lysine, proline, glycine, and alanine and higher in glutamine-glutamic acid, isoleucine, leucine, and phenylalanine (Pomeranz et al, 1973). As a result of these analyses, one might predict that germ and endosperm proteins are markedly dissimilar and that bran and starchy endosperm proteins also had some dissimilarities.

Storage protein in the endosperm of cereals and the cotyledons of dicots is found primarily in protein bodies (see reviews by Pernollet, 1978; Weber and Neumann, 1980). In some species, such as wheat, the protein bodies become degraded upon dehydration of the maturing seed, but in others, including oats, they persist.

A few studies on the intracellular location of oat storage proteins have been published. Fulcher et al (1972) used histochemical techniques to identify arginine-rich proteins in oat seeds. They found that protein bodies of the aleurone layer and scutellar parenchyma stained intensely with the Sakaguchi reaction for arginine and with alkaline fast green, which stains basic protein. They also found numerous, intensely stained protein bodies throughout the endosperm, which differed from their findings with wheat. These results agree with amino acid analyses: in wheat, aleurone protein had 12.3% arginine and starchy endosperm protein only 4.0% (Fulcher et al, 1972). Oats, in contrast, had 6.8 and 6.6% arginine in the bran and starchy endosperm, respectively (Pomeranz et al, 1973). Thus, the more uniformly high arginine level in oat protein led to staining in both tissues. Fulcher et al (1972) also reported that oat starchy endosperm protein bodies were dispersed in a protein matrix (stained by fast green at pH 2.0). Protein bodies of the aleurone layer and scutellum possessed unstained regions, presumably the globoids.

Two types of protein-containing structures were described in the oat caryopsis (Bechtel and Pomeranz, 1981; Peterson et al, 1985). Protein bodies of the aleurone layer had a central globoid crystal surrounded by a protein matrix. Within the protein matrix, "protein-carbohydrate bodies" were found. Bechtel and Pomeranz (1981) attempted to identify the chemical nature of these protein body components by digestion of fixed, sectioned material with various hydrolytic enzymes. Although the protein matrix was digested with trypsin and partially digested with pepsin and protease V and VI from *Streptomyces griseus*, the fact that these enzymes also digested the globoids and the protein-carbohydrate bodies causes one to interpret these results with caution. The phytic acid nature of the globoid has been confirmed (Buttrose, 1978; Fulcher et al, 1981). The interpretation of the carbohydrate nature of the "protein-carbohydrate body" rests largely on the analogy with barley (Jacobsen et al, 1971) and wheat (Morrison et al, 1975), where similar-appearing structures were stained by the silver-hexamine reaction. Microhistochemical reactions indicated that the protein-carbohydrate bodies represent the major site of niacin deposition (Fulcher et al, 1981). They also stained positively for carbohydrates, aromatic amines, and phenolic compounds. However, protein was not detected.

Protein bodies of the starchy endosperm are morphologically distinct from those of the aleurone layer (Bechtel and Pomeranz, 1981; Saigo et al, 1983; Peterson et al, 1985). In the mature kernel, they are variable in size from about 0.3 to 5.0 μm and are often irregular in shape. In appearance they have a uniform, electron-opaque matrix in which electron-lucent inclusions are embedded. The matrix is undoubtedly proteinaceous, but the composition of the inclusions is uncertain. They may be lipid, because of their lack of structure, their spherical shape, and a similar staining intensity as the lipid bodies. Or they could consist of "islands" of avenin embedded within the predominant globulin (Saigo et al, 1983). Endosperm protein bodies are most numerous in the subaleurone region of the starchy endosperm but are found throughout, scattered among the much larger starch grains (Fulcher et al, 1972; Pernollet and Mossé, 1980; Bechtel and Pomeranz, 1981). The development of aleurone grains and protein bodies is discussed in Section V.

Pernollet et al (1982) isolated a mixed aleurone and starchy endosperm protein body fraction by differential centrifugation. They found that avenin and the globulin-glutelin group represented two-thirds of the storage protein. Donhowe and Peterson (1983) devised techniques for isolating aleurone cell protoplasts free of any starchy endosperm cells. This was accomplished by digesting groat slices with cellulysin and macerase, which degraded starchy endosperm cell walls more rapidly than those of the aleurone layer. A slurry from starchy endosperm free of aleurone layer was also prepared by physical disruption. By centrifugation of the cellular contents of these two preparations on sucrose density gradients, they obtained relatively pure aleurone and starchy endosperm protein body preparations. The fractions were free of contaminating organelles, as shown by marker enzyme assays and observation of the electron micrographs, but structural modifications were apparent. The aleurone protein body preparation was 78% protein by micro-Kjeldahl analysis, and it contained 0.82% P (Table VIII). Eighty-eight percent of the P was accounted for by phytic acid. The remainder of the dry weight was not accounted for, but may be carbohydrate and cations bound to phytate, which were not determined. The protein body preparation from starchy endosperm was 98% protein with 0.69% P. No phytic acid was detectable. The protein analysis of the protein bodies seems unusually high, considering the amount of membranous material present in the preparations. Protein as determined from Kjeldahl analysis is overestimated by

TABLE VIII
Chemical Composition of Oat Aleurone Grains
and Protein Bodies[a]

Component	Aleurone Grains	Protein Bodies
N	12.56	15.72
Protein[b]	78.50	98.25
P	0.82	0.69
Phytic acid	2.56	0.00
Phytic acid-P[c]	87.80	0.00

[a] Source: Donhowe (1981); used by permission. Data based on percentage of dry weight.
[b] Protein values determined with conversion factor of 6.25 from total N values.
[c] Phytic acid-P as percentage of total P.

the AACC official conversion factor of 6.25. Tkachuk (1969) has determined that 5.5 is a more valid factor for oat grain, which would indicate that the actual protein concentration was 86%.

A primary objective of this work was to determine whether aleurone and starchy endosperm protein bodies contained similar or different protein. Earlier, some differences were suggested by amino acid analyses of crude bran and starchy endosperm fractions (Pomeranz et al, 1973). Amino acid analysis of the

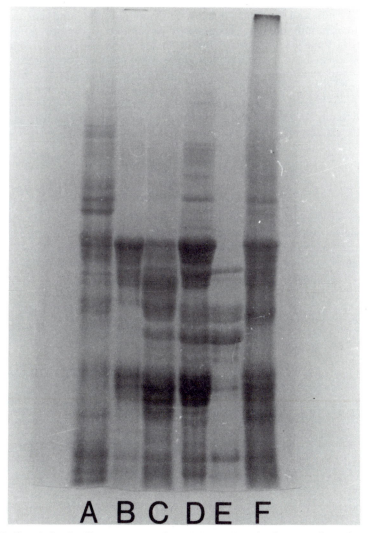

Fig. 12. Sodium dodecyl sulfate-polyacrylamide gel electrophoresis of extracts from aleurone and starchy endosperm protein bodies as compared with solubility fractions from mature oat groats, cultivar Froker. A, albumin; B, globulin; C, aleurone protein body extract; D, starchy endosperm protein body extract; E, prolamin; F, glutelin. Methods were similar to those of Laemmli (1970). (Reprinted, with permission, from Donhowe and Peterson, 1983)

aleurone and starchy endosperm protein body preparations (Donhowe, 1981) revealed considerable similarity of both fractions to the amino acid composition of oat globulin (Peterson, 1978). Only slight differences were found between the two preparations: glutamine-glutamic acid was lower and isoleucine and leucine were higher in the aleurone protein bodies. The differences between aleurone and starchy endosperm protein bodies for all other amino acids were within 0.2%.

Proteins of the two preparations were subjected to analysis by SDS-PAGE and compared with the four purified solubility fractions (Donhowe and Peterson, 1983) (Fig. 12). Both preparations contained the major globulin and prolamin bands, and they had some low-molecular-weight bands corresponding to certain albumin bands. The authors concluded that both globulin and prolamin storage proteins were found in aleurone and starchy endosperm protein bodies, but they could not determine whether there were separate globulin- and prolamin-containing populations of protein bodies within each tissue or whether both proteins were contained within the same protein bodies. Although the SDS-PAGE patterns were generally similar, certain differences indicated that aleurone and starchy endosperm protein bodies were not identical in their protein composition (Donhowe and Peterson, 1983).

V. DEVELOPMENT

A. Maturation

The protein percentage of oat groats has been shown by several investigators to increase during maturation (Brown et al, 1970; Peterson and Smith, 1976; Pomeranz et al, 1976; Welch et al, 1980). Differences among cultivars in protein percentage were established early in grain development and persisted throughout maturation (Peterson and Smith, 1976; Welch et al, 1980).

The amino acid balance of total crude protein changed significantly during maturation. Several reports agree that lysine, asparagine-aspartic acid, threonine, alanine, and isoleucine declined (as percentage of total protein),

TABLE IX
Changes in Amino Acid Composition with Maturation[a]

Amino Acid	Days after Anthesis					
	4	8	12	16	20	Mature
Declining trend						
Lysine	6.1	5.2	4.7	4.1	3.9	3.9
Aspartic acid	12.7	10.2	9.3	9.0	9.0	9.0
Threonine	4.5	4.1	3.7	3.6	3.5	3.3
Alanine	5.9	5.7	5.1	4.8	4.7	4.4
Valine	5.8	5.7	5.5	5.5	5.5	5.4
Isoleucine	4.4	4.4	4.2	4.2	4.1	4.0
Increasing trend						
Glutamic acid	15.8	17.2	19.0	20.2	20.8	20.9
Half-cystine	0.9	1.7	2.1	2.3	2.4	2.5
Tyrosine	2.8	3.4	3.5	3.6	3.6	3.5
Phenylalanine	4.1	4.8	5.0	5.1	5.2	5.1

[a] Source: Peterson and Smith (1976); used by permission of the Crop Science Society of America, Inc. Data based on g/100 g of amino acids recovered, mean values of six cultivars.

whereas glutamine-glutamic acid, proline, cystine, tyrosine, and phenylalanine increased (Table IX) (Brown et al, 1970; Peterson and Smith, 1976; Pomeranz et al, 1976). These changes in relative levels of amino acids reflect the differential synthesis of protein fractions early and late in maturity. Typical changes in the amounts of these solubility fractions are shown in Fig. 13. On a dry-weight basis, globulin showed the most rapid rate of increase. Albumin and prolamin also

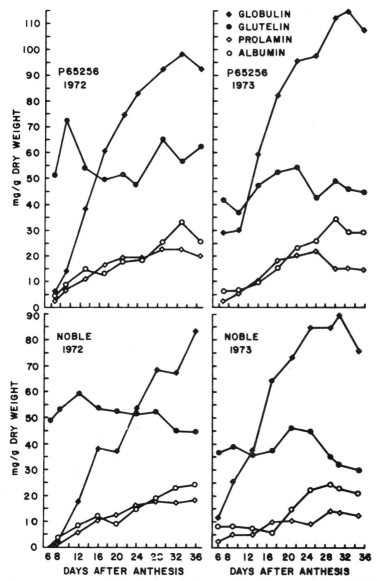

Fig. 13. Changes in protein solubility fractions with maturation of oat groats from P65256, a line from an F₅ plant selection from Diana × CI8320 (*Avena sterilis*), and from cultivar Noble in each of two years. (Reprinted from Ohn and Peterson, 1975, with permission of the Crop Science Society of America)

increased, whereas glutelin declined slightly (Noble) or showed no trend (P65256) (Ohm and Peterson, 1975). In another study, the solubility fractions were reported as percentages of total recovered protein (Peterson and Smith, 1976). The percentages of albumin and glutelin showed a declining trend, whereas globulin and prolamin percentages increased with maturation. Because these fractions have characteristic amino acid compositions (Table III), the overall amino acid balance is affected.

The separation of total protein extracts from maturing oat groats by SDS-PAGE (Fig. 8) was discussed in Section IV. The results confirmed and extended the information obtained by solubility fractionation. Whereas a wide array of protein bands was visible early in development, the later-formed proteins were primarily globulins.

The sequence of cell development in the subaleurone region of the starchy endosperm of Goodland oats, a high-protein cultivar, was followed by electron microscopy (Saigo et al, 1983). Particular attention was given to the origin and development of protein bodies, which developed from proteinaceous deposits within vacuoles of immature cells (Fig. 14). In the cytoplasm of these cells, small, densely stained deposits, probably proteinaceous, were observed within the cisternae of the RER (Fig. 15). Furthermore, the RER membranes were continuous with the tonoplast of the vacuole (Fig. 16). In more mature tissue, larger protein bodies nearly filled the vacuoles, which contained numerous small and large protein bodies (Fig. 17). From these micrographs, the following sequence of events is hypothesized. Protein, synthesized on the RER ribosomes, is inserted into the RER cisternae, where it may or may not condense. This protein migrates into vacuoles through the RER cisternae, where it condenses into insoluble bodies. These coalesce and enlarge, eventually filling the vacuole. Upon dehydration, these protein bodies become irregular in shape. Dictyosomes were infrequently observed (Saigo et al, 1983), indicating that this organelle did not play a major role in protein body formation, as has been suggested for several other species.

The formation of aleurone protein bodies within cells of the aleurone layer followed a different sequence (Peterson et al, 1985). Early in development, deposits of phytin were first seen within vacuoles. Somewhat later, protein appeared peripheral to the phytin deposit and separated from it by a membrane (Fig. 18). With maturation, the phytin deposit becomes solid and is called the globoid, while the proteinaceous material spreads around the globoid, eventually enclosing it (Fig. 19). In electron micrographs, the globoid often appears partially or completely empty. It is believed that this is an artifact of the preparation: the phytin was dissolved during fixation or is so brittle that it popped out during sectioning. The mechanism of protein deposition within the aleurone vacuoles is unknown. Rough ER was much less abundant in the aleurone cells than in the subaleurone, although free ribosomes were prominent. Connections between ER and the vacuoles were not observed, and dictyosomes were infrequent.

B. Germination

The effects of germination (sprouting or malting) on protein and amino acid composition of cereals are of interest because of the possibility of alteration of nutritional value. During the early phases of germination, hydrolytic enzymes are induced that degrade macromolecular substrates in the endosperm, enabling mobilization of their residues to the developing embryo. Sutcliffe and Baset (1973) showed that the rate of increase of protease activity peaked one to two days after imbibition, thereafter declining over the next five days. The rate of disappearance of endosperm protein closely paralleled the changes in enzyme activity (Fig. 7). These workers hypothesized that the rate of hydrolysis of substrate (protein) in oat endosperm is regulated by the rate of enzyme synthesis, under the assumption that only newly synthesized enzymes are active in the zone of hydrolysis.

Dalby and Tsai (1976) found that total protein (as percentage dry weight) of entire oat sprouts increased slightly during a five-day germination period (Fig. 20). This increase was attributed to declining dry weight resulting from respiration, rather than to actual increased synthesis of total protein. During this period the concentration of nutritionally poor prolamin declined, whereas that of the essential amino acids lysine and tryptophan increased.

The effects of germination on protein solubility fractions in the endosperm were examined in more detail by Kim et al (1979a). Using a sequential extraction procedure, they separated five protein fractions. The NaCl extraction step was performed at 5° C rather than at room temperature, which resulted in a very small "globulin" fraction. The different protein fractions declined at different rates (Fig. 21). On the basis of N content per endosperm, the residue fraction declined most rapidly over the first three days of germination, thereafter leveling off at about one-third of its original amount. The glutelin fraction was unaffected for 24 hr and then declined rapidly to negligible amounts after five days. Prolamin and globulin declined roughly in parallel, at a slower rate, but reaching negligible quantities at five days also. The albumins actually increased slightly and then declined slowly. Nonprotein N doubled within three days and then declined to its original level.

These workers extended their analyses of the proteins in germinating grain by electrophoresis on polyacrylamide gels either in $3M$ urea aluminum lactate buffer, pH 3.2, or in SDS Tris-borate buffer, pH 8.5. During the course of germination, protein bands were degraded at different rates and new bands appeared that may represent degradation products or newly synthesized proteins. This research lends credence to the idea that there is some specificity to the hydrolysis and remobilization of storage proteins. Unfortunately, little information is available regarding the induction, activity, and substrate specificity of proteases in germinating oat endosperm.

Robbins and Pomeranz (1971) malted several grains and compared the amino acid composition of the malt with that of the grain (Table X). They found a slight increase in lysine with malting, as was reported by Dalby and Tsai (1976). Threonine did not change, and most amino acids showed only slight differences. They also analyzed malt sprouts, a by-product of malting that consists primarily of rootlets. The sprouts were considerably elevated in N, lysine, and threonine, suggesting a superior feed value.

Germination increased the relative nutritional value (RNV) of oats from 85 to 91%, as determined by a *Tetrahymena* growth assay, but this increase was not statistically significant (Hamad and Fields, 1979). This protozoan has a requirement for essential amino acids similar to that of the rat, and *Tetrahymena*

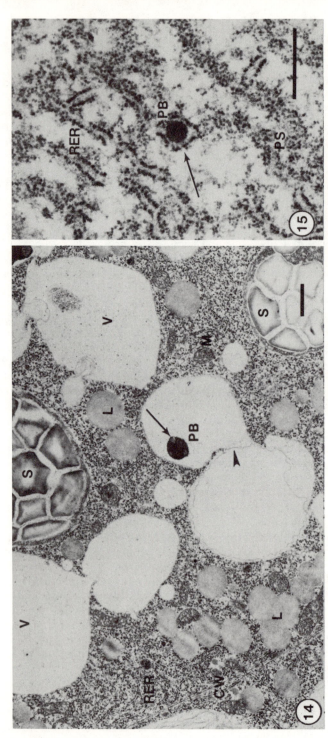

Figs. 14–15. Electron micrographs of developing subaleurone cells of Goodland oats. (14) Region of cell from 18-mg caryopsis. Large vacuoles connected by membranous profiles (arrowhead) contain small protein bodies with electron-lucent inclusions (arrow). Bar = 1 μm; CW = cell wall, L = lipid body, M = mitochondrion, PB = protein body, RER = rough endoplasmic reticulum, S = starch grain, and V = vacuole. (15) Cytoplasm of cell from 18-mg caryopsis, showing protein within cisternae of RER (arrow). Bar = 0.5 μm; PS = polyribosomes. (Reprinted, with permission, from Saigo et al, 1983)

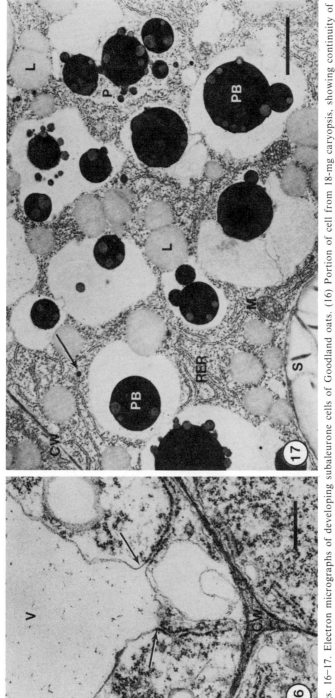

Figs. 16–17. Electron micrographs of developing subaleurone cells of Goodland oats. (16) Portion of cell from 18-mg caryopsis, showing continuity of endoplasmic reticulum with tonoplast of vacuoles (arrows). Bar = 0.5 μm; CW = cell wall and V = vacuole. (17) Cells of 25-mg caryopsis. Arrows show small protein bodies forming in rough endoplasmic reticulum (RER) cisternae. Bar = 2 μm; L = lipid body, M = mitochondrion, P = protein, PB = protein body, and S = starch grain. (Reprinted, with permission, from Saigo et al, 1983)

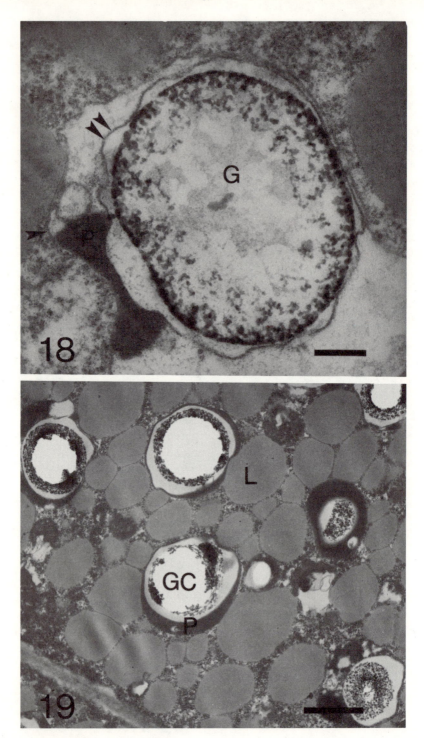

Figs. 18 and 19. Electron micrographs of developing aleurone protein bodies within the aleurone layer of Goodland oats. (18) Aleurone protein body in cell of caryopsis eight days after anthesis. The phytin globoid is enclosed within a unit membrane (double arrows). Protein is deposited in a bow-tie shape within a second unit membrane (single arrow) that completely surrounds the first. Bar = 0.2 μm; G = globoid. (19) Aleurone cell from caryopsis 14 days after anthesis, showing numerous aleurone grains and lipid bodies. Globoid cavities may have contained phytin in situ. Bar = 1.0 μm; GC = globoid cavity, L = lipid body, and P = protein. (Courtesy D. M. Peterson and R. N. Saigo)

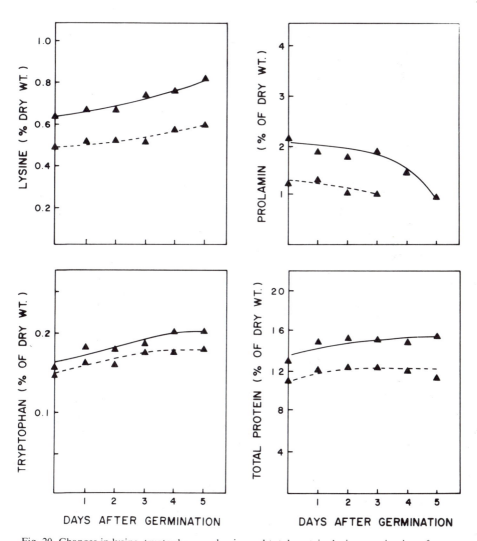

Fig. 20. Changes in lysine, tryptophan, prolamin, and total protein during germination of oats at 28°C. Solid line = cultivar Dal, broken line = cultivar Noble. (Reprinted, with permission, from Dalby and Tsai, 1976)

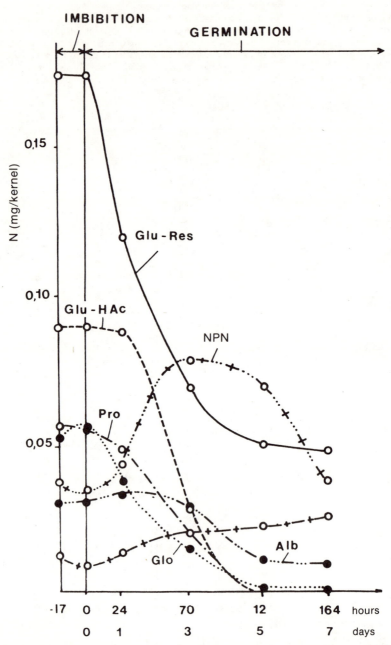

Fig. 21. Changes in protein fractions during germination of oats, cultivar Rhea, at 20°C. Alb = albumin, Glo = globulin, Glu-HAc = glutelin soluble in 0.1 N acetic acid, Glu-Res = residual glutelin, NPN = nonprotein N, and Pro = prolamin. (Reprinted, with permission, from Kim et al, 1979a)

RNV and rat protein efficiency ratio (PER) are highly correlated. The magnitude of increase in RNV with germination was less for oats than that for wheat, barley, and rice, all of which had a lower initial (ungerminated) RNV.

The above results are consistent with physiological processes known to occur during germination. The reserve proteins in the endosperm are hydrolyzed by proteases, and the amino acid residues are remobilized to the developing embryo. There they are used for the synthesis of proteins needed for growth of the shoot and root. Since these newly synthesized proteins have a different amino acid composition than the reserve proteins of the endosperm, metabolic conversion of amino acids probably occurs before their reincorporation into protein. The data suggest that germinated (sprouted) seed may be superior in nutritional value, at least in terms of amino acid balance and protein percentage. Whether the magnitudes of the changes are sufficient to have a measurable effect in feeding trials remains to be demonstrated. Certainly, the beneficial effect of sprouting oats would be less than that for grains with a more poorly balanced protein initially, such as corn.

VI. ENVIRONMENTAL EFFECTS

A. Fertility

Application of N fertilizer to soils has been shown to increase the protein concentration of oats (Portch et al, 1968; Ohm, 1976; Peterson, 1976; Welch and Yong, 1980). Portch et al (1968) found that 45 kg of N per hectare applied at

TABLE X
Nitrogen Concentration and Amino Acid Composition
of Oat Grain, Malt, and Malt Sprouts[a]

Amino Acid	Grain[b]	Malt[b]	Malt Sprouts[b]
Kjeldahl-N (%)	2.70	2.87	3.76
Lysine	3.6	3.8	6.4
Histidine	2.1	2.1	2.5
Ammonia	3.2	3.1	3.6
Arginine	6.9	6.8	5.3
Aspartic acid	9.8	10.1	18.3
Threonine	3.5	3.5	4.2
Serine	4.9	4.8	4.2
Glutamic acid	22.7	21.1	14.8
Proline	6.3	6.7	4.5
Half-cystine	0.7	0.5	0.2
Glycine	4.9	4.9	4.9
Alanine	4.7	5.0	6.2
Valine	5.4	5.6	6.1
Methionine	2.0	2.2	1.8
Isoleucine	4.1	4.2	4.2
Leucine	7.1	7.2	6.8
Tyrosine	2.8	2.7	2.4
Phenylalanine	5.5	5.5	3.7

[a] Source: Robbins and Pomeranz (1971); used by permission. Amino acid composition based on g/100 g of amino acids and ammonia recovered.
[b] Cultivar Florida 500.

seeding increased grain protein by 1.4 percentage points on well-drained soils and by 1.9 percentage points on imperfectly drained soils. Protein yield (in kilograms per hectare) was also increased significantly by N application. Ohm (1976) demonstrated that cultivars of oats responded differently to nitrogen at 110 kg/ha as a split application, 85 kg/ha at seeding, and 25 kg/ha as a foliar application of urea solution two to five days before heading. All cultivars from the fertilizer-treated plots had higher protein percentages than those from the controls, with increases ranging from 0.8 to 3.8 percentage points. The response to N was unrelated to the protein percentage of the cultivars in the control plots. Welch and Yong (1980) compared the effects of application of 125 kg/ha early (in the two- to three-leaf stage) versus late (at heading). The early application increased yield and slightly increased protein percentage. The late application increased grain protein percentage more than the early application but had no effect on yield. Double application (both early and late) increased both yield and protein percentage, thus giving the greatest protein yield.

The application of P and/or K to N-fertilized plots did not affect protein concentration of the grain, as compared with that from plots receiving N only (Portch et al, 1968). In a pot experiment, increasing levels of P had no consistent effect on grain protein percentage (Eppendorfer, 1978). Because of reduced grain yields, grain protein percentage was particularly high under conditions of K deficiency.

With increased N fertilization, Michael et al (1961) reported increased glutelin relative to the other fractions of oat protein. However, we consider their glutelin fraction to contain primarily globulin, as discussed in Section II. Völker (1975) found a slight increase in the globulin fraction relative to the others under high N fertility. Peterson (1976) reported that the increased grain protein resulting from higher N fertilization was accounted for primarily by increased globulin in AuSable and Rodney oats. In Clintland 64, however, a glutelin fraction showed the greatest response. Among samples of Nora oats ranging from 9.8 to 15.0% protein because of prior fertilizer treatments, globulin was most related to total protein, although prolamin and glutelin were also significantly correlated (Fig. 22). It is evident from these results that, under increasing grain protein percentages caused by N fertilization, the globulin fraction increases relative to other fractions.

The effects of soil fertility on amino acid composition have been considered (Steenbjerg et al, 1972; Eppendorfer, 1975, 1977, 1978; Peterson, 1976). Peterson (1976) was unable to distinguish differences for any amino acids as a consequence of N fertilization treatments, which yielded grain with protein concentrations ranging from 15.2 to 22.9%. In a pot experiment, Steenbjerg et al (1972) found an increase in glutamine-glutamic acid, histidine, and arginine with increased N per pot. Eppendorfer (1975, 1977, 1978), in a series of experiments, noted changes in amino acid levels with changes in N per pot. Results with the cultivar Selma showed that lysine, proline, glycine, methionine, and tryptophan decreased as N fertility increased, whereas glutamine-glutamic acid, asparagine-aspartic acid, and phenylalanine increased (Table XI). Particular attention was paid to lysine because of its nutritional importance (Eppendorfer, 1978). Although a negative regression of lysine on grain percentage of N was observed, the slope was not so severe as with wheat (Fig. 23). When considered as percentage of dry matter, lysine increased with increased fertility, as did all other amino acids. True

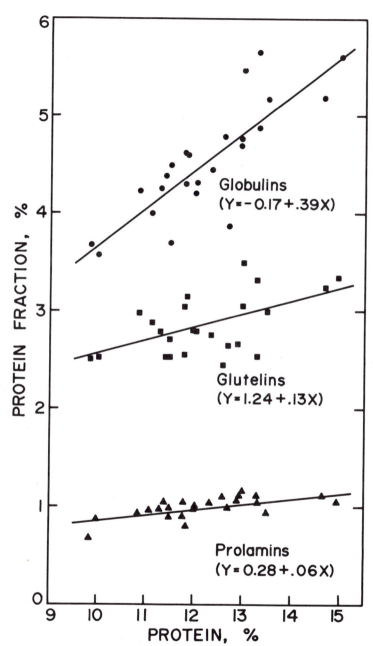

Fig. 22. Changes in protein fractions with increasing grain-N of oats, cultivar Nora, grown under various N fertilizer treatments. The correlation coefficients were 0.83***, 0.66***, and 0.55** for globulin, prolamin, and glutelin, respectively. Significance levels: ***, $P = 0.001$; **, $P = 0.01$ level. (Data from Peterson, 1976)

digestibility (TD) increased with increasing N concentration in grain, as was the case for other cereals (Eppendorfer, 1975). However, unlike other cereals, the biological value (BV) of which decreased at high N concentration, BV declined very little for oats. Consequently, net protein utilization (NPU = TD × BV / 100) increased nearly in parallel with the increase in TD (Fig. 24).

All data suggest that the practice of fertilization of oats to improve yield and protein percentage has a negligible effect on protein quality. This situation is unlike that of other cereals, in which a nutritionally poor prolamin fraction preferentially increases with fertilizer-induced higher protein levels. Furthermore, late applications of N at heading often can boost protein levels without the detrimental effects (such as excessive vegetative growth or lodging) of heavy applications at seeding.

B. Diseases

Two studies addressed the effect of crown rust (*Puccinia coronata*) on total protein percentage of oat groats from infected plants. Singleton et al (1979) investigated 12 cultivars of varying susceptibility in two years, one characterized by severe and the other by moderate infection. Control plots were maintained rust-free by spraying with a fungicide. The mean protein percentage was reduced 1.1% in 1971 (severe infection) and 0.4% in 1972 (moderate infection) (Table

TABLE XI
**Effects of Nitrogen on Grain Yield, Grain-N, and Amino Acid Composition
of Oats, Cultivar Selma[a]**

	Nitrogen per Pot (g)				
	1.0	**3.0**	**6.0**	**10.0**	**8.0 + 4.0[b]**
Grain yield, g	78	158	180	181	157
Grain N, %	1.02	1.26	1.91	2.24	2.60
Lysine ↓	4.6	4.4	4.2	4.1	4.0
Histidine	2.2	2.2	2.2	2.2	2.2
Ammonia	2.5	2.4	2.5	2.5	2.6
Arginine	7.0	7.0	7.1	7.3	7.0
Aspartic acid ↑	8.0	7.9	8.0	8.2	8.4
Threonine	3.5	3.5	3.4	3.6	3.4
Serine	4.7	4.6	4.5	4.6	4.5
Glutamic acid ↑	18.8	19.3	20.1	19.9	20.4
Proline ↓	6.6	6.5	6.4	6.1	6.2
Half-cystine	3.3	3.2	3.0	3.3	3.0
Glycine ↓	5.5	5.5	5.1	5.0	4.9
Alanine ↓	5.0	5.1	4.7	4.7	4.7
Valine	5.3	5.4	5.4	5.3	5.3
Methionine ↓	1.9	2.0	2.0	1.8	1.7
Isoleucine ↑	3.7	3.7	3.8	3.8	3.9
Leucine	7.3	7.3	7.3	7.4	7.4
Tyrosine	4.0	3.8	3.8	3.9	3.8
Phenylalanine ↑	4.8	4.9	5.0	5.1	5.2
Tryptophan ↓	1.5	1.5	1.4	1.3	1.3

[a] Recalculated from Eppendorfer (1977). Amino acid composition based on g/100 g of amino acids and ammonia recovered. Arrows indicate increasing or decreasing trend with increased nitrogen per pot.
[b] 4.0 g of N applied at flowering.

XII). The reduction was statistically significant for only some of the cultivars. Protein yield was more drastically affected than protein concentration, owing to the more severe effects of rust on yield. In the other study (Simons et al, 1979), 40 lines of oats were compared with fungicide-treated controls in a year of severe infection. Thirty-four lines did not differ significantly from their uninfected controls in protein percentage, five had increased levels of protein, and only one showed a significant decrease. The effects of rust on protein percentage were not correlated with inherent yield or inherent protein percentage, suggesting that selection in a plant breeding program for high yield or high protein can be done in conjunction with selection for crown rust resistance.

Barley yellow dwarf virus (BYDV) infection significantly increased percentage crude protein of Maldwyn oats, whereas powdery mildew (*Erysiphe graminis*) infection did not have a significant effect (Potter, 1980). However, grain yields were reduced 60 and 28% by BYDV and mildew, respectively. Potter (1980)

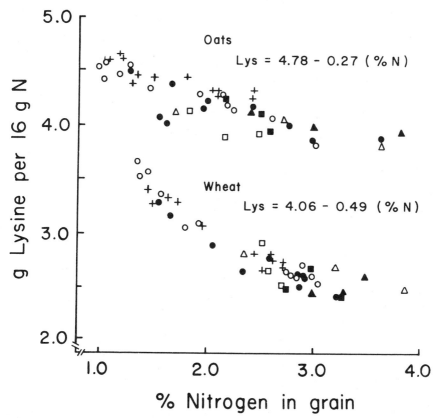

Fig. 23. Relationship between lysine and total N concentration of oats, cultivar Selma, and wheat, cultivar Kleiber, grown under varying treatments. Varying N, sandy soil for 1973 (o) and 1974 (●); varying P, sandy soil for 1973 (□) and 1974 (■); varying K, sandy soil for 1973 (△) and 1974 (▲); varying N, loam for 1973 (+). The correlation coefficients were significant at the $P = 0.001$ level. (Reprinted, with permission, from Eppendorfer, 1978)

speculates that the effect of BYDV on crude protein may result from a lower proportion of small, low-protein grains developing on the infected plants.

VII. FOOD USES AND FUNCTIONALITY

Cereal and legume proteins are rarely added to prepared foods in the purified, fractionated state. Oat proteins are no exception, with their major utilization coming in the forms of flour and flaked or rolled oats. Shukla (1975) has extensively reviewed the processing and uses of oats, including oat proteins.

Although apparently not yet in industrial use, wet- and dry-milling techniques have been devised for the preparation of oat protein concentrates and isolates from groats and flour. Cluskey et al (1973) obtained oat protein concentrates ranging from 59 to 89% protein from a pH 9 dilute alkali extract of dry-milled fractions. Such concentrates had good nitrogen solubility near pH 2.5 and above pH 8, good amino acid composition, a bland flavor, and other functional

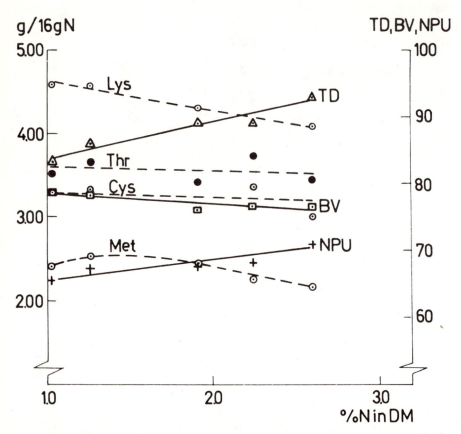

Fig. 24. Relationship between true digestibility (TD), biological value (BV), net protein utilization (NPU), lysine, threonine, cystine, and methionine and the concentration of N in oats, cultivar Selma. Plants were grown in pots with varying amounts of N. (Reprinted, with permission, from Eppendorfer, 1975)

properties (hydration capacity, emulsifying activity, and emulsion stability) equivalent to or better than those of soy protein isolate under acidic (pH 2.2) or basic (pH 8.6) conditions (Wu et al, 1973). An air-classification method (Wu and Stringfellow, 1973) for obtaining oat protein concentrates yielded a fraction that was only 2–5% of the total weight but consisted of 83–88% protein. Kjaergaard and Bruzelius (1979) have reported an aqueous extraction and precipitation procedure for obtaining products with 48–90% protein from dehulled, defatted oats. The use of density separation of protein from oat flour using mixtures of Freon and hexane (Cluskey et al, 1978) has resulted in oat protein concentrates as high as 70% protein. These concentrates had solubility (Fig. 25) and functional properties similar to those of oat protein concentrate from a wet-milling process. A 12,000 × g centrifugation of a water slurry of oat flour and bran has been reported to yield a layer of protein concentrate containing 50–57% protein (Youngs, 1974). Wu et al (1977) have prepared an oat protein isolate (>90% protein) by isoelectric precipitation (at pH 5) of a dilute alkali extract.

Although these oat protein concentrates and isolates suffer from low solubilities and emulsifying properties at pH 4–7, they have potential as nutritive, bland, and functional protein additives comparable to those derived from soybeans (Wu et al, 1973). Cluskey et al (1976) were able to formulate neutral and acidic beverages fortified with oat protein concentrates (up to 4%) having acceptable organoleptic qualities. The blending of a 50% oat protein concentrate at 5 and 10% levels with wheat flour yielded a bread that had a slightly decreased loaf volume but that gave very satisfactory taste panel scores (D'Appolonia and Youngs, 1978). Studies on the nutritive properties of similar oat protein preparations (Kjaergaard and Bruzelius, 1979) gave BV, TD, and NPU values of 72.7, 94.1, and 68.5, respectively, and a chemical score of 54 (compared with 43 for soy protein). In view of the current and potential market for functional and high-quality protein, further research on the utilization of oat proteins in formulated foods would seem to be in order.

Recently, a protein concentrate was prepared from high-protein oats by an alkaline extraction procedure (Ma, 1983). This concentrate was extensively characterized by chromatography, electrophoresis, and isoelectric focusing. The

TABLE XII
Effect of Crown Rust on Groat Protein Percentage and Total Protein Yield[a]

Cultivar	1971			1972		
	CRS[b] (%)	Groat Protein %[c]	Protein Yield[c]	CRS[b] (%)	Groat Protein %[c]	Protein Yield[c]
Tippecanoe	85	−10.4**[d]	−51.1**	57	−7.0**	−10.6
Coachman	85	−8.1**	−54.5**	50	−9.3**	−24.5**
Kota	40	0.0	−42.8**	18	0.0	−17.2**
Kelsey	30	−4.7	−46.3**	4	3.9	−14.8**
Ajax	27	−4.9*	−46.4**	5	−2.5	−11.5*
Minhafer	15	−11.5**	−37.0**	5	−3.6	0.6
Multiline M70	8	−4.1	−21.1**	6	−2.8	−2.1
Portage	5	−5.8*	−28.0**	1	0.6	1.1

[a] Source: Singleton et al (1979); used by permission.
[b] Crown rust severity (percentage of flag leaf infected).
[c] Percentage change due to rust.
[d] * and ** = Significant at the 5 and 1% levels, respectively.

TABLE XIII
Some Functional Properties of Oat Protein Concentrates and Selected Plant Proteins[a,]

Concentrate or Protein	EAI[b] (m²/g)	ESI[c] (min)	WHC[d] (ml/g)	FBC[e] (ml/g)	Foaming Capacity[f] (vol increase, %)	Foaming Stability[f] (vol remaining, %) 30 min	60 min
Hinoat protein concentrate	45.2	8.0	2.70	2.62	25	55	40
Defatted Hinoat protein concentrate	37.0	7.2	1.95	2.80	120	70	52
Sentinel protein concentrate	53.4	6.5	2.45	2.25	25	50	40
Defatted Sentinel protein concentrate	40.4	6.2	2.00	2.50	85	70	53
Wheat gluten	49.4	17.6	0.98	0.85	100	40	30
Soy protein isolate (Supro 610)	35.0	25.2	2.50	1.83	135	74	70

[a] Data from Ma (1983).
[b] Emulsifying activity index (pH 7.5) (Pearce and Kinsella, 1978).
[c] Emulsion stability index (Pearce and Kinsella, 1978).
[d] Water hydration capacity (Quinn and Paton, 1979).
[e] Fat-binding capacity (Lin et al, 1974).
[f] Data from Yasumatsu et al (1972).

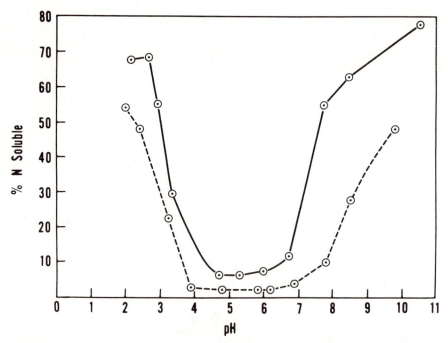

Fig. 25. Effect of pH on N solubility of oat protein concentrate from density separation in Freon-hexane mixtures (dotted line) as compared with oat protein concentrate from a wet-milling process (solid line). (Reprinted, with permission, from Cluskey et al, 1978)

patterns obtained for the concentrate by these techniques were complex, suggesting a mixture of the Osborne fractions with the possibility of interactions among polypeptides. It appeared that globulin was a major component, based on correspondence of major electrophoretic and isoelectric-focused bands. The pH solubility curve was similar to previous reports, with minimal solubility between pH 4 and 6. Prior hexane extraction of lipid slightly decreased the protein solubility.

Table XIII lists some functional properties of the oat protein concentrates in comparison with those of wheat gluten and soy protein isolate. The oat proteins had emulsifying activity comparable to that of wheat gluten and slightly higher than that of soy protein isolate. Emulsion stability was considerably lower for the oats. Water hydration capacity of the oat concentrates was similar to that of soy protein isolate and exceeded that of wheat gluten, and their fat-binding capacity was significantly higher that those of both gluten and soy protein isolates. Foaming properties of undefatted oats were poor but were improved by prior hexane extraction. The good water- and fat-binding properties of the oat protein concentrates suggested potential use as a meat binder or extender.

No studies have been conducted on the functional properties of the purified oat globulin protein, but it would seem a reasonable undertaking in light of its abundance relative to other oat proteins. An understanding of the basic properties of the globulin, some of which we have presented, might serve as a basis for a more scientific approach to the utilization of oats as a source of high-quality food protein. For example, our observations that the globulin β polypeptides are water-soluble when separated from the α polypeptides and contain significantly more lysine and methionine (Brinegar and Peterson, 1982a) suggest potential applications if a feasible large-scale procedure for their separation could be developed.

Consumer awareness has grown significantly concerning the nutritional benefits of whole grains and minimally processed grain products, and this may be the best way for many people to obtain oat protein. Aside from rolled oats, the average consumer is limited in the choice of unrefined oat products. In addition to the need for such products in the marketplace, there should be more of an effort by the food industry to stress the high protein content and quality of oats (as well as other nutritional advantages) and the versatility of oat products as regular components in the diet. The export of oat products for human consumption to developing countries lacking food high in protein and energy is another area that needs to be explored in detail. Currently only 5–10% of oats are used for human consumption. In any case, the promotion of oats for human food uses (as a protein additive or in a less refined form) appears to be necessary to expand their market and prevent the continuing decline in acreage and production that is threatening the role of oats as a major world crop.

VIII. SUMMARY

Oats contain a high percentage of protein (typically 15–20% groat protein) of superior amino acid balance as compared with that of other cereals. Lysine, threonine, and methionine are considered nutritionally limiting amino acids. Solubility fractionation shows that the primary storage protein is globulin. Although 50–55% of the total protein is salt-extractable globulin, its actual concentration may be as high as 75% because of the failure of the Osborne method to extract purified protein fractions quantitatively and preferentially.

Each solubility fraction has a distinctive amino acid composition. The prolamin (avenin) fraction is low in lysine and high in glutamine-glutamic acid, as are other cereal prolamins. The superior amino acid balance of oats results from the fact that the more balanced globulin fraction predominates. The globulin and avenin fractions have been purified and their chromatographic and electrophoretic characteristics determined. The native globulin is a hexamer of disulfide-linked α and β polypeptide—i.e., $(\alpha\beta)_6$. Both the acidic α polypeptides (pI 5.9–7.2, 32,500–37,500 mol wt) and the basic β polypeptides (pI 8.7–9.2, 22,000–24,000 mol wt) are heterogeneous, as shown by SDS-PAGE and IEF. The avenin fraction, as purified on Sephadex G-100, yielded eight bands by starch gel electrophoresis. Molecular weights ranged from 20,000 to 34,000.

Oat globulin polypeptides become major protein components about eight days after anthesis, and, along with avenin, are the predominant proteins synthesized thereafter. An isolated mRNA fraction synthesized polypeptides in vitro of about 60,000–64,000 mol wt that immunoprecipitated with antibodies raised against globulin. These polypeptides were about 2,000 mol wt larger than in vivo synthesized species, suggesting removal of a signal sequence. These globulin precursors are later cleaved into the disulfide-linked α and β polypeptides. The similarities between oat globulin and the 11S storage globulins of certain legumes in synthetic mechanism, structure, and chemical properties reflect a high degree of conservation among these evolutionarily diverse species.

Bran and starchy endosperm were shown to contribute about equally to oat groat protein, although protein concentration was higher in the bran. In the aleurone layer (a component of the bran), protein is concentrated with phytin in aleurone protein bodies. Within the starchy endosperm, protein bodies are more numerous in the subaleurone but persist throughout.

During groat development, protein concentration generally increases, and the proportion of the solubility fractions is altered markedly. Globulin is primarily synthesized beyond nine days after anthesis. The changing proportion of protein fractions is reflected in a changing amino acid balance of total crude protein. Germination improves the amino acid balance somewhat by replacing endosperm storage proteins with embryonic proteins of a different composition. The effect is less marked in oats than in high-prolamin seeds and has not been demonstrated to be nutritionally significant.

Protein bodies of the starchy endosperm were shown to develop within vacuoles, presumably from protein synthesized on polyribosomes of the RER, and transported to the vacuole within the RER cisternae. In the aleurone protein bodies, phytin was deposited first in the vacuoles and protein was then observed around the periphery of the phytin globoid. It is not known how the protein was transported into these vacuoles.

The increase in protein caused by increased N fertility was accounted for primarily by increased globulin. There was some alteration of amino acid composition with increased N, but the changes were not great. An increased TD offset the slight decline in BV, and NPU increased with higher grain protein concentration.

Several wet- and dry-milling processes that yield protein concentrates from oats have been demonstrated in the laboratory. These concentrates are characterized by low solubility between pH 4 and 6. Functional properties were generally similar to those of soy protein isolate or wheat gluten, except that emulsion stability was poor. We are unaware of any commercial utilization of these concentrates at the present time.

LITERATURE CITED

ADELI, K., ALLAN-WOJTAS, P., and ALTOSAAR, I. 1984. Intracellular transport and posttranslational cleavage of oat globulin precursors. Plant Physiol. 76:16-20.

ADELI, K., and ALTOSAAR, I. 1983. Role of endoplasmic reticulum in biosynthesis of oat globulin. Plant Physiol. 73:949-955.

ADELI, K., and ALTOSAAR, I. 1984. Characterization of oat vicilin-like polypeptides. Plant Physiol. 75:225-227.

ANDERSON, D. M., BELL, J. M., and CHRISTISON, G. I. 1978. Evaluation of a high-protein cultivar of oats (Hinoats) as a feed for swine. Can. J. Anim. Sci. 58:87-96.

BECHTEL, D. B., and POMERANZ, Y. 1981. Ultrastructure and cytochemistry of mature oat (*Avena sativa* L.) endosperm. The aleurone layer and starchy endosperm. Cereal Chem. 58:61-69.

BHATIA, C. R., and RABSON, R. 1976. Bioenergetic considerations in cereal breeding for protein improvement. Science 194:1418-1421.

BIETZ, J. A. 1982. Cereal prolamin evolution and homology revealed by sequence analysis. Biochem. Genet. 20:1039-1053.

BLOBEL, G., and DOBBERSTEIN, B. 1975. Transfer of proteins across membranes. I. Presence of proteolytically processed and unprocessed nascent immunoglobulin light chains on membrane bound ribosomes of murine myeloma. J. Cell Biol. 67:835-851.

BRESSANI, R., WILSON, D. C., CHUNG, M., BÉHAR, M., and SCRIMSHAW, N. S. 1963. Supplementation of cereal proteins with amino acids. VI. Effect of amino acid supplementation of rolled oats as measured by nitrogen retention in young children. J. Nutr. 81:399-404.

BRINEGAR, A. C. 1983. Isolation and characterization of oat seed globulin and synthesis of oat seed storage proteins. Ph.D. thesis, University of Wisconsin, Madison.

BRINEGAR, A. C., and PETERSON, D. M. 1982a. Separation and characterization of oat globulin polypeptides. Arch. Biochem. Biophys. 219:71-79.

BRINEGAR, A. C., and PETERSON, D. M. 1982b. Synthesis of oat globulin precursors. Analogy to legume 11S storage protein synthesis. Plant Physiol. 70:1767-1769.

BROHULT, S., and SANDEGREN, E. 1954. Seed proteins. Pages 487-512 in: The Proteins,

Vol. 2, Part A. H. Neurath and K. Bailey, eds. Academic Press, New York.

BROWN, C. M., WEBER, E. J., and WILSON, C. M. 1970. Lipid and amino acid composition of developing oats (*Avena sativa* L. cultivar 'Brave'). Crop Sci. 10:488-491.

BURGESS, S. R., SHEWRY, P. R., MATLASHEWSKI, G. S., ALTOSAAR, I., and MIFLIN, B. J. 1983. Characteristics of oat (*Avena sativa* L.) seed globulins. J. Exp. Bot. 34:1320-1332.

BUTTROSE, M. S. 1978. Manganese and iron in globoid crystals of protein bodies from *Avena* and *Casuarina*. Aust. J. Plant Physiol. 5:631-639.

CLUSKEY, J. E., WU, Y. V., INGLETT, G. E., and WALL, J. S. 1976. Oat protein concentrates for beverage fortification. J. Food Sci. 41:799-804.

CLUSKEY, J. E., WU, Y. V., and WALL, J. S. 1978. Density separation of protein from oat flour in nonaqueous solvents. J. Food Sci. 43:783-786.

CLUSKEY, J. E., WU, Y. V., WALL, J. S., and INGLETT, G. E. 1973. Oat protein concentrates from a wet-milling process: Preparation. Cereal Chem. 50:475-481.

COLYER, T. E., and LUTHE, D. S. 1984. Quantitation of oat globulin by radioimmunoassay. Plant Physiol. 74:455-456.

DALBY, A., and TSAI, C. Y. 1976. Lysine and tryptophan increases during germination of cereal grains. Cereal Chem. 53:222-226.

D'APPOLONIA, B. L., and YOUNGS, V. L. 1978. Effect of bran and high-protein concentrate from oats on dough properties and bread quality. Cereal Chem. 55:736-743.

DONHOWE, E. T. 1981. Isolation and characterization of oat aleurone and starchy endosperm protein bodies. M.S. thesis, University of Wisconsin, Madison.

DONHOWE, E. T., and PETERSON, D. M. 1983. Isolation and characterization of oat aleurone and starchy endosperm protein bodies. Plant Physiol. 71:519-523.

DRAPER, S. R. 1973. Amino acid profiles of chemical and anatomical fractions of oat grains. J. Sci. Food Agric. 24:1241-1250.

EGGUM, B. O. 1969. Evaluation of protein quality and the development of screening techniques. Pages 125-135 in: New Approaches to Breeding for Improved Plant Protein. Int.

Atomic Energy Agency, Vienna.

EPPENDORFER, W. H. 1975. Effects of fertilizers on quality and nutritional value of grain protein. Pages 249-263 in: Fertilizer Use and Protein Production. Proc. Colloq. Int. Potash Inst., 11th. Bornholm, Denmark. IPI, Berne, Switzerland.

EPPENDORFER, W. H. 1977. Nutritive value of oat and rye grain protein as influenced by nitrogen and amino acid composition. J. Sci. Food Agric. 28:152-156.

EPPENDORFER, W. H. 1978. Effects of nitrogen, phosphorus and potassium on amino acid composition and on relationships between nitrogen and amino acids in wheat and oat grain. J. Sci. Food Agric. 29:995-1001.

EWART, J. A. D. 1967. Amino acid analyses of cereal flour proteins. J. Sci. Food Agric. 18:548-552.

EWART, J. A. D. 1968. Fractional extraction of cereal flour proteins. J. Sci. Food Agric. 19:241-245.

FAO/WHO. 1973. Energy and protein requirements. FAO Nutr. Rep. Ser. 52. Food and Agric. Org. of the U.N., Rome. Also: Tech. Rep. Ser. 522. World Health Org., Geneva. 118 pp.

FREY, K. J. 1977. Protein of oats. Z. Pflanzenzuecht. 78:185-215.

FULCHER, R. G., O'BRIEN, T. P., and SIMMONDS, D. H. 1972. Localization of arginine-rich proteins in mature seeds of some members of the Gramineae. Aust. J. Biol. Sci. 25:487-497.

FULCHER, R. G., O'BRIEN, T. P., and WONG, S. I. 1981. Microchemical detection of niacin, aromatic amine, and phytin reserves in cereal bran. Cereal Chem. 58:130-135.

HAMAD, A. M., and FIELDS, M. L. 1979. Evaluation of the protein quality and available lysine of germinated and fermented cereals. J. Food Sci. 44:456-459.

HISCHKE, H. H., Jr., POTTER, G. C., and GRAHAM, W. R., Jr. 1968. Nutritive value of oat proteins. I. Varietal differences as measured by amino acid analysis and rat growth responses. Cereal Chem. 45:374-378.

JACOBSEN, J. V., KNOX, R. B., and PYLIOTIS, N. A. 1971. The structure and composition of aleurone grains in the barley aleurone layer. Planta 101:189-209.

KIES, C., PETERSON, M. R., and FOX, H. M. 1972. Protein nutritive value of amino acid supplemented and unsupplemented pre-cooked dehydrated oatmeal. J. Food Sci. 37:306-309.

KIM, S. I., CHARBONNIER, L., and MOSSÉ, J. 1978. Heterogeneity of avenin, the oat prolamin: Fractionation, molecular weight, and amino-acid composition. Biochim. Biophys. Acta 537:22-30.

KIM, S. I., and MOSSÉ, J. 1979. Electrophoretic patterns of oat prolamines and species relationships in *Avena*. Can. J. Genet. Cytol. 21:309-318.

KIM, S. I., PERNOLLET, J. C., and MOSSÉ, J. 1979a. Évolution des protéines de l'albumen et de l'ultrastructure du caryopse d'*Avena sativa* au cours de la germination. Physiol. Veg. 17:231-245.

KIM, S. I., SAUR, L., and MOSSÉ, J. 1979b. Some features of the inheritance of avenins, the alcohol soluble proteins of oat. Theor. Appl. Genet. 54:49-54.

KITAMURA, K., and SHIBASAKI, K. 1975. Isolation and some physicochemical properties of the acidic subunits of soybean 11S globulin. Agric. Biol. Chem. 39:945-951.

KJAERGAARD, L., and BRUZELIUS, E. 1979. Protein for human consumption from oats. Abstr. 3.4.2 in: Food Process Engineering 1979. Abstr. Int. Congr. Eng. Food, 2d. European Food Symp., 8th. P. Linko and J. Larinkari, eds. Helsinki Univ. Technol., Espoo, Finland.

LAEMMLI, U. K. 1970. Cleavage of structural proteins during the assembly of the head of bacteriophage T_4. Nature (London) 227:680-685.

LANDRY, J., SALLANTIN, M., BAUDET, J., and MOSSÉ, J. 1965. Extraction des protéines des graines. VI. Extraction exhaustive et fractionnement des protéines de la farine de blé. Electrophorese en gel d'amidon des fractions séparées. Physiol. Veg. 7:283-293.

LAPORTE, J., and TRÉMOLIÈRES, J. 1962. Action inhibitrice des farines de riz, d'avoine, de mais, d'ogre, de blé, de seigle, de sarrasin, sur certains enzymes protéolytiques du pancréas. C. R. Seances Soc. Biol. Ses Fil. 156:1261-1263.

LARKINS, B. A. 1981. Seed storage proteins: Characterization and biosynthesis. Pages 449-489 in: The Biochemistry of Plants: A Comprehensive Treatise, Vol. 6. Proteins and Nucleic Acids. A. Marcus, ed. Academic Press, New York.

LAURIÈRE, M., and MOSSÉ, J. 1982. Polyacrylamide gel-urea electrophoresis of cereal prolamins at acidic pH. Anal. Biochem. 122:20-25.

LIN, M. J. Y., HUMBERT, E. S., and SOSULSKI, F. W. 1974. Certain functional properties of sunflower meal products. J. Food Sci. 39:368-370.

LORENC-KUBIS, I. 1969. Bailka ziarniaków

niektórych gatunków rodziny Graminae i ich wlasności antytrypsynowe. Acta Soc. Bot. Pol. 38:165-175.

LUTHE, D. S., and PETERSON, D. M. 1977. Cell-free synthesis of globulin by developing oat (*Avena sativa* L.) seeds. Plant Physiol. 59:836-841.

MA, C.-Y. 1983. Chemical characterization and functionality assessment of protein concentrates from oats. Cereal Chem. 60:36-42.

MARUYAMA, K., SHANDS, H. L., HARPER, A. E., and SUNDE, M. L. 1975. An evaluation of the nutritive value of new high protein oat varieties (cultivars). J. Nutr. 105:1048-1054.

MATLASHEWSKI, G. J., ADELI, K., ALTOSAAR, I., SHEWRY, P. R., and MIFLIN, B. J. 1982. In vitro synthesis of oat globulin. FEBS Lett. 145:208-212.

MICHAEL, G., BLUME, B., and FAUST, H. 1961. Die Eiweissqualität von Körnern Verschiedener Getreidearten in Abhängigkeit von Stickstoffversorgung und Entwicklungszustand. Z. Pflanzenernahr. Bodenkd. 92:106-116.

MITRA, R. K., BHATIA, C. R., and RABSON, R. 1979. Bioenergetic cost of altering the amino acid composition of cereal grains. Cereal Chem. 56:249-252.

MOORE, S. 1963. On the determination of cystine as cysteic acid. J. Biol. Chem. 238:235-237.

MORRISON, I. N., KUO, J., and O'BRIEN, T. P. 1975. Histochemistry and fine structure of developing wheat aleurone cells. Planta 123:105-116.

OHM, H. 1976. Response of 21 oat cultivars to nitrogen fertilization. Agron. J. 68:773-775.

OHM, H. W., and PETERSON, D. M. 1975. Protein composition in developing groats of an *Avena sativa* L. cultivar and an *A. sativa* ×*Avena sterilis* L. selection. Crop Sci. 15:855-858.

OSBORNE, T. B. 1910. Die Pflanzenproteine. Ergeb. Physiol. 10:47-215.

PEARCE, K. N., and KINSELLA, J. E. 1978. Emulsifying properties of proteins: Evaluation of a turbidimetric technique. J. Agric. Food Chem. 26:716-723.

PENNING DE VRIES, F. W. T., BRUNSTING, A. H. M., and VAN LAAR, H. H. 1974. Products, requirements, and efficiency of biosynthesis: Quantitative approach. J. Theor. Biol. 45:339-377.

PERNOLLET, J. C. 1978. Protein bodies of seeds: Ultrastructure, biochemistry, biosynthesis and degradation. Phytochemistry 17:1473-1480.

PERNOLLET, J. C., KIM, S. I., and MOSSÉ, J. 1982. Characterization of storage proteins extracted from *Avena sativa* seed protein bodies. J. Agric. Food Chem. 30:32-36.

PERNOLLET, J. C., and MOSSÉ, J. 1980. Charactérisation des corpuscules protéiques de l'albumen des caryopses de Céréales par microanalyse élémentaire associée à la microscopie électronique à balayage. C. R. Hebd. Seances Acad. Sci. Ser. D 290:267-270.

PETERSON, D. M. 1976. Protein concentration, concentration of protein fractions, and amino acid balance in oats. Crop Sci. 16:663-666.

PETERSON, D. M. 1978. Subunit structure and composition of oat seed globulin. Plant Physiol. 62:506-509.

PETERSON, D. M., and SMITH, D. 1976. Changes in nitrogen and carbohydrate fractions in developing oat groats. Crop Sci. 16:67-71.

PETERSON, D. M., SAIGO, R. H., and HOLY, J. 1985. Development of oat aleurone cells and their protein bodies. Cereal Chem. 62:366-371.

POMERANZ, Y., SHANDS, H. L., ROBBINS, G. S., and GILBERTSON, J. T. 1976. Protein content and amino acid composition in groats and hulls of developing oats (*Avena sativa*). J. Food Sci. 41:54-56.

POMERANZ, Y., YOUNGS, V. L., and ROBBINS, G. S. 1973. Protein content and amino acid composition of oat species and tissues. Cereal Chem. 50:702-707.

PORTCH, S., MacKENZIE, A. F., and STEPPLER, H. H. 1968. Effect of fertilizers, soil drainage class and year upon protein yield and content of oats. Agron. J. 60:672-674.

POTTER, L. R. 1980. The effects of barley yellow dwarf virus and powdery mildew in oats and barley with single and dual infections. Ann. Appl. Biol. 94:11-17.

QUINN, J. R., and PATON, D. 1979. A practical measurement of water hydration capacity of protein materials. Cereal Chem. 56:38-40.

ROBBINS, G. S., and POMERANZ, Y. 1971. Amino acid composition of malted cereals and malt sprouts. Proc. Am. Soc. Brew. Chem., pp. 15-21.

ROBBINS, G. S., POMERANZ, Y., and BRIGGLE, L. W. 1971. Amino acid composition of oat groats. J. Agric. Food Chem. 19:536-539.

ROBERT, L. S., ADELI, K., and ALTOSAAR, I. 1985a. Homology among 3S and 7S globulins from cereals and pea. Plant Physiol. 78:812-816.

ROBERT, L. S., CUDJOE, A., NOZZOLILLO, C., and ALTOSAAR, I. 1983a. Total solubilization of storage protein in oat (*Avena sativa* L. cv. Hinoat): Evidence that glutelins are a minor component. J. Can. Inst. Food Sci. Technol. 16:196-200.

ROBERT, L. S., MATLASHEWSKI, G. J., ADELI, K., NOZZOLILLO, C., and ALTOSAAR, I. 1983b. Electrophoretic and developmental characterization of oat (*Avena sativa* L.) globulins in cultivars of different protein content. Cereal Chem. 60:231-234.

ROBERT, L. S., NOZZOLILLO, C., and ALTOSAAR, I. 1983c. Molecular weight and charge heterogeneity of prolamins (avenins) from nine oat (*Avena sativa* L.) cultivars of different protein content and from developing seeds. Cereal Chem. 60:438-442.

ROBERT, L. S., NOZZOLILLO, C., and ALTOSAAR, I. 1985b. Homology between legumin-like polypeptides from cereals and pea. Biochem. J. 226:847-852.

ROBERT, L. S., NOZZOLILLO, C., and ALTOSAAR, I. 1985c. Homology between rice glutelin and oat 12S globulin. Biochim. Biophys. Acta 829:19-26.

RODEN, L. T., MIFLIN, B. J., and FREEDMAN, R. B. 1982. Protein disulfide-isomerase is located in the endoplasmic reticulum of developing wheat endosperm. FEBS Lett. 138:121-124.

ROSSI, H. A., and LUTHE, D. S. 1983. Isolation and characterization of oat globulin messenger RNA. Plant Physiol. 72:578-582.

SAIGO, R. H., PETERSON, D. M., and HOLY, J. 1983. Development of protein bodies in oat starchy endosperm. Can. J. Bot. 61:1206-1215.

SCHEELE, G., and JACOBY, R. 1982. Conformational changes associated with proteolytic processing of presecretory proteins allow glutathione-catalyzed formation of native disulfide bonds. J. Biol. Chem. 257:12277-12315.

SHUKLA, T. P. 1975. Chemistry of oats: Protein foods and other industrial products. Crit. Rev. Food Sci. Nutr. 6:383-431.

SIBBALD, I. R. 1979. Bioavailable amino acids and true metabolizable energy of cereal grains. Poult. Sci. 58:934-939.

SIMONS, M. D., YOUNGS, V. L., BOOTH, G. D., and FORSBERG, R. A. 1979. Effect of crown rust on protein and groat percentages of oat grain. Crop Sci. 19:703-706.

SINGLETON, L. L., STUTHMAN, D. D., and MOORE, M. B. 1979. Effect of crown rust on oat groat protein. Phytopathology 69:776-778.

STEENBJERG, F., LARSEN, I., JENSEN, I., and BILLE, S. 1972. The effect of nitrogen and simazine on the dry-matter yield and amino acid content of oats and on the absorption and utilization of various plant nutrients. Plant Soil 36:475-496.

SUTCLIFFE, J. F., and BASET, Q. A. 1973. Control of hydrolysis of reserve materials in the endosperm of germinating oat (*Avena sativa* L.) grains. Plant Sci. Lett. 1:15-20.

TKACHUK, R. 1969. Nitrogen-to-protein conversion factors for cereals and oilseed meals. Cereal Chem. 46:419-423.

TKACHUK, R., and IRVINE, G. N. 1969. Amino acid compositions of cereals and oilseed meals. Cereal Chem. 46:206-218.

VÖLKER, T. 1975. Untersuchungen über den Einfluss der Stickstoffdürngung auf die Zusammensetzung der Weizen- und Hafer-proteine. Arch. Acker Pflanzenbau Bodenkd. 19:267-276.

WAHLSTROM, R. C., and LIBAL, G. W. 1975. Varying levels of high protein oats in diets for growing finishing swine. J. Anim. Sci. 41:809-812.

WAHLSTROM, R. C., and LIBAL, G. W. 1979. Effect of high-protein oats in diets for young weaned pigs. J. Anim. Sci. 48:1374-1378.

WALBURG, G., and LARKINS, B. 1983. Oat seed globulin. Subunit characterization and demonstration of its synthesis as a precursor. Plant Physiol. 72:161-165.

WALDSCHMIDT-LEITZ, E., and ZWISLER, O. 1963. Über die Proteine des Hafers. Z. Physiol. Chem. 332:216-224.

WEAVER, C. M., CHEN, P. H., and RYNEARSON, S. L. 1981. Effect of milling on trace element and protein content of oats and barley. Cereal Chem. 58:120-124.

WEBER, E., and NEUMANN, D. 1980. Protein bodies, storage organelles in plant seeds. Biochem. Physiol. Pflanz. 175:279-306.

WEBER, K., and OSBORNE, M. 1969. The reliability of molecular weight determinations by dodecyl sulfate-polyacrylamide gel electro-phoresis. J. Biol. Chem. 244:4406-4412.

WELCH, R. W., and YONG, Y. Y. 1980. The effects of variety and nitrogen fertiliser on protein production in oats. J. Sci. Food Agric. 31:541-548.

WELCH, R. W., YONG, Y. Y., and HAYWARD, M. V. 1980. The distribution of protein and non-structural carbohydrate in five oat varieties during plant growth and grain development. J. Exp. Bot. 31:1131-1137.

WIESER, H., SEILMEIER, W., and BELITZ, H. D. 1980. Vergleichende Untersuchungen über partielle Aminosäuresequenzen von Prolaminen und Glutelinen Verschiedener Getreidearten. I. Proteinfraktionierung nach Osborne. Z. Lebensm. Unters. Forsch. 170:17-26.

WIESER, H., SEILMEIER, W., and BELITZ,

H. D. 1981. Vergleichende Untersuchungen über partielle Aminosäuresequenzen von Prolaminen und Glutelinen Verschiedener Getreidearten. III. Amidegehalt der Proteinfraktionen. Z. Lebensm. Unters. Forsch. 173:90-94.

WU, Y. V., CLUSKEY, J. E., WALL, J. S., and INGLETT, G. E. 1973. Oat protein concentrates from a wet-milling process: Composition and properties. Cereal Chem. 50:481-488.

WU, Y. V., SEXSON, K. R., CAVINS, J. F., and INGLETT, G. E. 1972. Oats and their dry-milled fractions: Protein isolation and properties of four varieties. J. Agric. Food Chem. 20:757-761.

WU, Y. V., SEXSON, K. R., CLUSKEY, J. E., and INGLETT, G. E. 1977. Protein isolate from high protein oats: Preparation, composition, and properties. J. Food Sci. 42:1383-1386.

WU, Y. V., and STRINGFELLOW, A. C. 1973. Protein concentrates from oat flours by air classification of normal and high protein varieties. Cereal Chem. 50:489-496.

YASUMATSU, K., SAWADA, K., MORITAKA, S., MISAKI, M., TODA, J., WADA, T., and ISHII, K. 1972. Whipping and emulsifying properties of soybean products. Agric. Biol. Chem. 36:719-727.

YOUNGS, V. L. 1972. Protein distribution in the oat kernel. Cereal Chem. 49:407-411.

YOUNGS, V. L. 1974. Extraction of a high-protein layer from oat groat bran and flour. J. Food Sci. 39:1045-1046.

YOUNGS, V. L., and SENTURIA, J. 1976. Relationship in protein concentration between whole oats and oat groats. Crop Sci. 16:87-88.

ZARKADAS, C. G., HULAN, H. W., and PROUDFOOT, F. G. 1982. A comparison of the amino acid composition of two commercial oat groats. Cereal Chem. 59:323-327.

CHAPTER 8

OAT LIPIDS AND LIPID-RELATED ENZYMES

VERNON L. YOUNGS
Spring and Durum Wheat Quality Laboratory
U.S. Department of Agriculture
Agricultural Research Service
North Dakota State University
Fargo, North Dakota

I. INTRODUCTION

Among the cereals, oats are recognized for their protein quality and quantity, which have contributed to their importance as a livestock feed and as the leading hot cereal breakfast food. Oats are also notable for their lipid content: oat groats have the highest lipid concentration among the cereal grains. Free lipid percentages of several cereal grains (Morrison, 1978) are, for oat groats, 5.0–9.0; wheat, 2.1–3.8; rice, 0.8–3.1; millet, 4.0–5.5; maize, 3.9–5.8; barley, 3.3–4.6; rye, 2.0–3.5; and sorghum, 2.1–5.3.

In addition to being energy sources, oat lipids and other cereal lipids are important nutritionally because they are highly unsaturated and contain considerable amounts of linoleic acid. This essential fatty acid for humans is utilized in the synthesis of prostaglandins, chemicals that are found in all tissues of the body and that regulate smooth muscles like the heart (Woitas, 1981).

Oat lipids have created rancidity problems for processors because lipid-related enzyme systems are activated when oats are rolled or milled.

This review includes discussion of oat lipid components; the effects of maturity, environment, storage, and heritability on lipid composition; related enzyme systems; and the feasibility of oats as an oil crop.

II. LIPID EXTRACTION

Cereal lipid extraction requires attention to solvent-related variables such as the solvent used, extraction time, ratio of solvent volume to sample weight, temperature, and extraction equipment. Grain-related variables include moisture, particle size, and age of the ground grain. The last is important in oats because the enzyme lipase, which releases fatty acids, becomes very active once a sample is ground. All variables need to be identified in reporting data.

The most important factor in lipid extraction is the solvent. Predominantly free or unbound lipids are extracted from cereal grain by nonpolar solvents such as ethyl ether, hexane, or a petroleum ether. Since nonpolar solvents also vary in their polarity, the extractions differ; but generally the hydrocarbons, triglycerides, free fatty acids, sterols, and diglycerides are the major lipid classes removed. For total extraction, the lipids may be hydrolyzed first with hydrochloric acid and then extracted (American Association of Cereal Chemists, 1983). Without hydrolysis, bound lipids such as the glycolipids and phospholipids require polar solvents for extraction. Two commonly used solvent systems are water-saturated 1-butanol, which has been reviewed by Mecham (1971), and chloroform-methanol-water mixtures (Bligh and Dyer, 1959). Components other than lipids might be extracted with polar systems; the lipids are often reextracted with a less polar solvent, such as chloroform, or by washing procedures, such as the Folch procedure (Folch et al, 1957). If polar solvents are used to extract lipids from the original grain, total lipids are obtained; if the dried grain residue from a previous nonpolar solvent extraction is reextracted with polar solvents, bound lipids are obtained. The sum of the two fractions is total lipids. This practice is often used to establish amounts of free and bound lipids in a grain.

Morrison (1978) has reviewed methods of extracting lipids from cereal grains and has included pertinent comments about the various systems. Sahasrabudhe (1979) extracted lipids from Hinoat oat groats with seven different solvent systems. Total lipids extracted (as percentage of dry groat weight) ranged from 5.57% with *n*-hexane to 8.84% with ethanol.

III. LIPID CONTENT IN OATS

A considerable range exists in lipid content of oat groats. Table I reports lipid ranges from 11 studies conducted in seven countries. Two of these studies were

TABLE I
Lipid Concentration in Oats[a]

Source	Country	No. of Samples	Lipids (%)[b]		
			Free	Bound	Total
Hutchinson et al (1951)	England	1,000	4.5–11	...	...
Hutchinson and Martin (1955)	England, Scotland	54	5.0–10.1	about 2.5	...
Brown et al (1966)	United States	169	3.8–9.8	...	...
Johansson (1976)	Sweden	35	3.7–6.9	1.8–3.3	5.9–8.4
Frey and Hammond (1975)	United States	445	2.0–11.0	...	...
Brown and Craddock (1972)	United States	4,000	3.1–11.6	...	...
de la Roche et al (1977)	Canada	9	3.2–8.9 (oats)	...	4.5–10.3 (oats)
Martens et al (1979)	Africa and Middle East				
	(*A. abyssinica*)	36	5.4–7.3	...	...
	(*A. sterilis*)	215	5.7–8.9	...	...
Sahasrabudhe (1979)	Canada	12	4.2–11.8	...	...
Lasztity et al (1980)	Hungary	12	4.3–6.9	...	...

[a] All groats of *Avena sativa*, unless otherwise noted.
[b] Free lipids are extractable with nonpolar solvents such as hexane; bound-lipid extraction requires a more polar solvent or solvent system, such as water-saturated butanol.

quite extensive: Hutchinson et al (1951) analyzed 1,000 samples, and Brown and Craddock (1972) analyzed 4,000 samples. In the later study, 90% of the entries ranged from 5 to 9% lipid; 5 were over 11%, and 25 contained less than 4% lipid. The overall range reported from the 11 studies was 2.0–11.8%.

Youngs et al (1977) utilized a hand dissection technique to determine lipid distribution within oat kernels. Figure 1 is a drawing of an oat kernel showing the

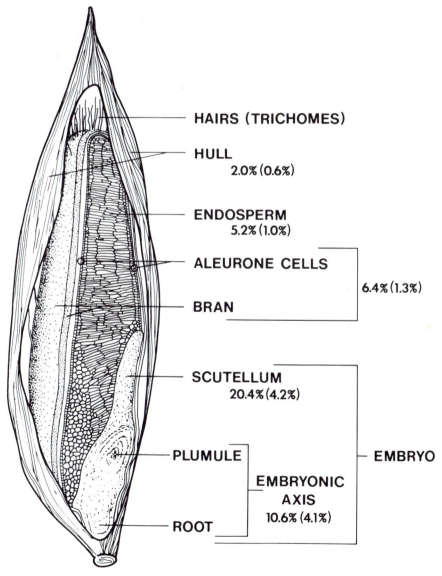

Fig. 1. Oat kernel, showing lipid concentrations found in the major parts of the cultivar Froker. First values are free lipids; values in parentheses are bound lipids.

distribution of lipids. In Table II, data are shown for Dal, a high-lipid, and Froker, a medium-lipid cultivar. Hulls contain the lowest percentage of total lipid. Although lipid percentages are highest in the embryonic axis and scutellum, the relative amounts of the total lipids that appear in these fractions are small because they represent only a small portion of the total groat weight (Table II, columns 3 and 6). Conversely, lipid percentages are small in the endosperm, but over 50% of the lipids in the groat appear in this fraction. In typical separations of oat bran from the endosperm, the layer of aleurone cells is removed with the bran. This layer is very rich in lipid (Fig. 2) and probably is a major source of bran lipids.

IV. LIPID CLASSES AND COMPONENTS

After lipids have been extracted from cereal grains, they are often separated by polarity into lipid classes. Column chromatography and sometimes thin-layer chromatography (TLC) are used; examples of the separation of oat lipids by classes are shown in Table III. Lipid compounds included within each class vary, and published results are sometimes hard to compare. Also, the solvent used to extract the lipids has a considerable effect on the relative amounts of each lipid class extracted. For instance, in Table III, results from Sahasrabudhe (1979) show 5.6% phospholipids in lipids extracted with hexane and 24.6% in lipids extracted with chloroform-methanol. Regardless of the extraction technique, the triglyceride fraction is the major class in oat lipids. Youngs et al (1977) reported the lowest value (41%). However, if percentage of triglycerides was based only on lipid components identified, the value would be 57%.

The isolation and identification of specific compounds within oat lipid classes have received less attention that those from corn and wheat lipids, yet several have been isolated.

TABLE II
Lipids in Groats and Oat Fractions from Two Oat Cultivars

	Dal Lipids			Froker Lipids		
Source	Free[a] (%)	Bound[a] (%)	Percentage of Total Lipids in Groat[b]	Free[a] (%)	Bound[a] (%)	Percentage of Total Lipids in Groat[b]
Groats	8.0	1.6	⋯	5.5	1.4	⋯
Hull	2.3	0.6	⋯	2.0	0.6	⋯
Embryonic axis	12.6	3.3	1.9	10.6	4.1	2.3
Scutellum	20.6	2.8	5.8	20.4	4.2	6.9
Bran	9.5	1.2	40.0	6.4	1.3	36.3
Starchy endosperm	6.8	1.0	52.1	5.2	1.0	54.4

[a] Data from Youngs et al (1977), expressed as percentage of source (db). Groats and hulls, average of three replications for two years; groat fractions, average of two replications for two years. Free lipids were removed by an 8-hr extraction in a Goldfisch extractor, and the bound lipids were extracted by shaking the residue for 30 min in water-saturated butanol.

[b] Calculated from total lipids reported here (free and bound) and weight distribution of fractions in groats as reported by Youngs (1972).

A. Sterol Esters and Free Sterols

Sahasrabudhe (1979) separated sterol esters from oat lipids by column chromatography and reported over 15% in hexane-extracted lipids and 2–3% in lipids extracted with chloroform-methanol. No identification was made of individual components. de la Roche et al (1977) reported only 0.3%. Price and Parsons (1975) did not quantitate sterol esters, but when seven different cereal lipids were separated by TLC, oat lipids produced the smallest sterol ester spot. Youngs et al (1977) were not able to quantitate sterol ester spots obtained by TLC separation because of the very small amounts. Individual sterol esters have been identified in wheat lipids (Gilles and Youngs, 1964).

Youngs et al (1977) reported that the total free sterols averaged about 1% of the total lipids, whereas Sahasrabudhe (1979) reported that they were 3.3% (Table III). The individual free sterols in oat lipids were studied by Knights (1965, 1968)

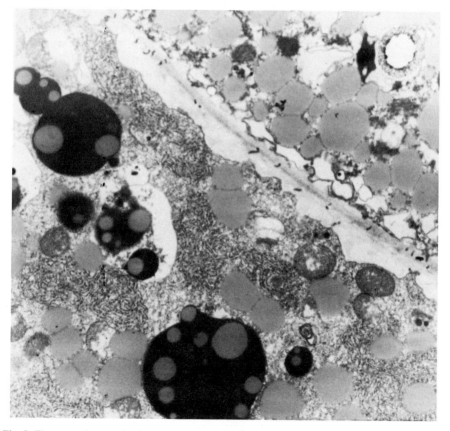

Fig. 2. Electron micrograph of part of Goodland oat aleurone cell (lower left), taken 14 days after flowering. Gray areas are lipid globules and dark circles are protein. (>10,000×) (Courtesy R. H. Saigo, University of Wisconsin, Eau Claire, and D. M. Peterson, USDA Oat Quality Laboratory, University of Wisconsin, Madison)

and Knights and Laurie (1967). They used gas-liquid chromatography (GLC) and reported values for eight sterols. Sitosterol was most abundant (39%), followed by Δ^5 avenasterol and Δ^7 avenasterol. Youngs et al (1977) combined TLC and GLC and analyzed the free sterols in one cultivar. Of the six sterols measured, the three most abundant were sitosterol, 69%; campesterol, 10%; and stigmasterol, 8%.

B. Triglycerides and Partial Glycerides

Major differences that would occur among oat triglycerides would be in the location of specific fatty acids on the glycerol molecule. This has been studied in corn (Brockerhoff and Yurkowski, 1966; de la Roche et al, 1971; Weber et al, 1971; Weber and Alexander, 1975) and other vegetable oils (Brockerhoff and Yurkowski, 1966; Evans et al, 1966). Corn triglycerides show high unsaturation in fatty acids linked to the beta (second) triglyceride carbon. In unpublished experiments performed at the Oat Quality Laboratory, Madison, Wisconsin, in 1977, fatty acids linked to the beta carbon of triglycerides extracted from the cultivars Dal and Froker showed low palmitic acid and high oleic and linoleic acid concentrations. This would agree with published data on corn. One could speculate that high unsaturation might also occur on carbon two of 1,2-diglycerides and that 1,3-diglycerides and 1-monoglycerides would be quite saturated. However, data are not available to support this. Youngs et al (1977) reported that about 2% of the total oat lipids were 1,3-diglycerides and 1% were 1,2-diglycerides.

TABLE III
Percentage Distribution of Oat Lipids in Different Classes

| Lipid Class | Sahasrabudhe (1979)[a] | | | de la Roche et al (1977)[b] Average of Nine Strains: CHCl₃-MeOH and Others | Price and Parsons (1975)[c] Chief: CHCl₃-MeOH-H₂O | Youngs et al (1977)[d] Avg. of Dal and Froker: Ethyl Ether, H₂O-BuOH |
	Hinoat: Hexane	Hinoat: CHCl₃-MeOH (2:1)	Avg. of Six Cultivars: CHCl₃-MeOH (2:1)			
Sterol esters	15.8	2.1	3.2	0.3	···	···
Triglycerides	64.8	53.7	51.1	77.8	79.9 (1–5)[e]	41
Free fatty acids	12.2 (3,4,5)[e]	10.0 (3,4,5)[e]	7.3	4.8	···	5
Partial glycerides	···	···	7.0	···	···	3
Free sterols	···	···	3.3	···	···	1
Glycolipids	1.5	9.6	8.3	8.6 (4,6)[e]	17.0	12
Phospholipids	5.6	24.6	19.7	9.0	10.1	10
Components not identified	···	···	···	···	···	28

[a] Oat groats; separation by silicic-acid column chromatography.
[b] Oats with hulls; separation by thin-layer chromatography.
[c] Oats with hulls; separation by silicic-acid column chromatography.
[d] Oat groats; separation by thin-layer chromatography, with components grouped to fit lipid classes. All spots were measured and those unidentified were reported as components not identified.
[e] Reported as combined classes, which are identified by numbers in parentheses.

C. Free and Esterified Fatty Acids

Considerable variation exists in the level of free fatty acids reported in the literature because of differences in analytical techniques. Free fatty acids as a lipid class may increase after the kernel is ground unless appropriate precautions are taken. This increase is due primarily to the enzyme lipase, which is very active in ground oats or groats; lipase is discussed later. Hence, the reported amount of free fatty acids may be an indication of the length of time between grinding and extracting a sample rather than of the actual free fatty acid content.

The individual fatty acids associated with the total oat lipids have received more attention than other lipid components. For analysis, the lipids are generally transesterified to form methyl esters of the fatty acids, and these are separated and quantitated by GLC. Data from five studies, each of which evaluated several oat cultivars, are reported in Table IV. Palmitic, oleic, and linoleic acids comprise about 95% of the fatty acids, with myristic, stearic, and linolenic acids occurring in lesser amounts. In each study the range in content is large for each major fatty acid. Palmitic acid is the major saturated fatty acid, comprising 18–20% of the fatty acids. The ranges of oleic and linoleic acids (both unsaturated) overlap, but a comparison of the averages reported in Table IV shows that percentages of linoleic acid are slightly higher than those of oleic.

Interesting relationships occur when individual fatty acid content is correlated with total groat lipids (Table V). As total lipids increase, palmitic and linoleic acids tend to decrease and oleic acid tends to increase. This may be explained partially by the data in Table VI, which show the fatty acid distribution in the three major lipid classes. The triglycerides have more oleic and less palmitic acid than the glycolipids or phospholipids. In oats, these polar lipids occur in much smaller amounts than do triglycerides; an increase in total lipid content would thus mean a relatively greater increase in amounts of triglycerides, hence greater amounts of oleic acid.

D. Glycolipids and Phospholipids

The bound lipids contain primarily glycolipids and phospholipids in amounts that have been reported as shown in Table III; the individual components found in these two lipid classes are reviewed in Table VII. In addition, Price and Parsons (1975) reported the presence of sulfatides. These components are not unique to oats and have been reported previously in other cereal lipids.

Limited quantitative data are shown in Table VII. The major glycolipid is digalactosyl diglyceride, and the major phospholipid is phosphatidylcholine. In addition, Tevekelev (1970) compared the phospholipids in wheat, barley, oats, rye, and millet and reported that phosphatidylcholine constituted 41–65% of the total phospholipids in all cereals except oat flakes, which contained 24–28%. The percentage of phosphatidylinositol in oats was about double that found in other cereals.

E. Distribution of Lipid Components Within the Oat Groat

Youngs et al (1977) used TLC to isolate and measure 12 components from four groat fractions dissected from two cultivars (Table VIII). Percentage of

TABLE IV
Range and Mean of Relative Fatty Acid Composition of Oat Cultivars from Five Studies

Fatty Acid	Youngs and Püskülcü (1976)[a] (%)	Frey and Hammond (1975)[b] (%)	de la Roche et al (1977)[c] (mole %)	Sahasrabudhe (1979)[d] (%)	Lasztity et al (1980)[e] (%)
Myristic	0.4–0.8 (0.6)	...	...	0.5–4.9 (2.1)	...
Palmitic	16.2–21.8 (18.9)	14–23	17.2–23.6 (19.6)	15.6–25.8 (20.5)	12.9–25.8 (17.5)
Stearic	1.2–2.0 (1.6)	1–4	0.8–1.8 (1.4)	1.6–3.9 (2.6)	1.1–11.0 (2.5)
Oleic	28.4–40.3 (36.4)	29–53	26.5–47.5 (38.7)	25.8–41.3 (34.9)	25.4–36.9 (31.4)
Linoleic	36.6–45.8 (40.5)	24–48	33.2–46.2 (38.9)	31.3–41.0 (37.0)	38.1–46.7 (42.0)
Linolenic	1.5–2.5 (1.9)	1–5	0.9–2.4 (1.5)	1.7–3.7 (2.6)	1.5–19.4 (3.9)

[a] Data from groats of 15 oat cultivars, three replications of each, grown at three locations in the United States during two crop years.
[b] Data from groats of 64 cultivars and collections.
[c] Calculated from data from nine cultivars of oats grown in Canada.
[d] Calculated from data for 12 cultivars of oat (groats) grown in Canada. Not shown here is "other" fatty acids (range 0–2.4%), which included C_{20} and C_{22} fatty acids.
[e] Data from oats of 12 cultivars grown in Hungary.

triglyceride was least in the bran and greatest in the two embryo fractions. Amounts of some glycolipids and phospholipids were too small to be measured in the embryo fractions.

The distribution of fatty acids found in free and bound lipids extracted from oat groats and groat fractions is given in Table IX. Less palmitic acid was found in the free lipids than in the bound lipids in all fractions, and the converse was true for oleic acid. This agrees with findings of Sahasrabudhe (1979) (Table VI). No other major differences are evident.

V. EFFECT OF MATURITY

Oat lipids are synthesized shortly after flowering (Fig. 3) (Beringer, 1966; Brown et al, 1970), when the kernel contains more than 50% moisture (Lindberg et al, 1964). Early in lipid synthesis, the percentage of linolenic acid is high, but thereafter it decreases and linoleic acid increases with kernel maturation (Lindberg et al, 1964; Beringer, 1966; Brown et al, 1970). Youngs (1978) investigated changes in 12 lipid components found in whole oat kernels during maturation. Although kernel moisture was still about 50%, triglycerides (the major component) increased and free fatty acids decreased (Fig. 4). These findings agree with those of Brown et al (1970). Also, levels of free sterols and phosphatidylcholine decreased with maturity, but the other lipid components listed in Table VIII changed very little.

TABLE V
Linear Correlation of Fatty Acid Concentration in Oats
with Total Lipid or Total Fatty Acid

Fatty Acid	De la Roche et al (1977)[a]	Youngs and Püskülcü (1976)[a]	Frey and Hammond (1975)[a]	Welch (1975)[b]
Palmitic	−0.89**[c]	−0.76**	−0.08	−0.74***
Stearic	...	...	0.03	0.33*
Oleic	0.98**	0.91**	0.37**	0.76***
Linoleic	−0.98**	−0.85**	−0.38**	−0.55***
Linolenic	...	...	−0.36**	−0.39**

[a] Correlation between percentage of fatty acid and total lipid.
[b] Correlation between percentage of fatty acid and total fatty acid content.
[c] *, **, *** = Significant at $P = 0.05, 0.01$, and 0.001, respectively.

TABLE VI
Fatty Acid Distribution in Major Oat Lipid Classes[a]

	Fatty Acids (mole %)							
	14:0	16:0	18:0	18:1	18:2	18:3	20:x[b]	22:x[c]
Triglycerides	1.5	14.8	2.2	43.3	35.0	2.0	0.5	0.1
Glycolipids	4.3	22.1	4.4	25.1	36.2	4.0	3.0	1.2
Phospholipids	2.2	28.1	4.2	21.3	38.1	2.8	2.0	0.3

[a] Source: Sahasrabudhe (1979); used by permission. Values are averages for six cultivars.
[b] Includes 20:1 to 20:5.
[c] Includes 22:0, 22:1, and 24:0.

TABLE VII
Glycolipids and Phospholipids Reported in Oats

Lipid Component[a]	Aylward and Showler (1962)[b]	Tevekelev (1970)	Sahasrabudhe (1979)[c]	Price and Parsons (1975)	Youngs et al (1977)[c]
Glycolipids					
MGMG			X (6.2%)		X (4%)
DGMG			X (4.8%)		
DGDG			X (41.5%)	X	X (7%)
MGDG			X (18.5%)	X	
Acyl-MGDG			X		
MG-ceramide			X		
SG			X (7.0%)		X (1%)
Others			X (22.0%)		
Phospholipids					
Acyl-LPE			X		
LPE			X		X (1%)
Acyl-PE			X		
PE	X	X	X (14.8%)		X (2%)
LPC			X (20.4%)[d]		X (2%)
PC	X	X	X (29.9%)	X[e]	X (5%)
PS			X (3.2%)	X[e]	
PA		X	X (3.9%)[e]	X[e]	
PI		X	X[e]		
PG			X (9.5%)		
Others			(19.0%)		

[a] MGMG = monogalactosylmonoglyceride; DGDG = digalactosyldiglyceride, etc.; SG = sterol glucoside; L = lyso; PE, PC, PS, PI, and PG = phosphatidylethanolamine, -choline, -serine, -inositol, and -glycerol, respectively; PA = phosphatidic acid.
[b] Aylward and Showler reported that PE and PC comprised 10% of the total lipids.
[c] Measured values in parentheses. Sahasrabudhe: percentage of measured glycolipids or of measured phospholipids; Youngs et al: percentage of total lipid.
[d] This value was reported for combined lysophospholipids LPC, LPS, and LPE, although LPS was not identified by thin-layer chromatography.
[e] Compounds listed separately but reported within the same thin-layer chromatography spot.

TABLE VIII
Distribution of Lipid Components in Oat Groat Fractions[a]

Component	Percentage of Total Lipids in			
	Bran	Endosperm	Scutellum	Embryonic Axis
Triglycerides	39	41	50	58
1,3-Diglycerides	2	2	1	2
1,2-Diglycerides	1	1	2	2
Free fatty acids	3	3	2	2
Sterols	1	1	1	1
Sterol glucosides	1	1	1	1
Monogalactosylmonoglycerides	4	5	···	···
Digalactosyldiglycerides	9	8	···	···
Phosphatidylethanolamine	3	2	1	1
Lysophosphatidylethanolamine	2	2	···	···
Phosphatidylcholine	4	4	3	3
Lysophosphatidylcholine	3	3	···	···
Components not measured (obtained by difference)	28	27	39	30

[a] Source: Youngs et al (1977); used by permission.

VI. EFFECT OF ENVIRONMENT

Growing temperature can have an effect on lipids in the mature seed. Beringer (1971) studied oats harvested from plants grown in a greenhouse at day temperatures of 12 and 28°C. The oats with hulls and oat fractions were extracted with a chloroform-methanol-water mixture (8:4:3) and analyzed (Table X). Grain from plants grown at day temperatures of 12°C showed higher lipid percentages and greater unsaturation in lipids in the embryo, bran, and endosperm.

The different growing temperatures also changed the relative amounts of free and bound lipids produced. Under cool temperatures, the lipids were 86.4% free and 10.7% bound; under warm temperatures, they were 79.6% free and 16.2% bound. In each case, the free lipids were more unsaturated than the bound lipids. Again, the effect of temperature is evident because cool temperatures caused greater unsaturation in both free and bound lipids (Table X). Since unsaturation in lipids promotes greater membrane fluidity, which is essential to development of the plant and seed, greater unsaturation would be expected in cooler environments.

Beringer (1966) stated that nitrogen fertilizer decreases lipid concentrations slightly but uniformly; no change was noted in the fatty acid pattern. This could be a relative change, because concentrations of lipids were expressed as a percentage of dry matter. Since protein percentage should increase as a result of the added nitrogen, lipid percentages might also decrease slightly.

VII. LIPID-PROTEIN RELATIONSHIPS

Although oats contain considerable amounts of both lipids and protein, relationships between the two have received little attention. Rohrlich and Niederauer (1968) isolated a proteolipid from wheat, rye, and oats that behaved like a uniform molecule during paper electrophoresis and TLC. The lipid portion

TABLE IX
Percentage of Fatty Acids in Free and Bound Lipids Extracted from Oat Groats and Fractions[a]

Fraction	Fatty Acid					
	Myristic	Palmitic	Stearic	Oleic	Linoleic	Linolenic
Ether extract (free lipids)						
Groats	0.4	18.8	2.2	39.4	37.9	1.3
Bran	0.4	18.1	1.9	38.4	39.6	1.6
Endosperm	0.6	18.9	2.3	37.4	39.4	1.4
Scutellum	0.6	21.1	1.2	34.5	39.7	2.8
Embryonic axis	0.9	21.6	1.9	28.8	42.5	4.1
WSB[b] extract (bound lipids)						
Groat	0.9	25.7	2.0	28.8	41.0	1.4
Bran	1.6	25.7	1.6	27.2	42.2	1.4
Endopserm	0.6	27.3	2.4	28.4	39.7	1.3

[a] Data from Youngs et al (1977). Average of two cultivars.
[b] Water-saturated 1-butanol.

was reported to contain three to four components similar to lecithin (phosphatidylcholine).

When protein was extracted from oat endosperm and bran by an aqueous extraction procedure, the high-protein fraction removed from bran contained

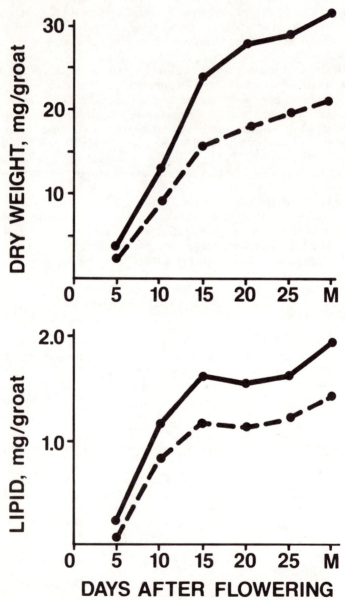

Fig. 3. Changes in groat dry weight and lipids (milligrams per groat) as the groat matures. Solid line = primary kernels, broken line = secondary kernels. (Selected from data published by Brown et al, 1970)

over 20% total lipid and the protein extracted from oat flour contained over 22% total lipid (Youngs, 1974). After this mixture was freeze-dried, a large portion of the lipids was still free—i.e., soluble in a nonpolar solvent. In contrast, in wheat flour subjected to a similar treatment, most of the triglycerides become bound (Youngs et al, 1970).

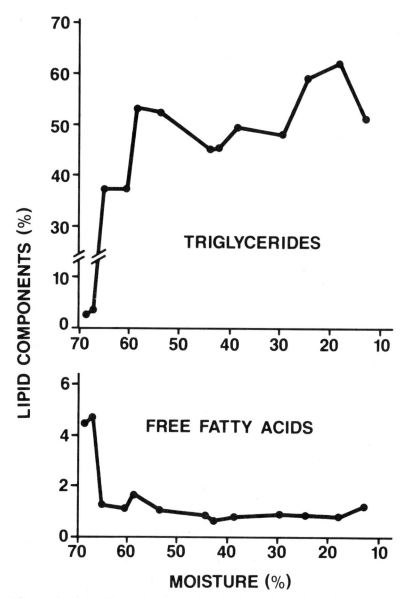

Fig. 4. Changes in triglycerides and free fatty acids related to moisture change in the oat kernel as it matures. Samples were harvested, frozen, then dried by freeze-drying; this may account for the very low values for free fatty acids.

Wu et al (1973) and Cluskey et al (1973) extracted high-protein fractions by mixing ground oat groats with $0.025N$ sodium hydroxide, passing the slurry through a bolting cloth, and separating the protein solution from the starch by centrifugation. The protein solution was adjusted to pH 6 and freeze-dried. Protein concentrates from two cultivars contained 17.7 and 10.1% lipids.

For several years, oat breeders and cereal chemists have made an effort to increase the protein levels in oats. The overall effect of this effort on lipid concentration in oats was not known, but the occurrence of the high-protein oat cultivar Dal, which was also high in lipids, caused some speculation that lipids would increase also. Although an increase in lipids would put extra energy in animal feed, it would be less desirable in human food because of the current interest in low-fat foods. Earlier, Brown et al (1966) reported a highly significant negative relationship ($r = -0.312**$) between protein and lipids in 129 spring oat strains, and a similar relationship existed between protein and lipids in 40 winter oat strains ($r = -0.477*$). Youngs and Forsberg (1979) utilized the data from the USDA-coordinated Uniform Early and Midseason Oat Performance Nurseries and correlated protein and lipids in oat genotypes by location across genotypes, by genotype across locations, and by genotype and location across years. Of 277 correlations calculated, only 36 were significant at the 5 or 1% level. Gullord (1980) also reported no correlation between oil and protein of 11 lines selected for high yield and protein content from crosses between *Avena sativa* and *A. sterilis* and four other established cultivars. Overall, there appears to be no trend that increasing protein levels also increase lipid concentration.

VIII. LIPIDS IN OAT STARCH

Efforts to extract high-protein fractions from oats have generally also produced fairly pure starch fractions. Lipid concentration of oat starch is low. Wu et al (1973) reported 0.8% lipid in starch extracted from the cultivar Wyndmere and 0.1% lipid in Garland starch. Extraction was described as by AACC official methods, but the specific method was not identified. Youngs

TABLE X
Influence of Temperature on Percentage of Oat Lipids[a]

Day Temp. (°C)	Fraction	Lipid[b]	Fatty Acids					
			14:0	16:0	18:0	18:1	18:2	18:3
12	Embryo	20.1	0.1	17.4	0.5	30.3	45.2	6.6
	Bran	7.5	0.3	17.2	1.2	36.4	41.9	3.1
	Endosperm	8.8	0.2	16.4	0.9	37.1	42.4	2.9
28	Embryo	12.6	0.3	19.0	0.8	28.9	45.5	5.6
	Bran	6.9	0.8	19.9	2.0	35.8	38.8	2.8
	Endosperm	7.3	0.5	17.9	2.1	37.8	39.8	3.0
12	Bound lipids	10.7[b]		20.2	1.6	18.8	56.4	2.8
	Free lipids	86.4		13.9	1.1	41.0	41.2	2.6
28	Bound lipids	16.2		24.5	1.7	22.8	47.2	3.0
	Free lipids	79.6		19.2	2.3	39.5	35.6	2.4

[a] Selected data from Beringer (1971).
[b] Lipids in groat fractions are expressed as percentage of fraction (db). The free and bound lipids do not equal 100% because they do not include free fatty acids or monoglycerides and diglycerides.

(1974) reported 0.8% free lipids (extracted by ethyl ether) and 0.8% bound lipids (extracted by water-saturated butanol) in Froker starch and 1.0% free and 0.3% bound lipids in Goodland starch. Acker and Becker (1971) studied lipids associated with several starches, including oat starch, and reported that 51.6% of the lipid factor was lysophosphatidylcholine, 5.1% lysophosphatidylethanolamine, 5.1% lysophosphatidylinositol, and 7.7% free fatty acids. They suggested that starch lipid resists oxidation, apparently because of the inclusion of the lipids within the amylose helix.

Morrison (1981) differentiated starch lipids as "starch surface lipids" and "internal starch lipids," the latter being true starch lipids. He stated that these were exclusively monoacyl lipids. Extraction studies showed that both water and heat are necessary to swell the starch granule sufficiently to permit the alcohol (methanol, propanol, or butanol) to extract the lipids (Morrison, 1981). Oat starch was not discussed, and it is probable that the true nature of oat starch lipids has not yet been determined.

IX. HERITABILITY

Since the variation in lipid concentration among oat cultivars is greater than would be expected from the environmental effects, high heritability is suggested (Baker and McKenzie, 1972; Brown and Aryeetey, 1973; Brown et al, 1974; Frey and Hammond, 1975; Frey et al, 1975). Frey and Hammond (1975) reported a tendency for high oil percentage to be partially dominant. However, Karow (1980) could not confirm that Dal, a high-lipid oat, carried dominant genes for high lipid in his studies of a Dal-Exeter (high lipid-low lipid) cross and a Dal-Sauk (high lipid-medium lipid) cross. He concluded that additive genes controlled oil percentage in the progeny of Dal-Exeter crosses and that heterotic gene effects controlled oil in the Dal-Sauk cross.

Individual fatty acid concentrations also exhibit wide ranges among cultivars (Table IV), and significant relationships exist between total fatty acid and total lipid concentrations and individual fatty acids (Table V). This suggests that fatty acid concentrations may also be heritable. Karow (1980) reported that additive genes controlled the palmitic, stearic, and linoleic acid concentrations of the progeny of the two crosses he studied and that oleic and linoleic acid concentration were influenced by partially dominant genes in both crosses. Broad-sense heritability estimates for the major fatty acids were similar in both studies—66.0% for oleic acid, 66.6% for palmitic, and 79.8% for linoleic. These values were lower than results reported by Youngs and Püskülcü (1976), whose study included 15 oat cultivars grown in three locations during two years. The results showed broad-sense heritabilities of 91.4, 98.6, and 95.7% for palmitic, oleic, and linoleic acids, respectively.

X. ANTIOXIDANT PROPERTIES

Oats have unique antioxidant properties, and some of the responsible agents are found in the lipids. Peters and Musher (1937) first proposed the use of oat flour as an antioxidant in foods and food packaging. Oat flour as an antioxidant was marketed as Avenex and Aveno. A hexane extract of oat flour, called Avenol, was also used (Caldwell and Pomeranz, 1973). These products were later replaced by butylated hydroxyanisole (BHA) and butylated hydroxytoluene (BHT).

Many compounds probably contribute to oat antioxidant properties. Several have been identified as glyceryl esters of hydroxycinnamic, ferulic, and caffeic acids (Daniels et al, 1963; Daniels and Martin, 1964, 1968), and these are discussed in some detail elsewhere in this monograph. Since the structures of these naturally occurring compounds are similar to those of BHA and BHT, one could hypothesize that they might contribute to the antioxidant properties of oats. Kalbasi-Ashtari and Hammond (1977) found that ferulic and caffeic acid esters could not be extracted with hexane but that they could be removed with a chloroform-methanol mixture. They reported that Dal oats contained ferulic acid at 22 μg/g and caffeic acid at 34 μg/g.

The tocopherols (vitamin E) are fat soluble and have strong antioxidant properties. α-Tocopherol is a major antioxidant component in crude oat lipids (105 μg/g), and β- and γ-tocopherols have also been detected in minor amounts (Kalbasi-Ashtari and Hammond, 1977); only α-tocopherol remained when the lipid was refined. Other reports of tocopherols in oats include that by Lasztity et al (1980), 2.98 mg/100 g of whole oats; Barr et al (1956), 1.20 mg/100 g of groats; and Herting and Drury (1969), 1.54 mg/100 g of groats. Lasztity et al (1980) also cited the relative distribution of 10 components making up the tocopherol fraction.

XI. EFFECT OF STORAGE

Lipids can undergo dramatic changes in oats. In whole oats or undamaged oat groats stored at normal temperatures and low moisture levels (less than 10%), lipids show little change. Hulless oats (*Avena nuda*) store as well as hulled cultivars unless the kernels are bruised (Welch, 1977). A temperature increase to 30°C does not significantly affect lipids in unground oats (Thomke, 1970), but free fatty acids increase at moisture levels above 10% (Sosedov et al, 1971; Frey and Hammond, 1975). In ground oats, free fatty acids increase rapidly during storage, and the increase is enhanced at 30°C (Thomke, 1970).

XII. RELATED ENZYMES

A. Lipase

Lipase (EC 3.1.1.3) and lipoxygenase (EC 1.13.11.12) are the major lipid-related enzymes; of the two, lipase has received the most attention. It is a hydrolytic enzyme that produces free fatty acids from triglycerides and partial glycerides. The oat kernel is distinctive because it has considerable lipase activity in the ungerminated seed. Sahasrabudhe (1982) reported that this activity is reduced 45–50% over two years. Frey and Hammond (1975) found a 20-fold variation in the lipase activity of 352 oat strains.

Lipase has been described as being located on the surface of the caryopsis (Hutchinson et al, 1951; Martin and Peers, 1953). Urquhart et al (1983) found that the percentage distribution of lipase in oats at 12% moisture was bran, 69.2%; first shorts, 5.1%; second shorts, 24.7%; and germ, 1%. Hence, lipase is probably associated with the aleurone layer.

Lipase activity in oats often is measured by comparing the amount of free fatty acids present in the original oat sample with the amount present after a specific treatment such as grinding. Lipids from the oat samples can be extracted and free fatty acids titrated as described in methods 02-01A and 02-02A of the American Association of Cereal Chemists (1983). The lipids can also be spotted on thin-layer silicic acid plates, developed, and visualized by acid charing, and the density of the free fatty acid spots can be compared. Sahasrabudhe (1982) has measured lipase activity in single kernels of oats. He soaked the kernel for 1 min in $1N$ hydrochloric acid, then ground and extracted it in a chloroform-heptane-methanol mixture (49:49:2, v/v). An aliquot of the extract was treated with a reagent of cupric nitrate and triethanolamine, and absorbance was read at 400 nm.

Matlashewski et al (1982) have cited several other methods for measuring lipase activity. In basic studies of the enzyme, they used a unique lipase activity assay system. The petroleum-ether defatted oat flour, which still retained the lipase activity, was allowed to react with glycerol $(1\text{-}^{14}C)$ trioleate for specific time periods. The reaction mixture was extracted with chloroform and spotted on thin-layer plates. The free fatty acid and triglyceride (trioleate) spots were removed and radioactivity was compared.

Lipase activity in oat flour is optimum at pH 7.4–7.5 and a temperature of 35–39°C (Martin and Peers, 1953; Matlashewski et al, 1982). In studies using triolein as a substrate, Michaelis-Menton kinetics were demonstrated with increasing concentration of the substrate, and the apparent Michaelis constant (K_m) was 3.5mM. When oat flour was extracted with petroleum ether, hexane, or acetone, the lipase activity was unaffected. However, extraction with a chloroform-methanol mixture (2:1) completely inactivated the enzyme (Matlashewski et al, 1982).

B. Lipoxygenase and Lipoperoxidase

Popov and Cheleev (1959) concluded that during storage the degree of saturation increased in free fatty acids because of increased lipoxygenase activity on free linoleic acid. That conclusion was based upon their finding that lipoxygenase oxidized hydrolyzed oat oil more extensively than nonhydrolyzed oil. When they added sodium linoleate to oat oil, oxidation decreased. Hence, the free acid form was more susceptible to oxidation. Conversely, the addition of sodium linoleate to corn oil increased oxidation. The peroxide numbers and oxidation of unsaturated compounds also increase as humidity increases (Sosedov et al, 1971).

Heimann and co-workers (1973) incubated oat enzymes and linoleic acid to study the products of linoleic acid oxidation. Their results showed that lipoxygenase catalyzes the formation of 9-hydroperoxyoctadecadienoic acid. Isomerase-controlled enzymatic and nonenzymatic processes, and secondary decomposition of the hydroperoxides by lipoperoxidase (EC 1.11.1.7), lead to the formation of several hydroxy acids: 9-hydroxy-*cis,trans*-octadecadienoic; 9-hydroxy-*cis*-12,13-epoxy-*trans*-10-octadecenoic; and 9,12,13-trihydroxy-*trans*-10-octadecenoic acids. They reported that lipoxygenase activity is greatest at pH 6.75.

Von Bierman et al (1980) incubated water suspensions of oat, soft wheat, and durum wheat flours for 4.5 hr at 38° C. Only the oat flour became intensely bitter. The freeze-dried samples of incubated flours were analyzed for free hydroxy acids with bitter taste. The following acids were found, per gram of oat flour: 750 µg of a mixture of 9-hydroxy-*trans,cis*-10,12-octadecadienoic acid and 13-hydroxy-*cis,trans*-9,11-octadecadienoic acid; and 62 µg of a mixture of 9,12,13-trihydroxy-*trans*-10-octadecenoic acid and 9,10,13-trihydroxy-*trans*-11-octadecenoic acid. Extraction of the lipids removed the bitterness from the oat flour, and the bitterness could be restored by adding back the hydroxy acids.

In summary, the action of lipid-related enzymes appears to take place in stages. First, lipase hydrolyzes the glycerides and releases free fatty acids. The unsaturated fatty acids are converted to hydroperoxides, which are changed to hydroxy acids by lipoxygenase, lipoperoxidase, and other enzymes and by nonenzymatic processes.

C. Enzyme Control of Unsaturation

It has been mentioned that oats grown in cool temperatures produce lipids with greater unsaturation than oats grown in warm climates. Studies of other crops have shown that this may be under the control of desaturase enzymes, which are more active in cool climates and hence lead to increased unsaturation (Woitas, 1981). Certain desaturase enzymes may be specific for certain changes—e.g., oleic to linoleic acid. The enzymes may have different temperature coefficients, and the change probably takes place in phospholipids and glycolipids rather than in the free fatty acids (Lem and Williams, 1981). One would assume that oat lipids would be affected in a similar way.

D. Industrial Control of Lipid-Related Enzyme Action

Lipase activity is not affected by freezing or freeze-drying (Urquhart et al, 1980), but the enzyme is heat labile. Industrially, groats are commonly steam-treated before rolling. This stops the series of reactions that lead to rancid by-products. However, damaged kernels that exist before heat treatment are subject to lipase activity.

Several other methods have been proposed for denaturing lipase, mostly for laboratory purposes. These were reviewed briefly by Youngs (1978).

XIII. UTILIZATION OF OAT LIPIDS AND LIPID-RELATED COMPOUNDS

Kalbasi-Ashtari and Hammond (1977) studied the refining capability and stability of oil extracted from Dal oats. Losses by degumming were 15%, and alkaline refining losses were 25–30%. The oat oil was bleached successfully with charcoal and deodorized. The stability of the oil was good, especially if citric acid was added. The flavor was also good.

Although a good oil product can be extracted from oats, the amount of oil present is probably a limiting factor. Frey and Hammond (1975) reported that for oats to be competitive with soybeans they should contain at least 21% protein and 17% oil. Obviously, these levels would vary with oat and soybean prices, but no oats have been reported with oil levels more than 12%. Hence, this criterion is impossible to meet at the present time. Also, other oil crops, such as sunflower, rape, buffalo gourd, and palm, are gaining in importance now.

If markets could be found for products of fractionated oats such as the starch and high-protein fractions, the oil would be an obvious by-product. A high-protein fraction extracted from oat groats can contain up to 20% oil. The oil is not bound, and it is easily extracted. Also, routine solvent extraction of oat flour before flaking, or solvent extraction of rolled oats, may not affect the products greatly, yet it may yield oat oil as a by-product.

XIV. SUMMARY

Oats contain more lipids than other cereal crops. Like other cereal lipids, they contain large amounts of triglycerides and are highly unsaturated. Lipid contents vary considerably among cultivars, and growing temperature can have a slight effect on the amount and level of saturation.

The heritability of oat lipids is high. It may also be possible to manipulate the relative distribution of the fatty acids genetically. Although high concentrations of linoleic acid would be good nutritionally, the stability of the oil would decrease because of the greater unsaturation.

Oat flour was used as an antioxidant several years ago. If compounds such as BHA and BHT come into disfavor with the public, utilization of the naturally occurring antioxidants in oat lipids may again become important. Concentration of these antioxidants may make them more usable in industry.

Lipase, the major enzyme in oats, causes rapid release of free fatty acids in damaged or milled oats. In industry, the action of lipase is controlled successfully by steam-heating the oats. Commercial use of this enzyme is questionable, but if it were used, oats should be considered as a source.

The quality of refined oat oil is good, but there is no great demand for it at present. This could change. Morgan and Schultz (1981), reviewing the fuels and chemicals made from novel seed oils, did not mention oat oil; but a big effort is being make to utilize vegetable oils as energy sources—e.g., as fuels for tractors and automobiles and for heating. This effort may open the door for more oat utilization. Oat oil would not be competitive in this market, but it may then become competitive as an oil for human consumption.

LITERATURE CITED

ACKER, L., and BECKER, G. 1971. Lipids of cereal starches. 2. Lipids of various types of starch and their binding to amylose. Staerke 23:419-424.

AMERICAN ASSOCIATION OF CEREAL CHEMISTS. 1983. Approved Methods of the AACC. Methods 02-01A and 02-02A, approved 1983; Method 30-10, approved April 1961, revised October 1975 and October 1981. The Association: St. Paul, MN.

AYLWARD, F., and SHOWLER, A. J. 1962. Plant lipids. IV. The glycerides and phosphatides in cereal grains. J. Sci. Food Agric. 13:429-496.

BAKER, R. J., and McKENZIE, R. I. H. 1972. Heritability of oil content in oats, *Avena sativa* L. Crop Sci. 12:201-202.

BARR, L., BROCKINGTON, S. F., BUDDE, E. F., BUNTING, W. R., CARROLL, R. W., GOULD, M. R., GROGG, B., HENSLEY, G. W., RUPP, E. G., STOUT, P. R., and WESTERN, D. E. 1956. Facts on oats. Quaker Oats Co., Chicago. 93 pp.

BERINGER, H. 1966. Einfluss von Reifegrad und N-Dungung auf Fettbildung und Fettsaurezusammensetzung in Haferkorner. Z. Pflanzenernaehr. Bodenkd. 114:117-127.

BERINGER, H. 1971. An approach to the

interpretation of the effect of temperature on fatty acid biosynthesis in developing seeds. Z. Pflanzenernaehr. Bodenkd. 128:115-122.

BLIGH, E. G., and DYER, W. J. 1959. A rapid method of total lipid extraction and purification. Can. J. Biochem. Physiol. 37:911-917.

BROCKERHOFF, H., and YURKOWSKI, M. 1966. Stereospecific analysis of several vegetable fats. J. Lipid Res. 7:62-64.

BROWN, C. M., ALEXANDER, D. E., and CARMER, S. G. 1966. Variation in oil content and its relation to other characters in oats. Crop Sci. 6:190-191.

BROWN, C. M., and ARYEETEY, A. N. 1973. Maternal control of oil content in oats (*Avena sativa* L.). Crop Sci. 13:120-121.

BROWN, C. M., ARYEETEY, A. N., and DUBEY, S. N. 1974. Inheritance and combining ability for oil content in oats (*Avena sativa* L.). Crop Sci. 14:67-69.

BROWN, C. M., and CRADDOCK, J. D. 1972. Oil content and groat weight of entries in the world oat collection. Crop Sci. 12:514-515.

BROWN, C. M., WEBER, E. J., and WILSON, C. M. 1970. Lipid and amino acid composition of developing oats (*Avena sativa* L. cultivar 'Brave'). Crop Sci. 10:488-491.

CALDWELL, E. F., and POMERANZ, Y. 1973. Industrial uses of cereals: Oats. Pages 393-411 in: Industrial Uses of Cereals. Y. Pomeranz, ed. Am. Assoc. Cereal Chem., St. Paul, MN.

CLUSKEY, J. E., WU, Y. V., WALL, J. S., and INGLETT, G. E. 1973. Oat protein concentrates from a wet-milling process: Preparation. Cereal Chem. 50:475-481.

DANIELS, D. G. H., KING, H. G. C., and MARTIN, H. F. 1963. Antioxidants of oats: Esters of phenolic acids. J. Sci. Food Agric. 14:385-390.

DANIELS, D. G. H., and MARTIN, H. F. 1964. Antioxidants in oats: Light-induced isomerization. Nature (London) 203: No. 4942, p. 299.

DANIELS, D. G. H., and MARTIN, H. F. 1968. Antioxidants in oats: Glyceryl esters of caffeic and ferulic acids. J. Sci. Food Agric. 19:710-712.

DE LA ROCHE, I. A., BURROWS, V. D., and McKENZIE, R. I. H. 1977. Variation in lipid composition among strains of oats. Crop Sci. 17:145-148.

DE LA ROCHE, I. A., WEBER, E. J., and ALEXANDER, D. E. 1971. Effects of fatty acid concentration and positional specificity on maize triglyceride structure. Lipids 6:531-536.

EVANS, C. D., McCONNELL, D. G., LIST, G. R., and SCHOLFIELD, C. R. 1966. Structure of unsaturated vegetable oil glycerides: Direct calculation from fatty acid composition. J. Am. Oil Chem. Soc. 46:421-424.

FOLCH, J., LEES, M., and SLOANE-STANLEY, G. H. S. 1957. A simple method for the isolation and purification of total lipids from animal tissue. J. Biol. Chem. 226:497-509.

FREY, K. J., and HAMMOND, E. G. 1975. Genetics, characteristics and utilization of oil in caryopses of oat species. J. Am. Oil Chem. Soc. 52:358-362.

FREY, K. J., HAMMOND, E. G., and LAWRENCE, P. K. 1975. Inheritance of oil percentage in interspecific crosses of hexaploid oats. Crop Sci. 15:94-95.

GILLES, K. A., and YOUNGS, V. L. 1964. Evaluation of durum wheat and durum products. II. Separation and identification of sitosterol esters of semolina. Cereal Chem. 41:502-513.

GULLORD, M. 1980. Oil and protein content and its relation to other characters in oats. Acta Agric. Scand. 30:216-218.

HEIMANN, W., DRESEN, P, and KLAIBER, V. 1973. Formation and decomposition of linoleic acid hydroperoxides in cereals. Quantitative determination of the reaction products. Z. Lebensm. Unters. Forsch. 153:1-5.

HERTING, D. C., and DRURY, J. E. 1969. Alpha-tocopherol content of cereal grains and processed cereals. J. Agric. Food Chem. 17:785-790.

HUTCHINSON, J. B., and MARTIN, H. F. 1955. The chemical composition of oats. I. The oil and free fatty acid content of oats and groats. J. Agric. Sci. 45:411-418.

HUTCHINSON, J. G., MARTIN, H. F., and MORAN, T. 1951. Location and destruction of lipase in oats. Nature (London) 167:758-759.

JOHANSSON, H. 1976. Lipid content and fatty acid composition in barley and oats. Sver. Utsaedesfoeren. Tidskr. 86:279-289.

KALBASI-ASHTARI, A., and HAMMOND, E. G. 1977. Oat oil: Refining and stability. J. Am. Oil Chem. Soc. 54:305-307.

KAROW, R. S. 1980. Oil composition in parental, F_1, and F_2 populations of two oat (*Avena sativa* L.) crosses. Master's thesis, Department of Agronomy, University of Wisconsin, Madison. 30 pp.

KNIGHTS, B. A. 1965. Identification of the sterols of oat seed. Phytochemistry 4:857-862.

KNIGHTS, B. A. 1968. Comparison of the grain sterol fractions of cultivated and wild oat species. Phytochemistry 7:2067-2068.

KNIGHTS, B. A., and LAURIE, W. 1967. Application of combined gas-liquid chromatography-mass spectrometry to the identification of sterols in oat seed. Phytochemistry 6:407-416.

LASZTITY, R., BERNDORFER-KRASZNER, E., and HUSZAR, M. 1980. On the presence and distribution of some bioactive agents in oat varieties. Pages 429-455 in: Cereals for Food and Beverages. Recent Progress in Cereal Chemistry. G. Inglett and L. Munck, eds. Academic Press, New York.

LEM, N. W., and WILLIAMS, J. P. 1981. Desaturation of fatty acids associated with monogalactosyl diacylglycerol: The effects of SAN 6706 and SAN 9785. Plant Physiol. 68:944-949.

LINDBERG, P., BINGEFORS, S., LANNER, N., and TANHUANPAA, E. 1964. The fatty acid composition of Swedish varieties in wheat, barley, oats and rye. Acta Agric. Scand. 14:3-11.

MARTENS, J. W., BAKER, R. J., McKENZIE, R. I. H., and RAJHATHY, T. 1979. Oil and protein content of *Avena* species collected in North Africa and the Middle East. Can. J. Plant Sci. 59:55-59.

MARTIN, H. F., and PEERS, F. G. 1953. Oat lipase. Biochem. J. 55:523-529.

MATLASHEWSKI, G. J., URQUHART, A. A., SAHASRABUDHE, M. R., and ALTOSAAR, I. 1982. Lipase activity in oat flour suspensions and soluble extracts. Cereal Chem. 59:418-422.

MECHAM, D. K. 1971. Lipids. Pages 393-451 in: Wheat: Chemistry and Technology, 2d ed. Y. Pomeranz, ed. Am. Assoc. Cereal Chem., St. Paul, MN.

MORGAN, R. P., and SCHULTZ, E. G. 1981. Fuels and chemicals from novel seed oils. Chem. Eng. News 59:69-77.

MORRISON, W. R. 1978. Cereal lipids. Pages 221-348 in: Advances in Cereal Science and Technology, Vol. 2. Y. Pomeranz, ed. Am. Assoc. Cereal Chem., St. Paul, MN.

MORRISON, W. R. 1981. Starch lipids: A reappraisal. Staerke 33:408-410.

PETERS, F. N., Jr., and MUSHER, S. 1937. Oat flour as an antioxidant. Ind. Eng. Chem. 29:146-151.

POPOV, M. P., and CHELEEV, D. A. 1959. Investigation of the lipoxidase of cereals in connection with the development of a bitter taste in groats. Biokhim. Zerna Khlebopech. 5:263-240.

PRICE, P. B., and PARSONS, J. G. 1975. Lipids of seven cereal grains. J. Am. Oil Chem. Soc. 52:490-493.

ROHRLICH, M., and NIEDERAUER, T. 1968. Fat-protein complexes in cereals. II.

Composition of the protein and lipid components. Fette Seifen Anstrichm. 70:58-62. (Chem. Abstr. 68:86195f.)

SAHASRABUDHE, M. R. 1979. Lipid composition of oats (*Avena sativa* L.). J. Am. Oil Chem. Soc. 56:80-84.

SAHASRABUDHE, M. R. 1982. Measurement of lipase activity in single grains of oat (*Avena sativa* L.). J. Am. Oil Chem. Soc. 59:354-355.

SOSEDOV, N. B., PECHAEV, L., SHVARTSMAN, M., and TERENT'EVA, G. N. 1971. Effect of moisture on the lipid complex of oat grain. Mukomolno-Elevat. Prom. 11:39-40. (Chem. Abstr. 78:56653t.)

TEVEKELEV, D. 1970. Phospholipids in foods. II. Cereals. Bulg. Akad. Nauk 9:5-19. (Chem. Abstr. 75:97395s.)

THOMKE, S. 1970. Determination of free fatty acids in grain. Z. Tierphysiol. Tierernaehr. Futtermittelkd. 17:31-35. (Chem. Abstr. 74:86552u.)

URQUHART, A., MATLASHEWSKI, G., and ALTOSAAR, I. 1980. Lipase activity in high-protein oats during maturation and germination. (Abstr.) Cereal Foods World 25:509.

URQUHART, A. A., ALTOSAAR, I., MATLASHEWSKI, G. J., and SAHASRABUDHE, M. R. 1983. Localization of lipase activity in oat grains and milled oat fractions. Cereal Chem. 60:181-183.

VON BIERMAN, V., WITTMANN, A., and GROSCH, W. 1980. Vorkommen bitterer Hydroxyfettsauren in Hafer und Weizen. Fette Seifen Anstrichm. 82:236-240.

WEBER, E. J., and ALEXANDER, D. E. 1975. Breeding for lipid composition in corn. J. Am. Oil Chem. Soc. 52:370-373.

WEBER, E. J., DE LA ROCHE, I. A., and ALEXANDER, D. E. 1971. Stereospecific analysis of maize triglycerides. Lipids 6:525-530.

WELCH, R. W. 1975. Fatty acid composition of grain from winter and spring sown oats, barley and wheat. J. Sci. Food Agric. 26:429-435.

WELCH, R. W. 1977. The development of rancidity in husked and naked oats after storage under various conditions. J. Sci. Food Agric. 28:269-274.

WOITAS, D. 1981. Linoleic acid—We need it, sunflowers got it. Sunflower 7:39-40.

WU, Y. V., CLUSKEY, J. E., WALL, J. S., and INGLETT, G. E. 1973. Oat protein concentrates from a wet-milling process: Composition and properties. Cereal Chem. 50:481-488.

YOUNGS, V. L. 1972. Protein distribution in the oat kernel. Cereal Chem. 49:407-411.

YOUNGS, V. L. 1974. Extraction of a high-protein layer from oat groat bran and flour. J. Food Sci. 39:1045-1046.

YOUNGS, V. L. 1978. Oat lipids. Cereal Chem. 55:591-597.

YOUNGS, V. L., and FORSBERG, R. A. 1979. Protein-oil relationships in oats. Crop Sci. 19:798-802.

YOUNGS, V. L., MEDCALF, D. G., and GILLES, K. A. 1970. The distribution of lipids in fractionated wheat flour. Cereal Chem. 47:640-649.

YOUNGS, V. L., and PÜSKÜLCÜ, H. 1976. Variation in fatty acid composition of oat groats from different cultivars. Crop Sci. 16:881-883.

YOUNGS, V. L., PÜSKÜLCÜ, M., and SMITH, R. R. 1977. Oat lipids. I. Composition and distribution of lipid components in two oat cultivars. Cereal Chem. 54:803-812.

CHAPTER 9

OAT PHENOLICS: STRUCTURE, OCCURRENCE, AND FUNCTION

F. W. COLLINS
Food Research Centre
Agriculture Canada
Ottawa, Ontario, Canada

I. INTRODUCTION

In recent years, it has become increasingly apparent that naturally occurring plant substances generically referred to as "phenolics" can and do contribute to the overall quality of cereal food products and ingredients. Such quality factors as color, odor, flavor, stability, nutrition, and safety are all potentially influenced, both by the types and amounts of native phenolic compounds originally present in the grain and by the degradation products of these phenolics formed during subsequent processing and preparation steps. The contribution of phenolics to food quality may be direct, or at least readily surmised. For example, the odor of the phenolic aldehyde 4-hydroxy-3-methoxybenzaldehyde (vanillin) can be detected by most individuals at a concentration of 1 part in 10 million. Phenolics may also contribute more indirectly to cereal grain quality. The inclusion of phenolic moieties in macromolecular complexes such as polysaccharides and proteins, whether biosynthetically or as a result of processing, can greatly alter the physicochemical and biological properties of the polymer. Such processes as the lignification of cell walls and the "tanning" of hide proteins brought about by the formation of intermolecular and intramolecular phenolic cross-linkages are examples. In these cases, the incorporation of phenolics into otherwise relatively labile macromolecules imparts increased physicochemical stability to the polymers and decreases their susceptibility toward enzymatic degradation by hydrolases and proteases.

The nature of phenolic functionality in the cereal grains and its consequences from the processing and nutritional standpoints are only now beginning to be formulated. However, progress is still hampered by a lack of knowledge concerning the precise structures and chemical forms of the phenolics present in the grains and their in vivo localization. Despite their long-established and widespread use as food sources, in-depth phytochemical examination of phenolics in cereal grains has not been pursued, at least to the same extent as in other food and beverage sources such as vegetables, fruits, and tubers. Indeed, few reviews have been written on cereal phenolics in general, and none, to my knowledge, on oat phenolics in particular. It is with the above insufficiencies in mind that this chapter dealing with oat phenolics has been prepared.

In compiling the review, I have exercised considerable liberty in drawing attention to phenolic constituents known to occur in other cereal crops and grasses but for which evidence of their absence or occurrence in oats has yet to appear. Although such inclusions may seem superfluous in some instances, they are intended to show the types and diversity of phenolics present in systematically related genera and to offer a broader prospective for future studies on oat phenolics.

The review deals primarily with the structures, occurrences, and physiological functions of free phenols, phenolic acids, flavonoids, aminophenolics and their esterified or otherwise conjugated forms that occur in oats and to a lesser extent in related cereals. In Sections II and III, the structure, occurrence, and function of free phenolic acids, simple phenols, and related phenolic aldehydes, quinones, and phenolic ether derivatives are presented. Section IV summarizes the major flavonoids in oats and related cereals and includes some recent findings in our laboratory on oat groat flavonoid phytochemistry and histochemistry. Section V deals with the amine-containing phenolics and related compounds. In each section, a brief outline of the biosynthetic relationships between various phenolics within these three classes is also given. In Section VI, the structure and occurrence of some of the phenolic esters, amides, and covalently linked phenolics are presented and possible functions of the covalent linking process are briefly discussed. The review concludes with a short summary and includes some future research needs in the field of phenolics.

II. PHENOLIC ACIDS AND RELATED COMPOUNDS

A. Structure

The structures of the major phenolic acids (primarily substituted cinnamic and benzoic acids) and related neutral phenolics are summarized in Fig. 1. The seemingly bewildering array of naturally occurring phenolics forms a more coherent picture when biogenetic considerations are taken into account, and this approach has been adopted in this and other figures in the text. Appreciation of the biosynthetic product-precursor relationships shown also allows a more logical prediction and interpretation of the occurrences and structural similarities exhibited by partially characterized phenolics. However, the biosynthetically active intermediates of cinnamic acids (and probably benzoic acids as well) consist of the labile coenzyme A thioesters rather than the free acids. Thus, for example, the absence of appreciable quantities of free phenolic acids does not necessarily reflect the biosynthetic capabilities of a cereal grain or the metabolic status of the acids in the tissues.

As shown in Fig. 1, the nonreversible first step in the allocation of carbon from primary metabolism to the phenolics involves the oxidative deamination of phenylalanine and, to a lesser extent, tyrosine, to form cinnamic acid (2) and *p*-coumaric acid (3), respectively. Thus, phenyl, benzyl, and phenylpropanoid carbons all appear to be derived from these two amino acids. Both cinnamate and *p*-coumarate can be considered to occupy central metabolic roles since they are substrates for a number of subsequent modifications (*meta* and *para*

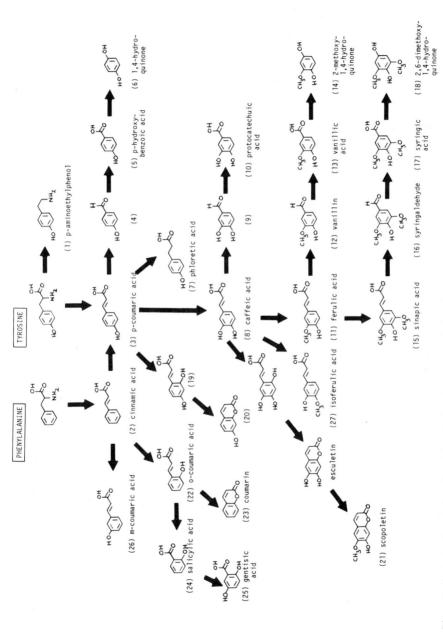

Fig. 1. Biosynthetic derivation of phenolic acids and phenols in oats and related cereals.

hydroxylation and methylation, reduction, oxidative decarboxylation) in the elaboration of substituted aldehydes, acids, and quinones. Hydroxylation *ortho* to the vinylic function of the appropriate cinnamic acid leads to lactone ring closure and the formation of substituted coumarins. Details of the biosynthesis of the phenolics in Fig. 1 can be found in the articles by Bolkart and Zenk (1968) and Krisnangkura and Gold (1979) and the reviews by Neish (1964), Brown (1966, 1979), Zenk (1966, 1979), Stafford (1974), Hanson and Havir (1979), Harborne (1980), and Gross (1981).

B. Occurrence

.Phenolic acids in the free form have been reported in alcoholic extracts of oat grains and a variety of other monocot cereal grains. In the cereals, the total free phenolic acid content rarely exceeds 0.02% of the dry weight of whole, mature, undamaged grain. The majority of these acids can be classified as either substituted benzoic or cinnamic acid types, the substituents consisting of hydroxyl functions and their corresponding methyl ethers. Following this classification, the hydroxybenzoic and hydroxycinnamic acids of oat and closely related cereal grains are summarized in Tables I and II, respectively.

Of the benzoic acid type, *p*-hydroxybenzoic (5), vanillic (13), and syringic (17) have been reported from oat hulls (Taketa, 1957; Vogel, 1961), and oat groats contain free vanillic acid (Durkee and Thivierge, 1977). In each of the above reports, identification of the acids was made by comparative paper chromatography of crude extracts and, because of the relatively low sensitivity and poor resolution of this technique, may represent a somewhat incomplete picture. All three acids have been found in whole-grain extracts of barley (*Hordeum vulgare*), triticale, and wheat (*Triticum aestivum*). Salicylic (24) and protocatechuic (10) acids have also been reported in these grains, whereas gentisic acid (25) was detected only in wheat and triticale (Maga and Lorenz, 1974). Recently, Sosulski et al (1982), using the more sensitive procedures of gas-liquid chromatography (GLC) and mass spectrometry (MS), have reported *p*-hydroxybenzoic, protocatechuic, vanillic, and syringic acids from debranned oat flour extracts. They found that the quantities of these acids were less than 2.5 ppm in the flour.

Comparable results for other cereal mill fractions are scarce. Maga and Lorenz (1974) also used GLC procedures to separate and determine free acids in several commercial wheat flours and Quadrumat Sr. mill fractions of triticale varieties. As seen in Table I, the concentration of each free acid was generally greatest in the bran fraction and least in the flour. The most prevalent acid was vanillic, occurring at concentrations as high as 37 ppm in triticale bran. Similar qualitative and quantitative results were obtained for barley fractions (Wall et al, 1961), indicating that the outer tissues (husks, pericarp, aleurone, etc.) of these grains constitute substantial but not exclusive sources of free benzoic acids. The germ may actually contain higher concentrations because, for example, in barley, about 20–25% of the total vanillic and *p*-hydroxybenzoic acids are located in the embryo (Van Sumere et al, 1972).

Because of the presence of a vinylic substitution in the cinnamic acid derivatives (Table II), both *trans* and *cis* isomers are possible (*trans* isomer shown). From an analytical standpoint, the *trans* and *cis* isomers of these acids are readily separated chromatographically and differ considerably in spectral characteristics. Commercially available standards (although this is usually not stated) are exclusively the *trans* isomers, which are rapidly converted to the *cis* forms in the presence of ultraviolet (UV) light or daylight (Hartley and Jones, 1975). This photolytic *trans-cis* conversion is also pH and solvent dependent, and the rates differ for each acid (Kahnt, 1967). Until recently, these factors have rarely been taken into account when reporting quantitative data. Since most of the original papers consulted in compiling Table II did not specify which isomers were analyzed, and since isomerization is virtually inevitable during extraction, the concentrations listed should be viewed as roughly indicative rather than strictly quantitative.

Free *p*-coumaric (3) and ferulic (11) acids have been detected in oats in both hulls and groats, and the hulls also contain sinapic acid (15) (Taketa, 1957; Vogel, 1961; Durkee and Thivierge, 1977) and dihydro-*p*-coumaric (4-hydroxyphenylpropionic) acid (7) (Vogel, 1961). Thus the hulls contain homologous pairs of benzoic and cinnamic acids with the same substitution patterns (5 and 3, 13 and 11, 17 and 15). Debranned oat flour contains free *p*-coumaric, ferulic, and caffeic (8) acids along with traces of sinapic acid and 4-hydroxyphenylacetic acid (Sosulski et al, 1982). As noted for the benzoic acids in barley and triticale mill fractions, the concentrations of each acid were highest in fractions rich in tissues from the outer portions of the grain and least in the flours. Ferulic and *p*-coumaric acids, the homologues of vanillic and *p*-hydroxybenzoic acids, respectively, are the major free cinnamic acids in wheat, barley, and triticale.

Despite the widespread occurrence of the more complex plant polyphenols, which usually contain catechol (1,2-dihydroxybenzene) or phloroglucinol (1,3,5-trihydroxybenzene) units, the simple phenolics themselves are rarely found in living plant tissues, especially in the free form. The relative infrequency of simple phenols is probably a consequence of their significant phytotoxicity when in the free form and of the rapidity with which they are detoxified, primarily to water-soluble β-glycosides, when administered to living cells (Towers, 1964; Pridham, 1965; Harborne, 1977). In the case of hydroxybenzoic and hydroxycinnamic acids, which contain both phenolic and carboxylic hydroxyl functions, both the ether and ester glucosides are formed when the free acids are fed (Harborne, 1977). In the mature cereal grain, the enzymatic complement necessary for phenol glycosylation is present and active. Wheat embryos, for example, contain two different phenol glucosyltransferases that are uridine diphosphate glucose (UDPG) dependent (Yamaha and Cardini, 1960a, 1960b). The first enzyme glucosylates simple phenols, especially dihydric phenols such as hydroquinones (1,4-dihydroxybenzenes; 28), to the mono-β-D-glucopyranosides (29), whereas the second catalyzes the transfer of a glucopyranosyl moiety to form the β-D-gentiobiosides (30), as shown in the reaction sequence below.

(28)
hydroquinone

(29)
hydroquinone–O–
β-D-glucoside

(30)
hydroquinone–O–
β-D-gentiobioside

TABLE I
Structure and Occurrence of Free Hydroxybenzoic Acids in Oats and Related Cereal Grains

Compound and Structure	Genus	Tissue or Material Examined	Concentration	Reference
 (5) p-Hydroxybenzoic acid (4-hydroxybenzoic acid)	*Avena sativa*	Hulls	...	Taketa (1957)
		Flour	0.7 ppm	Sosulski et al (1982)
	Hordeum vulgare	Whole grains	...	Slominski (1980)
		Whole grains	4 µg/100 grains	Van Sumere et al (1972)
		Whole grains	1.2 ppm	Wall et al (1961)
		Pearled grains	0.3 ppm	
		Pearlings (husks)	7.0 ppm	
		Husks		Van Sumere et al (1958)
		Embryos	1 µg/100 embryos	Van Sumere et al (1972)
	Triticale	Whole grains	6–11 ppm	Maga and Lorenz (1974)
		Bran	9–16 ppm	
		Shorts	5–8 ppm	
		Flour	5–8 ppm	
	Triticum aestivum	Flour	5–8 ppm	Maga and Lorenz (1974)
 (24) Salicylic acid (2-hydroxybenzoic acid)	*H. vulgare*	Whole grains	...	Slominski (1980)
	Triticale	Whole grains	5–7 ppm	Maga and Lorenz (1974)
		Bran	6–9 ppm	
		Shorts	5–7 ppm	
		Flour	3–5 ppm	
	T. aestivum	Flour	2–5 ppm	Maga and Lorenz (1974)
 (25) Gentisic acid (2,5-dihydroxybenzoic acid)	Triticale	Whole grains	2–4 ppm	Maga and Lorenz (1974)
		Bran	4–5 ppm	
		Shorts	3–5 ppm	
		Flour	1–2 ppm	
	T. aestivum	Flour	1–3 ppm	Maga and Lorenz (1974)

(10) Protocatechuic acid (3,4-dihydroxybenzoic acid)

Species	Plant part	Concentration	Reference
A. sativa	Flour	0.5 ppm	Sosulski et al (1982)
H. vulgare	Whole grains	...	Slominski (1980)
Triticale	Whole grains	2–4 ppm	Maga and Lorenz (1974)
	Bran	4–5 ppm	
	Shorts	3–5 ppm	
	Flour	1–2 ppm	
T. aestivum	Flour	1–3 ppm	Maga and Lorenz (1974)

(13) Vanillic acid (4-hydroxy-3-methoxy-benzoic acid)

Species	Plant part	Concentration	Reference
A. sativa	Hulls	...	Taketa (1957)
	Groats	...	Durkee and Thivierge (1977)
	Flour	0.7 ppm	Sosulski et al (1982)
			Slominski (1980)
H. vulgare	Whole grains	4.5 µg/100 grains	Van Sumere et al (1972)
	Whole grains	4.0 ppm	Wall et al (1961)
	Whole grains	1.0 ppm	
	Pearled grains	27.4 ppm	
	Pearlings (husks)	...	
	Husks		
Triticale	Embryo	1 µg/100 embryos	Van Sumere et al (1958)
	Whole grains	18–26 ppm	Van Sumere et al (1972)
	Bran	27–37 ppm	Maga and Lorenz (1974)
	Shorts	20–26 ppm	
	Flour	16–24 ppm	
T. aestivum	Flour	13–18 ppm	Maga and Lorenz (1974)
	Germ	0.6 ppm	Sosulski et al (1982)
			King (1962)

(17) Syringic acid (3,5-dimethoxy-4-hydroxy-benzoic acid)

Species	Plant part	Concentration	Reference
A. sativa	Hulls	...	Vogel (1961)
H. vulgare	Flour	...	Sosulski et al (1982)
	Whole grains	...	Slominski (1980)
	Husks		Van Sumere et al (1958)
Triticale	Whole grains	4–7 ppm	Maga and Lorenz (1974)
	Bran	5–14 ppm	
	Shorts	2–12 ppm	
	Flour	2–5 ppm	
T. aestivum	Flour	3–5 ppm	Maga and Lorenz (1974)
	Flour	0.5 ppm	Sosulski et al (1982)

TABLE II
Structure and Occurrence of Free Hydroxycinnamic Acids in Oats and Related Cereal Grains

Compound and Structure	Genus	Tissue or Material Examined	Concentration	Reference
 (3) *p*-Coumaric acid (4-hydroxy-*trans*-cinnamic acid)	*Avena sativa*	Hulls	...	Vogel (1961)
		Flour	0.7 ppm	Sosulski et al (1982)
		Groats	...	Durkee and Thivierge (1977)
	Hordeum vulgare	Whole grains	...	Slominski (1980)
		Whole grains	10.3 μg/100 grains	Van Sumere et al (1972)
		Whole grains	1.6 ppm	Wall et al (1961)
		Pearled grains	0.2 ppm	
		Pearlings (husks)	11.2 ppm	
		Husks	...	
	Triticale	Whole grains	18–28 ppm	Maga and Lorenz (1974)
		Bran	23–32 ppm	
		Shorts	16–25 ppm	
		Flour	15–24 ppm	
	Triticum aestivum	Whole grains	Trace	El-Basyouni and Towers (1964)
		Flour	17–24 ppm	Maga and Lorenz (1974)
			Trace	Sosulski et al (1982)
 (26) *m*-Coumaric acid (3-hydroxy-*trans*-cinnamic acid)	*H. vulgare*	Whole grains	...	Slominski (1980)

Compound	Species	Plant part	Amount	Reference
(22) o-Coumaric acid (2-hydroxy-*trans*-cinnamic acid)	*H. vulgare*	Whole grains	0.6 µg/100 grains (as coumarin)	Van Sumere et al (1972)
	Triticale	Whole grains	...	Slominski (1980)
		Husks	...	Van Sumere et al (1958)
		Whole grains	6–10 ppm	Maga and Lorenz (1974)
		Bran	10–15 ppm	
		Shorts	7–10 ppm	
		Flour	4–8 ppm	
	T. aestivum	Flour	4–8 ppm	Maga and Lorenz (1974)
(8) Caffeic acid (3,4-dihydroxy-*trans*-cinnamic acid)	*A. sativa*	Flour	1.0 ppm	Sosulski et al (1982)
(11) Ferulic acid (4-hydroxy-3-methoxy-*trans*-cinnamic acid)	*A. sativa*	Hulls	...	Taketa (1957)
		Flour	2.4 ppm	Sosulski et al (1982)
		Groats	...	Durkee and Thivierge (1977)
	H. vulgare	Whole grains	...	Slominski (1980)
		Whole grains	49 µg/100 grains	Van Sumere et al (1972)
		Whole grains	1.7 ppm	Wall et al (1961)
		Pearled grains	0.8 ppm	
		Pearlings (husks)	10.5 ppm	
		Husks	...	Van Sumere et al (1958)
		Embryo (germ)	3.3 µg/100 embryos	Van Sumere et al (1972)
	Triticale	Whole grains	19–22 ppm	Maga and Lorenz (1974)
		Bran	28–37 ppm	
		Shorts	14–20 ppm	
		Flour	15–18 ppm	
	T. aestivum	Whole grains	Trace	El-Basyouni and Towers (1964)
		Flour	15–18 ppm	Maga and Lorenz (1974)
		Flour	1.2 ppm	Sosulski et al (1982)
		Germ	...	King (1962)

(continued on next page)

TABLE II (continued)

Compound and Structure	Genus	Tissue or Material Examined	Concentration	Reference
(27) Isoferulic acid (3-hydroxy-4-methoxy-*trans*-cinnamic acid)	Triticale	Whole grains Bran Shorts Flour	4–9 ppm 5–15 ppm 5–7 ppm 2–5 ppm	Maga and Lorenz (1974)
	T. aestivum	Flour	1–3 ppm	Maga and Lorenz (1974)
(15) Sinapic acid (3,5-dimethoxy-4-hydroxy-*trans*-cinnamic acid)	*A. sativa*	Hulls	…	Vogel (1961)
	H. vulgare	Flour Whole grains	Trace …	Sosulski et al (1982) Slominski (1980)

In excised barley embryos, administration of ferulic acid resulted in the appearance of small quantities of the methanol-soluble ferulic acid β-glucoside ether, which was rapidly incorporated into more complex ester-linked, methanol-insoluble components. Neither the free acid nor the glucoside accumulated to any major extent. Similar results were also obtained with whole barley grains, although the uptake rate was somewhat slower (Van Sumere et al, 1972).

Table III summarizes the structure and occurrence of phenol glucosides so far reported from oats and related cereals. The glucoside of 2-methoxy-hydroquinone (31) has been isolated from the leaves of oat seedlings (Olsen, 1971). It is not present in mature leaves, nor is it found in the roots. Analysis of freshly germinated oats by Turner (1960) suggested that the glucoside was absent from hulls, endosperm, and root tissues, being confined to the coleoptile and apices of embryonic and rudimentary leaves. The glucoside was first reported from wheat germ by Bungenberg de Jong et al (1953), and a similar if not identical compound occurs in barley germ (Bungenberg de Jong et al, 1955; Mace and Hebert, 1963). Daniels (1959), however, failed to detect it in extracts of whole wheat and a variety of wheat milling fractions but did observe two uncharacterized, more highly glycosylated forms of the hydroquinone. The nature of the wheat germ glycosides was reexamined by Conchie et al (1961) in connection with their enzymatic synthesis. Chromatographic analysis showed the presence of four glycosides of the same substituted hydroquinone, in all probability 2-methoxyhydroquinone. Hydrolytic data suggested that the four glycosides consist of the mono-β-D-glucoside (31), the β-gentiobioside (32), and the β-D-gentiotri- and tetra-glucosides (33, 34).

Phenolic acid ether glucosides (35–38) have been reported in mature barley grains (Van Sumere et al, 1972), which also contain the ester glucosides (see Section VI). Durkee and Thivierge (1977) reported β-glucosidase-labile forms of ferulic acid in dehulled oats, and, although the procedure does not distinguish between ester and ether types since both are β-glucosidase labile (Harborne and Corner, 1961), the ether glucoside (38) is almost certainly present (F. W. Collins, *unpublished results*).

The coumarin glucoside scopolin (39) and its aglycone scopoletin (21), which contain the hydroxyl-methoxyl substitution pattern analogous to ferulic acid, have also been reported in oat seedlings (Goodwin and Pollock, 1954; Turner, 1960). Although mature grains were not examined, both coumarins are present shortly after germination in the differentiating and maturing zones but not the apical zones of seminal and crown roots from excised embryos (Turner, 1960).

Methoxyhydroquinone glucoside (31) is readily hydrolyzed both nonenzymatically by weak acid (Olsen, 1971) and enzymatically by the action of cereal embryo and yeast β-glucosidases (Bungenberg de Jong et al, 1953, 1955), as shown in the following reaction sequence:

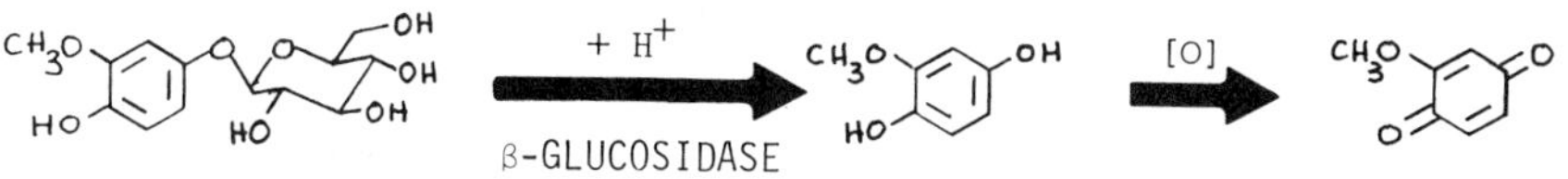

(31)
methoxyhydroquinone-
glucoside

β-GLUCOSIDASE

(31a)
methoxyhydro-
quinone

(31b)
methoxy-*p*-benzo-
quinone

TABLE III
Structure and Occurrence of Simple Phenolic Ether Glycosides in Oats and Related Cereal Grains

Structure	Name	Tissue or Material Examined	References
(31)	(31) 2-methoxyhydroquinone-4-*O*-β-D-glucoside	Oat seedling leaves Wheat germ	Olsen (1971) Bungenberg de Jong et al (1953, 1955)
(32, 33, 34)	(32) R = H; 2-Methoxyhydro-quinone-4-*O*-β-gentiobioside	Wheat germ	Conchie et al (1961)
	(33) R = β-D-glucosyl; 2-methoxyhydroquinone-4-*O*-β-gentiotrioside	Wheat germ	Conchie et al (1961)
	(34) R = O-β-gentiobiosyl; 2-methoxyhydroquinone-4-*O*-β-gentiotetraoside	Wheat germ	Conchie et al (1961)
(35)	(35) *p*-Hydroxybenzoic acid 4-*O*-β-glucoside	Barley grain	Van Sumere et al (1972)
(36)	(36) Vanillic acid 4-*O*-β-glucoside	Barley grain	Van Sumere et al (1972)
(37)	(37) *o*-Coumaric acid 2-*O*-β-glucoside	Barley grain	Van Sumere et al (1972)
(38)	(38) Ferulic acid 4-*O*-β-glucoside	Dehulled oats Barley grain	Durkee and Thivierge (1977) Van Sumere et al (1972)
(39)	(39) Scopolin (scopoletin-β-glucoside)	Oat roots seedling	Turner (1960)

The aglycone, methoxyhydroquinone (31a), is unstable in air and is autooxidized to methoxy-*p*-benzoquinone (31b).

C. Functionality

Although a great deal has been written about potential physiological and biochemical functions of phenolic acids and related phenolics in growth, development, and pathology, it seems unlikely that as a group they have one particular universal role. More likely, individual substances may possess significant activities in particular physiological or biochemical processes. For example, benzoquinones have been shown to exhibit effects on a wide variety of biochemical processes in germinating barley grains, such as decreasing oxygen consumption rates, uncoupling oxidative phosphorylation, and inhibiting amino acid activation (Van Sumere et al, 1975, and references cited therein). Olsen (1971) found methoxyhydroquinone or its oxidation product to be fungistatic at 100 ppm to *Ophiobolus graminis*, which causes take-all disease in wheat and barley seedlings but not in oats. The author concluded that the methoxyhydroquinone glucoside, which is present in oat leaves but not in those of wheat or barley, probably contributes to the resistance of oats against take-all disease. Both breakdown products, methoxyhydroquinone and methoxy-*p*-benzoquinone, could be formed after the fungus penetrated the cell walls of the host tissues and released or activated endogenous β-glucosidase. Both compounds have also been found to be fungicidal (100% inhibition of spore germination) at 5 ppm or slightly higher concentrations against loose smuts of wheat and barley grains (Mace and Hebert, 1963). In addition, methoxy-*p*-benzoquinone exhibits bacteriostatic activity (Bungenberg de Jong et al, 1955).

In the above example, the latent protective property of the free quinone is conserved in an inactive form as the hydroquinone glucoside and only becomes active following disruption or exposure of the cells to hydrolytic and oxidizing conditions. A somewhat similar control mechanism regulating longitudinal root growth in germinating oats has been ascribed to differences in the endogenous concentrations of scopoletin and its glucoside along the length of the root (Goodwin and Taves, 1950; Goodwin and Pollock, 1954; Pollock et al, 1954). Numerous further examples of the bacteriostatic, fungistatic, allelopathic, phytoalexic, phytohormonal, and allosteric activities of phenolic acids and related phenolics can be found in articles by Baranowski and Nagel (1982) and Herald and Davidson (1983) and in reviews by Swain (1977), Harborne (1977, 1980), and Brown (1981).

III. ALKYLPHENOLS AND RELATED PHENOLICS

A. Structure

Some free phenols that have been reported as naturally occurring in cereal grains, especially heat-processed ingredients, are now known to be enzymatic or thermal decarboxylation by-products derived from substituted cinnamic and benzoic acids. The structure and derivation of some of the more prevalent artifacts formed from phenolic acids are outlined in Fig. 2. Fiddler et al (1967) studied the formation of simple phenols during the thermal decomposition of

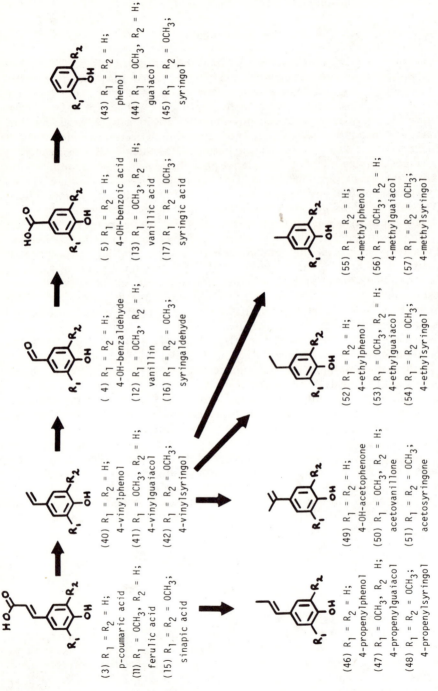

Fig. 2. Major pathways of formation of simple phenols from degradation of hydroxycinnamic acids.

solid ferulic acid in both air and nitrogen atmospheres over the temperature range from 20 to about 500°C. Two distinct stages in the decomposition were observed during a constant heating rate treatment of 6°C/min. The first stage commenced at about 200°C, with a maximum decomposition rate of about 5% per minute in both air and nitrogen at 280°C. The compound 4-vinylguaiacol (41) was the only product produced during this stage in both atmospheres. The second stage, commencing at about 340°C in air and 345°C in nitrogen, exhibited maximum rates of 2.1 and 2.5% per minute, respectively, at 380°C. In both atmospheres, the products (unsubstituted, 4-methyl-, and 4-ethylguaiacols) formed at the expense of 4-vinylguaiacol. Additionally, vanillin (12), acetovanillone (50), and vanillic acid (13) were produced only in the presence of air. In a model system simulating dry roasting of cereal grains, Tressl et al (1976) identified the major components present after thermal decarboxylation of several cinnamic and hydroxycinnamic acids (2 hr at 200°C) (Table IV). For all acids studied, the major component produced was the 4-vinylphenyl homologue following loss of carbon dioxide. The product ratios for the 4-vinyl derivatives were higher in the presence of air than in nitrogen for all hydroxyl-substituted

TABLE IV
Thermal Degradation Products of Cinnamic and Hydroxycinnamic Acids in Air and N$_2$[a]

Starting Compound	Structure[b]	Product	Relative Proportion in Degradation Mixture (%)	
			N$_2$	Air
Cinnamic acid		Vinylbenzene	99.2	97.5
		Benzaldehyde	0.2	1.2
		Others	0.6	1.3
p-Coumaric acid	(40)	4-Vinylphenol	40.6	69.1
	(52)	4-Ethylphenol	16.3	5.6
	(55)	4-Methylphenol	2.5	0.6
	(43)	Phenol	22.8	1.8
	(46)	4-Propenylphenol	13.8	8.6
	(49)	4-OH-acetophenone	···	1.9
		Others	4.0	12.4
Ferulic acid	(41)	4-Vinylguaiacol	56.2	79.9
	(53)	4-Ethylguaiacol	14.2	5.5
	(56)	4-Methylguaiacol	0.5	Trace
	(44)	Guaiacol	14.3	3.1
	(47)	4-Propenylguaiacol	12.5	2.5
	(12)	Vanillin	0.7	6.4
	(50)	Acetovanillone	···	2.6
		Others	1.6	···
Sinapic acid	(42)	4-Vinylsyringol	54.0	78.5
	(54)	4-Ethylsyringol	23.3	1.8
	(57)	4-Methylsyringol	0.3	Trace
	(45)	Syringol	18.3	4.5
	(48)	4-Propenylsyringol	3.1	0.7
	(16)	Syringaldehyde	···	13.4
	(51)	Acetosyringone	···	1.1
		Others	1.0	···

[a] Adapted from Tressl et al (1976).
[b] As shown in Fig. 2.

acids, whereas the 4-ethyl and 4-methyl homologues were generally more prevalent in nitrogen than in air atmospheres. Aldehydes and ketones (compounds 12, 16, 49, 50, and 51) were virtually absent from mixtures obtained under nitrogen.

The kinetics of the decarboxylation reaction in solution and the effects of pH have recently been examined by Pyysalo et al (1977) in connection with the formation of simple phenolic aroma chemicals produced during heat processing of foods. Table V summarizes the rate constants for the decarboxylation of several cinnamic and hydroxycinnamic acids commonly encountered in cereal grains. An indication of the relative stability after a 3-hr exposure to the various pH and temperature conditions has been included. The authors found that in solution the decarboxylation followed approximately first-order kinetics, with hydroxylation of the aromatic ring essential for decarboxylation under these conditions. Heating for 3 hr at 100°C resulted in decarboxylation of more than 50% of the *p*-coumaric, caffeic, and ferulic acids at pH 1.0, which corresponds to $0.1N$ HCl. With increasing pH, the rate diminished, although even at pH 6.0 almost 10% of the ferulic acid was decarboxylated. This obvious lability of the hydroxycinnamic acids to heat under acidic conditions should always be taken into account when determinations are made of esterified, conjugated, or otherwise "bound" forms of these acids following acid hydrolysis.

Unlike the short-chain alkylphenols discussed above, some phenols with extended alkyl side chains do occur naturally in cereal grains. Other than the terpenoid hydroquinones and chromanols (e.g., tocopherols, plastochromanols), which will not be discussed in this chapter, the most thoroughly studied groups are the 5-*n*-alkylresorcinols (*n*-alkyl-1,3-dihydroxybenzenes). Wenkert et al

TABLE V
Rate of Thermal Decarboxylation of Cinnamic and Hydroxycinnamic Acids
of Various pH and Temperature Conditions[a]

Compound	pH	Temperature (°C)	$10^3 \times$ Rate Constant (per min)	Decarboxylation After 3 hr (%)
Cinnamic acid	1.0	100	0	0
		110	0	0
o-Coumaric acid	1.0	100	0.78	13.0
		110	1.75	27.0
p-Coumaric acid	1.0	100	5.75	64.5
		110	14.50	92.6
	2.0	100	0.71	12.0
	4.0	100	0.31	5.5
	6.0	100	0.19	3.4
Caffeic acid	1.0	100	4.45	55.1
		110	7.60	74.5
Ferulic acid	1.0	100	6.00	66.3
		110	14.70	92.9
	2.0	100	0.95	15.7
	4.0	100	1.05	17.2
	6.0	100	0.55	9.4

[a] Adapted from Pyysalo et al (1977).

(1964) first reported the presence in wheat bran of a complex series of free phenols in which *n*-alkyl side chains of 17, 19, 21, 23, and 25 carbons were directly coupled to a resorcinol ring at the 5 position. Subsequently, Wieringa (1967) identified the animal growth-inhibiting factor from rye grains as a similar mixture of 5-*n*-alkylresorcinols, which also contained small quantities of 5-*n*-alkenyl analogues of unknown structure. The author showed that the rye alkylresorcinol also contained odd-number *n*-alkyl side chains with 15–25 carbons. Wieringa concluded that, as with the wheat bran mixtures, the rye alkylresorcinols were confined to the outer parts of the grain (i.e., pericarp) and were not present in the endosperm or embryo. The structure and approximate composition of typical wheat and rye alkylresorcinols (Wieringa, 1967) are shown in Table VI.

B. Occurrence

Steinke and Paulson (1964) reported the presence of free phenol (43) (Fig. 2), 4-vinylphenol (40), guaiacol (44), and 4-vinylguaiacol (41) in steam condensates from cornmeal "cooks." A number of fermented grain extracts were also found to contain these phenolics, as well as 4-ethylphenol (52), 4-methyl-phenol (55), 4-ethylguaiacol (53), and 4-methylguaiacol (56). The authors concluded that the alkylphenols were artifacts resulting from microbial (yeast and bacteria) and thermal decarboxylation of *p*-coumaric acid and that the alkylguaiacols were derived from ferulic acid decarboxylation. Crispbread prepared from rye was found to contain 4-vinylguaiacol along with small amounts of 4-propenylguaiacol (46), guaiacol, and 4-ethylphenol (von Sydow and Anjou, 1969). The formation of short-chain alkyl phenols in oats arising from heat processing (e.g., roasting, baking, toasting) has not been studied to date. That acidic conditions and/or elevated temperatures are prerequisite to their formation in oats seems evident from the recent work of Heydanek and McGorrin (1981a, 1981b). The authors analyzed total volatiles and purge trap headspace volatiles from dry and hydrated groats (heating at 55–60° C, vacuum distillation at 0.02 torr for 4 hr). Rancid oats prepared by boiling groats for 30 min followed by storage at ambient temperatures for one week were also

TABLE VI
Composition of *n*-Alkylresorcinol Mixtures from Wheat and Rye[a]

Structure HO — ⬡ — R OH	Percentage Composition by Weight of Component[b]	
	Wheat	Rye
R = *n*-$C_{15}H_{31}$	n.d.[c]	2
R = *n*-$C_{17}H_{35}$	4	27
R = *n*-$C_{19}H_{39}$	34	37
R = *n*-$C_{21}H_{43}$	48	24
R = *n*-$C_{23}H_{47}$	9	7
R = *n*-$C_{25}H_{51}$	5	3

[a] Adapted from Wieringa (1967).
[b] Mixture separated and quantified by gas chromatography after permethylation.
[c] Not detected.

analyzed. In dry and hydrated oats, numerous alkylbenzenes were detected but no substituted phenols were found. Benzaldehyde represented the sole aromatic carbonyl. In rancid oats, although the number of aromatic carbonyls increased dramatically, none of the components found corresponded to any of the decarboxylation products of the hydroxycinnamic acids known to be present in the grain. The compound 2-methylphenol (*o*-cresol) was the only simple phenol detected, and most, if not all, of the oxygenated aromatic compounds could be accounted for by lipid oxidation.

To date, oat grains have not been examined for the presence or absence of alkylresorcinols, although their presence in wheat, rye, millet, and barley suggests that they may also be present in oats. Musehold (1978) has used a rapid-screening technique employing thin-layer chromatography for qualitative analysis of nine genotypes of wheat, rye, and triticale. Considerable variation in the qualitative composition of alkylresorcinols among the various genotypes was encountered. Quantity of total alkylresorcinols expressed as 5-*n*-pentadecylresorcinol equivalents on a dry weight basis was also determined and found to vary from 657 to 926 ppm for wheat, 633 to 863 ppm for triticale, and 1,057 to 1,214 ppm for rye. Table VII summarizes the distribution and quantities of alkylresorcinols in several whole-cereal grains and Quadrumat Sr. mill fractions. Bran fractions show the highest concentration, whereas the flours exhibit only about 30% (rye and triticale) of the overall concentration of whole grains. Levels in the flour fraction for wheat may reflect contamination, under the milling conditions used, of flour with bran, which appears to contain high

TABLE VII
Occurrence and Quantitative Distribution of Alkylresorcinols in Several Cereal Grains

| Species | Cultivar | Alkylresorcinol Content (ppm dry wt) as 5-*n*-Pentadecylresorcinol Equivalent | | | | References |
		Whole Grain	Bran	Shorts	Flour	
Rye (*Secale cereale*)	Prolific	1,050	1,860	1,150	310	Verdeal and Lorenz (1977)
		970				Evans et al (1973)
Triticale	6-TA-204	820	2,130	870	260	Verdeal and
	Rosner	790				Lorenz (1977)
Wheat (*Triticum aestivum*)	Stewart 63	530				Evans et al (1973)
	Pitic 62	790				
	Manitou	600				
	Kenya Farmer	550				
	Neepawa	640				
	Selkirk	890				
	Chris	620	2,110	700	380	Verdeal and Lorenz (1977)
Barley (*Hordeum vulgare*)	Conquest	<100				Evans et al (1973)
Millet (*Setaria italica*)		<100				

concentrations (2,110 ppm). The physiological and biochemical functionality of these alkylresorcinols remains obscure.

IV. FLAVONOIDS

A. Structure

The structures and biosynthetic derivation of the major groups of cereal flavonoids are summarized in Fig. 3. Condensation of an activated cinnamoyl moiety (58) with three malonyl-CoA-derived C_2 units results in the formation of the C_{15} skeleton of the chalcones (59). Chalcone-flavanone interconversion (59–60) brings about closure of the pyrone ring, upon which subsequent oxidations and substitutions are dependent. Dehydrogenation of the flavanone produces the flavone group (61), which undergoes further substitution to give a wide variety of hydroxylated, methoxylated, and glycosylated derivatives (62). C-glycosylation substitution (31, R = glycosyl), however, precedes O-glycosylation and O-methylation in flavone substitution sequences and probably occurs before the flavanone-flavone oxidation step (Hahlbrock and Grisebach, 1975). Hydroxylation of the flavanone pyrone ring *ortho* to the carbonyl function generates the flavanonol group (63), the precursors of the flavonols (64, 65). Transformation of the flavanonol to yield the flav-3-en-3-ol intermediate (66) gives rise to an important branch point in the biosynthetic pathway. It leads, on the one hand, to anthocyanidin derivatives (68) via a flavylium cation (67); or, on the other, to the flavanol carbocation (69) and procyanidin derivatives (condensed tannins, 70) (cf. Stafford, 1983). Further information regarding these transformations can be found in reviews by Gross (1977, 1979) and Zenk (1979), and details of flavonoid biosynthesis as outlined in Fig. 3 are discussed by Hahlbrock and Grisebach (1975), Haslam (1975), and Grisebach (1979).

B. Occurrence

Despite the economic importance of the cereal grains, surprisingly little has been done to elucidate the major flavonoids present in the grain; such is the case particularly with oats. To some extent, other cereals closely related to oats have been examined, although to my knowledge no major reviews have been published. It is perhaps useful, therefore, to collate the available information on various flavonoid classes reported to occur in the mature grains of oats, wheat, barley, rye, corn, sorghum, and millet (i.e., *Pennisetum typhoides* Stafp. and Hubb.). The data are summarized in Tables VIII–X.

Proanthocyanidins and leucoanthocyanidins (Table VIII) have been detected in most of the cereal grains, although they remain poorly characterized and unquantitated in most cases. From sorghum seed coats, Haslam and Gupta (1978) have described the major polymeric component of sorghum "tannin" as a hexameric-heptameric polyflavan-3-ol (71) with an average molecular weight of 1,700–2,000. At maturity, the procyanidin polymer may constitute as much as 5% of the dry weight of the seed. It appears to be localized in the pericarp/testa tissues (Rooney et al, 1980). The monomeric flavan-4-ol luteoforal (78) and the flavan-3,4-diols leucopelargonidin (74), leucocyanidin (75), and leucofisetinidin

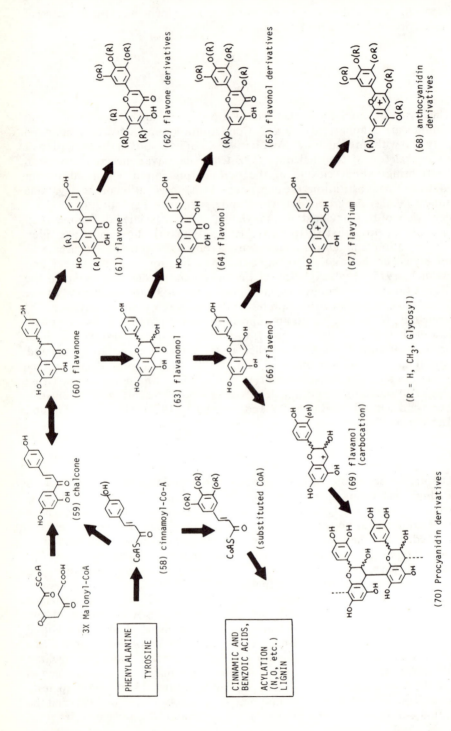

Fig. 3. Biosynthetic derivation of the major flavonoid and phenolic types in oats and related cereals.

TABLE VIII
Proanthocyanidins, Anthocyanidins, and Derivatives
Occurring in Oat and Related Cereal Grains

Structure	Genus	Tissue or Material Examined	References
		Proanthocyanidins	

(71)
Sorghum tannin (n = 4,5)

Sorghum vulgare — Seed coat — Haslam and Gupta (1978)

(72)
Procyanidin B-3 dimer

Hordeum vulgare — Aleurone cells — Jende-Strid (1981)

(73)
(+)-Catechin

H. vulgare — Aleurone cells — Jende-Strid (1981)

(74)
Leucopelargonidin

Zea mays — Aleurone tissues — Reddy (1964)
S. vulgare — Pericarp tissues — Yasumatsu et al (1965)

(75)
Leucocyanidin

Z. mays — Aleurone tissues — Reddy (1964)
H. vulgare — Whole grains — Metche and Urion (1961)

(76)
Leucofisetinidin

S. vulgare — Pericarp tissues — Yasumatsu et al (1965)

(*continued on next page*)

TABLE VIII (continued)

Structure	Genus	Tissue or Material Examined	References
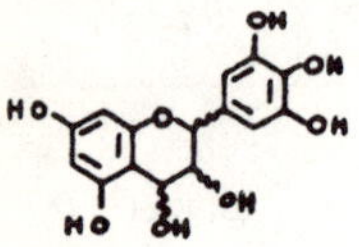 (77) Leucodelphinidin	*Avena sativa* *H. vulgare*	Hulls Aleurone cells	Vogel (1961) Jende-Strid (1981)
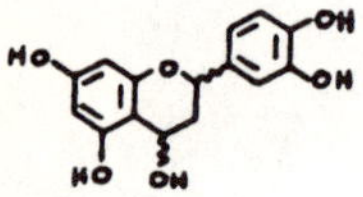 (78) Luteoforol	*Sorghum* spp.	Pericarp tissues	Bate-Smith (1969) Bate-Smith and Rasper (1969) Kambal and Bate-Smith (1976)

Anthocyanidins

Structure	Genus	Tissue or Material Examined	References
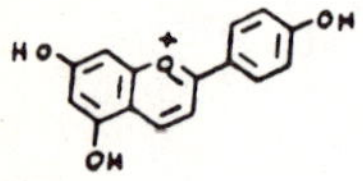 (79) Apigeninidin	*Sorghum* spp.	Pericarp tissues Glumes	Kambal and Bate-Smith (1976) Misra and Seshadri (1967)
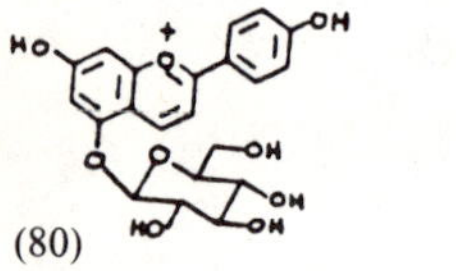 (80) Apigeninidin-5-glucoside	*Sorghum* sp.	Stylar area, epidermis, and tube cell layers of pericarp	Nip and Burns (1969, 1971)
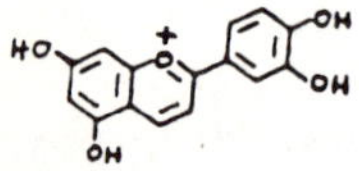 (81) Luteolinidin	*Sorghum* spp.	Pericarp tissues Glumes	Kambal and Bate-Smith (1976) Misra and Seshadri (1967)
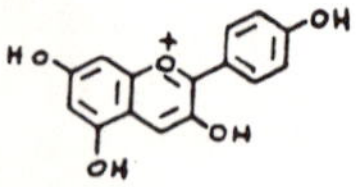 (82) Pelargonidin	*H. vulgare*	Pericarp and aleurone tissues; also occurs glycosylated	Mullick et al (1958)
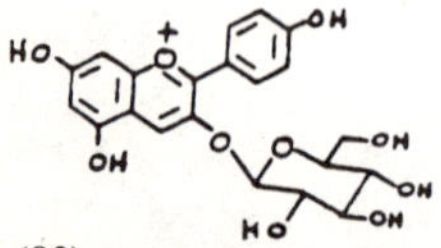 (83) Pelargonidin-3-glucoside	*Z. mays*	Aleurone tissues	Harborne and Gavazzi (1969)

(*continued on next page*)

TABLE VIII (continued)

Structure	Genus	Tissue or Material Examined	References
(84) Cyanidin	*H. vulgare*	Pericarp and aleurone tissues; also occurs glycosylated	Mullick et al (1958)
(85) Cyanidin-3-glucoside	*Z. mays*	Aleurone tissues Husks Cultured endosperm	Harborne and Gavazzi (1969) Sando et al (1935) Straus (1959)
(86) Peonidin-3-glucoside	*Triticum aestivum* *Secale cereale*	Pericarp tissues Pericarp tissues	Dedio et al (1972a) Dedio et al (1972b)
(87) Delphinidin	*H. vulgare*	Pericarp and aleurone tissues; also occurs glycosylated	Mullick et al (1958)
(88) Delphinidin-3-rutinoside	*S. cereale*	Aleurone tissues	Dedio et al (1969)

TABLE IX
Chalcones, Flavanones, and Flavonols Occurring in Oat and Related Cereal Grains

Structure	Genus	Tissue or Material Examined	References
Chalcones			
(89) 2′,3,4,4′,6-(OH)₅- Chalcone	*Sorghum bicolor*	Pericarp tissues	Kambal and Bate-Smith (1976)
(90) 2′,4,4′,6-(OH)₄-3- OCH₃-Chalcone	*Avena sativa*	Hulls	Vogel (1961)
Flavanones			
(91) Eriodictyol	*S. bicolor*	Pericarp tissues	Kambal and Bate-Smith (1976)
(92) Homo-eriodictyol	*A. sativa*	Hulls	Vogel (1961)
Flavonols			
(93) Kaempferol	*A. sativa*	Dehulled grains	F. W. Collins and R. G. Fulcher (unpublished results)
	Zea mays	Hydrolyzed aleurone extracts	Kirby and Styles (1970)
	S. bicolor	Stylar area, epidermis and tube cell layers; occurs as diglycoside	Nip and Burns (1969)
(94) Quercetin	*A. sativa*	Dehulled grains	F. W. Collins and R. G. Fulcher (unpublished results)
	Z. mays	Hydrolyzed aleurone tissue extracts husks; occurs as monoglucoside	Kirby and Styles (1970)

TABLE X
Flavone Derivatives Occurring in Oat and Related Cereal Grains

Structure	Genus	Tissue or Material Examined	References
		Flavones	
(95) Luteolin	*Avena sativa*	Dehulled grains	F. W. Collins and R. G. Fulcher (unpublished results)
(96) 7-*O*-Methyl-luteolin	*Sorghum durrha*	Glumes	Misra and Seshadri (1967)
(97) Chrysoeriol	*Hordeum vulgare*	Whole grains	Bhatia et al (1972)
(98) Tricin	*A. sativa*	Dehulled grains	F. W. Collins and R. G. Fulcher (unpublished results)
		Hulls	Vogel (1961)
	Triticum aestivum	Wheat flour	Chen and Geddes (1945)
		C-Glycosylflavones	
(99) Vitexin	*A. sativa*	Dehulled grains	F. W. Collins and R. G. Fulcher (unpublished results)
	Pennisetum typhoides	Whole and dehulled grains	Reichert (1979)
(100) X″-Glucosylvitexin (R = 3×H + 1×glucosyl)	*P. typhoides*	Whole and dehulled grains	Reichert (1979)

(*continued on next page*)

TABLE X (continued)

Structure	Genus	Tissue or Material Examined	References
(101) Vicenin-3	*T. aestivum*	Wheat germ (also occurs as ester with sinapic acid)	King (1962) Wagner (1974)
(102) Vicenin-1	*T. aestivum*	Wheat germ (also occurs as ester with sinapic acid)	King (1962) Wagner (1974)
(103) Shaftoside	*T. aestivum*	Wheat germ (also occurs as ester with sinapic acid)	King (1962) Wagner (1974)
(104) x″-Glucosylorientin	*P. typhoides*	Whole and dehulled grains	Reichert (1979)

(76) have been isolated from aleurone and pericarp tissues of sorghum, barley, and corn. In extracts of oat hulls, leucodelphinidin (77) was tentatively identified by Vogel (1961) as the major proanthocyanidin present.

Free and glycosylated anthocyanidins (79–88) have been reported from a variety of cereal grains, and their occurrence is widespread, especially in cultivars with red or purple seed coats. In some light-colored grains, the localization of anthocyanidin derivatives in the aleurone layer is at least in part responsible for characteristic bluish coloration of this tissue (Mullick et al, 1958; Dedio et al, 1969). Cultured corn endosperm cells apparently have the genetic capability to synthesize anthocyanidin glycosides, although this trait is not usually expressed because intact tissues and dark-grown cell cultures exhibit extremely low pigment levels. In the presence of light, however, the 3-*O*-glucosides of pelargonidin and cyanidin (83 and 85) are synthesized de novo.

This information suggests that the proanthocyanidin and anthocyanidin flavonoid types are confined to the peripheral tissues of monocot cereal grains. In this connection, Nip and Burns (1969) used the visible orange color and insolubility of the lead salts of 3-deoxyanthocyanidins to study their cellular distribution in sorghum grains. In all the red sorghum varieties studied, the 3-deoxyanthocyanidins were found in the stylar region, the epidermis, and the cross-cell/tube cell layers of the pericarp; the mesocarp (pericarp tissue between the outermost layer and the cross-cell/tube cell layer) was not pigmented. In white-seeded varieties, a yellow pigment—a glycoside of apigeninidin (79)—replaced the orange component of red varieties, but the flavonoid distribution pattern within the grain remained the same. It is therefore not surprising that proanthocyanidins were not detected by Durkee and Thivierge (1977) in the meal prepared from dehulled oats.

The chalcones, flavanones, and flavonol derivatives of oat and related cereal grains are summarized in Table IX. Kambal and Bate-Smith (1976) have isolated eriodictyol (91) and its related chalcone (89) from sorghum pericarp tissue. These compounds appear to be closely related to the more highly reduced leucoanthocyanidin luteoforol (78) and the anthocyanidin luteolinidin (81), with which they may form a direct biosynthetic sequence. In oat hulls, Vogel (1961) identified the flavanone homo-eriodictyol (92) and its related chalcone (90). The presence of a methoxy group at the 3'-position of the flavanone suggests the occurrence in oats of a methyltransferase capable of methylating the phenyl ring hydroxyl group at the flavanone-chalcone stage of biosynthesis.

Flavonol derivatives occur relatively rarely in the grass family (Harborne and Williams, 1976) and appear to be restricted to two subfamilies, Festucoideae and Panicoideae. Trace amounts of both kaempferol (93) and quercetin (94) have, however, recently been detected in dehulled oat extracts (F. W. Collins and R. G. Fulcher, *unpublished results*). In corn, Kirby and Styles (1970) found both kaempferol and quercetin in hydroxylates of aleurone tissue, and kaempferol-3-rutinoside-7-glucuronide has been tentatively identified in the pericarp/testa tissue of red sorghum varieties (Nip and Bruns, 1969). As with the procyanidins and anthocyanidins, chalcones, flavanones, and flavonols appear to be characteristically localized in the pericarp-aleurone regions of mature monocot cereal grains.

Flavone derivatives have been detected in cereal grains and their structures and occurrences summarized in Table X. Luteolin aglycone (95) occurs in the free form in dehulled oat grains along with tricin (98) (F. W. Collins and R. G. Fulcher, *unpublished results*), which has been reported from oat hulls (Vogel, 1961). Interestingly, tricin was first characterized as the resistance factor in rust-resistant cultivar Khapli wheat by Anderson (1933) and later reported by Chen and Geddes (1945) as a major coloring component along with other unidentified flavones in wheat grains. They found the highest concentration of flavones in the embryo (24.3–98.8 ppm, expressed as tricin equivalents) and the bran (4.4–14.7 ppm), whereas low concentrations were exhibited by endosperm preparations (0.6–0.8 ppm). Recently, the structures of several *C*-glycosyl-flavones from wheat germ and pearl millet have been elucidated. The flavones originally isolated by King (1962) from commercial wheat germ, where they occurred to the extent of 0.2–0.3% (2,000–3,000 ppm), were apparently a mixture of vicenin-3 (101), vicenin-1 (102), and shaftoside (103) (Wagner, 1974). Acylated forms of these glycosides involving an alkali-labile sinapic acid moiety attached to the sugar are also present. The 8-*C*-glucosyl derivative of apigenin, vitexin (99), has been detected chromatographically in oat groats (F. W. Collins and R. G. Fulcher, *unpublished results*) and occurs as well in pearl millet (Reichert, 1979). Along with vitexin, an *O"*-glucosyl-vitexin (100) and the corresponding glycosyl-orientin derivative (104) have been partially characterized from millet (Reichert, 1979). The three *C*-glucosyl-flavones are responsible for the intense yellow-green color of flour-water pastes at alkaline pH and appear to be located in the peripheral region of the endosperm. In terms of glucosyl-vitexin equivalents, whole millet grains contain about 0.1% *C*-glucosyl-flavones (1,000 ppm), but the concentration was reduced to about 20% of this value after half the kernel was removed with a laboratory pearler.

Considerable work has been done to elucidate the structural chemistry and physiology of flavonoid glycosides in the developing oat plant. The main flavonoid derivatives of aerial parts consist of a series of both *C*-glycosyl- and *O*-glycosyl-flavones, most of which exhibit particular organ-specific distributions within the plant (Table XI). Popovici and Weissenbock (1976) analyzed the flavonoid patterns of developing oat seedlings (cv. Gelbhafer) under standardized growth conditions and field cultivation. The primary leaf of phytotron-grown plants contained predominantly three *C*-glycosyl-flavones based on apigenin aglycone, viz., isoswertisin-2"-*O*-rhamnoside (108), vitexin-2"-*O*-rhamnoside (107), and isovitexin-2"-*O*-arabinoside (106), along with small amounts of isovitexin (105) (Chopin et al, 1977). Flavonoid (109) is probably identical with the isoswertisin-*O"*-rhamnoside isolated from *Avena sativa* cv. Shokan 1 by Nabeta et al (1977), whereas flavonoids A1 and A2 found by Harborne and Hall (1964) in leaves of cultivar Blenda correspond to flavonoids (107) and (106), respectively. Field-grown primary leaves contain these glycosides as well as an additional pair of glycosides based on luteolin, viz., isoorientin (110) and its 2"-*O*-rhamnoside (112). With rising insertion of the leaves, the flavonoid pattern of subsequent leaves remained qualitatively the same.

The stalk exhibited a broader spectrum of flavonoids containing not only all of the leaf flavonoids but also the 6,8-di-*C*-glucoside (108), the 7-*O*-glucoside of isoorientin (111), and a series of flavone-*O*-glycosides based on the aglycone tricin (compounds 116–121).

Inflorescences exhibited a full and complex pattern of more than 20 flavonoids. All of the leaf and stalk flavonoids were present, along with the 3'-methyl ether derivative of isoorientin (115) and several isoorientin glycosides, including two new, partially characterized biosides (113) and (114) (Chopin et al, 1977). The caryopses showed a similar high level of diversity in flavonoid biosynthetic capability. Within the caryopses, the *C*-glycosyl-flavone content of the kernel was low, with most of the components localized in the glumes. No further work was done on the mature grain.

Studies on the cellular distribution of the oat leaf flavonoids have revealed tissue-specific distributions for mesophyll and epidermal cells (Weissenbock et al, 1976; Effertz and Weissenbock, 1980). Mesophyll tissues showed quantitatively and temporally distinct patterns of *C*-glycosyl-flavone

TABLE XI
*C3- and O-*Glycosylflavones of the Oat Plant

Structure	Common and Systematic Name	Code	Distribution
(105)	Isovitexin Apigenin-6-C-β-D-glucoside	F₅	Leaves Stems Inflorescences
(106)	Isovitexin-2″-O-arabinoside Apigenin-6-C-(2″-O-α-L-arabinosyl)- β-D-glucoside	F₂	Leaves Stems
(107)	Vitexin-2″-O-rhamnoside Apigenin-8-C-(2″-O-α-L-rhamnosyl)β- D-glucoside	F₃	Leaves
(108)	Vicenin-2 Apigenin-6,8-di-C-β-D-glucoside	F_AD	Stems Inflorescences
(109)	Isoswertisin-2″-O-rhamnoside Apigenin-8-C-(2″-O-α-L-rhamnosyl)β-D- glucoside-7-O-methyl ether	F₁	Leaves Stems
(110)	Isoorientin Luteolin-6-C-β-D-glucoside	Fₑ	Leaves Inflorescences
(111)	Isoorientin-7-O-glucoside Luteolin-6-C, 7-O-di-β-D-glucoside	Fₑ	Stems
(112)	Isoorientin-2″-O-rhamnoside Luteolin-6-C-(2″-O-α-L-rhamnosyl)β-D- glucoside	Fₐ	Leaves Inflorescences

(*continued on next page*)

TABLE XI (continued)

Structure	Common and Systematic Name	Code	Distribution
(113,114)	R = glycosyl (two different triosides) Isoorientin-2"-*O*-glycosyl arabinosides Luteolin-6-*C*-(x"'-*O*-glycosyl-2"-*O*-α-L-arabinosyl)β-D-glycoside	F_c F_d	Inflorescences Inflorescences
(115)	Isoscoparin Luteolin-6-*C*-β-D-glucoside-3-*O*-methyl ether	F_f	Inflorescences
(98)	Tricin	T	Stems
(116)	Tricin-4'-*O*-α-L-arabinoside	T_2	Stems Inflorescences
(117)	Tricin-4'-*O*-β-D-glucoside	T_4	Stems Inflorescences
(118)	R = glucose or arabinose Tricin-7-*O*-monoglycoside	T_6	Inflorescences
(119–121)	R_1 = glucose or arabinose (three different dimonosides) R_2 = rhamnose or glucose Tricin-4',7-*O*-diglycoside	T_7 T_8 T_9	Inflorescences Inflorescences Inflorescences

accumulation, which differed substantially from those of the epidermal tissues and were greatly influenced by environmental growth regimes.

I am currently working on the phytochemical characterization of the flavonoids of dehulled oat grains. In conjunction with R. G. Fulcher of Agriculture Canada, I am also attempting to determine the location of specific phenolics within the grain using fluorescence microscopy. Some of the results have already been included in Tables IX and X to facilitate comparisons, where appropriate, between various other cereal tissues, but they will be considered here in more detail. To isolate the flavonoids, 80% methanol extracts of oat flours (from mechanically dehulled *A. sativa* cv. Hinoat and Sentinel) were partitioned with chloroform:methanol:water (45:45:10, v/v) to remove lipids and other nonpolar components. The aqueous methanol phase was then fractionated by Sephadex LH-20 column chromatography into high- and low-molecular-weight phenolic classes, using 50% isopropanol as solvent. A preliminary "mapping" was done on the low-molecular-weight fraction by two-dimensional high-resolution thin-layer chromatography (HRTLC) on Schleicher and Schuell Type F 1700 micropolyamide sheets. Such preliminary mapping was found extremely useful in comparative studies and offered an economical and rapid method for studying the complex mixtures of phenolics in oat grains. Detection of the phenolics on chromatographic plates was based on examination of fluorescence/absorption characteristics of the components under long-wavelength UV light before and after spraying with a 0.1% solution of 2-aminoethyl diphenylborinate (ADPB) in 95% ethanol. This sensitive reagent has been found particularly useful for detection and characterization of flavones, flavonols, hydroxycinnamic acids, and their glycosylated derivatives because of diagnostic fluorescent colors and color shifts induced upon complexation with these phenolics (Homberg and Geiger, 1980). When serial dilutions of standard flavones and flavonols spotted on polyamide sheets are used, as little as 1–50 pg/mm^2 can be readily detected (F. W. Collins, *unpublished results*).

A typical two-dimensional map of Hinoat oat phenolics from the low-molecular-weight fraction is shown in Plate 1. The upper portion (Plate 1A) shows the unsprayed HRTLC plate (actual size, approximately 7.5×7.5 cm) containing the phenolic mixture equivalent to about 50–100 mg of oat flour, and the lower portion (Plate 1B) shows the same plate after spraying with the ADPB reagent. These and subsequent photographs (Plates 2 and 3) were prepared from Kodak Ektachrome PR 64 (daylight) color slides. Exposures varied between 15 and 30 sec (f4), using types 2A and 85 glass filters (effective cutoff wavelength, 405 nm), with excitation provided by laterally positioned 15-W long-wavelength UV lamps (two on each side, about 30 cm from the plate). The unsprayed map is characterized by a series of purple, dark brown, gray, and bluish zones extending from the origin (in the lower right-hand corner) directly upward. Several black-absorbing spots, typical of glycosylated flavones and 3-*O*-substituted flavonols, are visible in the central and right-central area, along with a series of blue and bluish white autofluorescent spots. After spraying the same plate with ADPB (Plate 1B), a number of the dark-absorbing spots appear fluorescent yellow, orange yellow, yellow green, and green, diagnostic for flavones and flavonols containing a free hydroxyl function at C-5 and varying degrees of hydroxyl/methoxyl substitutions at C-7, 3', 4', and 5' positions. The series of zones at the right is clearly differentiated into a group of ADPB-positive components (i.e., flavonoids) co-chromatographing with several dark ADPB-negative zones. The central bluish white zone prominent in Plate 1A remained unaffected, whereas the faint blue-green fluorescence of several adjacent spots was greatly enhanced with ADPB. Some of the barely discernible dark-absorbing spots in the lower portion of Plate 1A are readily detected with the reagent, appearing as a group of yellow, orange yellow, and green fluorescent spots. In the particular two-dimensional solvent pair used here, a considerable number of phenolics remained at the origin. These components can readily be resolved, however, using different solvent-pair combinations. In fact, using several solvent combinations, over 150 chromatographically distinct phenolics have been consistently observed and mapped from the low-molecular-weight phenolic fraction.

To facilitate purification of the flavonoids, the low-molecular-weight fraction was subfractionated into two groups based on ionization characteristics of the phenolics at pH 7 in aqueous isopropanol. The low-molecular-weight mixture in 50% isopropanol was loaded onto a DEAE-Sephacel (Pharmacia) ion-exchange column, which had been converted to the formate and packed in 50% isopropanol. Elution of the column with several bed volumes of 50% isopropanol gave a "neutral-plus-cationic" subfraction containing the uncharged flavones, flavonols, and their glycosides along with neutral and cationic hydroxycinnamoyl and benzoyl esters, amides, and glycosides. Elution with the acidic solvent water:isopropanol:formic acid (50:45:5, volume %) gave an anionic subfraction containing the free hydroxycinnamic and benzoic acids, their ether glycosides, and a group of anionic *N*-acylaminobenzoic acids.

Analysis of the two subfractions by two-dimensional HRTLC revealed rich and diversified phenolic complements in each subfraction. Representative maps of the neutral-plus-cationic and anionic subfractions from the cultivar Hinoat, both before and after spraying with ADPB, are shown in Plates 2 and 3, respectively. Subfractionation allowed greater loads to be applied to the HRTLC plates, and the amounts spotted for Plates 2 and 3 represent the typical amounts corresponding to about 150–200 mg of oat flour. The unsprayed map (Plate 2A) of the neutral-plus-cationic subfraction was qualitatively similar to the entire low-molecular-weight fraction (Plate 1). The dull brown and dark-absorbing compounds extending upward from the origin and the absorbing spots in the lower central and right-hand halves were more easily discernible than in Plate 1A because of the greater amounts spotted. However, notably absent from the neutral-plus-cationic subfraction were some of the gray brown and whitish zones along the right-hand margin of the map. After spraying with ADPB (Plate 2B), the series of dull brown and black-absorbing compounds appeared yellow, yellow green, orange yellow, and green yellow fluorescent (T, A, K, L, Q), diagnostic for flavone and flavonol aglycones. A number of similarly fluorescent zones displayed considerably greater mobilities in the aqueous direction (first direction), typical of flavone and flavonol glycosides, and appeared across the lower half of the plate. In comparison with Plate 1B, it can be seen that most of the ADPB-positive compounds in the entire fraction were recovered in the neutral-plus-cationic subfraction.

A typical map of the anionic subfraction phenolics from oat flour (Hinoat) is shown in Plate 3. The unsprayed map (Plate 3A) showed a prominent series of poorly resolved gray, brown orange, and greenish brown zones along the right-hand side of the map. Once again, a considerable amount of phenolic material appears at the origin of this particular map, but it can be clearly resolved using different solvent pairs (at the expense of other phenolics). The anionic behavior of this series of compounds has been found to be the result of the free carboxylic function of at least 20 different *N*-acyl-2-aminobenzoic acids present in oat grains. The complete structure of some of these conjugates has been determined (Collins, 1983) and is discussed in Section VIE. A conspicuous blue-white autofluorescent zone, similar to the one seen in the neutral-plus-cationic subfraction, and several less distinct bluish phenolics are also present. After spraying with ADPB (Plate 3B), some of the low-mobility *N*-acylaminobenzoic acids became fluorescent yellow and blue white, although most remained unaffected. Subsequent to this work, it has been found that these acid derivatives become intensely fluorescent yellow, blue, orange, and green when sprayed with a 5% solution of ethanolamine in isopropanol. The prominent blue-white autofluorescent central zone remained unchanged with ADPB, but several of the neighboring blue fluorescent phenolics exhibited enhanced fluorescence typical of free hydroxycinnamic acids (*p*-coumaric, ferulic, sinapic, etc.). A number of ADPB-positive components that give colors similar to those seen in Plate 2B were also present in the anionic subfraction, although only in trace amounts.

The major flavonoids were purified using repeated preparative column chromatography of the neutral-plus-cationic subfraction on Sephadex LH-20, using the solvent systems consisting of methanol:water (50:50, 65:35, and 80:20, volume %) and methanol:water:formic acid (50:45:5 and 80:15:5). Purity of the aglycones (T, A, K, L, and Q in Plate 2B) was monitored by one- and two-dimensional HRTLC using a wide variety of solvent systems. A number of different glycoside derivatives of these aglycones were also purified by the same procedure (1A, 2A, 1Q, and 1K). All flavonoids purified were identified by comparison of UV spectra and the effects of diagnostic shift reagents (Mabry et al, 1970), direct-probe mass spectra, and co-chromatography with authentic samples. The major flavonoids from Hinoat oat flour identified to date are summarized in Table XII. The flavone class is represented by three major derivatives—apigenin (A), luteolin (L), and tricin (T)—containing mono-, di-, and trisubstituted phenol (B-ring) groups, respectively. The same three flavones have previously been detected in the form of glycosides in the aerial parts of the oat plant (see Table XI). The flavonols kaempferol (K) and quercetin (Q), containing B-ring substitution patterns analogous to apigenin and luteolin, are also present in the flour but are apparently not present in any other organ or tissue of the oat plant so far examined (Table XI). Several different glycosidic derivatives of these major aglycones are present, of which apigenin-6-*C*-glucoside and 8-*C*-glucoside (1A and 2A) as well as the 3-*O*-[L-rhamnosyl-α-1→6]-β-D-glucosides (i.e., rutinosides) of both quercetin and kaempferol (1Q and 1K) have been identified. In the case of *C*-substituted flavones, tentative identification as to the aglycone moiety is possible using the ADPB reagent by direct comparison with the aglycone on the TLC map, since *C*-substitution does not significantly alter the fluorescence characteristics of the ADPB-flavone complex (Homberg and Geiger, 1980). Some of these chromatographically separate but spectrally similar derivatives are clearly evident on the TLC plate. Work is in progress to elucidate the structures of some of these derivatives.

In collaboration with R. G. Fulcher, an attempt has been made histochemically to locate some of the flavonoids identified in the extracts within intact oat grain tissues, using the ADPB reagent and fluorescence microscopy. Hand-sectioned oat grains (cultivar Hinoat) were either preextracted with three changes of 80% methanol and treated with a 0.1% solution of ADPB in 100% methanol or were treated directly with the reagent. Control sections were treated with a 100% ethanol reagent blank lacking the ADPB. The sections were dried and observed by epi-illumination with excitation at approximately 360 nm and photographed using filter combinations I and II (as described in Chapter 3). The extract from preextracted sections was evaporated to dryness by rotary evaporation, taken up in 50% isopropanol, and fractionated on an LH-20 minicolumn, prepared using a pasteur pipette, into high- and low-molecular-weight phenolic fractions as previously described. The low-molecular-weight fraction was subsequently analyzed by two-dimensional HRTLC on polyamide plates.

Fluorescence microscopy of the sections treated directly with ADPB showed a localization of intensely yellow fluorescence around the entire endosperm in a more or less increasing gradient peripherally from the outer layers of the endosperm, through the subaleurone region, and into the aleurone layer. The yellow fluorescence was particularly striking in the area of the crease (Plate 4), whereas the embryo exhibited little if any reaction. As shown in Plate 4A, some yellow and orange autofluorescence could be detected in the control sections but appeared to be largely unaffected by the ADPB reagent. On the other hand, an extensive region embracing aleurone and subaleurone tissues exhibited the bright yellow fluorescence in the presence of ADPB (Plate 4B). When preextracted sections were treated with ADPB, virtually none of the yellow

TABLE XII
Structure and Ultraviolet (UV) Characteristics of Some Neutral Oat Flavonoids

Code	Name	R_1	R_2	R_3	R_4	R_5	Alone	+ADPB[b]
A	Apigenin-6-C-glucoside	H	H	H	H	H	Black abs.	Fluorescent blue green to brown
1A	Apigenin-6-C-glucoside	H	H	Glu	H	H	Black abs.	Fluorescent blue green to brown
2A	Apigenin-8-C-glucoside	H	Glu	H	H	H	Black abs.	Fluorescent blue green to brown
L	Luteolin	H	H	H	H	H	Black abs.	Fluorescent yellow
T	Tricin	H	H	H	OCH₃	OCH₃	Black abs.	Yellow green
K	Kaempferol	OH	H	H	H	H	Brown abs.	Fluorescent green
1K	Kaempferol-3-O-rutinoside	O-rut	H	H	H	H	Purple abs.	Green
Q	Quercetin	OH	H	H	H	OH	Orange brown	Fluorescent orange yellow
1Q	Quercetin-3-O-rutinoside	O-rut	H	H	H	OH	Purple abs.	Yellow

Characteristics in UV Light on Polyamide TLC Layers (Excitation ~360 nm)[a]

[a] TLC = thin-layer chromatography.
[b] ADPB = aminoethyl diphenylborinate.

fluorescence could be detected in any of the tissues, indicating that the constituents responsible had been removed (or destroyed) by preextraction with 80% methanol. Analysis of the 80% methanol extract by TLC revealed a pattern of yellow, greenish yellow, and orangy yellow fluorescent components similar to that shown in Plate 1B and, by direct comparison with the authentic standard, composed primarily of the five flavone and flavonol aglycones and their glycosides. It seems probable, therefore, that the flavones apigenin, luteolin, and tricin, the flavonols quercetin and kaempferol, and to a lesser extent their glycosides, are at least in part responsible for the in situ ADPB-induced fluorescence observed in the sections. If so, these flavonoids are for the most part confined specifically to the subaleurone and aleurone tissues of the oat grain and can be removed by simple extraction with aqueous methanol.

C. Functionality

The physiological function of specific flavonoids in the cereal grain remains largely unknown. Most observations on the effects of flavonoid derivatives on biological systems have been made using either in vitro analyses, such as alterations in specific enzyme kinetics, or in vivo modifications of physiological parameters (such as growth and germination) of model systems in the presence of exogenously applied flavonoids. In vitro, certain flavonoids inhibit enzyme systems including mitochondrial ATP formation, malate dehydrogenase, glutanate decarboxylase, pectin methylesterase, cellulases, β-amylase, and many others (see the review by Van Sumere et al, 1975). The amino acid activation step in protein synthesis in barley embryo is strongly inhibited by quercetin and related *ortho* dihydroxy phenolics capable of being oxidized to quinones (Van Sumere et al, 1975). Finally, chalcones, flavones, and flavonols, because of their strong absorption (log E between 4 and 4.5) in the 320–370 nm range, can act as UV screens and photodynamic pigments. Zenk (1967) has suggested a relationship between C-glycosyl-flavones and phototropic responses in *A. sativa* coleoptiles. A transverse light gradient in the coleoptile is formed by specific localization of flavones in epidermal and mesophyll cells (e.g., at 370 nm, 98% of the light is absorbed by flavones and less than 2% by other pigments), effectively shielding photoreceptors at critical action spectrum wavelengths.

In vivo, flavonoids undoubtedly exert regulatory control through direct modification of enzymatic processes such as oxidative uncoupling and inhibiting ATP formation, but also more indirectly through interaction with gibberellin- and auxin-mediated processes (Zinsmeister and Hollmuller, 1964; Corcoran et al, 1972; Jacobson and Corcoran, 1977); numerous other examples are given by McClure (1975). It is generally concluded in the case of auxin-mediated processes that certain flavonoids alter the levels of auxin by modifying the rate of destruction of IAA by peroxidase/IAA oxidase systems. Flavonoids containing a monohydroxylated phenyl ring (i.e., 4'-OH, as in kaempferol [93]) act as cofactors strongly promoting IAA oxidation, whereas the corresponding 3',4'-dihydroxy derivatives (e.g., quercetin [94]) are potent inhibitors. Modification of auxin-mediated responses may, therefore, be dependent more on relative than on absolute concentrations of particular homologous flavonoid pairs. However, their subcellular compartmentation and their ability to reach the in situ IAA-destroying systems are also crucial factors, which remain largely unexplored.

Specific *C*-glycosylflavonoids may also function in protective roles as insect-feeding deterrents or by displaying narrow-spectrum pesticidal activity. For example, maysin, the toxin factor toward the corn earworm larvae, *Heliothis zea*, which occurs in the silks of resistant maize varieties, has recently been shown to be 2″-*O*-α-L-rhamnosyl-6-*C*-(4″-ketofucosyl)-luteolin (Elliger et al, 1980). Tricin and unidentified *C*-glycosyl-flavones have been implicated as the aphid-feeding deterrents in wheat (Dreyer and Jones, 1981). Other examples are cited by Swain (1977).

V. AMINOPHENOLICS

A. Structure

In cereal grains in general and oats in particular, this group of amine-containing phenolics has received little attention. However, the structures and somewhat tentative biosynthetic derivations of several important aminobenzoic acids and aminophenolics are outlined in Fig. 4. As is the case with phenylalanine, tyrosine, and related aminoalkylphenols, the aromatic ring carbons of the amino-substituted benzoic acids are also derived from the shikimate pathway. Chorismic acid (122) occupies an important intermediate position, leading to both amino-substituted aryl derivatives and indolyl compounds, including the amino acid tryptophan. Transamination of the enolpyruvate side chain of chorismate utilizing glutamine yields the amides (123, 129). Aromatization (124, 125) involving *ortho*-ring substitution of (123), followed by elimination of the side chain, generates *o*-amino-substituted benzoic acid (126), also known as anthranilic acid. Comparable aromatization of (129), involving *para*-displacement and substitution (130, 131) with side-chain elimination, yields *p*-aminobenzoic acid (132), a component of the pteroyl function in folacin. Anthranilic acid (126) is a precursor in the synthesis of tryptophan and related indoles and probably the immediate precursor of the *o*-aminophenoxy group (128) via either 3-hydroxyanthranilate (127) or perhaps more directly through oxidative decarboxylation analogous to *p*-hydroquinone formation from the corresponding benzoic acid.

The substituted benzoxazinones (135) are also biosynthetically derived from shikimate. The benzene ring carbons of (135) are derived from ring-labeled shikimate or anthranilate and the amine nitrogen of anthranilic acid is incorporated into the oxazinone ring (Tipton et al, 1973). A possible route may involve aromatization of intermediate (124) followed by oxidation, decarboxylation, substitution, and glucosylation, but further work is clearly needed to elucidate the details of benzoxazinone biosynthesis. Additional biosynthetic aspects of some of the compounds shown in Fig. 4 are discussed by Ganem (1978) and Floss (1979).

B. Occurrence

Several phenolic compounds containing amine and substituted-amine functions are known to occur in oats and related cereals. For the purpose of this review, these phenolics are briefly discussed under the following general groupings: 1) aminoalkylphenols (including compounds containing primary, secondary, tertiary, and quaternary amino nitrogen); 2) aminobenzoic acid and aminophenol derivatives; and 3) benzoxazinone-glycosides and related compounds.

AMINOALKYLPHENOLS

With the exception of the phenolic amino acid tyrosine, aminoalkylphenols are not commonly reported as constituents of cereal grain tissues. In most cases, the reported occurrences have stemmed from follow-up studies of unusual ninhydrin-positive components detected during amino acid and/or protein hydrolysate analyses. Table XIII summarizes the occurrence of simple

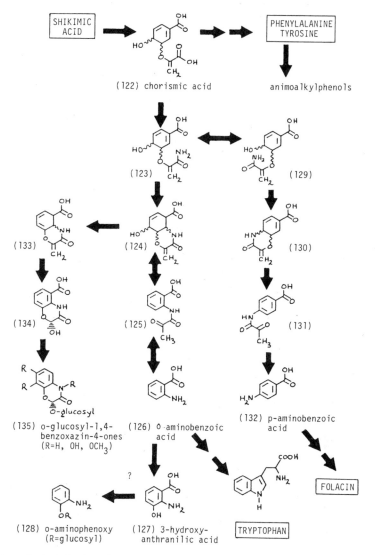

Fig. 4. Biosynthetic derivation of aminobenzoic acids and aminophenolics in oats and related cereals.

TABLE XIII

Structure and Occurrence of Aminoalkylphenols in Oats and Related Cereals

Structure	Name	Genus	Tissue or Material Examined	References
(136)	DOPA (dihydroxyphenylalanine)	*Avena sativa* *Triticum aestivum*	Whole grains	Hoeldtke et al (1972)
(137)	*p*-Aminoethylphenol (tyramine)	*Hordeum vulgare* *Panicum* sp.	Seedlings Hulls	Hegnauer (1963) Sato et al (1970)
(138)	*p*-Methylaminoethylphenol (*N*-methyltyramine)	*H. vulgare* *Panicum* sp.	Seedlings Hulls	Hegnauer (1963) Sato et al (1970)
(139)	*p*-Dimethylaminoethylphenol (*N,N*-dimethyltyramine)	*A. sativa* *H. vulgare* *Secale cereale* *Zea mays* *Panicum* sp.	Seedlings Hulls	Hegnauer (1963) Sato et al (1970)
(140)	*p*-Trimethylaminoethylphenol (*N,N,N*-trimethyltyramine; candicine)	*H. vulgare*	Seedlings	Hegnauer (1963)

Structure	Compound	Species	Plant part	Reference
(141)	*p*-Aminomethylphenol (4hydroxybenzylamine)	*H. vulgare*	Whole grains	Slaughter and Uvgard (1972)
		Fagopyrum esculentum (Polygonaceae)	Whole seeds	Koyama et al (1971)
(142)	*o*-Aminomethylphenol (salicylamine)	*F. esculentum* (Polygonaceae)	Whole seeds	Koyama et al (1971)
(143)	Serotonin (5-hydroxytryptamine)	*H. vulgare*	Seedlings	Schneider and Wightman (1973)
		Oryza sativa	Endosperm, shoots	Smith (1977)
(144)	*N*-Methylserotonin (*N*-methyl-5-hydroxytryptamine)	*H. vulgare*	Seedlings	Schneider and Wightman (1973)
(145,146)	(145) R = H; *N,N*-dimethyl-7-methoxytryptamine	*Phalaris aquatica*	Seedlings	Mulvena et al (1983)
	(146) R = OCH₃; *N,N*-dimethyl-5,7-dimethoxytryptamine			
(147)	(*S*)-Dhurrin	*Sorghum* spp.	Seedlings	Hegnauer (1963)
		Triticum spelta		Erb et al 1979
		Secale spp.		Erb et al (1981a, 1981b)

aminoalkylphenols in the cereal and forage crops. Hoeldtke et al (1972) have reported the occurrence of dihydroxyphenylalanine (DOPA, 136) in both oat and wheat grains. The authors concluded that a tyrosine hydroxylase was present and active in the grains, since pretreatment of the ground grains with proteolytic enzymes dramatically increased the DOPA content. Primary (137), secondary (138), tertiary (139), and quaternary (140) *p*-aminoethylphenols have been found in barley seedlings, and the tertiary amine hordenine (139) also occurs in seedlings of oats, rye, corn, and millet (Hegnauer, 1963). In the case of barley, the amines are apparently absent in the mature dry grain but appear sequentially with the onset of germination. They arise from the decarboxylation and subsequent stepwise N-methylation of tyrosine by enzyme systems produced in the embryo and later in the roots (Mann et al, 1963). However, in millet (*Panicum* sp.), these amines are already present in the mature grain, where they are localized in the hulls (Sato et al, 1970). Slaughter and Uvgard (1972) have isolated a tyramine homologue, *p*-aminomethylphenol (141), from ungerminated barley, but its localization within the grain was not studied. In this connection, both *p*-aminomethylphenol and *o*-aminomethylphenol (142) also occur in buckwheat seeds (*Fagopyrum esculentum* Moench; Polygonaceae). They occur not only in the free form (Koyama et al, 1971) but also as the β-*O*-glucopyranosides (Koyama et al, 1974) and as *N*-acyl conjugates with L-glutamic acid derivatives (Koyama et al, 1973).

Several hydroxylated and methoxylated tryptophan analogues, including serotonin (143), have been isolated from germinated barley (Schneider and Wightman, 1973) and *Phalaris* spp. (Mulvena et al, 1983). Like the tyramine derivatives, they are not found in mature grain but are produced shortly after germination in shoot and root tissues. Serotonin has been detected in the endosperm and shoot of rice seedlings along with 5-hydroxytryptophan (cited by Smith, 1977). The phenolic cyanogenic glucoside (S)-dhurrin (147) has been isolated from several species of sorghum, wheat, and rye seedlings. Similar but as yet unidentified cyanogenic glycosides are also present in oat seedlings and mature caryopses of oats, wheat, barley, rye, and corn (Erb et al, 1979, 1981a, 1981b).

AMINOBENZOIC ACIDS AND AMINOPHENOLS

Aminobenzoic acids, apart from *p*-aminobenzoic acid, are relatively rare in nature. Like the free phenols, they are not accumulated in the free form to any major extent in living tissues and generally occur in conjugated forms. For example, *p*-aminobenzoic acid is present in all cereal grains examined to date but occurs at least in part as the pteroyl function (148) of the ubiquitous B-vitamin

(148) Pteroyl function containing conjugated *p*-aminobenzoic acid

folic acid (folacin, N-pteroylglutamic acid). Neither qualitative nor quantitative analyses of either total or folate-conjugated *p*-aminobenzoate have been reported for oats to date. Data for *p*-aminobenzoate as folacin conjugates in wheat, rye, and triticale mill fractions (Calhoun et al, 1960; Michela and Lorenz, 1976) indicate that the high folacin content in bran and shorts fractions may result from localization of this component in the peripheral tissues of cereal grains.

Aminophenols containing a free primary amine attached directly to the phenol ring are perhaps the newest type of aminophenolic to be described. Mason et al (1973) and Mason and Kodicek (1973) reported the occurrence of an arylamine, identified as *o*-aminophenol, in alkali-hydrolyzed wheat bran extracts. Partial hydrolysis revealed the presence of *o*-aminophenol-*O*-glucoside and traces of oligomeric homologues, presumably the di- and triglucosides (Mason and Kodicek, 1973), suggesting the occurrence of *o*-aminophenoxy-substituted glycans in wheat bran. Fulcher et al (1981) have recently confirmed the presence of alkali-labile forms of *o*-aminophenol in 80% ethanol-soluble extracts of wheat bran; similar results were also obtained with oat and barley bran extracts. Chromatographic analyses of bran and flour extracts and fluorescence microscopy of sectioned oat, wheat, and barley grains were done, using, in both studies, the fluorophore *p*-dimethylaminobenzaldehyde to detect the primary arylamine function. Unhydrolyzed flour and bran extracts of oats, wheat, and barley did not contain free *o*-aminophenol, and only traces of the free arylamine were detected in the flour extracts even after hydrolysis. The unhydrolyzed bran extracts did, however, exhibit chromatographically complex arylamines, which reacted with the detection reagent in the same manner as *o*-aminophenol, indicating that the amine group was unsubstituted in these components. Fluorescence microscopy of the sectioned grains revealed a similar reaction that was localized exclusively in the aleurone protein bodies, in agreement with the chemical extraction analyses of arylamine components in bran but not in flour extracts of all three cereal grains.

Previous evidence from chromatographic and hydrolytic studies by Kodicek and collaborators (Kodicek and Wilson, 1960; Mason and Kodicek, 1973; Mason et al, 1973) and more recently by Koetz et al (1979) and Fulcher et al (1981) have shown a close biochemical and anatomical association of *o*-aminophenoxyglucosyl, hydroxycinnamoyl, and nicotinoyl moieties with a series of glycan- and/or glycopeptide-containing macromolecules. The components range in molecular weight from about 1,500 to 17,000 and vary considerably in qualitative composition with respect to *o*-aminophenoxy, feruloyl, sinapoyl, and nicotinoyl substituents. A general schematic model is shown in Fig. 5, updated from the one first proposed by Kodicek and Wilson (1960) to indicate the types of functional groups exhibited by these cereal grain constituents. In particular, the *o*-aminophenoxy moiety is shown in ether linkage to a main-chain or branch-substituted glycosyl residue through the phenolic hydroxyl. The amino group is unsubstituted, a prerequisite for reactivity in situ with *p*-dimethylaminobenzaldehyde (Suzuki et al, 1976).

BENZOXAZINONES

A series of 2-hydroxy-2H-1,4-benzoxazin-3(4H)-ones and their glucosides have been characterized from cereals, the aglycones of which have been shown to exhibit a wide spectrum of biological activities. The glucosides have been isolated from coleoptiles, leaves, stems, and roots of both seedlings and mature tissues of corn, wheat, and rye, but they are apparently also present in abundance in the endosperm of wheat and rye (Virtanen, 1964). To date, no work has been reported on oat grains, but the presence of benzoxazinones in wheat and rye endosperm would suggest that they may also be present in oats. The occurrence of the structurally known glucosides in the cereals is summarized in Table XIV. Recently the probable occurrences of two glucosides, (153) and (154) (based on analysis of the aglycones), in a number of cultivated and wild grasses and cereals, including wheat, have been reported by Zuniga et al (1983). These authors failed to detect either compound (i.e., aglycone) in mature wheat grains, but the concentrations of both increased rapidly following germination. In *Triticum durum*, for example, concentrations as high as 6.3mM/kg of fresh weight were found in seedlings after four days, and levels decreased gradually afterwards (Argandona et al, 1981; Zuniga et al, 1983). These compounds are, however, not present in barley or sorghum seedlings.

C. Functionality

In oats and other cereal grains, the physiological functions of the aminophenolics remain largely unexplored. Koyama et al (1971) have suggested that certain aminomethylphenols may play a secondary role in pericarp

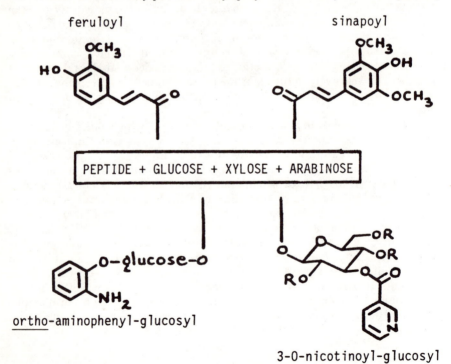

Fig. 5. Major aromatic functional groups in the structural model of "bound" niacin (niacytin) from wheat bran.

pigmentation and therefore alter the quality of light reaching underlying tissues during maturation of the grain. In vitro air oxidation of these aminophenolics results in the formation of progressively darker polymeric products, ranging from yellow to dark brown, and may occur in vivo during the latter stages of grain maturation. Enzymatic dimerization of tyramine (137) by peroxidase to give the yellow brown biphenyl ether derivative 3,3'-dityramine (157) (Gross and

(157) 2,2'-dihydroxy-5,5'-bis-(amino-ethyl) diphenyl (3,3'-dityramine)

Sizer, 1959) may exemplify the first step in comparable enzyme-mediated processes. A possible role as protective agents against herbivores can be ascribed to certain aminophenols such as DOPA (136), serotonin (143), and the tyramine derivatives. The production of some of these neutroactive and neutrotoxic

TABLE XIV
Structure and Occurrence of Benzoxazin-3-one Glucosides in Cereals

Structure (R = glucosyl)	Number	Source	Tissue or Material Examined	Reference
	(149)	*Zea mays* *Secale cereale*	Seedlings	Hofman and Hofmanova (1969)
	(150)	*Z. mays*	Seedlings	Hofman et al (1970b)
	(151)	*Z. mays*	Roots	Gahagan and Mumma (1967)
	(152)	*Z. mays*	Seedlings	Hofman and Masojidkova (1973)
	(153)	*Z. mays* *S. cereale* *Triticum aestivum*	Seedlings	Hofman and Hofmanova (1969) Hietala and Virtanen (1960)
	(154)	*Z. mays* *S. cereale* *T. aestivum*	Seedlings Leaves	Gahagan and Mumma (1967) Wahlroos and Virtanen (1959)
	(155)	*Z. mays* *T. aestivum*	Roots	Hofman et al (1970a)
	(156)	*Z. mays*	Coleoptiles Seedlings	Klun et al (1970) Hofman and Masojidkova (1973)

constituents in young seedling plants may allow early protection against predators without expenditure of vast amounts of metabolic energy. Undoubtedly the physiological role of cyanogenic phenolic glucosides is related to the ease with which toxic amounts of HCN can be generated by simply damaging the cells (Conn, 1981), thus repelling or inhibiting herbivores, pathogens, or competitors.

Similar roles for *o*-aminophenoxyglycosides in peripheral pigmentation and protection might be implicated if these components prove to be glucosidase labile. The aglycone *o*-aminophenol has an exceptionally high reduction potential and is readily oxidized in air (and on TLC plates!) (Gerber, 1968) to give deeply colored dimeric condensation products, including primarily 2-aminophenoxazin-3-one (158) (Gerber, 1968; Ayyangar et al, 1981). Enzyme-

o-aminophenol

(158)

2-amino-phenoxazin-3-one

mediated oxidative dimerization has also been effected in vitro by mammalian (Nagasawa et al, 1959) and plant systems (Nair and Vaidyanathan, 1964), and similar arylamine oxidases have long been known to occur in wheat and barley grains, as well as in malted wheat, barley, corn, and rice (Wallerstein et al, 1948). The dimer (158), also known as Questiomycin B, exhibits potent antibacterial and antifungal activity (Gerber and Lechevalier, 1964). Like the related actinomycin antibiotics, it inhibits DNA-dependent RNA synthesis by binding to double-stranded DNA (Chibber et al, 1973). This latent biological activity of *o*-aminophenoxy-substituted macromolecules combined with their localization in the aleurone layer may provide cereal grains with a peripheral barrier against penetration and catabolism by certain pathogens. In addition, they may function through endogenous hydrolysis and dimerization to enable localized self-regulation of enzyme synthesis during certain stages of germination by blocking gene transcription.

The physiological roles for the benzoxazinone glucosides in the cereals have focused on the biological activities of the aglycones. The native glucosides of general formula (159) are initially present in intact tissues as β-D-glucopyranosides (Hofman and Hofmanova, 1971; Willard and Penner, 1976) and are spatially isolated from endogenous β-glucosidase enzymes. On mechanical disruption of the plant tissue, they are rapidly hydrolyzed to the corresponding unstable aglycone (160), which rearranges (Bredenberg et al, 1962) to give a substituted benzoxazolin-2-one (161). Further treatment of the benzoxazolinone with mild alkali generates the corresponding *o*-aminophenol (162). The benzoxazolinones have been shown both in vitro and in vivo to

(159)   (160)   (161)   (162)

provide protection against bacterial, fungal, and parasitic invasion. Growth and development of the bacterium *Erwinia carotovora* are severely inhibited by benzoxazolinones from rye, wheat, and corn (Corcuera et al, 1978). In corn, resistance to corn leaf blight (*Helminthosporium turcicum*), and in wheat, resistance to stem rust (*Puccinia graminis* var. *tritici*), have been related to the potential of the host plant to produce the aglycones (Elnaghy and Linko, 1962; Molot and Anglade, 1968; Couture et al, 1971). Resistance against the European corn borer (*Ostrinia nubialis*) and aphids (*Rhopalosiphum maidis, Metopolophium dirhodum,* and *Schizaphis graminum*) have been shown to be highly correlated with natural and media-supplemented benzoxazolinone concentrations (Klun et al, 1967; Long et al, 1978; Argandona et al, 1980, 1981; Zuniga et al, 1983). At the molecular level, benzoxazolinones have recently been shown to inhibit energy transfer reactions (Queirolo et al, 1983) and to act as required cofactors in the enzymatic detoxification of S-triazine herbicides in corn, wheat, rye, oats, barley, and sorghum (Hamilton, 1964; Tipton et al, 1971). Finally, the possible growth regulatory functions of benzoxazolinones have been investigated by Venis and Watson (1978). Two of the aglycones in corn (6-methoxy and 6,7-dimethoxy derivatives) inhibited binding of auxin to both membrane-bound and solubilized auxin receptor sites in corn coleoptile. The same benzoxazolinones inhibited IAA and auxin-induced growth in oat coleoptiles in the concentration range of $1-10 \ \mu g/ml$. The 6-methoxy derivative also inhibited root elongation in wheat seedlings (Wilkins et al, 1974).

VI. PHENOLIC ACID ESTERS AND AMIDES

From early analytic studies of cereal grain phenolics, it became apparent that much greater quantities of phenolic acids were present in covalently and noncovalently bound forms than in the free acid form. To describe the diversity of uncharacterized molecular components yielding phenolic acids, a rather vague terminology has developed. It classifies the components into such categories as soluble bound, residue bound, insoluble, acid and alkali labile, and others. Although perhaps initially helpful, these operationally defined categories are nevertheless based on the combined solubility and stability properties of both phenolic and nonphenolic moieties rather than strictly on the structural properties or types of bonding. Coupled with variations in the types and sequences of solvent extraction and hydrolysis procedures used, valid comparison and interpretation of results for "bound" phenolic acids in different cell, tissue, and organ preparations are extremely difficult (cf. Durkee and Thivierge, 1977; Sosulski et al, 1982). Obviously, a clearer understanding will depend on continued efforts to elucidate the structural chemistry of these complex phenolic-containing grain components.

The structures of some covalently bound forms of phenolic acids in oat and related cereal grains are, however, known, and examples of these are discussed in this section. From a biosynthetic standpoint, both ester and amide linkages in the examples below appear to be formed through the participation of the same carboxyl-activated cinnamoyl CoA-thio ester intermediates (Zenk, 1979; Bird and Smith, 1983).

A. Phenolic Esters with α-Hydroxy Acids

Representatives of this group, such as the chlorogenic acids, are extremely widespread in the plant kingdom and are often encountered in substantial quantities in fruits and vegetables (Herrmann, 1967, 1978, and references cited therein). Since early reports of the presence of chlorogenic acid (163) in grasses

(163) chlorogenic acid (3-O-*trans*-caffeoyl-D(-)-quinic acid)

and cereals (summarized by Hegnauer, 1963), relatively few confirmatory studies have been made. Van Sumere et al (1958) reported acid-labile forms of chlorogenic acid in barley husks, but chlorogenic acid itself was not present. Maga and Lorenz (1974) also failed to detect chlorogenic acid in wheat and triticale mill fractions but found chlorogenic and isochlorogenic acids (i.e., mixtures of 3,4-, 3,5-, and 4,5-di-O-*trans*-caffeoyl-D(−)-quinic acid) in both cereals after hydrolysis. The above citations, however, should be viewed with some caution. Not only are *cis* and *trans* isomers encountered, but also positional isomers are known and often occur with the p-coumaroyl-, feruloyl-, sinapoyl-, and dicaffeoyl-quinic acid analogues (Herrmann, 1978). Lack of readily available standards, difficulty in purification, inherent instability, and confusion in nomenclature (e.g., *chlorogenic, neochlorogenic, cryptochlorogenic, Band 510, pseudochlorogenic, isochlorogenic*) have all led to misidentifications. Recent reevaluation of oat, wheat, rice, and corn flours (Sosulski et al, 1982) failed to detect any chlorogenic acid, although *cis* and *trans* isomers of p-coumaric, ferulic, and sinapic acids could readily be generated by mild alkaline hydrolysis of the "soluble esters" fractions from all four cereal flours. Rather than quinic acid esters, perhaps the hydroxycitric acid esters (164–166) isolated from germinated corn by Ozawa et al (1977) are widespread in the cereals.

(164) R=H, 2-O-*trans*-p-coumaroyl-hydroxycitric acid

(165) R=OH, 2-O-*trans*-caffeoyl-hydroxycitric acid

(166) R-OCH₃ 2-O-*trans*-feruloyl-hydroxycitric acid

B. Phenolic Esters with n-Alkanols

Daniels and Martin (1961) isolated a lipophilic phenolic antioxidant from oats, which yielded upon alkaline hydrolysis caffeic acid, ferulic acid, glycerol, and long-chain ω-hydroxy fatty acid. Further work (Daniels et al, 1963; Daniels and Martin, 1964, 1965, 1967, 1968) established that, in fact, a complex mixture was present in the flour, consisting of at least 24 phenolic components when separated by TLC. Using a combination of liquid-liquid extractions, precipitations, adsorption chromatography, and preparative thin-layer techniques, a number of pure substances were obtained. The structures are shown in Table XV, along with the reference numbers assigned by Daniels and Martin (1967) to facilitate cross-referencing to the original papers.

The known components can be divided into four distinct families based on long-chain monoalcohols, diols, ω-hydroxy fatty acids, and glycerol. In each family, the alcoholic hydroxyl group is esterified with caffeic and/or ferulic acids. All the phenolic esters so far described are freely soluble in diethyl ether, sparingly soluble or insoluble in petroleum ether, and insoluble in water. However, all are soluble in dilute aqueous alkali because of the ionization of the phenolic hydroxyl group and formation of water-soluble phenoxy anions. The melting points are relatively low (e.g., compounds [167], 70°C; [168], 118°C), and, owing to the ease with which these compounds autooxidize on heating and undergo light-induced isomerization *trans-cis*, extraction and isolation must be done at room temperature in subdued light or in darkness (Daniels and Martin, 1967). The pure substances are off-white or yellowish and exhibit bright bluish-green fluorescence in solution when exposed to long-wave UV light.

The first family of hydroxycinnamate esters with fatty alcohols (Table XV, compounds 167–169) are not restricted to oat grains. Ferulic and *p*-coumaric acid esters of C_{18}- to C_{28}-*n*-alkanols occur in the bark and heartwood of larch, pine, and spruce species (Nair and Von Rudloff, 1959; Rowe et al, 1969; Norin and Winell, 1972; Leont'eva et al, 1974). The *p*-coumaroyl esters of C_{18}- to C_{28}-fatty alcohols also occur in the composite *Artemesia campestris* L. (Vajs et al, 1975), and 1-O-feruloyl eicosan-1-ol (C_{20}) has been reported in rapeseed and linseed oils (Tamura et al, 1963).

In oat flour, the diesters (171–173) were found to be mixtures, in each case containing about 30% of the corresponding derivative of octoacosan-1,28-diol (Daniels and Martin, 1965). Diesters of this type have not been reported from natural sources other than oats.

Oat flour also contains the ferulic and caffeic acid esters (174–176) of the ω-hydroxy fatty acids related to the alcohol homologues described above. The only other documented occurrence of this type of compound is the methyl ester of 22-O-feruloyl-22-hydroxy-docosanoic acid, phellochrysein, isolated from native cork (*Quercus suber* L.) (Guillemonat and Traynard, 1963).

The family of esters containing glycerol, representing approximately 36% of the oat antioxidant mixture, consisted of inseparable mixed pairs of monoglycerides containing one mole of ω-hydroxy fatty acid with 22, 26, or 28 carbons. Three moles of hydroxycinnamoyl groups acylate the three aliphatic hydroxyl functions (Daniels and Martin, 1968), although the exact substitution pattern has yet to be verified. The authors concluded that, in addition to the dicaffeoyl-monoferuloyl derivatives (two positional isomers with each of three homologous monoglycerides), the monocaffeoyl-diferuloyl and triferuloyl series of analogues were also present in oats. Several other glyceride-phenolic esters were not structurally identified. Similar monoesters of glycerol have been reported from pineapple stems (Taketa and Scheuer, 1976). Both 1-feruloyl-3-*p*-

Plate 1. Polyamide two-dimensional separation by thin-layer chromatography of phenolics from the low-molecular-weight fraction of oat groats (cv. Hinoat). The solvent system for direction 1 was water, isopropanol, and formic acid (75:20:5) and for direction 2 was toluene, methanol, acetic acid, and water (65:20:14:1). The plate was photographed in long-wavelength ultraviolet light (~ 360 nm emission maximum) before (a) and after (b) spraying with the aminoethyl diphenylborinate reagent.

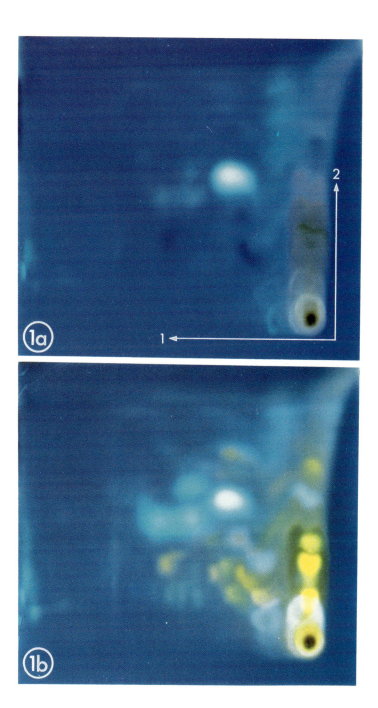

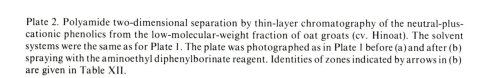

Plate 2. Polyamide two-dimensional separation by thin-layer chromatography of the neutral-plus-cationic phenolics from the low-molecular-weight fraction of oat groats (cv. Hinoat). The solvent systems were the same as for Plate 1. The plate was photographed as in Plate 1 before (a) and after (b) spraying with the aminoethyl diphenylborinate reagent. Identities of zones indicated by arrows in (b) are given in Table XII.

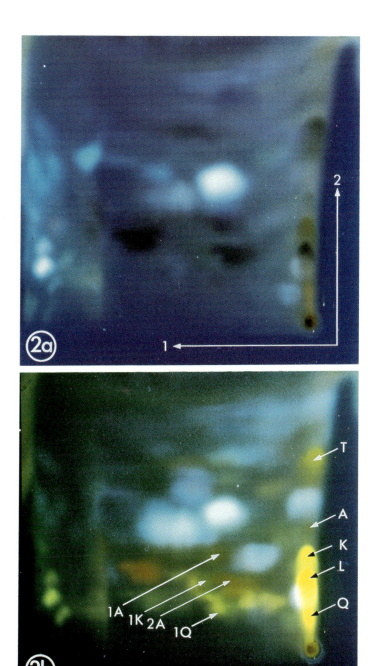

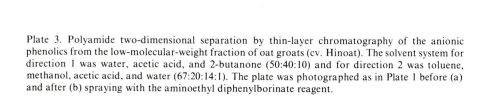

Plate 3. Polyamide two-dimensional separation by thin-layer chromatography of the anionic phenolics from the low-molecular-weight fraction of oat groats (cv. Hinoat). The solvent system for direction 1 was water, acetic acid, and 2-butanone (50:40:10) and for direction 2 was toluene, methanol, acetic acid, and water (67:20:14:1). The plate was photographed as in Plate 1 before (a) and after (b) spraying with the aminoethyl diphenylborinate reagent.

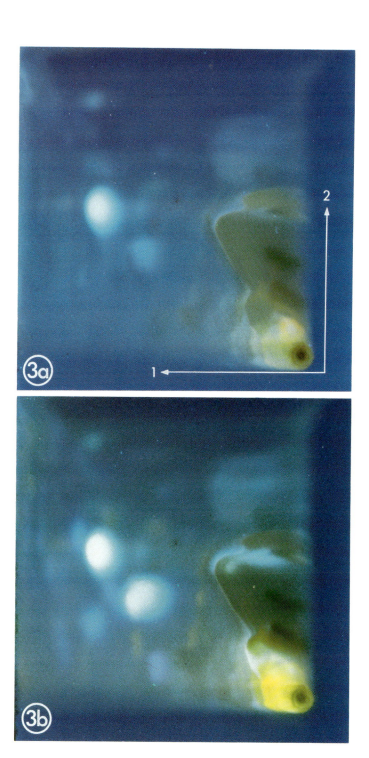

Plate 4. Fluorescence micrographs of hand-sectioned oat grains (cv. Hinoat), showing the vicinity of the crease after treatment with either a 100% methanol reagent blank (a) or the aminoethyl diphenylborinate reagent (b) consisting of 0.1% aminoethyl diphenylborinate in 100% methanol. Flavone and flavonol derivatives, which exhibit intense yellowish fluorescence after this treatment, occur primarily in the aleurone and subaleurone layers of the grain.

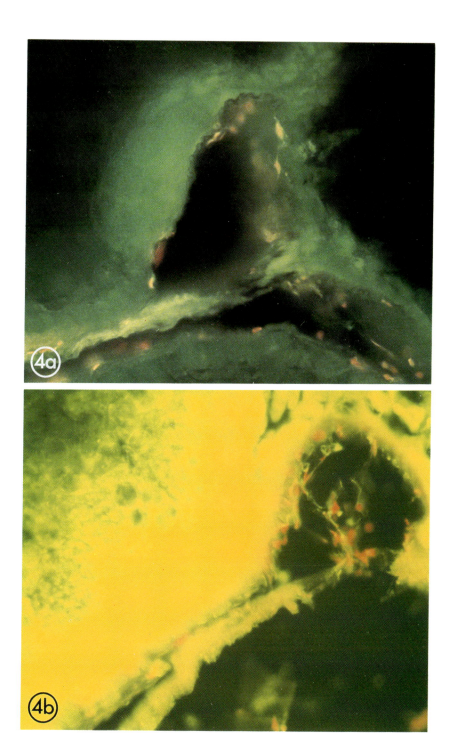

TABLE XV
Caffeoyl and Feruloyl Esters of *n*-Alkanols in Oat Groats

Structure	Name	Antioxidant Reference No. (Daniels and Martin, 1967)
Long-Chain *n*-Alkanols		
(167) R = CH₃ (168) R = H (169)	(167) 1-*O*-feruloyl-*n*-hexacosan-1-ol	1
	(168) 1-*O*-caffeoyl-*n*-hexacosan-1-ol	6 (in part)
	(169) 1-*O*-caffeoyl-*n*-octacosan-1-ol	6 (in part)
Long-Chain *n*-Alkanediols Mono-Esters		
(170)	(170) 1-*O*-feruloyl-*n*-hexacosan-1,26-diol	14
Di-Esters		
(171)	(171) 1,26-di-*O*-feruloyl-*n*-hexacosan-1, 26-diol	8 (C₂₈ homologue also present)

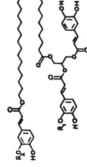

(172) 1-*O*-caffeoyl-26-*O*-feruloyl-*n*-hexacosan-1,26-diol — 15 (C$_{28}$ homologue also present)

(173) 1,26-di-*O*-caffeoyl-*n*-hexacosan-1,26-diol — 19 (C$_{28}$ homologue also present)

(172) R = CH$_3$
(173) R = H

ω-Hydroxy Alkanoic Acids

(174) 26-*O*-feruloyl-26-hydroxy-*n*-hexacosan-1-oic acid — 5

(175) 26-*O*-caffeoyl-26-hydroxy-*n*-hexacosan-1-oic acid — 11 (in part)

(174) R = CH$_3$
(175) R = H

(176) 28-*O*-caffeoyl-28-hydroxy-*n*-octacosan-1-oic acid — 11 (in part)

(176)

Glycerol

(177) 1,2-di-*O*-caffeoyl-3-*O*(26'-*O*-feruloyl)-*n*-hexacosanoyl-glycerol — 20 Mixed pairs:
64% C$_{26}$ (as shown)
31% C$_{28}$
5% C$_{22}$

(178) 1-*O*-caffeoyl-2-*O*-feruloyl-3-*O*-(26-*O*-caffeoyl)-*n*-hexacosanoyl-glycerol

(177) R$_1$ = H, R$_2$ = CH$_3$
(178) R$_1$ = CH$_3$, R$_2$ = H

coumaroyl- and 1,3-diferuloyl-2-acetylglycerol, along with other unidentified phenolic triglycerides, have been isolated from wheat roots (Popravko et al, 1982); the progenitor of wheat, *Aegilops* (Cooper et al, 1978); and poplar winter bud secretions (Asakawa and Wollenweber, 1976).

That other cereal grains contain compounds similar to those described from oats is not known. However analogous antioxidants are apparently present in wheat germ oil, although the exact structures have not been elucidated (D. G. H. Daniels and H. F. Martin, *unpublished data*, as quoted by MacMasters et al, 1971). As pointed out by Daniels et al (1963), various oat products have been patented for use as antioxidants, and most of the observed activity can be ascribed to the presence of these esters in the flour. Of the structurally known components, the highest antioxidant capacities are directly correlated with the quantity of caffeoyl-containing functions, as opposed to the feruloyl-substituted derivatives. They exhibit, on a weight basis, from one- to two-thirds the activity of propyl gallate (Daniels and Martin, 1967). The physiological functionality remains obscure.

The cutins, exocellular lipophilic polymers associated with the waxy cuticles of outer plant surfaces, have recently been shown to contain esterified *p*-coumaric and/or ferulic acid moieties. The basic monomeric components of all plant cutins consist of aliphatic and hydroxy-aliphatic fatty acids containing primarily 16 and, to a lesser extent, 18 carbons (Kolattukudy, 1977). The monomers are linked through esterification to form a linear macromolecular polyester with a small degree of branching. The *p*-coumaroyl and/or feruloyl groups are attached to free secondary alcoholic functions and account for not more than 1% of the weight of the cutin polymer. The physiological function of cutins in cereal grains has received attention, but little work has been done on oats. In wheat, the epidermal layer of the outer pericarp contains cutin and the cell walls of the testa (seed coat) adjacent to the nucellar epidermis are heavily cutinized (Shetlar, 1948; Bradbury et al, 1956). These cutinized layers create a hydrophobic, waxy covering around the outside of the aleurone layer over the entire grain except for the micropylar region. A similar localization of cutin occurs in rye (Simmonds and Campbell, 1976) and oat (R. G. Fulcher, *personal communication*) grains and no doubt functions as an effective water barrier in the control of imbibition and dehydration. The role of the phenolic acid moieties within the cutin polymers is not known.

C. Phenolic Esters with Triterpene Alcohols

The "oil" fraction of several cereal grains has been shown to contain ferulic acid esters of tetracyclic triterpene alcohols (i.e., sterols). Because of the alkali lability of the phenolic esters, their occurrence was overlooked during normal fractionation schemes involving saponification. In rice bran oil, up to 20% of the crude oil is in the form of ferulic acid esters (Akiya, 1962). Two sterol derivatives, oryzanol A and C, have been identified in this oil as the 3-*O*-feruloyl esters of cycloartenol and 24-methylenecycloartanol, respectively (Ohta and Shimizu, 1958). Tanaka et al (1964) have also characterized 3-*O*-feruloyl-β-sitosterol from the bran. In unsaponified corn oil, Nilsson et al (1968) recovered substantial quantities of ferulic acid esters, including 3-*O*-feruloyl-dihydro-γ-sitosterol, which also occurs in wheat germ oil (Tamura et al, 1961). As yet, similarly acylated sterols have not been reported for oats. However, two pentacyclic triterpene alcohol glycosides, avenacin A and B, have been isolated from germinated oat grains. Both derivatives contain the relatively rare phenolic acid *N*-methylanthranilic acid esterified to a tertiary hydroxyl group in the aglycone, rather than to one of the sugar hydroxyls (Burkhardt et al, 1964). The presence of the *N*-methylanthraniloyl moiety imparts a brilliant blue fluorescence to the avenacins when root tips or extracts are observed in long-wavelength UV light. Tschesche et al (1973) have determined the glycosidic linkages, and the partial structures for avenacin A (179) and B (180) are shown in the accompanying

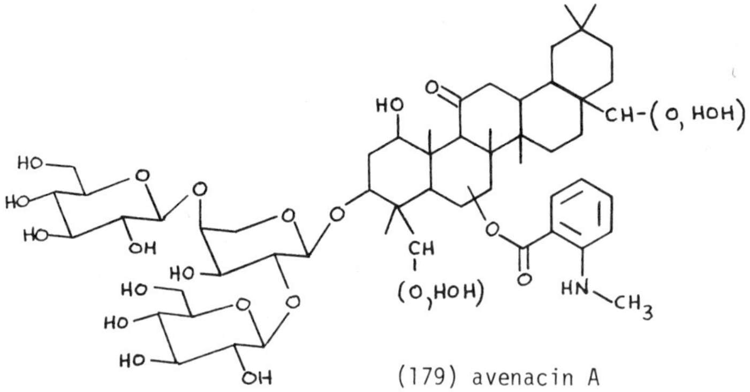

(179) avenacin A

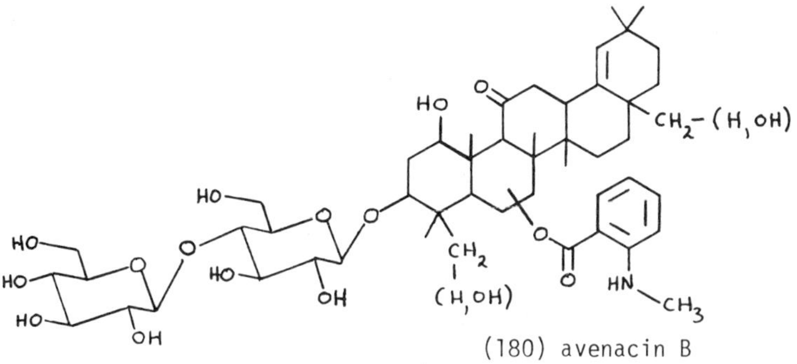

(180) avenacin B

diagrams. The avenacins are potent antifungal agents showing activity against *Ophiobolus graminis* and several nonpathogenic fungi (Maizel et al, 1964). Their occurrence in the mature grain has not been investigated to date.

D. Phenolic Esters with Sugars, Polysaccharides, and Lignin

The occurrence of phenolic acid esters with simple sugars is widespread in the plant kingdom, and almost all plant tissues that have been tested contain the enzyme systems required to synthesize sugar esters from exogenously supplied acids (Towers, 1964; Harborne, 1977; Herrmann, 1978). In actively metabolizing organs such as germinating embryos, young leaves, and maturing caryopses, low-molecular-weight phenolic esters, including presumably the ester glycosides, are rapidly turned over and appear to act as metabolic intermediates in the synthesis or breakdown of more complex macromolecules containing phenolic acids (Van Sumere et al, 1972). Administration of ^{14}C-labeled phenolic acids to various cereal tissues results in conversion of the free acids to both ester and ether glucosides (i.e., β-glucosidase-labile forms: Higuchi and Brown, 1963; Towers, 1964; Van Sumere et al, 1972).

Despite the ubiquitous occurrence of these sugar esters, little definitive work has been done to characterize those present in oat and related cereal grains. Invariably, the process involving extraction of the grain with aqueous alcohol, removal of free phenolic acids from the extract by lipophilic solvent extraction, and subsequent mild saponification of the soluble polar fraction yields substantial quantities of free benzoic and cinnamic acids along with concomitant increases in glucose and other sugars (El-Basyouni and Towers, 1964; Van Sumere et al, 1972; Durkee and Thivierge, 1977; Sosulski et al, 1982). Alternatively, β-glucosidase treatment of the polar fraction produces a similar spectrum of liberated acids, albeit usually in lesser quantities (Van Sumere et al, 1972; Durkee and Thivierge, 1977). Since both alkaline treatment and enzymatic hydrolysis will liberate the free acid from either the ester or the ether derivatives, differentiation between native forms of the phenolic acid requires additional work.

The occurrence of phenolic acids covalently linked to cereal cell wall polysaccharides by ester linkage has been well documented. Both lignified and nonlignified walls from oats, barley, wheat, rice, ryegrass, and corn release phenolic acids upon treatment with mild alkali (Guenzi and McCalla, 1966; Harris and Hartley, 1976; Hartley and Jones, 1977; Maniñgat and Juliano, 1982; Chesson et al, 1983; Tanner and Morrison, 1983). The predominant phenolic acids released include the *trans* isomers of ferulic and *p*-coumaric (approx. 90% *trans*, 10% *cis*), along with traces of vanillic and *p*-hydroxybenzoic acids (Hartley and Jones, 1976, 1977, 1978). In addition, small amounts of the oxidatively ring-coupled dimer, 5,5'-diferulic acid (181, 182), primarily in the *trans, trans*

(181)
trans, trans-5,5'-diferulic acid

(182)
cis, trans-5,5'-diferulic acid

form (181), have been reported (Hartley and Jones, 1976, 1977). Typical values for esterified acids in aerial parts of ryegrass, expressed as milligrams per gram of cell wall, range from 2.8 to 8.9 *trans*-ferulic acid (0.7–1.7 *cis*), 0.6–5.9 *trans*-*p*-coumaric acid (0.2–0.8 *cis*), and 0.09–0.36 *trans, trans*-diferulic acid (0.01–0.04 *cis, trans*) (Hartley and Jones, 1976, 1977). Similar values for *trans*-ferulic acid have been reported in wheat bran cell walls (Ring and Selvendran, 1980).

Evidence that these acids are attached via the carboxylic groups to the cell wall polysaccharides rather than to lignin has come from enzymatic and nonenzymatic degradation studies. Incubation of purified cell walls from developing wheat coleoptiles (Whitmore, 1974) and ryegrass (*Lolium* spp.) (Hartley, 1973; Hartley et al, 1976; Tanner and Morrison, 1983) with various fungal "cellulase" preparations resulted in the release of alkali-labile, water-soluble, phenolic acid carbohydrate esters ranging in molecular weight from several hundred to over 100,000. Purification of one of these components followed by alkaline hydrolysis showed it to contain *trans*-ferulic acid esterified to an oligosaccharide moiety containing xylose, arabinose, and glucose in the molar proportions 2.1:1.0:1.4 (Hartley, 1973). Tanner and Morrison (1983) found two types of complexes solubilized by cellulase treatment. Both types were based on $(1\rightarrow4)$-β-D-xylan chains with L-arabinofuranosyl and D-galactopyranosyl substituents and contained approximately one feruloyl ester group per 50–100 neutral sugar moieties. Treatment of the cellulase-resistant residue with mild acid released additional ferulic acid-galactoarabinoxylans. Both the cellulase-resistant and cellulase-susceptible polysaccharide-ferulate complexes contained closely associated $(1\rightarrow3)(1\rightarrow4)$-$\beta$-glucans as integral components. Within the cereal grain, phenolic acid esters of cell wall polysaccharides have also been shown to be associated with arabinoxylan components. In beeswing wheat bran cell wall preparations, ferulic and *p*-coumaric acids (4.8 and 0.14 mg/g of cell wall) occur in ester linkage, most probably to the arabinoxylan chains, which are differentially substituted with (arabinose) galactose, galacturonic acid, and 4-*O*-methylgalacturonic acid end groups (Ring and Selvendran, 1980). Fulcher et al (1972) observed a characteristic blue autofluorescence in UV light attributable to the feruloyl chromophore in the cell walls of wheat aleurone and scutellum. Ferulic acid was identified following mild alkaline hydrolysis of purified aleurone tissue. Aleurone cell walls of barley (Bacic and Stone, 1981), rice (Mod et al, 1981, Maniñgat and Juliano, 1982), and ryegrass (Harris and Hartley, 1976) also exhibit autofluorescence characteristics, primarily because of alkali-labile ferulate. (See Chapter 3 for details on oat cell wall autofluorescence.)

Structural evidence supporting arabinoxylan components as the major site of attachment of the feruloyl moieties in these cereal cell walls has recently been presented by Smith and Hartley (1983). Treatment of wheat bran cell wall preparations with *Oxyporus* spp. "cellulase" released a single ferulic acid-containing component identified as 2-*O*-(5′-*O*-*trans*-feruloyl-β-L-arabinofuranosyl)-D-xylopyranose (FAX) (183). This ester accounted for all the

(183)

2-0-(5'-0-*trans*-feruloyl- β -L-arabinofuranosyl)-D-xylopyranose (FAX)

aleurone cell wall ferulate released by alkaline hydrolysis. Similar enzymatic hydrolyses using cell wall preparations of wheat endosperm and the aerial parts of wheat, barley, and ryegrass also released FAX as the major ferulate ester, suggesting that this subunit may be a widespread component of cereal cell walls. From published quantitative analyses of wheat bran cell wall pentosan composition, these authors estimated the degree of ferulate substitution to be about one feruloyl moiety per 150 pentose residues.

Phenolic acids have also been recovered from the arabinoxylans of cereal endosperm (flour). Both soluble (Fausch et al, 1963) and insoluble (Geissmann and Neukom, 1973a) wheat endosperm pentosans contain small amounts of ferulic acid esterified to the arabinoxylan chains. This ferulate substitution has been implicated in the unique gelation properties exhibited by these pentosans in the presence of such oxidants as H_2O_2/peroxidase, sodium chlorite, ferricyanide, etc. (Painter and Neukom, 1968; Neukom and Markwalder, 1978). According to these authors, in the process of oxidative gelation, the aromatic rings of feruloyl esters on adjacent chains are oxidatively coupled to form diferulic diester cross-linkages between arabinoxylan polymers, with a resulting reduction in water solubility. Supporting this hypothesis, diferulic acid has been recovered from experimentally gelled, water-soluble arabinoxylans (Geissmann and Neukom, 1973b), and diferulic acid has in fact been found naturally occurring in the insoluble arabinoxylans (Markwalder and Neukom, 1976). This suggests that cross-linking or arabinoxylans by oxidative coupling of ferulic acid polysaccharide esters may occur in vivo during the deposition of cell wall hemicelluloses in the developing kernel. However, other hypotheses involving protein thiol side-chain coupling to the vinylic function of the ferulated subunit (Hoseney and Faubion, 1981) and ring coupling of protein tyrosyl to ferulate (Neukom and Markwalder, 1978) have been advanced to explain the gelation phenomena. Smith and Hartley (1983) recovered FAX (183) in cellulase hydrolysates of both soluble and insoluble wheat endosperm arabinoxylans. The quantities recovered accounted for all the alkali-labile ferulic acid present in the pentosans. The degree of feruloyl substitution based on recovery of this ferulated subunit was estimated at one feruloyl moiety per 20,000 sugar residues in the water-soluble pentosan and one feruloyl moiety per 1,140 residues in the insoluble pentosan.

Although structural aspects of noncellulosic polysaccharides of oat grains have received considerable attention, they have not been examined for phenolic acid esters. However, the overall similarities among oat, wheat, and related cereal cell walls for arabinoxylan composition (Burke et al, 1974) and ferulate autofluorescence localization within the grains suggest that at least some of the pentosan ferulate ester complexes described above are also present in oats. No doubt these complexes contributed to the sizable quantities of ethanol-insoluble, alkali-labile ferulic acid reported by Durkee and Thivierge (1977) in dehulled oats.

Potentially important physiological functions for phenolic acids as aromatic ester substituents in polysaccharides probably involve the manner in which they modify the physicochemical properties of the parent polymer. Esterification of soluble polysaccharides with hydrophobic acids such as ferulic acid renders the polymer less soluble (Geissmann and Neukom, 1973a), allowing it to act as a cementing agent at the site of cell wall deposition. Esterification would be expected to modify the tertiary and quaternary structures of polysaccharides such as the hemicelluloses, altering bonding interactions and close alignment between different types of polysaccharides (Ring and Selvendran, 1980). Phenolic acid ester subunits such as FAX may serve as important cross-linkage sites between adjacent pentosan chains and/or between pentosan and wall proteins (Neukom and Markwalder, 1978) and may serve as anchored free-radical-initiation sites for the oxidative polymerization of hydroxycinnamyl alcohols into lignin. Finally the esterification of polysaccharides imparts resistance toward enzymatic degradation (Fulcher et al, 1972; Chesson et al, 1983) and may play an important regulatory role in the ordered utilization of reserves during germination.

Phenolic acids attached by ester linkage to the lignins of grasses and cereals have been reported. Spectrophotometric and hydrolytic analyses have shown that lignins from wheat and other grasses contain *p*-coumaric and ferulic along with traces of *p*-hydroxybenzoic, vanillic, and syringic acids esterified directly to the lignin core (Sarkanen and Ludwig, 1971, and references cited therein). The major acids liberated by mild saponification of Klason lignins from a number of grasses consisted of *p*-coumaric (2.7–6.2% dry weight) and ferulic acid (1.3–3.9%) (Higuchi et al, 1967). In wheat grains, lignin (i.e., phloroglucinol-positive material) has been observed in the middle lamella of cross cells, particularly in that portion immediately adjacent to the testa cells, in the hairs, and in epidermal cell walls at the butt end of the grain, but it is absent from the aleurone (Bradbury et al, 1956). The same structural components of corresponding tissues and cell types of oat grains show comparable lignin localization (R. G. Fulcher, *personal communication*). Unless the lignins of oat and wheat grains differ, specifically with respect to these esters, from lignins in other parts of the plants, it might be anticipated that oat and other cereal grain lignins also contain esterified hydroxycinnamic acids. Clearly, qualitative and quantitative work is needed in this area. Rasper (1979) has determined crude lignin by the Southgate procedure for oat bran and oat hulls. Bran was found to contain about 2% lignin, and 6–8% by weight of the hulls represented lignin.

E. Phenolic Acids Conjugated with Amines

The occurrence of phenolic acids linked covalently to amine functions through "pseudopeptide" bonds is widespread in the plant kingdom, and the cereals, including oats, are no exception. The structures and occurrences of amide conjugates with hydroxycinnamic acids in oats and related cereals are shown in Table XVI. The series of ferulic, *p*-coumaric, and caffeic acid conjugates with alkyldi- and polyamines (184–190) appear to be widely distributed in meristematic tissues, reproductive organs (i.e., anthers, ovaries, and immature caryopses), and mature seeds throughout the grasses (Martin-Tanguy et al, 1978). They are apparently absent from cytoplasmic male sterile corn lines but are produced upon restoration of fertility (Martin-Tanguy et al, 1982). Their presence or absence in oats has not been established to date. However, oat grains do contain a series of at least 20 different phenolic conjugates with anthranilic acid derivatives, of which four (191–194) so far have been structurally identified (Collins, 1983). These conjugates (205) readily undergo cyclodehydration during

TABLE XVI
Structure and Occurrence of Phenolic Acid Conjugates of Amines, Amino Acids,
and Proteins in Oats and Related Cereals

Structure	Name	Species and Tissue or Material Examined	References
(184)	p-Coumaroylputrescine	Inflorescences of corn, wheat, millet	Martin-Tanguy et al (1978)
(185)	Feruloylputrescine	Inflorescences of corn, wheat, millet Corn embryo: 0.37 mg/gfw Endosperm: 0.16 mg/gfw	Martin-Tanguy et al (1978) Martin-Tanguy et al (1982)
(186)	Diferuloylputrescine	Wheat inflorescences Corn inflorescences Corn embryo: 2.24 mg/gfw Endosperm: 1.50 mg/gfw	Martin-Tanguy et al (1978) Martin-Tanguy et al (1982)
(187)	p-Coumaroyl-spermidine	Corn embryo: 1.53 mg/gfw Endosperm: 0.38 mg/gfw	Martin-Tanguy et al (1982)
(188)	Caffeoylspermidine	Millet inflorescences	Martin-Tanguy et al (1978)
(189)	Diferuloylspermidine	Corn inflorescences Corn embryo: 3.42 mg/gfw Endosperm: 0.84 mg/gfw	Martin-Tanguy et al (1982)
(190)	Diferuloylspermine	Corn inflorescences Corn embryo: 0.50 mg/gfw Endosperm: 0.39 mg/gfw	Martin-Tanguy et al (1982)

(*continued on next page*)

TABLE XVI (continued)

Structure	Name	Species and Tissue or Material Examined	References
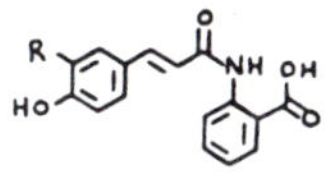 (191,192)	(191) R = H; *N-p*-coumaroylan- thranilic acid	Oat groats, hulls	Collins (1983)
	(192) R = OCH₃; *N*-feruloylan- thranilic acid	Oat groats, hulls	
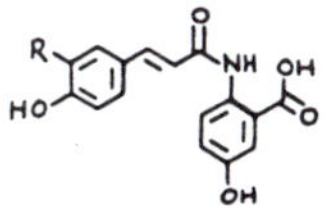 (193,194)	(193) R = H; *N-p*-coumaroyl-5- hydroxyanthranilic acid	Oat groats, hulls	Collins (1983)
	(194) R = OCH₃; *N*-feruloyl-5-hydroxy- anthranilic acid	Oat groats, hulls	
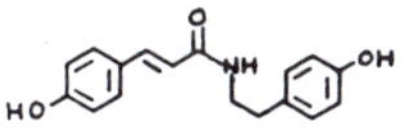 (195)	*p*-Coumaroyltyramine	Corn inflorescences	Martin-Tanguy et al (1978)
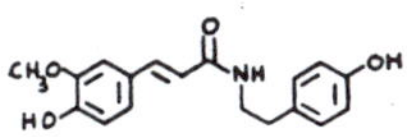 (196)	Feruloyltyramine	Corn inflorescences Corn embryo: 0.08 mg/gfw Endosperm: 0.20 mg/gfw	Martin-Tanguy et al (1978)
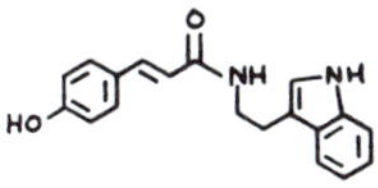 (197)	(197) R = H; *p*-couma- royltryptamine R = OCH₃; feru- loyltryptamine	*Zea mays* (whole grains)	Ehmann (1974)
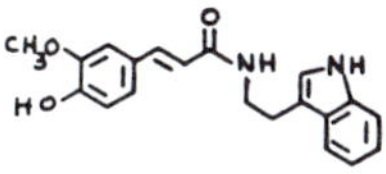 (198)	(198) R = H; *p*-couma- royltryptamine R = OCH₃; feru- loyltryptamine	*Z. mays* (whole grains)	Ehmann (1974)
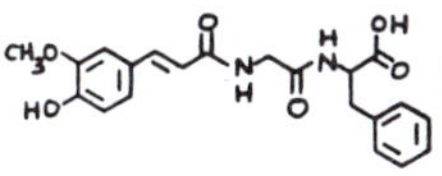 (199)	Feruloylglycyl- L-phenylalanine	*Hordeum vulgare* (seed globulin)	Van Sumere et al (1973)

(*continued on next page*)

TABLE XVI (continued)

Structure	Name	Species and Tissue or Material Examined	References
(200)	p-Coumaroylagmatine	*H. vulgare* (young shoots)	Stoessl (1965)
(201–203)	(201) R₁ = R₂ = H; hordatine A	*H. vulgare* (young shoots)	Stoessl (1970)
	(202) R₁ = H, R₂ = OCH₃; hordatine B		
	(203) R₁ = β-D-glucosyl, R₂ = H, OCH₃; hordatine M		

isolation or treatment with anhydrous solvents to form the correspondingly substituted 2-styryl-4H-3,1-benzoxazin-4-one derivatives (206). The naturally

(205)   −H₂O   +H₂O   (206)

occurring constituents probably exist as the *trans* isomers, but exposure to UV light promotes *trans-cis* isomerization, and mixtures of both isomers are usually encountered in extracts prepared in daylight. They are pale yellow or gray in solid form, very soluble in dilute aqueous alkali, appearing bright yellow-green to orange, and are decomposed above pH 10 to give reddish brown polymeric breakdown products. At neutral pH, they are practically insoluble in water but dissolve in acetone, methanol, and to some extent in chloroform. As a group, they constitute about 0.02–0.04% of the dry weight of bran-depleted oat flour and 0.02–0.08% of bran-rich mill fractions; they are also present in oat hulls.

Hydroxycinnamoyl conjugates of the decarboxylated amino acids tyramine and tryptamine (195–198) have been reported in corn kernels, and a conjugate (200) of agmatine (decarboxylated arginine) has been isolated from germinated barley (Stoessl, 1965; for reviews, see Stoessl, 1970, and Smith, 1977). The coupled dimers, hordatine A and B, along with an inseparable mixture of their β-D-glucosides, hordatine M (201–203), are also present in barley shoots. A recent survey (Smith and Best, 1978) found the hordatines to be present in

Hordeum vulgare, H. bulbosum, H. distichon, H. murinum, and *H. spontaneum* but absent in *H. jubatum,* corn, millet, oats, rice, rye, and wheat seedlings.

Few physiological roles have been ascribed to these low-molecular-weight conjugates. The *trans* isomers of the hordatines and their glucosides are potent antifungal agents completely inhibiting *Monilinia fructicola* spore germination at $7 \times 10^{-6} M$ concentrations, and hordatine A was similarly inhibitory to five other pathogenic fungi (Stoessl and Unwin, 1970). The conjugates of the alkyl amines appear to exhibit antiviral activity and are biochemical markers closely associated with pollen and ovule fertility (Martin-Tanguy et al, 1978, 1982). The functions of the oat grain benzoxazinones remain obscure, but their superficial similarity to the wheat, corn, and rye benzoxazolinones suggests that they may possess antifungal activity.

Several high-molecular-weight phenolic acid amine conjugates have been reported in cereal grains. The occurrence of ferulic acid covalently linked to the N-terminal amino acid of a barley seed globulin fraction was reported by Van Sumere et al (1973). From partial acid hydrolysis of the purified protein, the terminal dipeptide L-glycyl-L-phenylalanine containing the feruloyl moiety was recovered (Table XVI, 199). The same *N*-feruloyl dipeptide has recently been found in lucerne leaf protein and may be of widespread occurrence and function in plants (Van Sumere et al, 1980).

Yoshizawa et al (1970) have isolated a number of sugar-amino acid-phenolic acid conjugates from rice bran. Seven distinct compounds were recovered following column chromatography, and all were found to contain ferulic acid and at least one acidic amino acid (either aspartate or glutamate). All but one contained sugars (usually glucose), and three contained sinapic acid as well as the ferulate. One compound was further purified and its molecular composition determined to contain two molecules of ferulic acid, three of glucose, one glutamic acid moiety, and two calcium ions with a minimum molecular weight in excess of 1,100.

That phenolic acid conjugates involving low- and intermediate-molecular-weight glycopeptides and proteins are also present in oat grains has not been established. However, interpretation of recent evidence suggests that this is in fact the case. Percival and Bandurski (1976), using dehulled oats, studied the chemical composition of "bound IAA," which accounts for about 95% of the total IAA present in the seed (i.e., ~ 7.6 mg/kg of dry weight). Ester-linked forms of IAA were extracted with 50% aqueous acetone and subsequently fractionated by solvent and salt precipitation and ion-exchange gel-permeation chromatography. The authors reported that oat grain IAA was covalently linked to a heterogeneous mixture of IAA-phenolic acid-glucoprotein complexes, which differed in net charge and ranged in molecular weight from 5,000 to 20,000. In five individual complexes studied in detail, alkaline hydrolysis liberated essentially the same pattern of phenolic acids, which, although not identified by the authors, appear to be primarily of the ferulic-sinapic acid type, based on the spectral evidence given. One of the complexes appears to be of the albumin class, and the other belongs to the group of glutelins, based on solubility properties (Zimmermann, 1978). The site of attachment of the phenolic acid moieties remains unknown.

VII. SUMMARY

The information presented in the previous sections of this review demonstrates the wide biogenetic capability of the oat plant to elaborate simple phenolics. For the most part, they are all derived from the amino acids phenylalanine and tyrosine or their precursors. In the free form, they seldom accumulate in healthy tissues to any major extent but are themselves intermediates in biosynthetic routes leading to glycosides, esters, amides, and more complex cell constituents. Representatives of each major class of phenolic—including benzoic and cinnamic acids, quinones, flavones, flavonols, chalcones, flavanones, anthocyanidins, and the aminophenolics—have all been found in oats. The distribution of these phenolics within the oat plant and in closely allied cereals has also been presented, along with comments on their possible physiological functionality. It is hoped that such a broader perspective of form and function has provided insight into the relationship of oats to some of the other cereals in general and has pointed out gaps in our knowledge of oat phytochemistry in particular. As illustrated in the color plates, the oat grain contains a rich assortment of low-molecular-weight phenolics, of which only a few have been structurally identified. Clearly, more phytochemistry of oat phenolics is needed, using many different genetic sources, before a comprehensive picture of oat phenolics can emerge. Evident too is the lack of quantitative information on oat phenolics. However, until the basic chemistry has been unraveled and specific methodologies established, little meaningful purpose is served by generating quantitative data on uncharacterized constituents.

More importantly, perhaps, from the standpoint of phenolic functionality in oat-based foods and food ingredients, the phytochemical tables illustrate the diversity of molecular forms present in the grain that are known to contain phenolics. Taken collectively, these covalently bound phenolics amount to by far the greatest proportion of the total phenolics, and yet they are only trace components of some macromolecules. However, even as minor substituents of lipids, proteins, and polysaccharides, the phenolic moieties may drastically alter the physicochemical properties and functionality of the substrate. This is especially true for polymeric substances containing ester-linked hydroxycinnamoyl groups. In these macromolecules, phenolic hydroxyls are still readily dissociatable to give yellow-colored, highly soluble ionic groups, and the phenolic ring (and side-chain vinylic) carbons remain easily oxidizable through activated phenoxy radicals. Consequently the color and consistency of the substrate have become pH- and [O]-sensitive by phenolic ester substitution. In fact, a solution of galactomannan artificially ferulated to the extent of only one to two feruloyl groups per 1,200 sugar units will undergo oxidative gelation in vitro with trace amounts of hydrogen peroxide plus peroxidase (Geissmann and Neukom, 1971). Further studies to better control the rate, degree, and type of oxidative gelation using similar phenolic acid-substituted biopolymers may lead to new specialized food and nonfood uses.

It is beyond the scope of the present review to deal in depth with the functionality of phenolics in the processing, storage, preparation, and consumption of oats. In many instances, too little is yet known of specific structures present, their concentrations, their possible modes of interaction with other grain constituents, and the nutritional consequences of these mixtures to allow meaningful interpretation. Obviously, much work is needed in describing the processes of phenol interactions from the chemical as well as the nutritional standpoint. Nevertheless, several articles and general reviews of polyphenol functionality should be mentioned. Sensory properties and taste thresholds for many simple phenolics and their breakdown products are given by Maga and Lorenz (1973) and Maga (1978). Pratt (1976) has discussed the antioxidant properties of flavonoids in lipid mixtures, and Daniels et al (1963) have evaluated some of the oat alkanol hydroxycinnamates as antioxidants. The anticaries phenolics from oat bran and hull and the seeds of several other grains have been reviewed by Madsen (1981), and the activity of hydroxycinnamates as nitrosamine-formation inhibitors has been reported (Kikugawa et al, 1983, and references cited therein). Finally, Singleton (1981) has thoroughly reviewed health and nutritional aspects of a wide range of phenolics.

LITERATURE CITED

AKIYA, T. 1962. Components of unsaponifiable matter of rice bran oil. Agric. Biol. Chem. 26:180-186.

ANDERSON, J. A. 1933. The yellow colouring matter of Khapli wheat *Triticum dicoccum*. III. The constitution of tricin. Can. J. Res. 9:80-83.

ARGANDONA, V. H., LUZA, J. G., NIEMEYER, H. M., and CORCUERA, L. J. 1980. Role of hydroxamic acids in the resistance of cereals to aphids. Phytochemistry 19:1665-1668.

ARGANDONA, V. H., NIEMEYER, H. M., and CORCUERA, L. J. 1981. Effect of content and distribution of hydroxamic acids in wheat on infestation by the aphid *Schizaphis graminum*. Phytochemistry 20:673-676.

ASAKAWA, Y., and WOLLENWEBER, E. 1976. A novel phenolic acid derivative from buds of *Populus lasiocarpa* Oliv. Phytochemistry 15:811-812.

AYYANGAR, N. R., LUGADE, A. G., SANE, M. G., and SRINIVASAN, K. V. 1981. Separation of 2-amino-3H-phenoxazin-3-one impurity from o-aminophenol. J. Chromatogr. 209:113-116.

BACIC, A., and STONE, B. A. 1981. Isolation and ultrastructure of aleurone cell walls from wheat and barley. Aust. J. Plant Physiol. 8:453-474.

BARANOWSKI, J. D., and NAGEL, C. W. 1982. Inhibition of *Pseudomonas fluorescens* by hydroxycinnamic acids and their alkyl esters. J. Food Sci. 47:1587-1589.

BATE-SMITH, E. C. 1969. Luteoforol (3',4,4',5,7-pentahydroxyflavan) in *Sorghum vulgare* L. Phytochemistry 8:1803-1810.

BATE-SMITH, E. C., and RASPER, V. 1969. Tannins of grain sorghum: Luteoforol

(leucoluteolinidin), 3',4,4',5,7-pentahydroxy-flavan. J. Food Sci. 34:203-209.

BHATIA, I. S., KANSHAL, G. P., and BAJAJ, K. L. 1972. Chrysoeriol from barley seeds. Phytochemistry 11:1867-1868.

BIRD, C. R., and SMITH, T. A. 1983. Agmatine coumaroyltransferase from barley seedlings. Phytochemistry 22:2401-2403.

BOLKART, K. H., and ZENK, M. H. 1968. Biosynthesis of methoxylated phenols in higher plants. Z. Pflanzenphysiol. 59:439-444.

BRADBURY, D., MacMASTERS, M. M., and CULL, I. M. 1956. Structure of the mature wheat kernel. II. Microscopic structure of the pericarp, seed coat, and other coverings of the endosperm and germ of hard red winter wheat. Cereal Chem. 33:342-360.

BREDENBERG, J. B-Son., HONKANEN, E., and VIRTANEN, A. I. 1962. The kinetics and mechanism of the decomposition of 2,4-dihydroxy-1,4-benzoxazin-3-one. Acta Chem. Scand. 16:135-141.

BROWN, S. A. 1966. Biosynthesis of coumarins. Pages 15-24 in: Biosynthesis of Aromatic Compounds. G. Billek, ed. Proc. 2d meeting FEBS, Vienna, 21-24 April 1965. Pergamon Press, Oxford.

BROWN, S. A. 1979. Biochemistry of the coumarins. Recent Adv. Phytochem. 12:249-281.

BROWN, S. A. 1981. Coumarins. Pages 269-300 in: The Biochemistry of Plants. Vol. 7. Secondary Plant Products. E. E. Conn, ed. Academic Press, New York.

BUNGENBERG DE JONG, H. L., KLAAR, W. J., and VLIEGENTHART, J. A. A. 1953. A precursor of methoxy-parabenzoquinone in wheat germ. Nature (London) 172:402-403.

BUNGENBERG DE JONG, H. L., KLAAR, W. J., and VLIEGENTHART, J. A. 1955.

Glycosides and their importance in the wheat germ. Proc. Int. Bread Congr. 3rd., Sec. I, 29-33. Arbeitsgemeinschaft Getreide forschung, Detmold, W. Germany.

BURKE, D., KAUFMAN, P., McNEIL, M., and ALBERSHEIM, P. 1974. The structure of plant cell walls. VI. A survey of the walls of suspension-cultured monocots. Plant Physiol. 54:109-115.

BURKHARDT, H. J., MAIZEL, J. V., and MITCHELL, H. K. 1964. Avenacin, an antimicrobial substance isolated from *Avena sativa*. II. Structure. Biochemistry 3:426-428.

CALHOUN, W. K., HEPBURN, F. N., and BRADLEY, W. B. 1960. The distribution of the vitamins of wheat in commercial mill products. Cereal Chem. 37:755-761.

CHEN, K.-T., and GEDDES, W. F. 1945. Studies on the wheat pigments. Master's thesis, University of Minnesota, St. Paul.

CHESSON, A., GORDON, A. H., and LOMAX, J. A. 1983. Substituent groups linked by alkali-labile bonds to arabinose and xylose residues of legume, grass and cereal straw cell walls and their fate during digestion by rumen microorganisms. J. Sci. Food Agric. 34:1330-1340.

CHIBBER, B. A. K., BALASUBRAMANIAN, A., and VISWANATHA, T. 1973. Binding of actinomycin chromophore analogues to DNA. Can. J. Biochem. 51:204-217.

CHOPIN, J., DELLAMONICA, G., BOUILLANT, M. L., BESSET, A., POPOVICI, G., and WEISSENBOCK, G. 1977. C-glycosylflavones from *Avena sativa*. Phytochemistry 16:2041-2043.

COLLINS, F. W. 1983. Oat phenolics: Isolation and structural elucidation of substituted benzoxazinones. (Abstr.) Cereal Foods World 28:560.

CONCHIE, J., MORENO, A., and CARDINI, E. C. 1961. A trisaccharide glycoside from wheat germ. Arch. Biochem. Biophys. 94:342-343.

CONN, E. E. 1981. Cyanogenic glycosides. Pages 479-500 in: The Biochemistry of Plants. Vol. 7, Secondary Plant Products. E. E. Conn, ed. Academic Press, New York.

COOPER, R., GOTTLIEB, H. E., and LAVIE, D. 1978. New phenolic diglycerides from *Aegilops ovata*. Phytochemistry 17:1673-1675.

CORCORAN, M. R., GEISSMAN, T. A., and PHINNEY, B. O. 1972. Tannins as gibberellin antagonists. Plant Physiol. 49:323-330.

CORCUERA, L. J., WOODWARD, M. D., KELMAN, A., HELGESON, J. P., and UPPER, C. D. 1978. 2,4-dihydroxy-7-methoxy-2H-1,4-benzoxazin-3(4H)-one, an inhibitor from *Zea mays* with differential activity against soft rotting *Erwinia* species.

Plant Physiol. 61:791-795.

COUTURE, R. M., ROUTLEY, D. G., and DUNN, G. M. 1971. Role of cyclichydroxamic acids in monogenic resistance of maize to *Helminthosporium turcicum*. Physiol. Plant Pathol. 1:515-521.

DANIELS, D. G. H. 1959. The estimation and location of methoxyhydroquinone glycosides in the wheat grain. Cereal Chem. 36:32-41.

DANIELS, D. G. H., KING, H. G. C., and MARTIN, H. F. 1963. Antioxidants in oats: Esters of phenolic acids. J. Sci. Food Agric. 14:385-390.

DANIELS, D. G. H., and MARTIN, H. F. 1961. Isolation of a new antioxidant from oats. Nature (London) 191:1302.

DANIELS, D. G. H., and MARTIN, H. F. 1964. Structures of two antioxidants isolated from oats. Chem. Ind. London 1964:2058.

DANIELS, D. G. H., and MARTIN, H. F. 1965. Antioxidants in oats: Diferulates of long-chain diols. Chem. Ind. London 1965:1763.

DANIELS, D. G. H., and MARTIN, H. F. 1967. Antioxidants in oats: Mono-esters of caffeic and ferulic acids. J. Sci. Food Agric. 18:589-595.

DANIELS, D. G. H., and MARTIN, H. F. 1968. Antioxidants in oats: Glyceryl esters of caffeic and ferulic acids. J. Sci. Food Agric. 19:710-712.

DEDIO, W., HILL, R. D., and EVANS, L. E. 1972a. Anthocyanins in the pericarp and coleoptiles of purple wheat. Can. J. Plant Sci. 52:977-980.

DEDIO, W., HILL, R. D., and EVANS, L. E. 1972b. Anthocyanins in the pericarp and coleoptiles of purple-seeded rye. Can. J. Plant Sci. 52:981-983.

DEDIO, W., KALTSIKES, P. J., and LARTER, E. N. 1969. The anthocyanins of *Secale cereale* L. Phytochemistry 8:2351-2352.

DREYER, D. L., and JONES, K. C. 1981. Feeding deterrency of flavonoids and related phenolics towards *Schizaphis graminum* and *Myzus persicae*: Aphid feeding deterrents in wheat. Phytochemistry 20:2489-2493.

DURKEE, A. B., and THIVIERGE, P. A. 1977. Ferulic acid and other phenolics in oat seeds (*Avena sativa* L. var. Hinoat). J. Food Sci. 42:551-552.

EFFERTZ, B., and WEISSENBOCK, G. 1980. Tissue specific variation of C-glycosylflavone patterns in oat leaves as influenced by the environment. Phytochemistry 19:1669-1672.

EHMANN, A. 1974. *N*-(*p*-coumaryl)-tryptamine and *N*-(ferulyl)-tryptamine in kernels of *Zea mays*. Phytochemistry 13:1979-1983.

EL-BASYOUNI, S., and TOWERS, G. H. N. 1964. The phenolic acids in wheat. I. Changes during growth and development. Can. J. Biochem. 42:203-210.

ELLIGER, C. A., CHAN, B. G., WAISS, A. C., Jr., LUNDIN, R. E., and HADDON, W. F. 1980. C-glycosylflavones from *Zea mays* that inhibit insect development. Phytochemistry 19:293-297.

ELNAGHY, M. A., and LINKO, P. 1962. The role of 4-O-glucosyl-2,4-dihydroxy-7-methoxy-1,4-benzoxazin-3-one in resistance of wheat to stem rust. Physiol. Plant. 15:764-771.

ERB, N., ZINSMEISTER, H. D., LEHMANN, G., and MICHELY, D. 1981b. Prussic acid content of cereals in temperate climate regions. Z. Lebensm. Unters. Forsch. 173:176-179.

ERB, N., ZINSMEISTER, H. D., LEHMANN, G., and NAHRSTED, A. 1979. A new cyanogenic glycoside from *Hordeum vulgare*. Phytochemistry 18:1515-1517.

ERB, N., ZINSMEISTER, H. D., and NAHRSTEDT, A. 1981a. The cyanogenic glycosides of *Triticum, Secale* and *Sorghum*. Planta Med. 41:84-89.

EVANS, L. E., DEDIO, W., and HILL, R. D. 1973. Variability in the alkyl-resorcinol content of rye grain. Can. J. Plant Sci. 53:485-488.

FAUSCH, H., KUENDIG, W., and NEUKOM, H. 1963. Ferulic acid as a component of a glycoprotein from wheat flour. Nature (London) 199:287.

FIDDLER, W., PARKER, W. E., WASSERMAN, A. E., and DOERR, R. C. 1967. Thermal decomposition of ferulic acid. J. Agric. Food Chem. 15:757-761.

FLOSS, H. G. 1979. The shikimate pathway. Recent Adv. Phytochem. 12:59-89.

FULCHER, R. G., O'BRIEN, T. P. O., and LEE, J. W. 1972. Studies on the aleurone layer. I. Conventional and fluorescence microscopy of the cell wall with emphasis on phenol-carbohydrate complexes in wheat. Aust. J. Biol. Sci. 25:23-34.

FULCHER, R. G., O'BRIEN, T. P., and WONG, S. I. 1981. Microchemical detection of niacin, aromatic amine, and phytin reserves in cereal bran. Cereal Chem. 58:130-135.

GAHAGAN, H. E., and MUMMA, R. O. 1967. The isolation of 2-(2-hydroxy-7-methoxy-1,4-benzoxazin-3-one)-β-D-glucopyranoside from *Zea mays*. Phytochemistry 6:1441-1448.

GANEM, B. 1978. From glucose to aromatics: Recent developments in natural products of the shikimic acid pathway. Tetrahedron 34:3353-3383.

GEISSMANN, T., and NEUKOM, H. 1971. Vernetzung von Phenolcarbonsaureestern von Polysacchariden durch oxydative phenolische Kupplung. Helv. Chim. Acta 54:1108-1112.

GEISSMANN, T., and NEUKOM, H. 1973a. A note on ferulic acid as a constituent of the water-insoluble pentosans of wheat flour. Cereal Chem. 50:414-416.

GEISSMANN, T., and NEUKOM, H. 1973b. On the composition of the water soluble wheat flour pentosans and their oxidative gelation. Lebensm. Wiss. Technol. 6:59-62.

GERBER, N. N. 1968. Phenoxazinones by oxidative dimerization of aminophenols. Can. J. Chem. 46:790-792.

GERBER, N. N., and LECHEVALIER, M. P. 1964. Phenazine and phenazinones from *Waksmania aerata* sp. nov. and *Pseudomonas iodina*. Biochemistry 3:598-602.

GOODWIN, R. H., and POLLOCK, B. M. 1954. Studies on roots. I. Properties and distribution of fluorescent constituents in *Avena* roots. Am. J. Bot. 41:416-520.

GOODWIN, R. H., and TAVES, C. 1950. The effect of coumarin derivatives on the growth of *Avena* roots. Am. J. Bot. 37:224-231.

GRISEBACH, H. 1979. Selected topics in flavonoid biosynthesis. Recent Adv. Phytochem. 12:221-248.

GROSS, A. J., and SIZER, I. W. 1959. The oxidation of tyramine, tyrosine and related compounds by peroxidase. J. Biol. Chem. 234:1611-1614.

GROSS, G. G. 1977. Biosynthesis of lignin and related monomers. Recent Adv. Phytochem. 11:141-184.

GROSS, G. G. 1979. Recent advances in the chemistry and biochemistry of lignin. Recent Adv. Phytochem. 12:177-220.

GROSS, G. G. 1981. Phenolic Acids. Pages 301-316 in: The Biochemistry of Plants. Vol. 7, Secondary Plant Products. E. E. Conn, ed. Academic Press, New York.

GUENZI, W. D., and McCALLA, T. M. 1966. Phenolic acids in oats, wheat, sorghum and corn residues and their phytotoxicity. Agric. J. 58:303-304.

GUILLEMONAT, A., and TRAYNARD, J.-C. 1963. Sur la constitution chimique du liege. IV. Structure de la phellochryseine. Bull. Soc. Chim. Fr., 1963:142-144.

HAHLBROCK, K., and GRISEBACH, H. 1975. Biosynthesis of flavonoids. Pages 866-915 in: The Flavonoids. J. B. Harborne, T. J. Mabry, and H. Mabry, eds. Chapman and Hall, London.

HAMILTON, R. H. 1964. Tolerance of several grass species to 2-chloro-*s*-triazine herbicides in relation to degradation and content of benzoxazinone derivatives. J. Agric. Food Chem. 12:14-17.

HANSON, K. R., and HAVIR, E. A. 1979. An

introduction to the enzymology of phenyl-propanoid biosynthesis. Recent Adv. Phytochem. 12:91-137.

HARBORNE, J. B. 1977. Variation in and functional significance of phenolic conjugation in plants. Recent Adv. Phytochem. 12:457-474.

HARBORNE, J. B. 1980. Plant Phenolics. Pages 329-402 in: Encyclopedia of Plant Physiology (new series). Vol. 8, Secondary Plant Products. E. A. Bell and B. V. Charlwood, eds. Springer-Verlag, New York.

HARBORNE, J. B., and CORNER, J. J. 1961. Plant polyphenols. IV. Hydroxycinnamic acid-sugar derivatives. Biochem. J. 81:242-250.

HARBORNE, J. B., and GAVAZZI, G. 1969. Effects of Pr and pr alleles on anthocyanin biosynthesis in *Zea mays*. Phytochemistry 8:999-1002.

HARBORNE, J. B., and HALL, E. 1964. Plant polyphenols. XII. The occurrence of tricin and of glycoflavones in grasses. Phytochemistry 3:421-428.

HARBORNE, J. B., and WILLIAMS, C. A. 1976. Flavonoid patterns in leaves of the Gramineae. Biochem. Syst. Ecol. 4:267-280.

HARRIS, P. J., and HARTLEY, R. D. 1976. Detection of bound ferulic acid in cell walls of the Gramineae by ultraviolet fluorescence microscopy. Nature (London) 259:508-510.

HARTLEY, R. D. 1973. Carbohydrate esters of ferulic acid as components of cell-walls of *Lolium multiflorum*. Phytochemistry 12:661-665.

HARTLEY, R. D., and JONES, E. C. 1975. Effect of ultraviolet light on substituted cinnamic acids and the estimation of their *cis* and *trans* isomers by gas chromatography. J. Chromatogr. 107:213-218.

HARTLEY, R. D., and JONES, E. C. 1976. Diferulic acid as a component of cell walls of *Lolium multiflorum*. Phytochemistry 15:1157-1160.

HARTLEY, R. D., and JONES, E. C. 1977. Phenolic components and degradability of cell walls of grass and legume species. Phytochemistry 16:1531-1534.

HARTLEY, R. D., and JONES, E. C. 1978. Phenolic components and degradability of the cell walls of the brown midrib mutant bm₃ of *Zea mays*. J. Sci. Food Agric. 29:777-789.

HARTLEY, R. D., JONES, E. C., and WOOD, T. M. 1976. Carbohydrates and carbohydrate esters of ferulic acid release from cell walls of *Lolium multiflorum* by treatment with cellulolytic enzymes. Phytochemistry 15:305-307.

HASLAM, E. 1975. Natural procyanidins. Pages 505-559 in: The Flavonoids. J. B. Harborne, T. J. Mabry, and H. Mabry, eds. Chapman and Hall, London.

HASLAM, E., and GUPTA, R. K. 1978. Plant proanthocyanidins. V. Sorghum polyphenols. J. Chem. Soc. Perkin Trans. 1:892-896.

HEGNAUER, R. 1963. Chemotaxonomie der Pflazen. Vol. 2, Monocotyedoneae. Birkhauser Verlag, Basel.

HERALD, P. J., and DAVIDSON, P. M. 1983. Antibacterial activity of selected hydroxy-cinnamic acids. J. Food Sci. 48:1378-1379.

HERRMANN, K. 1967. Uber Hydroxy-zimtsauren und ihre Bedeutung in Lebensmitteln. Z. Lebensm. Unters. Forsch. 133:158-178.

HERRMANN, K. 1978. Hydroxyzimtsauren und hydroxybenzosauren enthaltende Naturstoffe in Pflanzen. Fortschr. Chem. Org. Naturst. 35:73-132.

HEYDANEK, M. G., and McGORRIN, R. J. 1981a. Gas chromatography-mass spectroscopy investigations on the flavor chemistry of oat groats. J. Agric. Food Chem. 29:950-954.

HEYDANEK, M. G., and McGORRIN, R. J. 1981b. Gas chromatography-mass spectroscopy identification of volatiles from rancid oat groats. J. Agric. Food Chem. 29:1093-1095.

HIETALA, P. K., and VIRTANEN, A. I. 1960. Precursors of benzoxazolinones in rye plants. II. Precursor I, the glucoside. Acta Chem. Scand. 14:502-504.

HIGUCHI, T., and BROWN, S. A. 1963. Studies of lignin biosynthesis using isotopic carbons. XI. Reactions relating to lignification in young wheat plants. Can. J. Biochem. Physiol. 41:65-76.

HIGUCHI, T., ITO, Y., SHIMADA, M., and KAWAMURA, I. 1967. Chemical properties of milled wood lignin of grasses. Phytochemistry 6:1551-1556.

HOELDTKE, R., BALIGA, B. S., ISSENBERG, P., and WURTMAN, R. J. 1972. Dihydroxy-phenylalanine in rat food containing wheat and oats. Science 175:761-762.

HOFMAN, J., and HOFMANOVA, O. 1969. 1,4-benzoxazine derivatives in plants. Sephadex fractionation and identification of a new glucoside. Eur. J. Biochem. 8:109-112.

HOFMAN, J., and HOFMANOVA, O. 1971. 1,4-benzoxazine derivatives in plants: Absence of 2,4-dihydroxy-7-methoxy-2H-1,4-benzoxazin-3(4H)-one from uninjured *Zea mays* plants. Phytochemistry 10:1441-1444.

HOFMAN, J., HOFMANOVA, O., and HANUS, V. 1970a. 1,4-benzoxazine derivatives in plants. A new type of glucoside from *Zea mays*. Tetrahedron Lett., pp. 3213-3214.

HOFMAN, J., HOFMANOVA, O., and HANUS, V. 1970b. 1,4-benzoxazine derivatives in plants. A new glucosidic derivative from *Zea mays*. Tetrahedron Lett., pp. 5001-5002.

HOFMAN, J., and MASOJIDKOVA, M. 1973. 1,4-benzoxazine glucosides from *Zea mays*. Phytochemistry 12:207-208.

HOMBERG, H., and GEIGER, H. 1980. Fluorescenz und struktur von vlavonen. Phytochemistry 19:2443-2449.

HOSENEY, R. C., and FAUBION, J. M. 1981. A mechanism for the oxidative gelation of wheat flour water-soluble pentosans. Cereal Chem. 58:421-424.

JACOBSON, A., and CORCORAN, M. R. 1977. Tannins as gibberellin antagonists in the synthesis of α-amylase and acid phosphatase by barley seeds. Plant Physiol. 59:129-133.

JENDE-STRID, B. 1981. Characterization of mutants in barley affecting flavonoid synthesis. Hoppe-Seyler's Z. Physiol. Chem. 362.12-13.

KAHNT, G. 1967. *Trans-cis*-equilibrium of hydroxycinnamic acids during irradiation of aqueous solutions at different pH. Phytochemistry 6:755-758.

KAMBAL, A. E., and BATE-SMITH, E. C. 1976. A genetic and biochemical study on pericarp pigments in a cross between two cultivars of grain sorghum, *Sorghum bicolor*. Heredity 37:413-416.

KIKUGAWA, K., HAKAMADA, T., HASUNUMA, M., and KURECHI, T. 1983. Reaction of *p*-hydroxycinnamic acid derivatives with nitrite and its relevance to nitrosamine formation. J. Agric. Chem. 31:780-785.

KING, H. G. C. 1962. Phenolic compounds of commercial wheat germ. J. Food Sci. 27:446-454.

KIRBY, L. T., and STYLES, E. D. 1970. Flavonoids associated with specific gene action in maize aleurone, and the role of light in substituting for the action of a gene. Can. J. Genet. Cytol. 12:934-940.

KLUN, J. A., TIPTON, C. L., and BRINDLEY, T. A. 1967. 2,4-dihydroxy-7-methoxy-1,4-benzoxazin-3-one (DIMBOA), an active agent in the resistance of maize to the European corn borer. J. Econ. Entomol. 60:1529-1533.

KLUN, J. A., TIPTON, C. L., ROBINSON, J. R., OSTREM, D. L., and BEROZA, M. 1970. Isolation and identification of 6-7-dimethoxy-2-benzoxazolinone from dried tissues of *Zea mays* (L.) and evidence of its cyclic hydroxamic acid precursor. J. Agric. Food Chem. 18:663-665.

KODICEK, E., and WILSON, P. W. 1960. The isolation of niacytin, the bound form of nicotinic acid. Biochem. J. 76:27-28.

KOETZ, R., AMADO, R., and NEUKOM, H. 1979. Nature of the bound nicotinic acid in wheat bran. Lebensm. Wiss. Technol. 12:346-349.

KOLATTUKUDY, P. E. 1977. Lipid polymers and associated phenols, their chemistry, biosynthesis, and role in pathogenesis. Recent Adv. Phytochem. 11:185-246.

KOYAMA, M., AIJIMA, T., and SAKAMURA, S. 1974. *o*- and *p*-(β-D-glucopyranosyloxy) benzylamine from buckwheat (*Fagopyrum esculentum* Moench). Agric. Biol. Chem. 38:1467-1469.

KOYAMA, M., OBATA, Y., and SAKAMURA, S. 1971. Identification of hydroxybenzyl-amines in Buckwheat seeds (*Fagopyrum esculentum* Moench). Agric. Biol. Chem. 35:1870-1879.

KOYAMA, M., TSUJIZAKI, Y., and SAKAMURA, S. 1973. New amides from buckwheat seeds (*Fagopyrum esculentum* Moench). Agric. Biol. Chem. 37:2749-2753.

KRISNANGKURA, K., and GOLD, M. H. 1979. Peroxidase catalyzed oxidative decarb-oxylation of vanillic acid to methoxy-*p*-hydroquinone. Phytochemistry 18:2019-2021.

LEONT'EVA, V. G., GROMOVA, A. S., LUTSKII, V. I., MODONOVA, L. D., and TYUKAVKINA, N. A. 1974. Alkyl ferulates from the bark and wood of plants of the Pinaceae family. Khim. Prir. Soedin. 10:240-241. (In Russian) (Chem. Nat. Compounds 10:224-245)

LONG, B. J., DUNN, G. M., and ROUTLEY, D. G. 1978. Relationship of hydroxamate concentration in maize and field reaction to *Helminthosporium turcicum*. Crop Sci. 18:573-575.

MABRY, T. J., MARKHAM, K. R., and THOMAS, M. B. 1970. The Systematic Identification of Flavonoids. Springer-Verlag, Berlin.

MACE, M. E., and HEBERT, T. T. 1963. Naturally occurring quinones in wheat and barley and their toxicity to loose smut fungi. Phytopathology 53:692-700.

MacMASTERS, M. M., HINTON, J. J. C., and BRADBURY, D. 1971. Microscopic structure and composition of the wheat kernel. Pages 51-113 in: Wheat: Chemistry and Technology. Y. Pomeranz, ed. Am. Assoc. Cereal Chem., St. Paul, MN.

MADSEN, K. O. 1981. The anticaries potential of seeds. Cereal Foods World 26:19-25.

MAGA, J. A. 1978. Simple phenol and phenolic compounds in food flavor. Crit. Rev. Food Sci. Nutr. 10:323-372.

MAGA, J. A., and LORENZ, K. 1973. Taste threshold values for phenolic acids which can influence flavor properties of certain flours, grains and oilseeds. Cereal Sci. Today 18:326-328.

MAGA, J. A., and LORENZ, K. 1974. Phenolic

acid composition and distribution in wheat flours and various triticale milling fractions. Lebensmitt. Wiss. Technol. 7:273-278.

MAIZEL, J. V., BURKHARDT, H. J., and MITCHELL, H. K. 1964. Avenacin, an antimicrobial substance isolated from *Avena sativa*. I. Isolation and antimicrobial activity. Biochemistry 3:424-426.

MANIÑGAT, C. C., and JULIANO, B. O. 1982. Composition of cell wall preparations of rice bran and germ. Phytochemistry 21:2509-2516.

MANN, J. D., STEINHART, C. E., and MUDD, S. H. 1963. Alkaloids and plant metabolism. V. The distribution and formation of tyramine methylpherase during germination of barley. J. Biol. Chem. 238:676-681.

MARKWALDER, H. U., and NEUKOM, H. 1976. Diferulic acid as a possible crosslink in hemicelluloses from wheat germ. Phytochemistry 15:836-837.

MARTIN-TANGUY, J., CABANNE, F., PERDRIZET, E., and MARTIN, C. 1978. The distribution of hydroxycinnamic acid amides in flowering plants. Phytochemistry 17:1927-1928.

MARTIN-TANGUY, J., PERDRIZET, E., PREVOST, J., and MARTIN, C. 1982. Hydroxycinnamic acid amides in fertile and cytoplasmic male sterile lines of maize. Phytochemistry 21:1939-1945.

MASON, J. B., GIBSON, N., and KODICEK, E. 1973. The chemical nature of the bound nicotinic acid of wheat bran: Studies of nicotinic acid-containing macromolecules. Br. J. Nutr. 30:297-311.

MASON, J. B., and KODICEK, E. 1973. The identification of *o*-aminophenol and *o*-aminophenyl glucose in wheat bran. Cereal Chem. 50:646-654.

McCLURE, J. 1975. Physiology and functions of flavonoids. Pages 970-1055 in: The Flavonoids. J. B. Harborne, T. J. Mabry, and H. Mabry, eds. Chapman and Hall, London.

METCHE, M., and URION, E. 1961. Isolement et identification d'anthocyanosides des enveloppe d'orge. C. R. Hebd. Seances Acad. Sci. 252:356-357.

MICHELA, P., and LORENZ, K. 1976. The vitamins of triticale, wheat, and rye. Cereal Chem. 53:853-861.

MISRA, K., and SESHADRI, T. R. 1967. Chemical components of *Sorghum durra* glumes. Indian J. Chem. 5:209-210.

MOD, R. R., NORMAND, F. L., ORY, R. L., and CONKERTON, E. J. 1981. Effect of hemicellulose on viscosity of rice flour. J. Food Sci. 46:571-573.

MOLOT, P.-M., and ANGLADE, P. 1968.

Resistance commune des lignees de mais a l'helminthosporiose (*Helminthosporium turcicum* Pass.) et a la pyrale (*Ostrinia nubilalis* Hebn.) en relation avec la presence d'une substance identifiable a la 6-methoxy-2(3)-benzoxazolinone. Ann. Epiphyt. 19:75-95.

MULLICK, B. D., FARIS, D. G., BRINK, V. C., and ACHESON, R. M. 1958. Anthocyanins and anthocyanidins of the barley pericarp and aleurone tissues. Can. J. Plant Sci. 38:445-456.

MULVENA, D. P., PICKER, K., RIDLEY, D. D., and SLAYTOR, M. 1983. Methoxylated gramine derivatives from *Phalaris aquatica*. Phytochemistry 22:2885-2886.

MUSEHOLD, J. 1978. Dunnschichtchromatographische Trennung von 5-alkyl-resorcin-homologen aus Getreidekornern. Z. Pflanzenzuecht. 80:326-332.

NABETA, K., KADOTA, G., and TANI, T. 1977. O″-rhamnosylisoswertisin from oats (*Avena sativa*). Phytochemistry 16:1112-1113.

NAGASAWA, H. T., GUTMANN, H. R., and MORGAN, M. A. 1959. The oxidation of *o*-aminophenols by cytochrome c and cytochrome oxidase. II. Synthesis and identification of oxidation products. J. Biol. Chem. 234:1600-1604.

NAIR, G. V., and VON RUDLOFF, E. 1959. The chemical composition of the heartwood extractives of tamarack [Larix laricina (Du Roi) K. Koch]. Can. J. Chem. 31:1608-1613.

NAIR, P. M., and VAIDYANATHAN, C. S. 1964. Isophenoxazine synthetase. Biochem. Biophys. Acta 81:507-516.

NEISH, A. C. 1964. Major pathways of biosynthesis of phenols. Pages 295-359 in: Biochemistry of Phenolic Compounds. J. B. Harborne, ed. Academic Press, New York.

NEUKOM, H., and MARKWALDER, H. U. 1978. Oxidative gelation of wheat flour pentosans: A new way of cross-linking polymers. Cereal Foods World 23:374-376.

NILSSON, J. L. G., REDALIEU, E., NILSSON, I. M., and FOLKERS, K. 1968. On the protection against infarction by corn oil. Acta Chem. Scand. 22:97-105.

NIP, W. K., and BURNS, E. E. 1969. Pigment characterization in grain sorghum. I. Red varieties. Cereal Chem. 46:490-495.

NIP, W. K., and BURNS, E. E. 1971. Pigment characterization in grain sorghum. II. White varieties. Cereal Chem. 48:74-80.

NORIN, T., and WINELL, B. 1972. Extractives from the bark of common spruce, *Picea abies* L. Acta Chem. Scand. 26:2289-2304.

OHTA, G., and SHIMIZU, M. 1958. A new triterpene alcohol, 24-methylenecy-cloartanol, as the ferulate ester, from rice bran oil. Chem. Pharm. Bull. 6:325-326.

OLSEN, R. A. 1971. Methoxyhydroquinone, a growth inhibitor of *Ophiobolus graminis*, in leaves of oat seedlings. Physiol. Plant. 24:34-39.

OZAWA, T., NISHIKIORI, T., and TAKINO, Y. 1977. Three new substituted cinnamoyl hydroxycitric acids from corn plant. Agric. Biol. Chem. 41:359-367.

PAINTER, T. J., and NEUKOM, H. 1968. The mechanism of oxidative gelation of a glycoprotein from wheat flour. Evidence from a model system based upon caffeic acid. Biochim. Biophys. Acta 158:363-381.

PERCIVAL, F. W., and BANDURSKI, R. S. 1976. Esters of indole-3-acetic acid from *Avena* seeds. Plant Physiol. 58:60-67.

POLLOCK, B. M., GOODWIN, R. H., and GREENE, S. 1954. Studies on roots. II. Effects of courmarin, scopoletin and other substances on growth. Am. J. Bot. 41:521-529.

POPOVICI, G., and WEISSENBOCK, G. 1976. Changes in the flavonoid pattern during the development of the oat plant (*Avena sativa* L.). Ber. Dtsch. Bot. Ges. 89:483-489.

POPRAVKO, S. A., SOKOLOV, I. V., and TORGOV, I. V. 1982. New natural phenolic triglycerides. Khim. Prir. Soedin. 1982:169-173.

PRATT, D. E. 1976. Role of flavones and related compounds in retarding lipid-oxidative flavour changes in foods. Pages 1-13 in: Phenolic, Sulfur and Nitrogen Compounds in Food Flavours. G. Charalambous and I. Katz, eds. Am. Chem. Soc. Symp. 26 (Jan.).

PRIDHAM, J. B. 1965. Low molecular weight phenols in higher plants. Annu. Rev. Plant Physiol. 16:13-32.

PYYSALO, T., TORKKELI, H., and HONKANEN, E. 1977. The thermal decarboxylation of some substituted cinnamic acids. Lebensm. Wiss. Technol. 10:145-147.

QUEIROLO, C. B., ANDREO, C. S., NIEMEYER, H. M., and CORCUERA, L. J. 1983. Inhibition of ATPase from chloroplasts by a hydroxamic acid from the Gramineae. Phytochemistry 22:2455-2458.

RASPER, V. G. 1979. Chemical and physical characteristics of dietary cereal fiber. Pages 93-115 in: Dietary Fibers: Chemistry and Nutrition. G. E. Inglett and S. I. Falkehag, eds. Academic Press, New York.

REDDY, G. M. 1964. Genetic control of leucoanthocyanidin formation in maize. Genetics 50:485-489.

REICHERT, R. D. 1979. The pH-sensitive pigments in pearl millet. Cereal Chem. 56:291-294.

RING, S. G., and SELVENDRAN, R. R. 1980. Isolation and analysis of cell wall material from Beeswing wheat bran (*Triticum aestivum*). Phytochemistry 19:1723-1730.

ROONEY, L. W., BLAKELY, M. E., MILLER, F. R., and ROSENOW, D. T. 1980. Factors affecting the polyphenols of sorghum and their development and localization in the sorghum kernel. Pages 25-35 in: Polyphenols in Cereals and Legumes. Publ. 145e. Int. Dev. Res. Cent., Ottawa.

ROWE, J. W., BOWER, C. L., and WAGNER, E. R. 1969. Extractives of jack pine bark: Occurrence of *cis*- and *trans*-pinosylvin dimethyl ether and ferulic acid esters. Phytochemistry 8:235-241.

SANDO, C. E., MILNER, R. T., and SHERMAN, M. B. 1935. Pigments of the Mendelian colour types in maize. Chrysanthemin from purple-husked maize. J. Biol. Chem. 109:203-211.

SARKANEN, K. V., and LUDWIG, C. H., eds. 1971. Lignins. Wiley Interscience, New York.

SATO, H., SAKAMURA, S., and OBATA, Y. 1970. The isolation and characterization of *N*-methyltyramine, tyramine and hordenine from Sawa millet seeds. Agric. Biol. Chem. 34:1254-1255.

SCHNEIDER, E. A., and WIGHTMAN, F. 1973. Amino acid metabolism in plants. V. Changes in basic indole compounds and the development of tryptophan decarboxylase activity in barley (*Hordeum vulgare*) during germination and seedling growth. Can. J. Biochem. 52:698-705.

SHETLAR, M. R. 1948. Chemical study of the mature wheat kernel by means of the microscope. Cereal Chem. 25:99-110.

SIMMONDS, D. H., and CAMPBELL, W. P. 1976. Morphology and chemistry of the rye grain. Pages 63-110 in: Rye: Production, Chemistry, and Technology. W. Bushuk, ed. Am. Assoc. Cereal Chem., St. Paul, MN.

SINGLETON, V. L. 1981. Naturally occurring food toxicants: Phenolic substances of plant origin common in foods. Adv. Food Res. 27:149-242.

SLAUGHTER, J. C., and UVGARD, A. R. A. 1972. The identification of *p*-hydroxybenzyl-amine in barley and malt. Phytochemistry 11:478-479.

SLOMINSKI, B. A. 1980. Phenolic acids in the meal of developing and stored barley seeds. J. Sci. Food Agric. 31:1007-1010.

SMITH, M. M., and HARTLEY, R. D. 1983. Occurrence and nature of ferulic acid substitution of cell-wall polysaccharides in graminaceous plants. Carbohydr. Res. 118:65-80.

SMITH, T. A. 1977. Recent advances in the biochemistry of plant amines. Prog. Phytochem. 4:27-81.

SMITH, T. A., and BEST, G. R. 1978. Distribution of the hordatines in barley.

Phytochemistry 17:1093-1098.

SOSULSKI, F., KRYGIER, K., and HOGGE, L. 1982. Free, esterified and insoluble-bound phenolic acids. III. Composition of phenolic acids in cereal and potato flours. J. Agric. Food Chem. 30:337-340.

STAFFORD, H. A. 1974. The metabolism of aromatic compounds. Annu. Rev. Plant Physiol. 25:459-489.

STAFFORD, H. A. 1983. Enzymic regulation of procyanidin biosynthesis; lack of a flav-3-en-3-ol intermediate. Phytochemistry 22:2643-2646.

STEINKE, R. D., and PAULSON, M. C. 1964. The production of steam-volatile phenols during the cooking and alcoholic fermentation of grain. J. Agric. Food Chem. 12:381-387.

STOESSL, A. 1965. The antifungal factors in barley. III. Isolation of *p*-coumaroylagmatine. Phytochemistry 4:973-976.

STOESSL, A. 1970. Antifungal compounds produced by higher plants. Recent Adv. Phytochem. 3:143-180.

STOESSL, A., and UNWIN, C. H. 1970. The antifungal factors in barley. V. Antifungal activity of the hordatines. Can. J. Bot. 48:465-470.

STRAUS, J. 1959. Anthocyanin synthesis in corn endosperm tissue cultures. I. Identity of the pigments and general factors. Plant Physiol. (Lancaster) 34:536-541.

SUZUKI, H., MATSUMOTO, T., and NOGUCHI, M. 1976. Identification of *p*-aminophenyl-α-D-glucose from *Hydrangea macrophylla*. Phytochemistry 15:555.

SWAIN, T. 1977. Secondary compounds as protective agents. Annu. Rev. Plant Physiol. 28:479-501.

TAKETA, F. J. 1957. Studies of various nutritional factors in relation to dental caries in the rat. Ph.D. thesis, University of Wisconsin, Madison.

TAKETA, R. H., and SCHEUER, P. J. 1976. Isolation of glyceryl esters of caffeic and *p*-coumaric acids from pineapple stems. Lloydia 39:409-411.

TAMURA, T., HIBINO, T., YOKOYAMA, K., and MATSUMOTO, T. 1961. On the occurrence of dihydro-γ-sitosterylferulate in wheat oil. Chem. Abstr. 55:4889.

TAMURA, T., TAKEDA, M., MURAKAMI, K., YOSHIDA, A., and MATSUMOTO, T. 1963.Ferulates. Chem. Abstr. 59:5230.

TANAKA, A., KATO, A., and TSUCHIYA, T. 1964. The isolation of β-sitosteryl ferulate from rice bran oil. Yakugaku Zasshi 13:260.

TANNER, G. R., and MORRISON, I. M. 1983. Phenolic-carbohydrate complexes in the cell walls of *Lolium perenne*. Phytochemistry 22:1433-1439.

TIPTON, C. L., HUSTED, R. R., and TSAO, F. H.-C. 1971. Catalysis of simazine hydrolysis by 2,4-dihydroxy-7-methoxy-1,4-benzoxazin-3-one. J. Agric. Food Chem. 19:484-486.

TIPTON, C. L., WANG, M.-C., TSAO, F. H.-C., TU, C.-C. L., and HUSTED, R. R. 1973. Biosynthesis of 1,4-benzoxazin-3-ones in *Zea mays*. Phytochemistry 12:347-352.

TOWERS, G. H. N. 1964. Metabolism of phenolics in higher plants and microorganisms. Pages 249-294 in: Biochemistry of Phenolic Compounds. J. B. Harborne, ed. Academic Press, New York.

TRESSL, R., KOSSA, T., RENNER, R., and KOPPLER, H. 1976. Gas chromatographic-mass spectroscopic investigations on the formation of phenolic and aromatic hydro-carbons in food. Z. Lebensm. Unters. Forsch. 162:123-130.

TSCHESCHE, R., CHANDRA JHA, H., and WULFF, G. 1973. Uber triterpene. XXIX. Zur Struktur des Avenacins. Tetrahedron 29:629-633.

TURNER, E. M. C. 1960. The nature of the resistance of oats to the take-all fungus. III. Distribution of the inhibitors in oat seedlings. J. Exp. Bot. 11:403-412.

VAJS, V., JEREMIC, D., STEFANOVIC, M., and MILOSAVLJEVIC, S. 1975. *p*-Coumaric esters and fatty alcohols from *Artemisia campestris*. Phytochemistry 14:1659-1660.

VAN SUMERE, C. F., ALBRECHT, J., DEDONDER, A., DE POOTER, H., and PE, I. 1975. Plant proteins and phenolics. Pages 221-264 in: The Chemistry and Biochemistry of Plant Proteins. J. B. Harborne and C. F. Van Sumere, eds. Academic Press, New York.

VAN SUMERE, C. F., COTTENIE, J., DE GREEF, J., and KINT, J. 1972. Biochemical studies in relation to the possible germination regulatory role of naturally occurring coumarin and phenolics. Recent Adv. Phytochem. 4:166-221.

VAN SUMERE, C. F., DE POOTER, H., ALI, H., and DEGRAUW-VAN BUSSEL, M. 1973. *N*-Feruloylglycyl-L-phenylalanine: A sequence in barley proteins. Phytochemistry 12:407-411.

VAN SUMERE, C., HILDERSON, H., and MASSART, L. 1958. Coumarins and phenolic acids of barley and malt husk. Naturwissenschaften 45:292.

VAN SUMERE, C. F., HOUPELINE-DE COCK, H. J., VINDEVAGHEL-DE BACQUER, Y. C., and FOCKENIER, G. E. 1980. *N*-Feruloylglycyl-L-phenylalanine isolated by partial hydrolysis of bulk leaf protein of lucerne, *Medicago sativa* cv.

Europe. Phytochemistry 19:704-705.

VENIS, M. A., and WATSON, P. J. 1978. Naturally occurring modifiers of auxin-receptor interaction in corn: Identification as benzoxazolinones. Planta 142:103-107.

VERDEAL, K., and LORENZ, K. 1977. Alkylresorcinols in wheat, rye and triticale. Cereal Chem. 54:475-483.

VIRTANEN, A. I. 1964. Primary plant substances and decomposition reactions in crushed plants, exemplified mainly by studies on organic sulfur compounds in vegetables and fodder plants. Biochem. Inst., Helsinki, pp. 3-6.

VOGEL, J. J. 1961. Studies on the diet in relation to dental caries in the cotton rat. Ph.D. thesis, University of Wisconsin, Madison.

VON SYDOW, E., and ANJOU, K. 1969. The aroma of rye crispbread. Lebensm. Wiss. Technol. 2:15-18.

WAGNER, H. 1974. Flavonoid-Glykoside. Fortschr. Chem. Org. Naturst. 31:185.

WAHLROOS, O., and VIRTANEN, A. I. 1959. The precursors of 6-methoxybenzoxazolinone in maize and wheat plants, their isolation and some of their properties. Acta Chem. Scand. 13:1906-1908.

WALL, J. S., SWANGO, L. C., TESSARI, D., and DIMLER, R. J. 1961. Organic acids of barley grain. Cereal Chem. 38:407-422.

WALLERSTEIN, J. S., HALE, M. G., and ALBA, R. T. 1948. Oxidizing enzymes in brewing materials. IV. *o*-Phenylenediamine oxidase in rice, wheat and corn. Wallerstein Lab. Commun. 11:221-226.

WEISSENBOCK, G., PLESSER, A., and TRINKS, K. 1976. Flavonoid content and activities of corresponding enzymes in chloroplasts isolated from primary leaves of *Avena sativa* L. Ber. Dtsch. Bot. Ges. 89:467-472.

WENKERT, E., LOESER, E.-M., MAHAPATRA, S. N., SCHENKER, F., and WILSON, E. M. 1964. Wheat bran phenols. J. Org. Chem. 29:435-439.

WHITMORE, F. W. 1974. Phenolic acids in wheat coleoptile cell walls. Plant Physiol. 53:728-731.

WIERINGA, G. W. 1967. On the Occurrence of Growth-Inhibiting Substances in Rye. H. Weeman en Zonen, N.V., Wageningen, pp. 41-46.

WILKINS, H., BURDEN, R. S., and WAIN, R. L. 1974. Growth inhibitors in roots of light- and dark-grown seedlings of *Zea mays*. Ann. Appl. Biol. 78:337-338.

WILLARD, J. I., and PENNER, D. 1976. Benzoxazinones: Cyclic hydroxamic acids found in plants. Residue Rev. 64:67-76.

YAMAHA, T., and CARDINI, C. E. 1960a. The biosynthesis of plant glycosides. I. Monoglucosides. Arch. Biochem. Biophys. 86:127-132.

YAMAHA, T., and CARDINI, C. E. 1960b. The biosynthesis of plant glycosides. II. Gentiobiosides. Arch. Biochem. Biophys. 86:133-137.

YASUMATSU, K., NAKAYAMA, T. O. M., and CHICHESTER, C. O. 1965. Flavonoids of sorghum. J. Food Sci. 30:663-667.

YOSHIZAWA, K., KOMATSU, S., TAKAHASHI, I., and OTSUKA, K. 1970. Phenolic compounds in the fermented products. I. Origin of ferulic acid in sake. Agric. Biol. Chem. 34:170-180.

ZENK, M. H. 1966. Biosynthesis of C_6-C_1 compounds. Pages 45-60 in: Biosynthesis of Aromatic Compounds. G. Billek, ed. Proc. 2d meeting FEBS, Vienna, 21–24 April 1965. Pergamon Press, Oxford.

ZENK, M. H. 1967. Untersuchungen zum Phototropismus der *Avena*-Koleoptile: II. Pigmente. Z. Pflanzenphysiol. 56:122-140.

ZENK, M. H. 1979. Recent work on cinnamoyl CoA derivatives. Recent Adv. Phytochem. 12:139-176.

ZIMMERMANN, H. 1978. Investigations on the extraction of indoleacetic acid proteins from oat. Z. Pflanzenphysiol. 89:115-118.

ZINSMEISTER, H. D., and HOLLMULLER, W. 1964. Gerbstoffe und Wachstum. II. Die Wirkung einiger Gerbstoffblausteine auf das Wachstum von Getreidekoleoptilen. Planta 63:133-145.

ZUNIGA, G. E., ARGANDONA, V. H., NIEMEYER, H. M., and CORCUERA, L. J. 1983. Hydroxamic acid content in wild and cultivated Gramineae. Phytochemistry 22:2665-2668.

CHAPTER 10

NUTRITION OF OATS

HAINES B. LOCKHART
H. DAVID HURT
John Stuart Research Laboratory
Quaker Oats Company
Barrington, Illinois

I. INTRODUCTION

The basic tenet that serves to unite this chapter with the food chemistry chapters on oats is that food and nutrition issues cannot be separated—food is the carrier of nutriment. To fully appreciate the advances of our understanding of the chemistry of oats, we must be cognizant of the eventual utility of oats to humans, that being its nutritional value. This chapter evaluates the relationship between the current nutritional needs of the U.S. public and the role that cereals, and in particular oats, play in meeting these needs.

The current nutrition scenario in this country is not one of seeking to eradicate nutritional voids and deficiencies among our population but rather one of attempting to solve the nutritional abuses that have resulted from overnutrition. These problems have been amply expressed in the joint recommendations set forth by the U.S. Department of Agriculture and the Department of Health and Human Service's "Dietary Guidelines for Healthy Americans." By way of review, these recommendations are to 1) eat a variety of foods; 2) maintain ideal body weight; 3) avoid too much fat, saturated fat, and cholesterol; 4) eat foods with adequate starch and fiber; 5) avoid too much sugar; 6) avoid too much sodium; and 7), if you drink, do so in moderation.

Adoption of these guidelines presents significant advantages to cereal-based foods. Cereals represent an economical alternative to animal products as the principal meal item. Nutritionally, these guidelines and others speak of the need to increase our intake of fiber and complex carbohydrate while decreasing consumption of refined sugar, animal fat, and protein. Cereal-based foods will attain greater acceptance as the health benefits and economic considerations are fully realized. The major task that lies ahead is to assimilate these basic ingredients into products that appeal to the sophisticated palates of our modern society. The beneficial role that cereal-based products play in the U.S. diet is not fully appreciated. If the food consumption patterns of the U.S. consumer were changed to meet the U.S. Dietary Guidelines, cereal consumption would increase 70–140%.

Presently, oats rank well below other cereals—corn, wheat, and rice—as an important food ingredient for human consumption. In the United States, only 8–10% of the oat crop is directed toward this purpose. Under these circumstances, there exist significant opportunities to expand the usage of oats by applying the knowledge of the chemical properties and nutritional value of oats.

Although not widely appreciated, oats have an interesting place in human culinary history. The first reference to human consumption of oats was by Germanic tribes of the first century, who knew oats well and made "their porridge of nothing else." The grain, however, found widespread acceptance only in Ireland and Scotland, where the Highlanders used it in a variety of porridges and baked goods. Oats are believed to have reached the American continent in 1602 through the Elizabeth Islands in Massachusetts. Their use as a human food in this country, however, was slow in coming. Until Ferdinand Schumacher, the Oatmeal King, promoted oats to the American people as a healthful addition to the diet, it was sold primarily by druggists for relief of stomach discomforts and for the infirm. History may have reported that oats are an orphan cereal, but hot porridge on a cold morning remains a time-tested reality with sound nutritional truths.

II. PROTEIN

Recent dietary surveys clearly indicate that the protein nutriture of the U.S. consumer is in excellent shape. The latest survey data indicate that 102 g or 150% of the recommended dietary allowance (RDA) is consumed per person per day. Average protein intake of individuals in all age-sex groups, although slightly lower than in similar surveys conducted in the mid-1960s, remains well above the requirement. The health effects of excessive intakes of protein, and in particular animal protein, have been suggested to be associated with such chronic disease conditions as cancer and coronary heart disease. Certainly, economic considerations question the efficacy of continued disproportionate consumption of animal rather than plant protein. In contrast, the major populations of the world rely heavily on cereal sources for their protein nutriture.

The relative value of cereals as a protein source is discounted primarily because of the lower quantity of protein and because of its marginal quality. The commonly consumed cereals range from 6 to 18% protein, with oats generally recognized as having the highest protein concentration of cereals routinely consumed by people of developed countries throughout the world (Fig. 1). Even though the concentration of protein is relatively low, people who consume enough cereal to meet their energy requirements will generally ingest enough protein to also meet these needs, according to Jansen (1972). Unpublished data by Graham et al suggest that protein needs of infants can be met with oats when fed at 66% of their caloric needs. The limiting nutritional concern, however, is directly related to the relative quality of cereal protein.

The quality of cereal protein is a reflection of the relative amino acid makeup of the total protein of the cereal, as well as its ability to be digested by the organism for which it is intended. The essential amino acid composition of the major cereals is illustrated in Table I. Cereal proteins are generally considered to be limiting in lysine, with methionine, threonine, and isoleucine secondary limiting amino acids. The amino acid composition of oats is remarkably constant over a wide range of protein contents. It has been reported that only a slight negative correlation exists between total protein and the lysine percentage. Of the oat cultivars surveyed, the lysine ranged between 3.2 and 5.2% of the total protein. Other amino acids demonstrate similar response to genetic alteration. However, oat breeders should pay attention not only to the protein content but to that of lysine as well as the other limiting amino acids, methionine and threonine.[1]

Results of human metabolic studies indicate that oat protein has a higher nutritive value than most other cereal proteins. The effect of supplementing oat

[1]D. J. Schrickel and W. L. Clark. Status of protein and quality improvement in oats. Paper presented at the CIMMYT-Purdue University International Symposium on Protein Level in Maize. El Batan, Mexico, 1972.

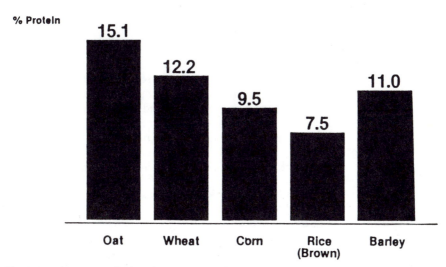

Fig. 1. Protein content of selected cereals. (Data from Lockhart and Nesheim, 1978)

TABLE I
Essential Amino Acid Profiles of Eggs and of Cereal Grains[a]

Amino Acid	Eggs	Oats	Wheat	Corn	Barley	FAO Standard
Lysine	6.4	3.7	2.9	2.7	3.5	4.2
Histidine	2.4	2.1	2.3	2.7	2.1	...
Arginine	6.6	6.3	4.6	4.2	4.7	...
Threonine	5.0	3.3	2.9	3.6	3.3	2.8
Valine	7.4	5.1	4.4	4.8	5.0	4.2
Methionine	3.1	1.7	1.5	1.9	1.7	2.2
Isoleucine	6.6	3.8	3.3	3.7	3.6	4.2
Leucine	8.8	7.3	6.7	12.5	6.7	4.8
Phenylalanine	5.8	5.0	4.5	4.9	5.1	2.8
Tryptophan	1.6	1.3	1.1	0.7	1.5	1.4

[a]Data from FAO (1970). Data given as grams per 100 g of protein.

protein with amino acids to achieve nitrogen balance has been studied. These studies with young children suggest threonine to be the limiting amino acid. Kies et al (1972) similarly evaluated the protein value of precooked oatmeal when fed to humans and determined methionine to be the first limiting amino acid in an oatmeal product.

The protein of oats is uniquely different from that of other cereals in that the major protein fraction is represented by the salt-soluble globulin fraction (Table II); little protein is present as glutelin. The high percentage of globulin is probably the primary reason for the better protein nutritive value of oats. This is believed to be the result of the higher level of lysine in the globulin fraction than in the glutelin and prolamine fractions.

A comparison of the biological value of oat protein and the protein of other cereal grains is given in Table III. The protein quality of oats is second only to that of the genetically modified *opaque-2* maize. Oats also have the highest level of protein of any of the grains.

III. LIPIDS

The kind and amount of fat that is included in the diet have received a good deal of attention from those interested in what we eat, as it may affect such chronic ailments as heart disease. Since 1960, there has been a steady decline in the consumption of butter and other animal fats and a simultaneous increase in polyunsaturated vegetable oils, especially oil expressed from corn.

The fat content of various cereal grains is presented in Table IV. Note that whole-grain oats have the greatest percentage of fat of the major cereals. For grains such as corn and wheat, the lipid fraction is concentrated in the germ fraction of the kernel, thus greatly facilitating extraction of the lipid. However,

TABLE II
Protein Fractions of Cereal Grains[a]

Cereal	Albumin	Globulin	Prolamine	Glutelin	Prolamine and Glutelin
Oat	1.0	78.0	16.0	5.0	21
Wheat	3.5	10.0	69.0	16.0	85
Corn	4.0	2.0	55.0	39.0	94
Rice	5.0	10.0	5.0	80.0	85
Barley	13.0	12.0	52.0	23.0	75

[a] Data from Osborne (1909).

TABLE III
Protein Quality of Cereal Grains (PER)[a]

Grain	Protein Efficiency Ratio	Grain	Protein Efficiency Ratio
Maize (normal)	1.2	Rye	1.6
Maize (*opaque-2*)	2.3	Oats	1.9
Pearl millet	1.8	Wheat	1.5
Rice (polished)	1.7	Triticale	1.6

[a] Source: Lockhart and Nesheim (1978); used by permission.

oat lipids are distributed throughout the grain rather than being concentrated in the germ.

Although there have been few (if any) nutritional studies on the value of oat lipid, the high content of oleic and linoleic acid, the resulting favorable polyunsaturated-to-saturated fatty acid ratio of 2.2 (the recommended ratio is at least 1.0), and the accompanying lipid-soluble antioxidant properties suggest that oats would readily lend themselves to diets designed to lower blood cholesterol levels.

The fact that the lipid content and composition of oats appear to be greatly influenced by varietal differences suggests that the grain may yet represent a valuable source of dietary lipid.

IV. CARBOHYDRATE AND FIBER

The U.S. Department of Agriculture's Dietary Guidelines have recognized the importance of increasing the consumption of complex carbohydrate and fiber as derived primarily from cereal-based products. Chronic disease conditions, such as diabetes, atherosclerosis, and digestive diseases, appear to be improved by inclusion of more complex carbohydrate and fiber in the diet. In this regard, the nutritional focal point of oat carbohydrate rests primarily with the high β-glucan or gum content, which is water soluble yet resistant to human digestive processes. The important nutritional attributes of oats as related to diabetes and control of blood cholesterol levels have been well documented by Anderson and Chen (1979).

Forsythe et al (1978) studied the laxation effect of oat bran and oat flour on laboratory rats. The oat products were fed to test animals in addition to other fiber sources, such as wheat bran, wheat middlings, cellulose, psyllium seeds,

TABLE IV
Fat Contents of Selected Cereal Grains[a]

Grain	Percentage of Fat (Moisture-Free Basis)
Oats	
Whole grain	5.4
Groats	7.6
Hulls	0.6
Corn	
Whole grain	4.4
Germ	22.0
Wheat	
Whole grain	2.1
Germ	15.0
Rice	
Whole grain	2.2
Polished	0.8
Bran	20.0
Rye	
Whole grain	1.8
Sorghum	
Whole grain	3.4
Barley	
Whole grain	2.1

[a] Data from Weber (1973).

beet pulp, and soy hulls. Results of the four-week study are presented in Table V.

Oat bran had an effect on dry fecal weight equal to that of wheat bran and slightly lower than that of wheat middlings. Even though oat flour supplied less cellulose and lignin to the diet than the other fiber sources tested, it produced a positive effect on fecal dry weight.

Meyer and Calloway (1977) fed diets containing oat bran, wheat bran, oat gum, or raffinose to adult humans to evaluate the laxation effect of various dietary fibers (Table VI). Frequency of bowel movements was highest when diets containing wheat bran and the indigestible sugar raffinose were fed; diets containing oat bran and oat gum resulted in only a slight increase in bowel frequency. Diets that contained wheat bran as the fiber source clearly produced increased fecal volume. Oat bran and oat gum feeding resulted in increased fecal output when compared with the control diet. Wheat bran and raffinose-containing diets, which produced discomfort and gaseous distress, were not well tolerated. Oat bran caused little or no problem.

Calloway and Kretsch (1978) investigated the effects on the fecal output of the test subjects of a fiber-free liquid formula diet and the same diet supplemented with toasted or plain oat bran. Each diet was fed to six young men for a period of 15 days. The addition of about 45 g of oat bran to the liquid formula diet approximately doubled the output of fecal matter. It did not seem to make any difference whether the oat bran was plain or toasted.

Thus, oat fiber, like cereal grain fibers in general, increases the fecal bulk. The increase consists of a change in both dry matter and total fecal weight; the percentage of moisture remains fairly constant. In most cases, frequency of defecation increases as a result of consuming cereal grain fibers. Transit time (the

TABLE V
Bulking Effect of Fiber (Rats)[a]

Diet	Fecal Dry Weight (g/day)
Control	0.9 ± 0.1
Wheat bran	3.7 ± 0.3
Wheat middlings	5.8 ± 0.2
Oat flour	1.4 ± 0.1
Oat bran	3.7 ± 0.1
Cellulose	2.2 ± 0.1
Psyllium seed	1.6 ± 0.1
Sugar beet pulp	2.5 ± 0.1
Soy hull fiber	2.3 ± 0.1

[a] Data from Forsythe et al (1978).

TABLE VI
Laxation Effect of Fiber (Humans)[a]

Food	Dry Fecal Weight (g)	No. of Bowel Movements/Day
Control diet	13.7 ± 3.0	1.4 ± 0.3
Oat bran	31.1 ± 8.5	1.8 ± 0.4
Wheat bran	38.1 ± 10.4	3.4 ± 1.1
Oat gum	22.0 ± 7.4	1.8 ± 0.4
Raffinose	14.3 ± 6.8	3.0 ± 0.9

[a] Data from Meyer and Calloway (1977).

time it takes food to travel through the digestive system) is frequently decreased. The laxation property of cereal grain fiber appears to be related to the insoluble fiber portion. At present, the mechanism of laxation is unknown, but possible factors are 1) the water-binding nature of the fiber, 2) physical bulk, 3) interference with normal digestion, and 4) products within the fiber itself or produced via bacterial fermentation that affect the motility of the intestinal system.

V. MINERALS

Although there are no recognizable micronutrient deficiencies in this country, the inexactness of the science of nutrition, expressed as our inability to characterize and detect the impact of suboptimal or marginal nutritional status, coupled with increased reliance on processed and formulated foods, has served in part as the basis for the dietary recommendation that the consumption of whole-grain cereals be increased.

Cereal-based food products are worthy contributors to our mineral and vitamin needs, provided the grains have not been subjected to extensive milling or other processes (Table VII). The mineral and vitamin content of cereal grains is affected by a number of environmental conditions and agricultural practices, such as soil condition, climate, state of maturity, cultivar, and so forth. The data of Morgan (1968) presented in Table VIII illustrate the effect of a number of these agricultural practices, as reflected in the degree of variation determined in

TABLE VII
Typical Vitamin/Mineral Contents of Rolled Oats[a]

Vitamin/Mineral	Content (per 100 g)	Vitamin/Mineral	Content (per 100 g)
Thiamine (mg)	0.67	Iron (mg)	3.81
Riboflavin (mg)	0.11	Phosphorus (mg)	450
Niacin (mg)	0.80	Manganese (mg)	4
Vitamin E (I.U.)	3	Magnesium (mg)	141
Vitamin B$_6$ (mg)	0.21	Zinc (mg)	3
Folic acid (μg)	104	Copper (mg)	0.33
Biotin (μg)	13	Sodium (mg)	4
Calcium (mg)	50	Potassium (mg)	370

[a] Source: Unpublished data from Quaker Oats Company, Barrington, IL.

TABLE VIII
Mineral Content of 171 Cross-Section Samples of Oats[a]

Mineral	Mean	Range	C.V. (%)
Calcium (%)	0.11	0.07–0.18	18.2
Phosphorus (%)	0.38	0.29–0.59	10.5
Sodium (%)	0.02	0.004–0.06	47.8
Potassium (%)	0.47	0.31–0.65	17.0
Magnesium (%)	0.13	0.10–0.18	13.1
Copper (ppm)	4.7	3.0–8.2	17.0
Cobalt (ppm)	0.05	0.02–0.17	53.0
Manganese (ppm)	45	22–79	28.9
Zinc (ppm)	37	21–70	27.0

[a] Data from Morgan (1968).

the micronutrient content of 171 random samples of oats raised in the relative confines of a small area of Wales. However, as a result of modern harvesting, distribution, and food manufacturing techniques, which all facilitate randomization of specific commodity lots, the natural variability of the nutrient components is greatly diminished and becomes of little practical significance in evaluating their relative nutritional value to humans.

Peterson et al (1974) at the Oat Quality Laboratory and Department of Agronomy at the University of Wisconsin, Madison, have determined the mineral contents of several oat cultivars adapted to the north central United States. To express this information in terms that more readily reflect the potential of oats to meet human nutritional requirements, the averages from the six cultivars tested are summarized in Table IX in terms of percentage of the RDA for the adult male.

As noted from this information, one ounce of oats (a typical serving) provides a significant quantity of a number of essential minerals. Oats are a good source of manganese, magnesium, and iron, as well as calcium, zinc, and copper. As will be mentioned later, the relative value of the phosphorus component to human nutrition has been questioned by nutritionists. Analytical values for several essential trace minerals in oats are not available. Although nutrition studies have identified the need for trace nutrients, such as chromium, nickel, cobalt, vanadium, silicon, and tin, there has been little research conducted to determine the amounts contained in oats or other cereals.

The micronutrient content of oats is not evenly distributed throughout the groat. To determine the relative distribution of these constituents, Peterson et al (1975) carefully hand-dissected samples of groats into their component parts and assayed each part for its mineral content. Their results (Table X) demonstrate that the mineral components of the groat are concentrated in the outer bran fraction. Any physical process that would alter the relative distribution of the component parts of the groat, i.e., milling, could be expected to alter the relative content of the mineral components. Fortunately, oats are generally consumed as the whole grain, including the bran fraction.

A review of the mineral content of cereals in general would not be complete without consideration of a negative nutritional concern related to the increased use of cereals in our diet. The major form of phosphorus in cereals exists as phytic

TABLE IX
Mineral Composition of Oats[a]

Nutrient	Amount (mg/oz)	Percentage of RDA[b] or SADDI[c]
Calcium	15.4	2
Phosphorus	155	19
Potassium	120	6
Magnesium	52	15
Iron	1.3	13
Zinc	1.0	7
Manganese	1.3	52
Copper	0.13	7

[a] Data from Peterson et al (1975).
[b] Recommended daily allowance.
[c] Safe and adequate daily dietary intakes.

acid, and the phytic acid content of cereals is highly correlated with total phosphorus. The following relationships between these variables have been noted: for wheat, the phytic acid concentration (PAC) = (phosphorus concentration [PC] − 0.096)/0.30, $r = 0.97$; for barley, PAC = (PC − 0.092)/0.504, $r = 0.96$; for oats, PAC = (PC − 0.153)/0.29, $r = 0.91$.

In mature grains, phytate accounts for 60–80% of the total phosphorus. Germination of the grains rye, barley, and corn completely eliminates phytate phosphorus, whereas, at the end of the same period of time, oats still have more than half the original phytate phosphorus. However, rye and wheat have high phytase activities, whereas in oats the enzyme has a much lower activity under comparable conditions.

Phytic acid is nutritionally important because it can bind essential minerals, such as calcium, zinc, and magnesium, and keep them from being absorbed by the body. Mineral deficiencies have been noted in humans and monogastic animals whose diets consist predominantly of whole grains or legumes high in phytic acid.

The practical implication of the effects of the phytic acid content of oats when they are consumed as part of a normal meal was evaluated by Sharpe (1950). Groups of 10 adolescent boys were fed diets containing a constant amount of reduced radioisotope of iron: either Fe^{55} (half-life of four years) or Fe^{59} (half-life of 44 days). The effects of various foods on the total amounts of iron absorbed are shown in Table XI.

Milk alone reduced the absorption of iron by one-third; this could not have been a result of the action of phytates, since milk contains none. Rolled oats and milk depressed absorption of iron by two-thirds. Thus, rolled oats interfered no

TABLE X
Relative Distribution of Minerals Within the Groat[a]

Element	Bran	Endosperm
Phosphorus (%)	1.02	0.26
Calcium (%)	0.11	0.10
Potassium (%)	1.00	0.16
Magnesium (%)	0.38	0.07
Manganese (ppm)	88	31
Iron (ppm)	90	18
Zinc (ppm)	58	24

[a] Data from Peterson et al (1975).

TABLE XI
Effect of Oatmeal on the Absorption of Radioactive Iron[a]

Food	Percentage of Iron Absorbed
Water	26
Milk	17
Milk + oatmeal (173 g)	9
Milk + oatmeal (285 g)	8
Complete meal (high bread)	5
Complete meal	4
Milk + Na phytate	1

[a] Data from Sharpe et al (1950).

more than an equal weight of milk. No correlation was found between the phytate content of rolled oats and the reduction in iron absorption. Rather, there appears to be an inverse correlation of iron absorption with the solids content of the test meals.

VI. VITAMINS

Although no vitamin deficiencies are identifiable in this country, the consumption of a balanced diet derived from a variety of foods will collectively provide the level of nutriment needed by humans. The vitamin content of groats and rolled oats expressed as milligrams per 100 g (as is) and percentage of RDA are presented in Table XII. As noted, oats and oat-based products contribute a small but significant amount of vitamins to the human diet. Rolled oats represent a good source of thiamine and pantothenic acid.

Little research has been done on the relative distribution of vitamins in the oat kernel. Holman and Godden (1947) determined that the hull and endosperm of oats contained little thiamine, whereas the major portion of the vitamin content was located in the outer bran fraction (Table XIII). Separation and isolation of the various fractions by milling or other processes alters the relative nutritional value of the resulting product. Since oats are generally eaten as a whole grain, the consumer gets the benefits of the nutrients in the bran.

Concern has been growing regarding the nutritional worth of the nutrients in grains because there is a question about how "bioavailable" they actually are. Studies by Frigg (1976) to estimate the availability of vitamin biotin in corn,

TABLE XII
Vitamin Content of Groat and Rolled Oats[a]

Vitamin	Groat		Rolled Oats	
	Content[b]	% RDA/oz	Content	% RDA/oz
Thiamine (B₁)	0.77	15	0.67	14
Riboflavin (B₂)	0.14	2	0.14	2
Niacin	0.97	2	0.98	2
Pantothenic Acid	1.36	10	1.48	10
Pyridoxine (B₆)	0.12	2	0.13	2
Folic acid	0.06	4	...	...
α Tocopherol	...	...	1.94	5

[a] Data from Shukla (1975).
[b] Given as milligrams per 100 g.

TABLE XIII
Distribution of Thiamine in the Various Subfractions
of the Groat (var. Early Miller)[a]

Fraction	Thiamine (mg/100 g dry wt)	Thiamine Contributed by Fraction (%)
Hull	0.0	0
Embryo	6.3	17
Scutellum	8.3	24
Bran	2.7	42
Endosperm	0.05	6

[a] Data from Holman and Godden (1947).

TABLE XIV
Biological Availability of Biotin of Various Cereals[a]

Grain	Biotin (μg/kg)	Percentage Available to Chick
Corn	45	100
Wheat	104	0
Barley	144	20
Oats	208	32
Milo	290	20

[a] Data from Frigg (1976).

wheat, barley, and oats are presented in Table XIV. The biological availability of biotin in cereals, compared with that of added biotin, was determined by a chick growth bioassay. Although the biotin content of corn was determined to be low, the relative availability was excellent. In milo, barley, and oats, the biotin availability was reduced to 20, 22, and 32%, respectively. Wheat gave no significant growth response, suggesting a low availability for the biotin. Although these data may not be directly applicable to humans, it is important to recognize the differences in utilizable nutrients that exist among various foodstuffs. These data should clearly indicate a continued need to evaluate carefully the effect of processing on mineral and vitamin content and to establish the relative biological availability of other minerals and vitamins as provided by oats and other cereals, especially for those micronutrients for which little or no information is available.

In summary, with the growing awareness of the importance of good nutrition to good health, cereals will undoubtedly play a much bigger role in feeding the consumer in the future. Our challenge rests in using our technological base of knowledge to develop acceptable products for the sophisticated consumer. Oats have a number of unique nutritional as well as functional qualities; consequently, it is puzzling that they are greatly underutilized for human consumption. Oats have the potential for making a major contribution to the human diet, benefits that are attainable if we are smart enough to guide the way.

LITERATURE CITED

ANDERSON, J. W., and CHEN, W. J. 1979. Plant Fiber. Carbohydrate and lipid metabolism. Am. J. Clin. Nutr. 32:346-363.

CALLOWAY, D. H., and KRETSCH, M. J. 1978. Protein and energy utilization in men given a rural Guatamalan diet and egg formulas with and without added oat bran. Am. J. Clin. Nutr. 31:1118-1126.

FAO. 1970. Amino acid content of foods and biological data on proteins. Food and Agric. Org. of the U.N., Rome.

FORSYTHE, W. A., CHENOWTH, W. L., and BENNICK, M. R. 1978. Laxative and serum cholesterol in rats fed plant fibers. J. Food Sci. 43:1470-1476.

FRIGG, M. 1976. Bio-availability of biotin in cereals. Poult. Sci. 55:2310-2318.

HOLMAN, W. I. M., and GODDEN, W. 1947. Aneurine content of oats. J. Agric. Sci. 37:51-59.

JANSEN, G. R. 1972. Seeds as a source of protein for humans. Pages 19-38 in: Symposium: Seed Proteins. G. E. Inglett, ed. Avi Publishing Co., Westport, CT.

KIES, C., PETERSON, R., and FOX, V. M. 1972. Protein nutritive value of amino acid supplemented and unsupplemented processed oatmeal. J. Food Sci. 37:306-309.

LOCKHART, H. B., and NESHEIM, R. O. 1978. Nutritional quality of cereal grains. Pages 201-221 in: Cereals '78: Better Nutrition for the World's Millions. Y. Pomeranz, ed.

Am. Assoc. Cereal Chem., St. Paul, MN.

MEYER, S., and CALLOWAY, D. H. 1977. Gastro-intestinal response to oat and wheat milling fractions in older women. Cereal Chem. 54:110-119.

MORGAN, D. E. 1968. Note on variations in the mineral composition of oat and barley grown in Wales. J. Sci. Food Agric. 19:393-395.

OSBORNE, T. D. 1909. The Vegetable Proteins. Longmans and Green, London.

PETERSON, D. M., SENTURIA, J., YOUNGS, V. L., and Schrader, L. E. 1975. Elemental composition of oat groats. J. Agric. Food Chem. 23:9-13.

SHARPE, L. M., PEACOCK, W. C., COOKE, R., and HARRIS, R. S. 1950. The effect of phytate and other food factors on iron absorption. J. Nutr. 41:433-446.

SHUKLA, T. P. 1975. Chemistry of oats: Protein foods and other industrial products. Crit. Rev. Food Sci. Nutr. 6:383-431.

WEBER, E. J. 1973. Structure and composition of cereal components as related to their potential industrial utilization. Pages 161-206 in: Industrial Uses of Cereals. Y. Pomeranz, ed. Am. Assoc. Cereal Chem., St. Paul, MN.

CHAPTER 11

CHOLESTEROL-LOWERING PROPERTIES OF OAT PRODUCTS

JAMES W. ANDERSON
WEN-JU LIN CHEN
Medical Service
Veteran's Administration Medical Center
University of Kentucky
Lexington, Kentucky

I. INTRODUCTION

Hypercholesterolemia is one of the three major risk factors for ischemic heart disease (Gotto et al, 1977). In animal experiments, cholesterol feeding produces hypercholesterolemia that is followed by development of antherosclerosis; usually the degree of atherosclerosis parallels the rise in serum cholesterol (Grundy, 1977). In humans, the association between serum cholesterol elevation and ischemic heart disease is well established (Gotto et al, 1977). Cholesterol may be involved in the final common pathway for atherosclerosis, since experimental animals usually do not develop atherosclerosis unless they are fed cholesterol. Among the cholesterol-rich lipoproteins, the low-density lipoproteins are considered to be atherogenic and the high-density lipoproteins are considered to be antiatherogenic (Miller and Miller, 1975; Carew et al, 1976).

Plant fibers of certain types lower serum cholesterol concentrations, and a generous intake of plant fiber may reduce the likelihood of ischemic heart disease. In 1972, Trowell (1972) postulated that fiber-depleted diets contribute to ischemic heart disease; subsequently, Morris et al (1977) reported that a high level of cereal fiber intake was correlated with a lower frequency of coronary heart disease. Water-soluble fibers are especially effective in lowering serum cholesterol (Anderson and Chen, 1979). Since oat products contain generous amounts of a water-soluble oat gum, a β-glucan, we investigated the cholesterol-lowering properties of oat products. This review will focus on studies that have been reported since our last review (Anderson and Chen, 1979). Current proposals regarding the mechanism for the cholesterol-lowering effects of oat fiber are discussed.

309

II. FIBER DEFINITIONS

A. Plant Fibers

Plant fibers are the portions of plant foods, consisting of polysaccharides and lignins, that are not digested in the human small intestine. These polysaccharides are branched and unbranched polymers of hexoses, pentoses, and sugar derivatives such as uronic acid. Plant fiber polysaccharides are fermented by bacteria in the colon to various organic products, hydrogen, carbon dioxide, and water. Some polysaccharides such as pectin are virtually completely hydrolyzed in the colon, whereas others such as cellulose are less completely hydrolyzed; the extent of fermentation depends on the type of food, the residence time in the colon, and other factors. Plant fibers can be extracellular, such as gums exuded from the surface; cell wall components, such as cellulose, hemicelluloses, lignin, and pectins; or intracellular, such as the storage polysaccharides of legumes.

B. Dietary Fiber

Dietary fiber has been defined in several different ways. In general, it is considered to be the portions of plant cell walls, including polysaccharides and lignin, that are not digested in the human small intestine. At present there is disagreement about whether indigestible cell wall constituents such as phytates, minerals, silica, lipids (e.g., cutins and waxes), and proteins should be included under the dietary fiber umbrella. When dietary fiber is defined as cell wall material not digested in the small intestine, the storage polysaccharides of legumes may not be included in this definition.

C. Purified Fiber Polysaccharides

Oat gum, pectins, guar, and psyllium are fibers that are extracted from plant foods and purified. Many of these fiber polysaccharides, such as the β-glucan of oat gum, have been defined chemically. Studies using these products may provide insight into the mechanism of action of plant fibers; however, the effects of a purified fiber may be quite different from those of the same type of fiber in its natural, unextracted state. When purified fibers are used, their source, physical properties, and chemical characteristics should be reported.

D. Unavailable Carbohydrates

Plant fiber polysaccharides, undigestible starch, and certain simple carbohydrates (e.g., lactalose, raffinose, and stachyose) are not digested in the human small intestine. Most of these sugars and polysaccharides are fermented in the colon.

E. Available Carbohydrates

Most carbohydrates in the diet can be digested in the human small intestine. These available carbohydrates include simple sugars (oligosaccharides) and polysaccharides such as starches, dextrins, and glycogen. Under certain circumstances, some of these carbohydrates are not completely hydrolyzed and absorbed; then they are readily fermented by colon bacteria.

F. Dietary Fiberlike Substances

Certain compounds of nonplant origin, like chitins and chitosans, have physiological properties that resemble plant fiber polysaccharides. Undoubtedly, other natural substances and synthetic products may have properties similar to plant fiber polysaccharides.

III. FIBER ANALYSIS

A. Total Fiber Content

Recent methods (Southgate, 1969; Chen and Anderson, 1981; Englyst et al, 1982) for measuring both water-soluble and water-insoluble fibers can best estimate total fiber content of plant foods. Chen and Anderson (1981) used the extraction methods of Southgate and measured the carbohydrate content by a gas-liquid chromatography technique; this method yields information about the monosaccharides present and may provide clues to the component polysaccharides. Measurement of plant fiber is difficult because of several factors: the complicated nature of complex branched and unbranched polysaccharides containing a variety of different sugars; the difficulty of separating fiber polysaccharides from digestible polysaccharides (e.g., starch); the difficulty of extracting delicate polysaccharides from fibrous cell walls without altering their chemical structure; and other reasons (Southgate, 1969; James and Theander, 1981; Englyst et al, 1982).

B. Neutral Detergent Residue Fiber

The neutral detergent residue (NDR) fiber technique (Goering and Van Soest, 1970) accurately measures the water-insoluble fiber content and gives a fairly good estimate of the cell-wall fiber content for most foods. The NDR method is less tedious, easier to automate, and easier to standardize than the Southgate method. Because the NDR method greatly underestimates the soluble fiber content, it has limitations in estimating fiber content for products such as oats, oat bran, and legumes, which are rich in soluble fibers.

C. Crude Fiber

This method, which was developed in the 19th century to prevent the dishonest "lacing" of cattle feed with sawdust, grossly underestimates the fiber content of most food. This archaic method should not be used to measure the fiber content of human foods. Table I shows that the ratio of total plant fiber to crude fiber ranges from 2.3 to 17.4.

Table I compares the total, insoluble, soluble, and crude fiber content of some selected foods. Most foods contain substantial amounts of soluble fibers; soluble fiber values average 41% of the total fiber content and range from 8 to 59%. The NDR method would measure about 59% of the total fiber content of these foods, with values ranging from 41 to 92% of total fiber content. The archaic crude fiber method measures only 23% of the total fiber content (range of 6–44%) and has little predictive value regarding total fiber content. For example, the crude fiber method measures less than 6% of the total fiber of cornflakes but measures almost 22% of the fiber of wheat bran. Neither the ratios of total fiber to crude fiber (ranging from 2.3 to 17.4) nor the ratios of insoluble or NDR fiber to crude fiber (ranging from 1.4 to 7.1) provide confidence that crude fiber values will be useful in predicting the total fiber content of plant foods.

Apparently two different fiber analysis techniques are required. First, an accurate method is needed to provide reference information about the soluble fiber components (pectins, gums, and storage polysaccharides), the insoluble polysaccharides (cellulose and noncellulosic polysaccharides), and the lignin content of foods. Second, a practical method is needed for routine use by food producers to estimate the total fiber content of food products. Modifications of the Southgate procedure may meet the need for reference values for foods and food products. Currently, an acceptable method for the practical estimation of fiber content in large numbers of samples is not available. The method of Asp et al (1983) is being evaluated by the Association of Official Analytical Chemists.

IV. PHYSIOLOGICAL EFFECTS

Plant fibers greatly influence the utilization of food by the body. High-fiber foods, compared with low-fiber ones, differ in appearance and usually take

TABLE I
Fiber Content of Various Foods[a]

Food	Total Plant Fiber	Insoluble Fiber	Soluble Fiber	Crude[b] Fiber	Ratios	
					Total Plant Fiber to Crude Fiber	Insoluble Fiber to Crude Fiber
Wheat bran	42.2	38.9	3.3	9.1	4.6	4.3
Oat bran	27.8	13.8	14.0	2.7	10.3	5.1
Oats (rolled oats)	13.9	6.2	7.7	1.2	11.6	5.2
Cornflakes	12.2	5.0	7.2	0.7	17.4	7.1
Grapenuts	13.0	7.4	5.6	1.5	8.7	4.9
Pinto bean	10.5	6.0	4.5	1.5	7.0	4.0
White bean	8.7	4.0	3.7	1.5	5.8	2.7
Kidney bean	10.2	5.5	4.7	1.5	6.8	3.7
Lima bean	9.7	6.4	3.3	1.8	5.4	3.6
Corn	3.3	1.5	1.8	0.7	4.7	2.1
Sweet potato	2.5	1.4	1.1	1.0	2.5	1.4
Kale	2.6	2.0	0.6	0.9	2.9	2.2
Asparagus	1.6	1.1	0.5	0.7	2.3	1.6
Cucumber	0.9	0.5	0.4	0.3	3.0	1.7
Apple	2.0	1.1	0.9	0.6	3.3	2.8
Orange	2.0	1.4	0.6	0.5	4.0	2.8
Banana	1.8	1.0	0.8	0.5	3.6	2.0
Peach	1.4	1.1	0.3	0.6	2.3	1.8

[a] Source: Chen and Anderson (1981): © Am. J. Clin. Nutr., Am. Soc. for Clinical Nutrition; used by permission. Data given as grams per 100 g (wet weight).
[b] Data taken from USDA (1963).

longer to eat. Fibers affect gastric emptying time, the rate of absorption of nutrients from the intestine, fecal bulk, and frequency of bowel movements. Thus, from mouth to anus, fibers profoundly influence food utilization. Furthermore, fibers directly or indirectly influence pancreatic hormone secretion, hepatic glucose, and lipid metabolism, and they may affect the peripheral metabolism of glucose and fats. At the present time, there is very limited information about the specific effects of fiber on the human gastrointestinal system. Table II, however, which is based on available data, outlines some potential effects of fiber on gastrointestinal physiology.

A. Gastric Effects

Pectin and other water-soluble fibers delay gastric emptying, thereby retaining foods for longer times in the stomach. Certain soluble fibers have been used therapeutically to prevent rapid gastric emptying and reactive hypoglycemia (Jenkins et al, 1977a). The higher degree of satiety associated with high-fiber food may be related in part to these gastric effects. Some fibers may have buffering capacity and may alter gastric acidity (Lennard-Jones et al, 1968; Jalan et al, 1979).

B. Intestinal Effects

Fibers may either increase or decrease intestinal transit times. Wheat bran and similar items appear to create "intestinal hurry" and reduce transit times (Kirwan et al, 1974). Many water-soluble fibers, however, appear to slow intestinal transit and increase transit time; this may be related to the gel-formation properties of

TABLE II
Gastrointestinal Effects of Plant Fiber

Site	Action
Stomach	Alter emptying time
	Affect intragastric pressure
	Affect gastric acidity
Small intestine	Alter transit time
	Affect absorption rate by:
	Altering transit time
	Forming gel-filtration system
	Ion exchange properties
	Binding of organic compounds
	Modify the secretion or activity
	of digestive enzymes
	Influence hormone secretion
Colon	
Right side	Fermentation effects
	Influence bile acid metabolism
	Antitoxic effect
	Antioxidant effect
Left side	Alter transit time
	Affect colonic pressure
	Influence amount and content of feces

these soluble fibers (Jenkins et al, 1978; Holt et al, 1979).

The chromatographylike properties of fibers are of considerable interest. Plant fibers can behave like gel filtration, ion exchange, or adsorption-type columns in the gastrointestinal tract (Eastwood and Kay, 1979). Most water-soluble fibers form gelatinous globules when hydrated and could act like molecular sieves in the small intestine. Large molecules could pass rapidly through the system, but smaller molecules would be trapped in the various pores for variable lengths of time. These gelatinous globules may slow down the digestion and absorption of various nutrients by decreasing the interaction between nutrients and digestive enzymes. Elsehans et al (1980) proposed that these types of fibers may alter the unstirred water layer and thus influence absorption rates.

Mineral absorption is influenced by many dietary factors, including plant fibers. Fiber polysaccharides with free carboxyl groups can bind minerals by ionic charge. Short-term studies in humans have documented that an increased intake of semipurified or purified fibers is accompanied by increased fecal loss of certain minerals (especially iron, zinc, and calcium) and negative mineral balance (Kelsay, 1981). Because of the adverse effects of chronic zinc and iron deficiency on Middle Eastern children who obtain a large portion of their energy from an unleavened high-fiber bread, this potential problem needs to be kept in mind. However, in Western people there is no firm evidence that a higher-than-average fiber intake is associated with long-term adverse effects on mineral balance. When we increased the plant fiber intake of diabetic patients from approximately 20 to 50 g/day and followed these patients over a two-year period, we detected no evidence of mineral depletion (Anderson et al, 1980). Several patients were followed for more than five years using these diets without evidence of mineral derangements being observed. However, further studies are required to ascertain the long-term effects of high-fiber diets or increased intake of high-fiber foods on mineral balance.

Absorption or affinity properties of plant fibers are still being defined. Many fibers bind bile acids and may result in the delivery of more bile acids to the colon or to the feces. The role of increased fecal bile acid excretion associated with fiber intake is discussed below. Fibers also have the capacity to bind many other organic compounds and may protect the colon from potential carcinogens. Preliminary studies (Kasper et al, 1979) indicate that fiber intake influences the absorption rate of certain drugs; this exceedingly important area has not been examined critically. Thus, the potential influence of high-fiber foods or fiber supplements on the absorption rates or extent of absorption of many drugs, organic compounds, and other substances remains a fertile area for investigation.

Digestive enzyme activities may be greatly influenced by fiber intake. Not only do fibers partition nutrients from their digestive enzymes but they also appear to alter the activities of certain pancreatic enzymes. In vitro studies (Dunaif and Schneeman, 1981) indicate that wheat bran and oat bran decrease amylase and chymotrypsin activities, whereas pectins increase the activities of these enzymes. Whether fibers alter secretion rates of pancreatic enzymes has not been determined.

Hormone secretion from the gut and pancreas is altered by fiber intake. After fiber-supplemented meals, insulin secretion is significantly lower than after low-fiber meals providing the same quantities of nutrients (Jenkins et al, 1976b, 1977a, 1977b). Plasma glucagon concentrations are also significantly lower on high-fiber diets than on low-fiber diets (Miranda and Horwitz, 1978). Furthermore, fiber ingestion reduces the secretion of gastrointestinal hormones such as gastric inhibitory polypeptide (GIP), vasoactive intestinal polypeptide (VIP), and gut glucagon (Goulder et al, 1978; Morgan et al, 1979). The decrease in portal concentration of glucagon and gut hormones that have similar hepatic actions would foster improved hepatic metabolism of glucose.

C. Colon Effects

Simplistically, the function of the human colon can be separated into the fermentation effects in the cecum and ascending colon and the stool-packaging function of the left side of the colon. Although fibers are essentially undigestible in the human small intestine, they are choice nutrients for colon bacteria. Virtually all fibers, except lignin, are almost completely digested in the right side of the colon (Holloway et al, 1978; Holloway et al, 1980). The products of bacterial fermentation are methane, carbon dioxide, water, and short-chain fatty acids (predominantly acetate, propionate, and butyrate). These short-chain fatty acids are almost completely absorbed and may affect hepatic metabolism of glucose, cholesterol, and fatty acids (Anderson and Bridges, 1981).

Fiber provides an adsorptive surface in the colon. Lignin, a nonpolysaccharide component of fiber, has the greatest bile acid adsorption capacity. Adsorption of bile acid by lignin may prevent bile acid degradation and reabsorption from the colon. Thus fiber may influence the types and proportions of primary and secondary bile acids returning to the liver through the enterohepatic circulation. Eastward and Kay (1979) have proposed an additional function for lignin in the colon. They suggested that it may act as a free radical scavenger, an antioxidant, and may alter the formation of carcinogens in the digestive tract. Adsorption of bacteria by fiber may alter bacterial metabolism: for example, the presence of dietary fiber in the colon prevents the bacterial transformation of cyclamate to its toxic metabolite cyclohexamine.

The left side of the colon is responsible for stool formation. In this colon section, undigestible residue, water, electrolytes, bacteria, gas, and other materials are assembled and packaged before stool evacuation. Water-insoluble fibers such as wheat bran appear to increase fecal bulk by increasing fecal content of water and bacteria. Coarse wheat bran has especially good bulking properties because of the trapping of water in its interstices (Kirwan et al, 1974). The increase in the dry weight of stools associated with high fiber intake is largely related to increased quantities of bacteria (Cummings, 1981). Water-soluble fibers such as pectin do not promote stool bulk because they are degraded almost completely by bacteria in the colon. Colonic transit time is inversely related, in a curvilinear fashion, to fecal weight (Burkitt et al, 1972; Stasse-Wolthuis et al, 1978). Cereal bran increases stool weight and accelerates colonic emptying (Payler et al, 1975).

Ingenious studies will be required to fully delineate the effects of high-fiber foods and fiber polysaccharides on human intestinal function. The specific effects of fibers on gastric emptying times must be determined in order to define their effect on intestinal transit time. The actual physiochemical nature of fiber polysaccharides in the small intestine must be ascertained, as well as the interactions between fiber polysaccharides, nutrients, digestive enzymes, the unstirred water layer, gut hormone secretion, and intestinal motility. Investigators do not know how fibers affect the quantity of nutrients and other substances delivered to the colon. Studies of patients with ileostomies will be useful, but they will have to be interpreted with caution because of bacterial colonization of the ileum. The interactions of fiber and bacterial metabolism in the right colon will need to be evaluated directly, since fecal samples are too far removed temporally and anatomically to allow valid conclusions to be drawn. Samples of bacteria must be obtained from the cecum to determine whether fibers influence the types and amounts of bacteria. In essence, investigators have just scratched the surface of this enormous area of fiber effects on gastrointestinal function.

V. HYPOCHOLESTEROLEMIC EFFECTS

A. Animals

DeGroot and colleagues (1963) demonstrated that whole oats had a hypocholesterolemic effect in cholesterol-fed rats. Other investigators (Fisher and Griminger, 1967; Chenoweth and Bennink, 1976; Hamilton and Carroll, 1976) had reported that whole oats lowered serum and liver cholesterol concentrations in chicks and rabbits. Similar results were subsequently reported for rats (Kelley et al, 1981).

Various fractions of whole oats were tested to determine which component had the greatest cholesterol-lowering activity. Whole oats and defatted oats exerted similar hypocholesterolemic effects in chicks, whereas defatted-defibered oats and defatted-defibered-degummed oats had no hypocholesterolemic effects (Chenoweth and Bennink, 1976). Fisher and Griminger (1967) reported that whole oats, oat hulls, or dehulled oats exerted a cholesterol-lowering effect in chicks fed hypercholesterolemic diets. Oat hulls were the most effective, whereas oat starch and oat oil had no cholesterol-lowering activity. Two studies (Forsythe et al, 1978; O'Donnell et al, 1981) showed that oat bran was ineffective in lowering serum cholesterol in rats and gerbils.

The differences between oat bran and oat hull effects led us to examine the effects of oat bran on the cholesterol metabolism of rats. Since oat gum is a component of oat bran, a second study was conducted to examine the effects of oat gum on the cholesterol metabolism of rats.

The first study was conducted with diets supplemented with oat bran. Male Sprague-Dawley rats were divided into four groups of 10 rats each, with each group receiving one of four different fiber-containing diets. All diets contained 62.5% carbohydrate, 15% protein, 0.3% DL-methionine, 6% fat, 4% salt mix, 1% vitamin fortification, 0.2% cholic acid, 1% cholesterol, and 10% plant fiber. The four fiber components were cellulose, pectin, guar, and oat bran. Oat bran, a coarse, branny fraction of oat groat, contained 22.1% protein, 43.2% carbohydrate, 6.9% fat, and 27.8% total plant fiber. Thus, 36.5% oat bran provided 10% oat bran fiber. Figure 1 presents a schematic flow chart for oat bran production. After three weeks, the rats were killed, and serum samples were analyzed for total cholesterol and high-density-lipoprotein (HDL) cholesterol concentrations. Liver samples were analyzed for total cholesterol concentrations.

In the second study, diets supplemented with oat gum were evaluated. The design and diet formulations for this experiment were the same as noted above, and the four different fiber components were cellulose, pectin, oat gum, and oat bran. Oat gum (which contained 66% β-glucan) was extracted and isolated from the oat bran fraction.

The oat-bran supplement experiments showed that serum total cholesterol concentrations were significantly lower in rats fed oat bran, guar, or pectin than in those fed cellulose (Fig. 2). Values were 15, 28, and 37% lower in the oat bran, guar, and pectin groups, respectively. On the other hand, plasma HDL cholesterol concentrations were significantly higher in rats fed oat bran, guar, or pectin. Oat bran and pectin feeding increased HDL cholesterol values by 38%; guar feeding increased HDL cholesterol values by 68%. The ratios of total:HDL cholesterol for the oat bran, pectin, and guar gum groups ranged from 3.3 to 4.5; these were significantly lower than the ratios for the cellulose group, or 7.3 ($P < 0.05$).

Liver cholesterol concentrations followed a similar pattern to that observed for plasma cholesterol levels: rats fed oat bran, guar, or pectin had lower liver

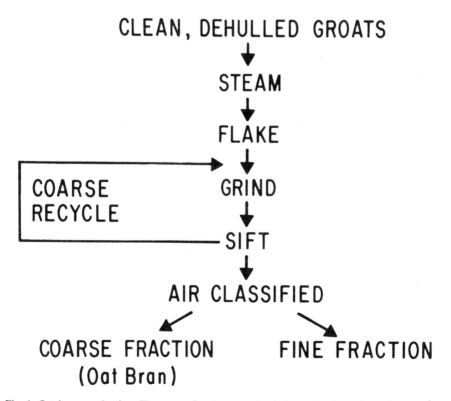

Fig. 1. Oat bran production. The coarse fraction contained about 45% by weight of the original groats, and the fine fraction contained about 55%.

cholesterol concentrations than rats fed cellulose. Values for the oat bran group were similar to those for the pectin or guar group (Fig. 2).

Similarly, the oat-gum supplement diet series showed that serum total cholesterol concentrations were significantly lower in the oat bran, oat gum, or pectin groups than in the cellulose group (Fig. 3). Values were 40% lower in the oat gum and pectin group and 24% lower in rats fed oat bran. On the other hand, serum HDL cholesterol concentrations were higher in oat bran, oat gum, and pectin groups. The largest increase was seen in the oat gum group, where HDL cholesterol values were 76% higher than values for the cellulose group. The serum total:HDL cholesterol ratios were substantially reduced from values for the cellulose group (6.7) to the lowest values for rats fed oat gum (2.2, $P<0.01$).

The oat bran, oat gum, and pectin groups also had a significant reduction in

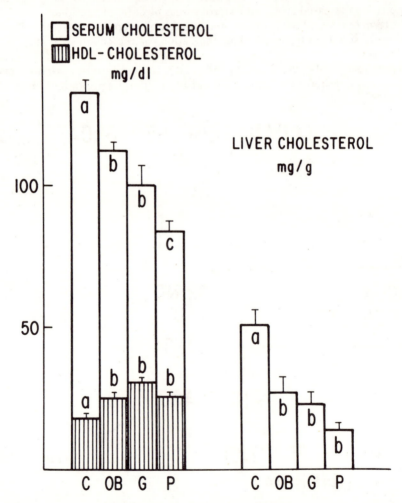

Fig. 2. Effect of oat bran on serum and liver cholesterol concentrations in rats. C = cellulose, OB = oat bran, G = guar gum, and P = pectin; values with different letters differ significantly ($P<0.05$).

liver cholesterol concentrations when compared with values for the cellulose group. Values for the oat gum and pectin groups were significantly lower than values for the oat bran groups (Fig. 3).

Oat gum thus appears to be the active cholesterol-lowering ingredient of oat bran. Our studies indicate that oat bran selectively lowers serum total cholesterol while raising HDL cholesterol concentrations; the cholesterol-lowering properties of oat bran appear related to its water-soluble gum content.

B. Humans

The hypocholesterolemic effect of rolled oats in cholesterol-fed rats prompted DeGroot et al (1963) to conduct an experiment with human volunteers. When

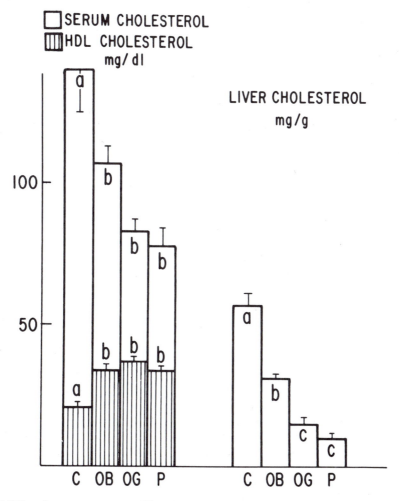

Fig. 3. Effect of oat gum on serum and liver cholesterol concentrations in rats. C = cellulose, OB = oat bran, OG = oat gum, and P = pectin; values with different letters differ significantly ($P < 0.05$).

they fed 140 g of rolled oats (dry weight) to healthy young men, serum cholesterol concentrations declined 11% (from 251 to 233 mg/dl) in three weeks. When intake was reduced to 50 g per day, a small cholesterol-lowering effect was still seen; but eight weeks were required on the test diet before the effect could be demonstrated (Luyken et al, 1965). Recently, Judd and Truswell (1981) also demonstrated a reduction in serum cholesterol levels, averaging 8% when 125 g of rolled oats were consumed daily for three weeks. However, Kretsch et al (1979) were unable to detect changes in serum cholesterol levels when healthy men were fed 0.6 g of oat bran per kilogram of body weight for two weeks. This failure to demonstrate an oat bran effect could be related to the fact that these subjects had normal serum cholesterol levels when they entered the study, to the lower intake of oat bran, or to the shorter experimental period.

Oat bran is a palatable cereal that is rich in soluble plant fiber and has twice as much soluble fiber as rolled oats (Table I). Many studies in humans (Anderson and Chen, 1979) have demonstrated that the ingestion of soluble plant fibers such as pectin, guar, or Bengal gram is associated with a significant reduction in serum cholesterol levels. Based on this information, we initiated a program to examine the effects of the addition of 100 g of oat bran to the diets of hypercholesterolemic patients.

We studied eight men whose serum total cholesterol concentrations in serum had exceeded 260 mg/dl. None of these patients had received cholesterol-lowering drugs in the previous three months. Two patients were obese, and one did not have hypercholesterolemia at the time of the study.

The two solid diets, control and oat bran, were composed of commonly available foods and were formulated to be identical in energy, carbohydrate, protein, and fat content (Table III). The subjects ate each diet for at least 10 days. Previous studies (Anderson and Ward, 1979; Anderson et al, 1980) had indicated that 10 days was an adequate length of time to determine the direction of change in cholesterol-rich lipoproteins, although it was not adequate to quantitate the maximal effects. The test diets differed only in the inclusion of 100 g of oat bran per day provided in muffins and hot cereals. Patients were fed control and oat bran diets in an alternating sequence.

TABLE III
Mean Daily Composition of Foods Consumed[a]

Component	Control Diet	Oat Bran Diet
Kilocalories	$1,954 \pm 84$[b]	$1,954 \pm 98$
Protein (g)	95 ± 5	93 ± 5
Carbohydrate (g)	209 ± 9	211 ± 11
Fat, total (g)	82 ± 3.8	82 ± 3.8
Saturated	30 ± 3.8	32 ± 3.8
Nonsaturated	30 ± 1.3	29 ± 1.5
Polyunsaturated	18 ± 2.4	17 ± 2.2
Cholesterol (mg)	435 ± 11	424 ± 12
Total fiber (g)	20 ± 1.3	43 ± 1.3
Soluble fibers (g)	7 ± 0.6	19 ± 0.7
Oat bran (g)	0	94 ± 2.3

[a] Adapted from Kirby et al (1981).
[b] Mean ± standard error.

Serum total cholesterol concentrations were measured daily after a 10-hr fast; serum HDL and LDL cholesterol concentrations were measured during the last two or three days of each dietary period.

The results showed that serum total, LDL, and HDL cholesterol concentrations were stable for the control diet period, whereas a progressive reduction was observed in seven men fed the oat bran diet. The oat bran diet lowered serum total cholesterol by 13% and serum LDL cholesterol by 14% ($P<0.05$) and did not alter HDL cholesterol concentrations (Fig. 4). The ratio of HDL to LDL cholesterol concentration was higher with the oat bran diet than with the control diet. Previously, using different experimental conditions, we observed that oat-bran feeding reduced serum LDL cholesterol concentrations by 36% but increased serum and HDL cholesterol concentrations by 82% in humans (Gould et al, 1980). Although this current oat bran diet failed to increase serum HDL cholesterol concentrations, the ratio of serum HDL to LDL cholesterol concentrations was altered favorably. These short-term studies, feeding oat bran for a period of 10 days, suggest that oat bran supplements might be useful in lowering serum cholesterol concentrations over a longer period of time.

To further evaluate the effects of oat bran, nine hypercholesterolemic patients

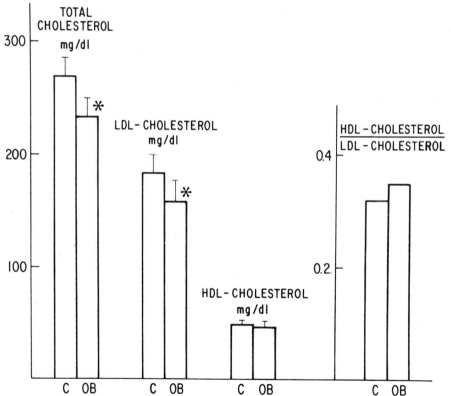

Fig. 4. Effect of oat bran on serum cholesterol and lipoprotein cholesterol concentrations in hypercholesterolemic patients. C = control and OB = oat bran; * = values significantly different from control ($P<0.01$).

were fed diets containing 100 g of oat bran per day for 21 days. Patients were fed weight-maintaining control diets for seven days before starting the test diet. The oat bran diets produced average reductions in serum total cholesterol concentrations of 22%; LDL cholesterol levels were 27% lower, and HDL cholesterol values were not changed. The hypocholesterolemic effect of feeding oat bran for 21 days was almost twofold greater than the effect of feeding oat bran for 10 days. Thus, in all of these studies, oat bran ingestion (100 g daily) was accompanied by a significant reduction in serum total cholesterol and LDL cholesterol, whereas HDL cholesterol either increased or remained stable.

The hypocholesterolemic effects of oat bran are very similar to the effects observed with purified soluble fibers such as pectin or guar (Anderson and Chen, 1979). Diets utilizing approximately 25 g of pectin per day produced average reductions in serum cholesterol of 13%. Similarly, when approximately 24 g of guar per day was fed to humans, a 16% reduction in serum cholesterol concentration was observed. Recently, Khan et al (1981) demonstrated that 9.0 g of guar gum daily lowered serum total cholesterol by 16.6% and LDL cholesterol by 25.6%. Thus, either pectin or guar clearly lowers serum cholesterol concentration in humans. However, both of these purified, water-soluble fibers often caused troublesome to intolerable nausea and vomiting. Oat bran, on the other hand, is palatable as a hot cereal and can be incorporated into muffins, breads, and other prepared foods. None of our patients experienced any gastrointestinal disturbance after long-term oat bran consumption, and the acceptability was generally good. Currently, we are using 50 g/day in outpatient studies with good acceptance and compliance by patients.

Based on these human and rat studies, oat bran has effective hypocholesterolemic properties and may have long-term usefulness for selected hypercholesterolemic patients. Furthermore, oat bran intake may have good preventive properties with respect to ischemic heart disease and may be useful for many individuals with disease risks.

VI. PROPOSED MECHANISMS FOR CHOLESTEROL-LOWERING EFFECTS

Oat products and other sources of soluble fiber have prominent hypocholesterolemic effects in humans and in experimental animals. These observations have important therapeutic implications, since oat products selectively lower atherogenic serum LDL cholesterol concentrations while either raising protective or antiatherogenic HDL cholesterol concentrations, or at least raising the ratio of HDL:LDL cholesterol concentrations. Most studies have been descriptive and have not looked at underlying mechanisms. Currently, the biochemical or physiological basis for fiber-induced changes in cholesterol metabolism is not understood. The following section briefly reviews the available data and contains some speculations about the hypocholesterolemic effects of soluble plant fibers.

A. Fecal Bile Acid and Cholesterol Excretion

The effect of plant fiber on serum cholesterol concentrations is generally believed to be largely mediated by enhanced fecal bile acid and neutral sterol excretion. Neutral sterol measurements include cholesterol and coprostanol. The effects of various fibers on fecal bile acids and neutral sterols in animals and humans are shown in Tables IV and V, respectively.

In animal studies, wheat bran had little effect on fecal bile acid and neutral sterol excretion (Chang et al, 1979; Reddy et al, 1980). However, one study observed that wheat bran feeding decreased fecal bile acid excretion (Nomani et al, 1979). There are inconsistencies regarding the effect of pectin on fecal bile acid and neutral sterol excretion (Leveille and Sauberlich, 1966; Nomani et al, 1979; Reddy et al, 1980). Reddy et al (1980) observed that pectin increased both bile acid and sterol excretion; Nomani et al (1979) observed decreased bile acid excretion and increased sterol excretion; and Leveille and Sauberlich (1966) observed increased bile acid excretion and no change in neutral sterol excretion. Diets supplemented with oat and rice bran increased fecal bile acid excretion but did not alter fecal neutral sterols. Cellulose, bagasse, alfalfa, carrageenan, and pulses all significantly increased bile acid and sterol excretion (Table IV).

In human studies, there are also inconsistencies in the reported effects of wheat bran on fecal bile acid and sterol excretion (Table V); some studies observed no change and one observed a decrease. Pectin significantly increased bile acid and sterol excretion in all studies. Cellulose, bagasse, guar, psyllium, Bengal gram, oat bran, and mixed plant fibers increased fecal bile acid excretion under various conditions (Table V).

Plant fibers appear to increase bile acid and neutral sterol excretion by binding these sterols and preventing their reabsorption. Many fibers have distinct bile acid binding properties in vitro (Story and Kritchevsky, 1976a, 1976b).

TABLE IV
Effects of Various Plant Fibers on Fecal Bile Acid and Neutral Sterol Excretion in Animals

Source of Fiber	No. of Studies	Bile Acid Changes (%)	Neutral Sterol Changes (%)	References
Alfalfa	3	+107 (37–162)[a]	+152 (57–247)	Horlick et al, 1967; Reddy et al, 1980; Kelley et al, 1981
Bagasse	1	+218	+ 44	Morgan et al, 1974
Bran				
Wheat	2	0	0	Chang et al, 1979; Reddy et al, 1980
	1	− 32	0	Nomani et al, 1979
Rice	1	+242	0	Vijayagopal and Kurup, 1973
Cellulose	3	+115 (20–220)	+ 22 (12–32)	Morgan et al, 1974; Jayakumari and Kurup, 1979; Kelley et al, 1981
Carrageenan	1	+250	+ 50	Reddy et al, 1980
Oats	3	+172 (166–228)	0	Fisher and Griminger, 1967; Chenoweth and Bennink, 1976; Kelley et al, 1981
Pectin	1	+ 60	+119	Reddy et al, 1980
	1	+ 32	0	Leveille and Sauberlich, 1966
	1	− 28	+ 70	Nomani et al, 1979
Pulses				
Black gram	2	+ 77 (38–116)	+ 61 (38–83)	Devi and Kurup, 1973; Jayakumari and Kurup, 1979
Bengal gram	1	+ 45		Mathur et al, 1964
Mixed source	1	+284	+ 89	Balmer and Zilversmit, 1974

[a]Numbers are mean values with ranges in parentheses.

Ordinarily, between 95 and 99% of bile acids are reabsorbed and reenter the enterohepatic circulation (Dowling, 1973). The liver may react to reduced reabsorption by diverting cholesterol from lipoprotein synthesis into bile acid synthesis and thereby secreting less lipoprotein cholesterol. Although this may represent a mechanism contributing to the hypocholesterolemic effects of plant fibers, two flaws prevent us from accepting this as the major mechanism. First, the liver has the capacity to increase bile acid production by almost 10-fold under conditions of biliary diversion where all bile acids are lost and with intensive cholestyramine treatment. Pectin intake, for example, significantly lowers serum cholesterol concentrations with only small increases in bile acid excretion (Table V). A 35% increase in bile acid excretion represents only an increase in fecal bile acid excretion of 100–150 mg/day; the liver should be able to compensate for this loss easily. Second, the in vitro binding of bile acids by fibers is not well correlated with their hypocholesterolemic effects. Thus, we cannot conclude that bile acid binding by fibers and subsequent fecal excretion play the major role in lowering serum cholesterol concentrations.

Some fibers increase fecal excretion of cholesterol. When fibers bind bile acids, they interfere with micelle formation and may decrease intestinal absorption of cholesterol. Some fibers that bind cholesterol can extract it from formed micelles in vitro (Vahouny et al, 1978; Story, 1980). Although many fibers do not increase fecal excretion of cholesterol (Table IV), they may deliver cholesterol to the

TABLE V
Effect of Various Plant Fibers on Fecal Bile Acid and Neutral Sterol Excretion in Humans

Source of Fiber	No. of Studies	Bile Acid Changes (%)	Neutral Sterol Changes (%)	References
Wheat	5	0	0	Eastwood et al, 1973; Jenkins et al, 1975; Walters et al, 1975; Kay and Truswell, 1977; McLean-Baird et al, 1977
	2	+ 39 (36–41)[a]	+37 (33–40)	Stasse-Wolthuis et al, 1980; Cummings et al, 1976
	1	− 49	−34	Tarpila et al, 1978
Bagasse	2	+ 50	0	Walters et al, 1975; McLean-Baird et al, 1977
Cellulose	1	0		Eastwood et al, 1973
	2	+ 35 (25–45)		Shurpalekar et al, 1971; Stanley et al, 1973
Psyllium	2	+102 (70–134)		Forman et al, 1968; Stanley et al, 1973
Oats	1	+ 35	0	Judd and Truswell, 1981
Oat bran	2	+ 83 (54–111)		Kretsch et al, 1979; Kirby et al, 1981
Pectin	5	+ 36 (6–51)	+13 (6–20)	Jenkins et al, 1976a; Kay and Truswell, 1977; Miettinen and Tarpila, 1977; Stasse-Wolthuis et al, 1980; Ross and Leklem, 1981
Guar gum	1	+ 84		Jenkins et al, 1976a
Bengal gram	1	+ 59		Mathur et al, 1968
Mixed source	2	0		Antonis and Bersohn, 1962;
	1	+ 36	+50	Raymond et al, 1977; Ullrich et al, 1981

[a] Numbers are mean values with ranges in parentheses.

colon, where it is degraded by bacteria. Thus, even though apparent fecal cholesterol loss is negligible under many fiber-feeding circumstances, derangements in intestinal cholesterol absorption may contribute to the cholesterol-lowering effects of these fibers.

B. Altered Bile Acid Metabolism

Plant fibers interrupt the enterohepatic circulation of bile acids by increasing fecal loss of bile acids. This interruption stimulates hepatic synthesis of primary bile acids such as cholic and chenodeoxycholic acids (McDogually et al, 1978). With fiber-depleted diets, bile acids are excreted less rapidly, bile acid return in the portal circulation is increased, hepatic bile acid synthesis is decreased, and more cholesterol is secreted into the serum as cholesterol-rich lipoproteins. Reduced synthesis of primary bile acids coincident with continuing bacterial degradation of primary to secondary acids (deoxycholic and lithocholic) would reduce the ratio of primary to secondary bile acids (Kritchevsky, 1978).

When rats were fed fiber-containing chow diets, biliary cholic acid secretion was threefold greater and biliary total bile acid secretion was twofold greater than values for rats fed fiber-free diets. Consequently, the rats fed fiber-free diets had longer half-lives for cholic acid and larger cholic acid pool sizes (Portman, 1960). Rats fed a high-fiber diet had higher biliary ratios of primary to secondary bile acids than those on a low-fiber diet, which reflects a higher rate of bile acid synthesis (Kritchevsky et al, 1974, 1975).

Biliary bile acid levels in humans also are influenced by plant fiber intake. Bran feeding increased chenodeoxycholate and decreased deoxycholate concentrations in bile (Pomare and Heaton, 1973; Pomare et al, 1974). Nigerians who habitually consume a high-fiber diet have more cholic acid and less deoxycholic acid in their bile than do people living in Western communities (Falaiye, 1974). Strict vegetarians also have a high ratio of primary to secondary bile acids in their feces (Aries et al, 1971).

An increase in fecal bile acid output will increase the rate of bile acid synthesis from cholesterol. Activity of cholesterol 7-α-hydroxylase, the rate-limiting enzyme in the conversion of cholesterol to bile acids, is increased in rats fed fiber-supplemented diets or high-fiber diets (Morgan et al, 1974; Story et al, 1977; Brydon et al, 1980). Further studies are needed to correlate the activity of this key bile acid synthesis enzyme with the hypocholesterolemic effect of certain fibers.

C. Influence of Colonic Metabolites
on Lipid Metabolism

Plant fibers, excepting lignin, are almost completely fermented to short-chain fatty acids (SCFA) in the colon (Cummings, 1981). These SCFA, including acetate (approximately 60% of the total SCFA), propionate (approximately 24%), and butyrate (approximately 16%), are virtually completely absorbed from the colon (Schmitt et al, 1976; Cummings, 1981). When the intake of plant fibers is increased from 20 to 70 g per day, 30–35 g of additional SCFA may be absorbed into the portal vein for metabolism by the liver and periphery (Cummings, 1981). The absorbed SCFA can influence hepatic cholesterol and fatty acid metabolism. Anderson and Bridges (1981) evaluated these effects by incubating isolated liver cells with physiological concentrations of SCFA. They measured rates of cholesterol and fatty acid synthesis and found that propionate reduced the cholesterol synthesis rate by 45% and the fatty acid synthesis rate by 84%. Further studies are required to evaluate the physiological significance of these observations.

Cholesterol biosynthesis is regulated through multiple mechanisms, which appear to converge on the activity of 3-hydroxy-3-methylglutary-CoA (HMG-CoA) reductase, the rate-limiting enzyme in cholesterol synthesis. In general, rates of cholesterol synthesis closely parallel changes in HMG-CoA reductase activity. When chicks are fed oat- or barley-based diets, HMG-CoA reductase activity is 59–79% lower than values for corn-fed chicks (Qureshi et al, 1980). On the other hand, HMG-CoA reductase activity is increased by the feeding of pectin (Reiser et al, 1977), alfalfa, or wheat bran (Story et al, 1977). Obviously, further studies are required to correlate changes in serum cholesterol concentrations with the activity of this key rate-limiting enzyme.

D. Altered Lipoprotein and Apoprotein Metabolism

Cholesterol is an integral part of all major lipoprotein particles. Of the three lipoproteins contributing significantly to the serum cholesterol pool, low-density lipoproteins (LDL) carry most of the serum cholesterol and are considered to be the most atherogenic. Very-low-density lipoproteins (VLDL), carrying primarily triglycerides, appear to be less atherogenic (Tzagournis, 1978). High-density lipoprotein (HDL), antiatherogenic lipoproteins, make a modest and relatively fixed contribution of cholesterol to the total serum cholesterol pool (Miller, 1978; Barboriak et al, 1979).

The effects of plant fiber intake on serum LDL and HDL levels are not well characterized. When rabbits were fed hypercholesterolemic diets, serum HDL as well as total cholesterol concentrations were lowered by pectin supplements (Berenson et al, 1975). In sharp contrast, cholesterol-fed rats had higher HDL cholesterol values and lower total serum cholesterol values when their diets were supplemented with oat bran, oat gum, pectin, or guar (Figs. 2 and 3). Furthermore, diets supplemented with soluble fiber increased the ratios of HDL to total serum cholesterol concentrations (Chen and Anderson, 1979).

Fiber-supplemented diets also favorably alter the ratio of serum HDL to LDL cholesterol concentrations in humans. When normal human subjects were fed pectin for three weeks, the reduction in serum total cholesterol concentration was largely related to a reduction in serum LDL cholesterol levels (Durrington et al, 1976). In hypercholesterolemic men, we observed that oat bran supplements selectively reduced serum LDL cholesterol concentrations without significantly altering HDL values (Fig. 4); consequently, the ratios of serum HDL and LDL cholesterol concentrations were 15% higher on oat bran than on control diets.

The mechanism by which plant fibers lower serum LDL cholesterol and raise HDL cholesterol concentrations is unclear. Plant fibers may modify serum lipoprotein profiles by altering the nature of chylomicrons and VLDL secreted into the lymph and portal vein circulation from the intestine. Plant fibers may alter the composition of lipoproteins, apoprotein synthetic rates, LDL receptors, and LDL catabolic rates. Plant fibers increase bile acid excretion and deplete hepatic cholesterol pools, which could lead to more efficient hepatic removal of free cholesterol from HDL. If HDL cholesterol removal capacity increased significantly, LDL receptor synthesis might be stimulated, causing an increase in the catabolic rate of LDL and lower serum LDL levels (Hopkins and Williams, 1981).

E. Other Possible Mechanisms

Other factors such as alterations in intestinal transit, colonic microflora, glucose metabolism, insulin secretion, and glucagon secretion could contribute to the cholesterol-lowering properties of certain fibers.

Plant fiber may alter the colonic flora and thus influence bile acid metabolism to alter cholesterol metabolism (Schoenfield et al, 1973). Many types of plant fibers modulate glucose absorption, resulting in reduced postprandial levels of glucose, insulin (Anderson, 1980; Jenkins, 1980), and glucagon (Miranda and Horwitz, 1978) and reducing the secretion of gastrointestinal hormones such as GIP, VIP, and gut glucagon (Goulder et al, 1978; Morgan et al, 1979). Insulin has been reported to increase hepatic cholesterol synthesis (Bhathena et al, 1974) and VLDL synthesis (Robertson et al, 1973; Reaven and Bernstein, 1978). Thus, changes in serum glucagon and gastrointestinal hormones might also influence hepatic metabolism of fatty acids and lipoproteins (Miranda and Horwitz, 1978; Munoz et al, 1979).

VII. CONCLUSIONS

Plant fibers can be defined as the portions of plant food, including polysaccharides and lignin, that are not digested in the human small intestine. Assessing the total fiber content of foods is tedious and time-consuming, and currently there are no practical ways of accurately estimating the sum of water-soluble and water-insoluble fibers. Soluble fibers, in general, have important effects on glucose and lipid metabolism, whereas insoluble fibers have greater effects on intestinal transit time and stool production.

The intake of certain plant fibers lowers serum cholesterol concentrations in humans. Several water-soluble fibers such as pectin, guar, and Bengal gram have a hypocholesterolemic effect, whereas most water-insoluble fibers such as cellulose and cellulose-rich products (e.g., wheat bran) do not lower serum cholesterol concentrations of humans and animals (Anderson and Chen, 1979). Oat bran is a palatable cereal, rich in water-soluble fiber. Because of the hypocholesterolemic effects of other water-soluble fibers, we have examined the effects of feeding 100 g of oat bran per day.

A diet supplemented with oat bran significantly reduced the serum cholesterol concentrations of hypercholesterolemic patients. The decrease in serum total cholesterol concentrations observed with oat bran diets occurred primarily through a reduction in LDL cholesterol concentrations. Oat-bran feeding produced a significant and selective reduction in serum LDL cholesterol concentration, whereas HDL either increased or remained stable. Supplementing diets with a palatable and inexpensive high-fiber food such as oat bran may have a valuable role in the treatment of certain patients with hypercholesterolemia.

Oat bran diets also improved the glucose and insulin metabolism of nondiabetic and diabetic patients (Kirby et al, 1981). The oral glucose tolerance tests of eight hypercholesterolemics were significantly improved by the oat bran as compared with the control diet. One of the patients was a diabetic receiving 20 units of insulin per day. We were able to reduce his insulin requirements progressively to zero after 10 days on the oat bran diet while maintaining his fasting plasma glucose concentration within an acceptable range (Gould et al, 1980). Oat bran intake may have long-term beneficial effects on glucose metabolism of nondiabetic and diabetic individuals.

With oat bran diets, fecal weight was slightly higher than values on control diets. Oat bran seems to have fecal-bulking effects, which could improve laxation. Earlier investigators (Meyer and Calloway, 1977) observed that oat bran or oat gum reduced intestinal transit time and increased fecal weight and the frequency of defecation in older women. Recent studies with oat supplements indicate that they reduced intestinal transit time and increased fecal weights (Anderson, 1980).

Thus, our studies indicate that oat products have beneficial effects on cholesterol metabolism, glucose metabolism, and fecal bulk. Studies from other laboratories substantiate these observations. Diets supplemented with oat bran very effectively lower serum cholesterol concentrations and favorably alter the ratio of antiatherogenic to atherogenic lipoproteins for hypercholesterolemic individuals. Oat bran is well tolerated physiologically and is a good ingredient for muffins and other bakery products. Because of their general health-promoting properties, oat bran and other oat products may offer general health benefits for most adults.

LITERATURE CITED

ANDERSON, J. W. 1980. Dietary fiber in diabetes. Pages 193-221 in: Medical Aspects of Dietary Fiber. G. A. Spiller and R. M. Kay, eds. Plenum Press, New York.

ANDERSON, J. W., and BRIDGES, S. R. 1981. Plant fiber metabolites alter hepatic glucose and lipid metabolism. Diabetes 30:133A.

ANDERSON, J. W., and CHEN, W. L. 1979. Plant fiber: Carbohydrate and lipid metabolism. Am. J. Clin. Nutr. 32:346-363.

ANDERSON, J. W., CHEN, W. L., and SIELING, B. 1980. Hypolipidemic effects of high-carbohydrate, high-fiber diets. Metabolism 29:551-558.

ANDERSON, J. W., FERGUSON, S. K., KAROUNOS, D., O'MALLEY, L., SIELING, B., and CHEN, W. L. 1980. Mineral and vitamin status on high-fiber diets: Long-term studies of diabetic patients. Diabetes Care 3(1):38-40.

ANDERSON, J. W., and WARD, K. 1979. High carbohydrate, high fiber diets for insulin-treated men with diabetes mellitus. Am. J. Clin. Nutr. 32:2312-2321.

ANTONIS, A., and BERSOHN, I. 1962. The influence of diet on fecal lipids in South African white and Bantu prisoners. Am. J. Clin. Nutr. 11:142-155.

ARIES, V. C., CROWTHER, J. S., DRASAR, B. S., HILL, M. J., and ELLIS, F. R. 1971. The effect of a strict vegetarian diet on the faecal flora and faecal steroid concentration. J. Pathol. 103(1):54-56.

ASP, N. G., JOHANSSON, C. G., HALLMER, H., and SILJESTROM, M. 1983. Rapid enzymatic assay of insoluble and soluble dietary fiber. J. Agric. Food Chem. 31:476-482.

BALMER, J., and ZILVERSMIT, D. B. 1974. Effects of dietary roughage on cholesterol absorption, cholesterol turnover and steroid excretion in the rat. J. Nutr. 104:1319-1328.

BARBORIAK, J., ANDERSON, A. J., RIMM, A. A., and KING, J. F. 1979. High density lipoprotein cholesterol and coronary artery occlusion. Metabolism 28:735-738.

BERENSON, L. M., BHANDARU, R. R., BADHAKRISHNAMURITY, V., SHINIVASAN, S. R., and BERENSON,

G. S. 1975. The effect of dietary pectin on serum lipoprotein cholesterol in rabbits. Life Sci. 16:1533-1543.

BHATHENA, S. J., AVIGAN, J., and SCHREINER, M. E. 1974. Effect of insulin on sterol and fatty acid synthesis and hydroxy methylglutaryl CoA reductase activity in mammalian cells grown in culture. Proc. Natl. Acad. Sci. U.S.A. 71(6):2174-2178.

BRYDON, W. G., TADESSE, K., and EASTWOOD, M. A. 1980. The effect of dietary fibre on bile acid metabolism in rats. Br. J. Nutr. 43:101-106.

BURKITT, D. P., WALKER, A. R. P., and PAINTER, N. S. 1972. Effect of dietary fibre on stools and transit times, and its role in the causation of disease. Lancet 2:1408-1412.

CAREW, T. E., KOSCHINSKY, T., HAYES, S. B., and STEINBERG, D. 1976. Mechanism by which high-density lipoproteins may slow the atherogenic process. Lancet 1:1315-1317.

CHANG, M. L. W., JOHNSON, M. A., and BAKER, D. 1979. Effects of whole wheat flour and mill-fractions on lipid metabolism in rats. Proc. Soc. Exp. Biol. Med. 160:88-93.

CHEN, W. J. L., and ANDERSON, J. W. 1979. Effects of plant fiber in decreasing plasma total cholesteral and increasing high density lipoprotein cholesterol. Proc. Soc. Exp. Biol. Med. 162:310-313.

CHEN, W. L., and ANDERSON, J. W. 1981. Soluble and insoluble plant fibers in selected cereals and vegetables. Am. J. Clin. Nutr. 34:1077-1082.

CHENOWETH, W. L., and BENNINK, M. R. 1976. Hypocholesterolemic effect of oat fiber. Fed. Proc. Fed. Am Soc. Exp. Biol. 35:Abstr. 1598.

CUMMINGS, J. H. 1981. Short chain fatty acids in the human colon. Gut 22:763-779.

CUMMINGS, J. W., HILL, M. J., JENKINS, D. J. A., PERSON, J. R., and WIGGINS, H. S. 1976. Changes in fecal composition and colonic function due to cereal fiber. Am. J. Clin. Nutr. 29:1468-1473.

DEGROOT, A. P., LUYKEN, R., and PIKAAR, N. A. 1963. Cholesterol-lowering effect of rolled oats. Lancet 2:303-304.

DEVI, K. S., and KURUP, P. A. 1973. Hypolipidaemic activity of the protein and polysaccharide fraction from *Phaseolus mungo* (black gram) in rats fed a high-fat-high cholesterol diet. Atherosclerosis 18:389-397.

DOWLING, R. H. 1973. The enterohepatic circulation of bile acids as they relate to lipid disorders. J. Clin. Pathol. 26 (Suppl. 5): 59-67.

DUNAIF, G., and SCHNEEMAN, B. O. 1981. The effect of dietary fiber on human pancreatic enzyme activity in vitro. Am. J. Clin. Nutr. 34:1034-1035.

DURRINGTON, P. N., MANNING, A. P., BOLTON, C. H., and HARTOG, M. 1976. Effect of pectin on serum lipids and lipoproteins, whole-gut transit time and stool weight. Lancet 2:394-396.

EASTWOOD, M. A., and KAY, R. M. 1979. An hypothesis for the action of dietary fiber along the gastrointestinal tract. Am. J. Clin. Nutr. 32:364-367.

EASTWOOD, M. A., KIRKPATRICK, J. R., MITCHELL, W. D., BONE, A., and HAMILTON, T. 1973. Effects of dietary supplements of wheat bran and cellulose on feces and bowel function. Br. Med. J. 392-394.

ELSEHANS, B., SUFKE, U., BLUME, R., and CASPARY, W. F. 1980. The influence of carbohydrate gelling agents on rat intestinal transport of monosaccharides and neutral amino acids in vitro. Clin. Sci. 59:373-380.

ENGLYST, H., WIGGINS, H. S., and CUMMINGS, J. H. 1982. Determination of the non-starch polysaccharides in plant foods by gas-liquid chromatography of constituent sugars. Analyst 107:307-318.

FALAIYE, J. M. 1974. Bile-salt patterns in Nigerians on a high-fibre diet. Lancet 1:1002-1008.

FISHER, H., and GRIMINGER, P. 1967. Cholesterol-lowering effects of certain grains and of oat fractions in the chick. Proc. Soc. Exp. Biol. Med. 126(1):108-111.

FORMAN, D. T., GARVIN, J. E., FORESTNER, J. E., and TAYLOR, C. B. 1968. Increased excretion of fecal bile acids by an oral hydrophilic colloid. Proc. Soc. Exp. Biol. Med. 127(4):1060-1063.

FORSYTHE, W. A., CHENOWETH, W. L., and BENNIK, M. R. 1978. Laxation and cholesterol metabolism in rats fed plant fibers. J. Food Sci. 43:1470-1476.

GOERING, K. H., and VAN SOEST, P. J. 1970. Forage fiber analysis. U.S. Dep. Agric. Agric. Handb. 379.

GOTTO, A. M., GORRY, G. A., THOMPSON, J. R., COLE, J. S. TROST, R., YOSHURUN, D., and DEBAKEY, M. E. 1977. Relationship between plasma lipid concentrations and coronary artery disease in 496 patients. Circulation 56:875-883.

GOULD, M. R., ANDERSON, J. W., and O'MAHONY, S. 1980. Biofunctional properties of oats. Pages 447-460 in: Cereals for Food and Beverages. G. G. Inglett and L. Munck, eds. Academic Press, New York.

GOULDER, T. J., MORGAN, L. M., MARKS, V., SMYTHE, P., and HINKS, L. 1978. Effects of guar on metabolic and hormonal responses to meals in normal and diabetic subjects. Diabetologia 15:235.

GRUNDY, S. M. 1977. Treatment of hypercholesterolemia. Am. J. Clin. Nutr. 30:985-992.

HAMILTON, R. M. G., and CARROLL, K. K. 1976. Plasma cholesterol levels in rabbits fed low fat, low cholesterol diets: Effects of dietary proteins, carbohydrates and fibre from different sources. Atherosclerosis 24:47-62.

HOLLOWAY, W. D., TASMAN-JONES, C., and LEE, S. P. 1978. Digestion of certain fraction of dietary fiber in humans. Am. J. Clin. Nutr. 31(6):927-930.

HOLLOWAY, W. D., TASMAN-JONES, C., and BELL, E. 1980. The hemicellulose component of dietary fiber. Am. J. Clin. Nutr. 33:260-263.

HOLT, S., HEADING, R. C., CARTER, D. C., PRESCOTT, L. F., and TOTHILL, P. 1979. Effect of gel fibre on gastric emptying and absorption of glucose and paracetamol. Lancet 1:636-639.

HOPKINS, P. N., and WILLIAMS, R. R. 1981. A simplified approach to lipoprotein kinetics and factors affecting serum cholesterol and triglyceride concentrations. Am. J. Clin. Nutr. 34:2560-2590.

HORLICK, L., COOKSON, F. B., and FEDOROFF, S. 1967. Effect of alfalfa feeding on the excretion of fecal neutral sterols in the rabbit. Circulation 36(Suppl. II):18.

JALAN, K. N., MAHALANABIS, D., MAITAN, T. K., and AGARWAL, S. H. 1979. Gastric acid secretion rate and buffer content of the stomach after a rice and a wheat-based meal in normal subjects and patients with duodenal ulcer. Gut 20:389-393.

JAMES, W. P. T., and THEANDER, D. 1981. The Analysis of Dietary Fiber in Food. Marcel Dekker, New York.

JAYAKUMARI, N., and KURUP, P. A. 1979. Dietary fiber and cholesterol metabolism in rats fed a high cholesterol diet. Atherosclerosis 33:41-47.

JENKINS, D. J. A. 1980. Dietary fiber and carbohydrate metabolism. Pages 175-192 in: Medical Aspects of Dietary Fiber. G. A. Spiller and R. M. Kay, eds. Plenum Press, New York.

JENKINS, D. J. A., GASSULL, M. A., LEEDS, A. R., METZ, G., DILAWARI, J. B., SLAVIN, B., and BLENDIS, L. M. 1977a. Effect of dietary fiber on complications of gastric surgery: Prevention of postprandial hypoglycemia by pectin. Gastroenterology 72:215-217.

JENKINS, D. J. A., HILL, M. J., and

CUMMINGS, J. H. 1975. Effect of wheat fiber on blood lipids, fecal steroid excretion and serum iron. Am. J. Clin. Nutr. 28:1408-1411.

JENKINS, D. J. A., LEEDS, A. R., GASSULL, M. A., COCHET, B., and ALBERTI, K. G. M. M. 1977b. Decrease in postprandial insulin and glucose concentrations by guar and pectin. Ann. Intern. Med. 86(1):20-23.

JENKINS, D. J. A., LEEDS, A. R., GASSULL, M. A., HOUSTON, H., DOFF, D. V., and HILL, M. J. 1976a. The cholesterol lowering properties of guar and pectin. Clin. Sci. Mol. Med. 51:8-9.

JENKINS, D. J. A., LEEDS, A. R., GASSULL, M. A., WOLEVER, T. M. S., GOFF, D. V., ALBERTI, K. G. M. M., and HOCKADAY, T. D. R. 1976b. Unabsorbable carbohydrates and diabetes: Decreased postprandial hyperglycemia. Lancet 2:172-174.

JENKINS, D. J. A., WOLEVER, T. M. S., LEEDS, A. R., GASSULL, M. A., HAISMAN, P., DILAWARI, J., GOFF, D. V., METZ, G. L., and ALBERTI, K. G. M. M. 1978. Dietary fibres, fibre analogues and glucose tolerance: Importance of viscosity. Br. Med. J. 1:1392-1394.

JUDD, P. A., and TRUSWELL, S. A. 1981. The effect of rolled oats on blood lipids and fecal steroid excretion in man. Am. J. Clin. Nutr. 34:2061-2067.

KASPER, H., ZILLY, W., FASSE, H., and FEHLE, F. 1979. The effect of dietary fiber on postprandial serum digoxin concentration in man. Am. J. Clin. Nutr. 32:2436-2443.

KAY, R. M., and TRUSWELL, A. S. 1977. The effect of wheat fiber on plasma lipids and faecal steroid excretion in man. Br. J. Nutr. 37:227-235.

KELLEY, M. J., THOMAS, J. N., and STORY, J. A. 1981. Modification of spectrum of fecal bile acids in rats by dietary fiber. Fed. Proc. Fed. Am. Soc. Exp. Biol. 40:Abstr. 3497.

KELSAY, J. L. 1981. Effect of diet fiber level on bowel function and trace mineral balances of human subjects. Cereal Chem. 58:2-5.

KHAN, A. R., KHAN, G. Y., MITCHEL, A., and QADEER, M. A. 1981. Effect of guar gum on blood lipids. Am. J. Clin. Nutr. 34:2446-2449.

KIRBY, R. W., ANDERSON, J. W., SIELING, B., REES, E. D., CHEN, W. J. L., MILLER, R. E., and KAY, R. M. 1981. Oat bran intake selectively lowers serum low-density lipoprotein cholesterol concentrations of hypercholesterolemic men. Am. J. Clin. Nutr. 34:824-829.

KIRWAN, W. O., SMITH, A. N.,

McCONNELL, A. A., MITCHELL, W. D., and EASTWOOD, M. A. 1974. Actions of different bran preparations on colonic function. Br. Med. J. 4:187-189.

KRETSCH, M. J., CRAWFORD, L., and CALLOWAY, D. H. 1979. Some aspects of bile acid and urobilinogen excretion and fecal elimination in men given a rural Guatemalan diet and egg formulas with and without added oat bran. Am. J. Clin. Nutr. 32:1492-1496.

KRITCHEVSKY, D. 1978. Influence of dietary fiber on bile acid metabolism. Lipid 13:982-985.

KRITCHEVSKY, D., TEPPER, S. A., and STORY, J. A. 1974. Isocaloric isogravic diets in rats. III. Effect of non-nutritive fiber (alfalfa or cellulose) in cholesterol metabolism. Nutr. Rep. Int. 9:301-308.

KRITCHEVSKY, D., TEPPER, S. A., and STORY, J. A. 1975. Nutritional perspectives and atherosclerosis non-nutritive fiber and lipid metabolism. J. Food Sci. 40:8-11.

LENNARD-JONES, J. E., FLETCHER, J., and SHAW, D. G. 1968. Effect of different foods on the acidity of the gastric contents in patients with duodenal ulcer. III. Effect of altering the proportions of protein and carbohydrate. Gut 9:177-182.

LEVEILLE, G. A., and SAUBERLICH, H. E. 1966. Mechanism of the cholesterol-depressing effect of pectin in the cholesterol-fed rat. J. Nutr. 88(2):209-214.

LUYKEN, R., WIZN, J. F., PIKAAR, N. A., and VANSTAVEREN, W. A. 1965. De involved van Havermont op HET SERUM-cholesterolgehalte van HET Bloed. Voeding 26:229.

MATHUR, K. S., KHAN, M. A., and SHARMA, R. D. 1968. Hypocholesterolemic effect of bengal gram: A long-term study in man. Br. Med. J. 1:30-31.

MATHUR, K. S., SINGHAL, S. S., and SHARMA, R. D. 1964. Effect of bengal gram on experimentally induced high levels of cholesterol in tissues and serum in albino rats. J. Nutr. 84:201-204.

McDOUGALL, R. M., YAKYMYSHYN, L., WALKER, K., and THURSTON, O. G. 1978. Effect of wheat bran in serum lipid lipoproteins and biliary lipids. Can. J. Surg. 21:433-435.

McLEAN-BAIRD, I., WALTERS, R. L., DAVIES, P. S., HILL, M. J., DRASAR, B. S., and SOUTHGATE, D. A. T. 1977. The effect of two dietary fiber supplements on gastrointestinal transit, stool weight and frequency, and bacteria flora, and fecal bile acids in normal subjects. Metabolism 26:117-127.

MIETTINEN, T. A., and TARPILA, S. 1977. Effect of pectin on serum cholesterol, fecal bile acids and biliary lipids in normolipidemic and hyperlipidemic individuals. Clin. Chim. Acta 79:471-477.

MEYER, S., and CALLOWAY, D. H. 1977. Gastrointestinal response to oat and wheat milling fraction in older women. Cereal Chem. 54:110-119.

MILLER, G., and MILLER, N. E. 1975. Plasma-high-density-lipoprotein concentrations and development of ischemic heart-disease. Lancet 1:16-19.

MILLER, N. E. 1978. The evidence for the antiatherogenicity of HDL in man. Symposium: HDL structure, function analysis, clinical epidemiological and metabolic aspects of HDL. Lipids 13:914-919.

MIRANDA, P. M., and HORWITZ, D. L. 1978. High fiber diets in the treatment of diabetes mellitus. Ann. Intern. Med. 88:482-486.

MORGAN, B., HEALD, M., ATKIN, S. A., and GREEN, J. 1974. Dietary fibre and sterol metabolism in the rat. Br. J. Nutr. 32:447.

MORGAN, L. M., GOULDER, T. J., TSIOLAKIS, D., MARKS, V., and ALBERTI, K. G. M. M. 1979. The effect of unabsorbable carbohydrate on gut hormones. Diabetologia 17:85-89.

MORRIS, J. N., MARR, J. W., and CLAYTON, D. G. 1977. Diet and heart: A postscript. Br. Med. J. 19:1307-1314.

MUNOZ, J. M., SANSTEAD, M. M., and JACOB, R. A. 1979. Effects of dietary fiber on glucose tolerance of normal men. Diabetes 28:496-502.

NOMANI, M. Z. A., BRADAC, C. J., LAI, H. D., and WATNE, A. L. 1979. Effect of dietary fiber fractions on the fecal output and steroids in the rat. Nutr. Rep. Int. 20:269-278.

O'DONNELL, J. A., LEE, H. S., and HURT, H. D. 1981. Effect of plant fiber on plasma and liver lipids of adult male gerbils. Fed. Proc. Fed. Am. Soc. Exp. Biol. 40:Abstr. 3539.

PAYLER, D. K., POMARE, E. W., HEATON, K. W., and HARVEY, R. F. 1975. The effect of wheat bran on intestinal transit time. Gut 16:209-213.

POMARE, E. W., and HEATON, K. W. 1973. Alteration of bile salt metabolism by dietary fiber (bran). Br. Med. J. 4:262-264

POMARE, E. W., HEATON, K. W., LOWBEER, T. S., and WHITE, C. 1974. Effect of wheat bran on bile salt metabolism and bile composition. Gut 15:824-825.

PORTMAN, O. W. 1960. Nutritional influence on the metabolism of bile acids. Am. J. Clin.

Nutr. 8:462-470.

QURESHI, A. A., BURGER, W. C., PRENTICE, N., BIRD, H., and SUNDE, M. 1980. Regulation of lipid metabolism in chicken liver by dietary cereals. J. Nutr. 110(3):388-393.

RAYMOND, T. L., CONNOR, W. E., LIN, S., WARNER, S., FRY, M. M., and CONNER, S. L. 1977. The interaction of dietary fiber and cholesterol upon the plasma lipids and lipoproteins, sterol balance, and bowel function in human subjects. J. Clin. Invest. 60:1429-1437.

REAVEN, G. M., and BERNSTEIN, R. M. 1978. Effect of obesity on the relationship between very low density lipoprotein production rate and plasma triglyceride concentration in normal and hypertrigly-ceridemic subjects. Metabolism 27:1047-1054.

REDDY, B. S., WATANABE, K., and SHEINFIL, A. 1980. Effect of dietary wheat bran, alfalfa, pectin, and carrageenan on plasma cholesterol and fecal bile acid and neutral sterol excretion in rats. J. Nutr. 110:1247-1254.

REISER, R., HENDERSON, G. R., O'BRIEN, B. C., and THOMAS, J. 1977. Hepatic 3-hydroxy-3-methylglutaryl coenzyme A reductase of rats fed semipurified and stock diets. J. Nutr. 107:453-457.

ROBERTSON, R. P., GAVARESKI, J. D., HENDERSON, J. D., PORTE, D., and BIERMAN, E. L. 1973. Accelerated trigly-ceride secretion. A metabolic consequence of obesity. J. Clin. Invest. 52:1620-1626.

ROSS, J. K., and LEKLEM, J. E. 1981. The effect of dietary citrus pectin on the excretion of human fecal neutral and acid steroids and the activity of 7-alpha-dehydroxylase and beta-glucuronidase. Am. J. Clin. Nutr. 34:2068-2077.

SCHMITT, M. G., SOERGEL, K. H., and WOOD, C. M. 1976. Absorption of short chain fatty acids from the human jejunum. Gastroenterology 70:211-215.

SCHOENFIELD, L. J., BONORRIS, G. G., and GANZ, P. 1973. Induced alterations in the rate-limiting enzymes of hepatic cholesterol and bile acid synthesis in the hamster. J. Lab. Clin. Med. 82:858-868.

SHURPALEKAR, K. S., DORAISWAMY, T. R., SUNDARAVALLI, O. E., NARAYANA, R. M., and ROA, M. 1971. Effect of inclusion of cellulose in an "atherogenic" diet in the blood lipids of children. Nature 232:554-555.

SOUTHGATE, D. A. T. 1969. Determination of carbohydrate in foods. II. Unavailable carbohydrate. J. Sci. Food Agric. 20:331-335.

STANLEY, M. M., PAUL, D., GACKE, D., and MURPHY, J. 1973. Effects of choles-tyramine, metamucil, and cellulose on fecal bile salt excretion in man. Gastroenterology 65:889-894.

STASSE-WOLTHUIS, M., ALBERS, H. F. F., VAN JEVEREN, J. G. C., WIL DE JONG, J., HAUTVAST, J. G. H. J., HERMUS, R. J. J., KATAN, M. B., BRYDON, W. B., and EASTWOOD, M. A. 1980. Influence of dietary fiber from vegetables and fruits, bran and citrus pectin on serum lipids, fecal lipids and colonic function. Am. J. Clin. Nutr. 33:1745-1756.

STASSE-WOLTHUIS, M., KATAN, M. B., and HAUNTVAST, J. G. A. J. 1978. Fecal weight, transit time, and recommendations for dietary fiber intake. Am. J. Clin. Nutr. 32:909-910.

STORY, J. A. 1980. Dietary fiber and lipid metabolism: Update. Pages 137- in: Medical Aspects of Dietary Fiber. G. A. Spiller and R. M. Kay, eds. Plenum Press, New York.

STORY, J. A., and KRITCHEVSKY, D. 1976a. Dietary fiber and lipid metabolism. Pages 171-184 in: Fiber in Human Nutrition. G. A. Spiller and R. J. Amen, eds. Plenum Press, New York.

STORY, J. A., and KRITCHEVSKY, D. 1976b. Comparison of the binding of various bile acids and bile salts in vitro by several types of fiber. J. Nutr. 106(9):1292-1294.

STORY, J. A., TEPPER, S. A., and KRITCHEVSKY, D. 1977. Influence of dietary alfalfa, bran or cellulose on cholesterol metabolism in rats. Artery 3:154-163.

TARPILA, S., MIETTINEN, T. A., and METSARANTA, L. 1978. Effects of bran on serum cholesterol, faecal mass, fat, bile acids and neutral sterols, and biliary lipids in patients with diverticular disease of the colon. Gut 19:137-145.

TROWELL, H. 1972. Fiber: A natural hypocholesterolemic agent. Am. J. Clin. Nutr. 25:464-465.

TZAGOURNIS, M. 1978. Triglycerides in clinical medicine: A review. Am. J. Clin. Nutr. 31:1437-1452.

ULLRICH, I. H., LAI, H. Y., VONA, L., REID, R. L., and ALBRINK, M. J. 1981. Alterations of fecal steroid composition induced by changes in dietary fiber composition. Am. J. Clin. Nutr. 34:2054-2060.

USDA. 1963. Composition of Foods. U.S. Dep. Agric. Agric. Handbk. 8.

VAHOUNY, G. V., ROY, T., GALLO, L., STORY, J. A., KRITCHEVSKY, D., CASSIDY, M., GRUND, B. M., and

TREADWELL, C. R. 1978. Dietary fiber and lymphatic absorption of cholesterol in the rat. Am. J. Clin. Nutr. 31:S208-S212.

VIJAYAGOPAL, P., and KURUP, P. A. 1973. Hypolipidemic principle of the husk and bran of paddy. Atheroclerosis 18:379-387.

WALTERS, R. L., McLEAN BAIRD, I., DAVIES, P. S., HILL, M. J. K., DRASER, B. S., SOUTHGATE, D. A. T., and MORGAN, B. 1975. Effect of two types of dietary fiber on fecal steroid and lipid excretion. Br. Med. J. 2:536-538.

CHAPTER 12

OAT FLAVOR CHEMISTRY:
PRINCIPLES AND PROSPECTS

MENARD G. HAYDANEK, JR.
ROBERT J. McGORRIN[1]
Quaker Oats Company
Barrington, Illinois

I. INTRODUCTION

Although cereal grains are important food staples throughout the world, their flavor chemistry is rather poorly documented and understood, especially in their raw or uncombined state. Cereals represent most of the diet of people in many countries, and their desirable sensory properties therefore contribute to the maintenance of the nutritional health of large numbers of the world's population through consumption of good-tasting cereal food products. Given this enormous consumption of cereal grains, one would assume that the study of flavor in these systems would be of intense interest. It is rather surprising that very little modern flavor chemistry is being reported on cereal grains. This omission may have an economic basis, since the availability of cereals has not been a general problem and their relative cost is low. In contrast, the intense interest in more scarce or expensive flavored foods such as coffee, cocoa, meats, etc., may be prompted by the need for substitutes or extenders to keep the price and availability within the means of the average consumer. No such conditions are present with cereal grains, and therefore they represent one of the areas of flavor chemistry that could be described as a new horizon of research.

This chapter and associated research focuses on the flavor chemistry of cereal grains as it relates to processed whole grains and, in particular, to oats. It is beyond our scope to discuss other grain-based foods such as bread, ready-to-eat cereals, beer, and fermented foods.

II. CURRENT STATUS OF CEREAL GRAIN
FLAVOR CHEMISTRY

An excellent review of the available literature on cereal volatiles through early 1977 has been written (Maga, 1978). Most of the components identified by early

[1]Present address: Kraft, Inc., Research and Development, Glenview, IL.

workers in the field are various aliphatic carbonyls. Many aldehydes and ketones are found to be common to wheat, rice, barley, oats, and corn. However, these compounds do not appear to provide all the flavor sensations associated with these grain sources. In fact, Maga points out that one of the most complete studies on rice, by Bullard and Holguin (1977), was done in an effort to formulate a better rat bait composed of rice volatiles. An interest in roasted barley flavor can probably be attributed to its importance to the brewing industry. Wang et al (1969) reported that volatile pyrazines are present in the basic fraction of roasted barley flavor isolates. This is one of the few early reports implicating flavor components other than carbonyls in grain flavors.

Buttery et al (1978, 1980) have described the volatiles from corn kernels, husks, and tassels as possible corn earworm attractants. By using vacuum steam distillation isolation and capillary gas chromatography-mass spectrometry (GC-MS), they identified nonan-2-ol, heptan-2-ol, hept-4-en-2-ol, and undecan-2-ol as the major volatile components of the corn kernel. The raw kernels were not removed from the cob, however, so the origin of the volatiles with reference to the corn kernel is in doubt. Raw corn kernels and cobs did contain low levels of hexanal, nonanal, and 2,4 decadien-1-al isomers in addition to several terpenoid components. Over 75% of the 0.1 ppm volatile oil isolated from the cobs and kernels was composed of aliphatic alcohols.

Recent reports on rice have also implicated components other than carbonyls as important to grain flavor. Yajima et al (1978) studied volatile flavor components from cooked rice. In the steam-distilled volatiles of cooked polished rice, 100 constituents were identified. Of these compounds, 92 were identified for the first time in cooked rice. Among them, 4-vinyl phenol, 4-vinyl guaiacol, methyl-substituted pyrazines, and pyridines are potentially flavor active. The authors felt that the oxygenated neutral fraction was the major contributor to cooked rice flavor and that these volatiles originated by lipid oxidation during the cooking process.

Buttery et al (1983) identified 2-acetyl-1-pyrroline in cooked aromatic varieties of rice favored in Southeast Asia and India. The authors believe that the "popcornlike" odor of this pyrroline derivative is principally responsible for the favorable aroma attributes of these rice varieties. The threshold of 2-acetyl-1-pyrroline was determined as 0.1 ppb, and its concentration was measured in various rice varieties at 6–90 ppb. Sensory attributes correlated very well with the absolute amount of the pyrroline present. It was also of interest that 45% of the judges used a descriptor of "cooked oatmeal" when asked to describe the aroma of a 0.05-ppm solution of the pyrroline in water.

Wild rice (*Zizania aquatica*) volatile flavor analysis was first reported by Withycombe et al (1978). Unlike some other grains, this gourmet item undergoes a series of postharvest steps that alter the flavor of the original grain. These include mild fermentation and a parching step that produces a toasted flavor quality. Volatiles isolated by vacuum steam distillation and identified by GC-MS numbered 112, the majority of them pyrazine and pyridine derivatives. The toasted flavor notes were attributed to the pyrazine fraction, and the "tealike" aroma of wild rice was attributed to the alkylpyridine components. The compounds identified in wild rice contrast dramatically with those in unprocessed white and cooked rice. Few aldehydes and ketones of lipid origin were found, and many of the neutral carbonyl components identified in wild rice can be attributed to the parching process since they are found in many other heat-processed, carbohydrate-rich foods.

The authors also present "aromagrams" obtained by smelling a portion of the capillary column effluent. Two peaks, labeled as wild-ricelike and smokey, corresponded to the 1-phenyl-2-propanone and methylindole identified at the respective retention index for these odors. However, the authors did not draw any conclusions as to their overall importance to wild rice flavor.

An extensive review of rice product volatiles including raw rice, rice bran, Soong-Neung, and rice cake has been compiled by Maga (1984).

Studies on the volatile flavor components of wheat are very few in number. An extensive review of bread flavor is available (Maga, 1974), but the complex fermentation and baking processes involved preclude extrapolation back to the flavor of the whole grain or flour. Twenty-four volatile compounds have been characterized in studies of whole-grain wheat or flour (Maga, 1978), most of them short-chain alkyl aldehydes and ketones that would not be expected to exhibit the total sensory character of whole-grain wheat. Since wheat is widely consumed as a breakfast cereal, a great deal of more fundamental investigation is needed to understand its desirable flavor attributes.

Buttery et al (1985) have compared wheat leaf and stem volatiles to those in oat and barley. Major volatiles noted were (Z)-3-hexenyl acetate, (Z)-3-hexanol, (E)-β-ocimene, (E)-2-hexenal, and caryophyllene. Although the work was directed toward these compounds as possible insect attractants, it is conceivable that they may also be associated with wheat flavor.

Several reports describing the volatile composition of buckwheat flour have appeared from Japanese laboratories interested in the flavor of buckwheat noodles. Of the 45 compounds identified by Aoki et al (1981), most could logically be expected to originate from lipid oxidation processes. These authors also measured volatiles as a function of flour fractions obtained by roller-milling. The best flavor scores were attributed to fractions having high contents of testa and embryo and little pericarp. These flavor fractions had relatively high hexanal and nonanal contents, but no firm flavor correlations were established.

Yajima et al (1983) described 209 compounds isolated from boiled buckwheat flour. The characteristic boiled aroma was obtained by using a Likens and Nickerson type isolation. Further separation into acidic, basic, and neutral fractions revealed characteristic buckwheat aromas in the basic and neutral fractions. The basic fraction had a nutty and slightly fishlike odor. The pleasant nutty aroma was attributed to several pyrazines, including 2-(1'-ethoxyethyl)-pyrazine and 2-(1'-ethoxyethyl)-5-methyl pyrazine. Both of these compounds appear to be unique to boiled buckwheat flour volatiles. The neutral fraction had a characteristic aroma of boiled buckwheat flour but with a somewhat green overtone. Extensive fractionation on silica gel separated the green odor into a fraction comprised mainly of alkanals; another fraction contained a characteristic aroma of boiled buckwheat flour but contained 55 components, none of which individually had the above aroma based upon GC exit port sniffing evaluation. The compound 1-octen-3-ol was a major one in a musty odor fraction that resembled freshly milled buckwheat flour in aroma. Yajima et al (1983) concluded that boiled buckwheat flour aroma is very complex and is composed of the combined aroma effects of many components.

Sparse literature is available on rye and triticale. Common aldehydes and ketones similar to those of other grains are reported by Maga (1978). Kaminski et al (1964) studied methods of isolating volatiles from rye, roasted rye malt, and whole-meal rye bread. The isolation techniques described were combined with GC odor effluent assessments and resulted in aromagrams of the volatiles present in the starting materials. No identifications were presented, but significant areas of roasted cereal aroma were noted, indicating areas of further effort needed to establish the components responsible for this desirable attribute.

Oats, which represent a somewhat unique and characteristic flavor in the cereal grain family, are consumed as a rolled, flaked breakfast cereal throughout the world. Their nutty, pleasant grain character is obviously highly acceptable to a great number of consumers. Additionally, oat flour is used as a primary ingredient in many ready-to-eat breakfast cereals. The rolled flakes were the first "unofficial" meat extender when mixed with hamburger, and they are also extensively used in the cookie industry.

Hrdlicka and Janicek (1964a, 1964b) studied the nutlike flavor of toasted oat flakes. Until recently, these two brief reports, which identified typical carbonyls and unusual amines, were the sum total of literature relating to volatile flavor compounds in oats. Table I lists the compounds they identified; unfortunately, the isolation method and toasting conditions were not reported. The authors stated that these compounds cannot be regarded as the sole source of toasted oat flavor. Heydanek and McGorrin (1980, 1981a) reported systematic studies on various aspects of oat flavor chemistry, which are presented with the main body of data of this paper to maintain completeness and continuity.

Flavor stability and potential off-flavorants in oat systems have been studied by several approaches. The flavor stability of oatmeal is markedly good, provided its moisture content remains above 5%. However, certain oat cereals treated severely with heat are much more prone to oxidative deterioration (Martin, 1958). Oats are probably more labile to oxidative deterioration that

TABLE I
Volatiles Reported in Toasted Oat Flakes Before 1980[a]

Type and Compound	Type and Compound
Alcohols	Bases
2-Aminoethanol	Methyl amine
Carbonyls	Ethyl amine
Formaldehyde	Propyl amine
Acetaldehyde	i-Propyl amine
Propanal	Butyl amine
2-Methylpropanal	i-Butyl amine
Butanal	Pentyl amine
2-Methylbutanal	Dimethyl amine
3-Methylbutanal	1,3-Diaminopropane
Hexanal	1,4-Diaminobutane
Octanal	1,5-Diaminopentane
Diacetyl	N,N'-di-1,4-diaminobutane
2-Pentanone	
2-Hexanone	
2-Octanone	
Furfural	
5-Hydroxymethylfurfural	

[a] Data from Hrdlicka and Janicek (1964a, 1964b).

other cereal grains because their fat content is considerably higher. Oat oil in unheated oatmeal is highly resistant to oxidation, probably because of high levels of natural antioxidant with potency comparable to that of propyl gallate and butylated hydroxytoluene (BHT) (Daniels and Martin, 1961, 1965). As a matter of fact, a specially fine-ground oat flour (Avenex) has been used as an antioxidant to improve the keeping qualities of many foods.

Native oat enzymes also have the potential to cause improperly processed oat products to have marked flavor instability. Oat lipase activity is high enough to render them objectionably soapy unless inactivated during processing (Shukla, 1975). Tyrosinase activity is used as a criterion of the adequacy of heat treatment for lipase inactivation. Oxidative deterioration via the action of oat lipoxygenase and peroxidase can have two effects on flavor quality. First, because of the high content of unsaturated fatty acids in oat oil, action by the above enzymes can lead to volatile aldehydes and ketones with varying types of flavor properties. Heimann et al (1975) showed that volatile aldehydes including (E)-3-nonen-1-al were produced by incubation of linoleic acid with oat lipoxygenase.

The second flavor effect of oat enzymes on lipids is the oxidation of nonvolatile bitter components. Bitterness in stored oat flour has been attributed to formation of specific hydroxy monoglycerides (Heimann and Schreier, 1970; Biermann and Grosch, 1979). A distinct bitter taste was found in a mixture containing 9-hydroxy-(E,Z)-10,12-octadecadienoic-1'-monoglyceride and 13-hydroxy-(Z,E)-9,11-octadecadienoic-1'-monoglyceride. Authentic compounds were prepared, and their taste thresholds were 1−8 μmol/ml in water with a pure bitter quality comparable to that of caffeine. These compounds could be prepared by incubation of monolinoleate with lipoxygenase and borohydride reduction of the resulting hydroperoxides.

III. CHEMICAL ANALYSIS OF OAT FLAVOR VOLATILES

A. Experimental Methodology

The analytical methodology used to develop the data on oat flavor is described in detail elsewhere (Heydanek and McGorrin, 1981a, 1981b). Combined techniques of vacuum steam distillation, headspace trapping with Tenax GC, capillary GC/MS/computer analysis, and sensory comparisons were all used in attempting to understand the chemical components responsible for oat flavor.

All oat sources used in the flavor isolation work reported were obtained from commercial production streams and represent items of commerce. Dried oat groats were dehulled oats that have been dried to ~7.5% moisture, representing the commercial starting material for the manufacture of oatmeal, oat flakes, and oat flour. The oatmeal used was the 5-min cooking variety, which was analyzed from commercial packages at less than two months of age. Any other treatments of these starting materials are discussed in the context of the experiments presented.

B. Oatmeal and Its Precursors

Any isolation technique and comprehensive study of volatiles must be assessed in terms of the fidelity between the flavor of the isolated volatiles and the flavor of the starting material. Intuitively, it seems that vacuum steam distillation should provide a suitable isolation for oat systems because of its low temperature and use of water as a distillation transfer medium. A convenient way to evaluate the isolated flavor volatiles is simply to taste the distillate water directly from a spoon. In terms of sensory assessment of the distillate flavor quality, this method is far more convenient and distinct that smelling the total extract in a concentrated form on a perfume blotter. However, blotter analysis was used in cases where CH_2Cl_2 extracts had to be checked for flavor and aroma character. All sensory judgments reported are the responses of three to five trained laboratory panelists who are familiar with oat products and accustomed to assessing flavor quality of isolated volatile fractions. Table II compares the flavor quality of vacuum-steam distilled isolates with the sensory properties of the starting oat source. Essentially, maintenance of odor and flavor character is observed after volatile isolation from three different oat sources. The different sources are distinguished by flavor, and their corresponding isolates can also be differentiated from each other by similar flavor characteristics. It is critical that these comparisons be made before chemical analysis so that confidence in the interpretation of the chemical data and its sensory consequences is high. Our interpretation of the sensory comparisons of Table II lead us to believe that the isolates analyzed represented oat flavor in its isolated volatile form.

Also included in Table II are the various flavor attributes associated with the separation of the basic fraction from the total distillate. As one would expect, most of the nutty, browned flavor character was found in the heat-processed systems of oatmeal. These would implicate the basic fraction commonly containing pyrazines and other nitrogen heterocycles in the nutty flavor impressions found in high-quality cooked oatmeal flavor. The acid/neutral fraction contained green-grain type flavor notes. This simple transformation effectively allows study of the two major flavor characteristics of oats in separate fractions.

The sensory properties of the oat systems detailed in Table II show that a progression of flavor development can be noted as additional heat treatments are applied during oatmeal manufacture and consumption. When groats are rolled and flaked to make oatmeal, the steaming step, which also completes enzyme inactivation, provides the first development of true oat flavor. The 5-min

TABLE II
Sensory Assessments of Oat Flavor Isolates

| Source | Initial Flavor | Vacuum-Steam Distillate Flavor | Separated Vacuum-Steam Isolate | |
			Acids/Neutrals	Basic
Groats	Raw oat Weedy-hay Grassy	Raw grain Hay-feedy Grassy	Green-weedy Mashy grain	Weak/none
Oatmeal	Mild oat Hay-weedy Browned	Mild oat Raw-green Brown-nutty	Green-weedy Mashy grain	Brown-nutty Pleasant
Cooked oatmeal	Oat-nutty Browned-burnt Weak weedy	Oaty, slightly burnt Nutty-browned Weedy-grain	Mashy grain Malty Caramelized slightly oat	Nutty-pecan Raw potato Harsh-chalky

cooking step required to prepare oatmeal completes the development of its flavor character. To study the fundamental flavor chemistry processes occurring during these heat-treatment steps, complete knowledge is needed of the flavor volatiles associated with the native groat. In an effort to develop this basic data on oat flavor chemistry, total volatile analysis and purge-trap headspace analysis were both used to study dried oat groat volatiles.

Total volatile analysis was used to evaluate materials with a GC Kovat's Retention Index (K_I) of >600, and the headspace technique was used to identify the lower-boiling components. Kovat's index values were measured using pure alkanes from carbon numbers 6 to 15 as GC reference retention standards. During our studies, a certain variability in the types and amounts of volatiles was noted in the system when excess water was used in vacuum steam distillation isolation. For comparison purposes, oat groat volatiles were isolated under high vacuum without added water. The results showed that the volatile profile changed dramatically when this technique was used. Figure 1 illustrates the differences in total volatiles noted between dry vacuum isolation and vacuum steam distillation. The relative concentrations of isolated components vary widely between the two chromatograms: a major peak such as limonene (peak E) in the dry vacuum isolate represents a 10-ppb concentration; in contrast, hexanal (peak B) in the hydrated isolation represents 3–5 ppm.

Table III lists the components identified by GC-MS analysis of both total volatile isolates. The dry vacuum isolate, which represented the weak grainy, haystraw odor of dry groats, contains mostly hydrocarbon materials. The principal components are $C_{10}H_{16}$ terpenes, alkyl benzenes, and some oxygenated constituents. No single component appears responsible for the entire odor.

The major oxygenated flavor compound present is hexanal (peak B), and its level is below 50 ppb. The presence of a large number of alkyl benzenes suggests that they occur via contamination of the groats in the natural environment during the normal process of getting them to an oat mill. All of these compounds are present at <50 ppb, and they would not be expected to be of importance to the overall flavor of oat groats or their subsequent products. In essence, dried oat groats derive their weak hay, grassy odor from very low levels of monoterpenes and hexanal. These would be expected to make little contribution to oat flavor in cooked, hydrated, or processed systems, especially at the concentrations observed.

In contrast, volatiles isolated by vacuum steam distillation showed a much higher level of oxygenated compounds, as well as a completely different component profile (Fig. 1, bottom). The distillates had a more green cereal-type odor and flavor. The major components observed were 3-methyl-1-butanol, 1-pentanol, 1-hexanol, hexanal, 1-octen-3-ol, (E,E)- and (Z,E)-3,5-octadien-2-one, and nonanal.

In addition, many other components associated with enzymatic activity, such as alcohols and aldehydes, were found; those most notable from a flavor standpoint include 3-methylbutanal, 2,4-decadienal, and benzaldehyde. Traces of the major components from the dry vacuum isolate were also found in the hydrated isolate. It appears that when the isolation step is done in the presence of water, enzymatic activity results from rehydration of the dried oat groat. The observed volatiles are produced and superimposed on the volatiles inherent in the dry groat.

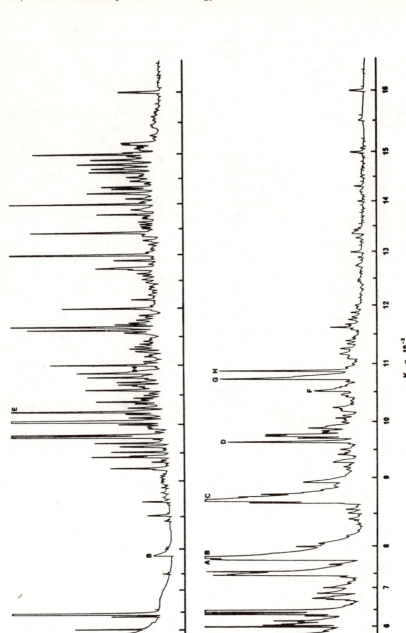

Fig. 1. SE-30 capillary gas chromatography-flame ionization detector responses of oat groat total volatile isolates as a function of Kovat's retention indexes. Dry isolation (top) and hydrated isolation (bottom) are recorded at ×256 and ×1, respectively. Peak identifications correspond to those listed in Table III.

TABLE III
Compounds Identified in the Total Volatile Fraction of Oat Groats
by Gas Chromatography-Mass Spectrometry

Kovat's Retention Index	Compound	Standard Match[b]	Dry[c]	Hydrated[c]
605	Ethyl acetate	++		+
619	2-Methyl-1-propanol	++		+
628	3-Methylbutanal	++		+
641	Benzene	++	+	+
651	Cyclohexane	+	+	+
669	Pentanal	++		+
689	3-Pentanol	+		+
699	2-Pentanol	+		+
720	Pyridine	++		+
743	3-Methyl-1-butanol	++	+	+
748	Toluene	++	+	+
754 (A)	1-Pentanol	++		+
774 (B)	Hexanal	++	+	+
782	Ethyl butyrate	++		+
798	2-Methylpyrazine	++		+
800	Octane	++	+	+
814	Furfural	++	+	+
845	Ethylbenzene	++	+	
854	*m/p*-Xylene	++	+	
858 (C)	1-Hexanol	++		+
860	γ-Butyrolactone	+		+
867	2-Heptanone	++	+	+
869	Styrene	++	+	
875	*o*-Xylene	++	+	+
878	Heptanal	++	+	+
883	2,5-Dimethylpyrazine	++	+	+
895	*n*-Pentyl acetate	++		+
900	Nonane	++	+	
910	*i*-Propyl benzene	++	+	
930	α-Pinene	++	+	
931	Benzaldehyde	++		+
937	Dimethylpyridine	++		+
940	*n*-Propyl benzene	++	+	
943	Camphene	++	+	
947	*m*-Ethyl toluene	++	+	+
948	2,5-Diethyl furan	+		+
949	*p*-Ethyl toluene	++	+	+
956	1,3,5-Trimethylbenzene	++	+	+
961	*t*-Butyl benzene	++	+	
964	*o*-Ethyl toluene	++	+	+
970	β-Pinene	++	+	
971 (D)	1-Octen-3-ol	++		+
972	2-Octanone	++	+	
978	2-*n*-Pentyl furan	++	+	+
979	1,2,4-Trimethylbenzene	++	+	+
982	Myrcene	++	+	
986	3-Octanol	+		+
988	Dichlorobenzene	+	+	
994	2-Octanol	+		+
997	*sec*-Butyl benzene	+	+	
1000	Decane	++	+	
1004	Δ-3-Carene	++	+	+
1007	1,2,3-Trimethylbenzene	+	+	
1011	*p*-Cymene	++	+	
1016	*o*-Methyl styrene	+	+	
1018	3-Octene-2-one	++	+	+

(*continued on next page*)

TABLE III (continued)

$K_1{}^a$	Compound	Match[b]	Dry[c]	Hydrated[c]
1021 (E)	Limonene	++	+	
1023	4-Methyldecane	+	+	
1035	*o*-Diethyl benzene	+	+	
1042	Propyl toluene	+	+	
1044	3-Ethyl-1,2-dimethylbenzene	+	+	
1046 (F)	(Z,E)-3,5-Octadien-2-one	++	+	+
1049	γ-Terpinene	++	+	
1053	*n*-Propyl toluene	+	+	
1055	2,5 Dimethyl-3-ethyl pyrazine	++		+
1062	3,5 Dimethyl-2-ethyl pyrazine	++		+
1065	Dimethylethylbenzene	+	+	
1069 (G)	(E,E)-3,5-Octadien-2-one	++	+	+
1071	Dimethylethylbenzene	+	+	
1072	Dimethylstyrene	+	+	
1079	Allo-ocimene	+	+	
1084 (H)	Nonanal	++	+	+
1091	Dimethylethylbenzene	+	+	
1097	Isophorone	+	+	
1100	Undecane	++	+	
1101	2-Phenyl ethanol	++		+
1103	Dimethylethylbenzene	+	+	
1106	Dimethylethylbenzene	+	+	
1108	2-Formyl imidazole	+	+	+
1113	4-Decanone	+	+	
1122	Ethyl styrene	+	+	
1156	Borneol	++		+
1160	Diethyl fumarate	+		+
1162	Naphthalene	++	+	+
1172	$C_{13}H_{28}$	+	+	
1174	2-Decanone	++		+
1193	Benzothiazole	+		+
1200	Dodecane	++	+	
1216	$C_{13}H_{28}$	+	+	
1274	2-Methyl naphthalene	++	+	
1289	1-Methyl naphthalene	++	+	
1293	2,4-Decadien-1-al	++		+
1300	Tridecane	++	+	
1324	γ-Nonalactone	++	+	+
1380	$C_{15}H_{32}$	+	+	
1400	Tetradecane	++	+	+
1411	$C_{15}H_{24}$ Sesquiterpene	+	+	
1422	$C_{15}H_{24}$ Sesquiterpene	+	+	
1434	α-Cedrene	+	+	
1446	$C_{15}H_{26}$	+	+	
1458	Biphenyl	+	+	
1463	$C_{16}H_{34}$	+	+	
1471	$C_{15}H_{22}$	+	+	
1479	$C_{15}H_{24}$ Sesquiterpene	+	+	
1488	Pentadecene	++	+	
1500	Pentadecane	++	+	+
1507	Methyl laurate	+		+
1509	$C_{15}H_{24}$ Sesquiterpene	+	+	
1515	α-Cadiene	+	+	
1559	Diethyl phthlate	++		+
1600	Hexadecane	++	+	+

[a] K_1 index and peak identification corresponding to Fig. 1.
[b] + = Matches published mass spectra (MS) value and has appropriate K_1; ++ = matches authentic compound K_1 and MS value.
[c] + = Present in the isolated volatiles from this isolation procedure.

The type of volatiles produced, C_5 and C_6 alcohols and aldehydes, indicates that lipoxygenase and aldehyde oxidoreductase activity is present in the dried oat groats. Since these volatiles comprise about 80% of the total isolated volatiles from the hydrated system, their specific production seems indicative of these enzymes. Similar production of C_5 and C_6 alcohol and aldehyde components has been shown to arise from these enzyme systems in green beans and seeds (De Lumen et al, 1978), soybeans and corn (Leu, 1974), and alfalfa seeds (Esselman and Clagett, 1974).

Oats are known to contain lipoxygenase (Heimann et al, 1975) and, in general, a very complex enzyme system (Shukla, 1975). It is thus not surprising to find enzymatic activity; however, it is critical to a flavor study to know whether enzymes are active and functional during a flavor isolation process. These data also suggest that since flavor volatiles can form in oat groats at high moisture contents, they could be a source of additional desirable flavor in certain end products as well as a source of off-flavor if the enzyme activity is unwanted or unknown.

Examination of the more highly volatile components by Tenax trapping of headspace volatiles from oat groats, both dry and hydrated, showed results similar to those obtained in the total volatile analysis. Figure 2 compares the headspace profiles obtained by purging dry (top) and hydrated (bottom) groats, and Table IV lists the components identified by GC-MS analysis of these volatiles. Most notable in the hydrated sample is the previously observed increase in C_5 and C_6 alcohols and a general increase in the level of oxygenated components.

Although the changes noted in Fig. 2 do not appear as dramatic in a 1-hr incubation at 60°C by headspace analysis, the Tenax trapping method is not ideally suited for retaining alcohols that boil at lower temperatures. Therefore, we are not sure whether all of the alcohol present was retained during trapping, but distinct increases in these components were noted. Again, the two chromatograms represent widely differing component concentrations. In the dry isolates, limonene (Fig. 2, peak E; see Table IV) represents 10 ppb, whereas the hexanal content (Fig. 2, peak B) of the hydrated samples compares favorably with a 3-ppm standard of hexanal added to the hydrated groat sample and immediately purged.

The most abundant components noted in the dry purge system are 1,2-dichloroethane, hexanal, toluene, and 2-pentyl furan. In the lower-boiling fraction eluting before hexanal, no components other than dimethylsulfide and dimethyldisulfide appear to be potentially significant for flavor. In contrast, the hydrated isolate contained 2-methylpropanal, ethanol, acetone, 2-butanone, 3-methylbutanal, 3-pentanone, and pentanal in the lower-boiling fraction. These results again suggest residual enzymatic activity present in the hydrated groat.

The combined use of total volatile analysis and headspace trapping effectively allowed analysis of the entire spectrum of volatiles arising from oat groats. The presence of large amounts of alcohols in both hydrated systems caused some aberration in the chromatographic systems as a result of tailing and column overloading. However, the profiles obtained illustrate the fact that these alcohols are abundant in the hydrated system, offering a clue to the processes taking place. The fact that residual enzyme activity may be present in dried cereal grains must

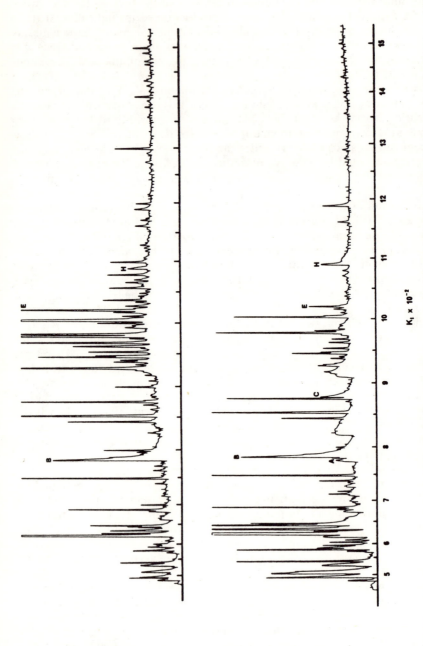

Fig. 2. SE-30 capillary gas chromatography-flame ionization detector responses of Tenax-trapped oat groat headspace volatiles as a function of Kovat's retention indexes. Groats purged dry (top) and hydrated (bottom) are recorded at the same attenuation scale. Peak identifications correspond to those listed in Table IV.

be taken into account whenever flavor isolations are conducted. Two completely different conclusions as to the type of flavor components present in an oat groat would be drawn depending on the isolation technique used. Therefore, further studies on the flavor of oat systems would be expected to show that the cooking and processing methods are highly important to the ultimate flavor.

In the context of the effect of heat processing on flavor development, the presence of small but detectable amounts of 2-methyl, 2,5-dimethyl, and C_4-substituted pyrazines in oat groats suggests that some browning takes place

TABLE IV
Headspace Volatiles (K_1 <800) Identified from Oat Groats
Using Gas Chromatography-Mass Spectrometry

Kovat's Retention Index[a]	Compound	Standard Match[b]	Dry[c]	Hydrated[c]
478	Acetone	++	+	+
485	Ethanol	++	+	+
493	Dimethylsulfide	+	+	+
506	1,2-Propanediol	++	+	+
538	2-Methylpropanal	++	+	+
565	2,3-Butanedione	++	+	+
572	1-Propanol	++	+	+
575	2-Butanone	++	+	+
593	2-Methylfuran	+	+	+
605	Chloroform	++	+	+
607	Ethyl acetate	++	+	+
612	2-Methyl-1-propanol	++		+
623	1,2-Dichloroethane	++	+	+
630	1,1,1-Trichloroethane	++	+	+
634	3-Methylbutanal	++	+	+
643	3-Pentanone	++	+	+
644	Benzene	++	+	+
649	Carbon tetrachloride	++	+	+
679	Pentanal	++	+	+
683	Trichloroethylene	++	+	+
688	2-Ethyl furan	+	+	+
700	Heptane	++	+	+
716	Methylcyclohexane	+	+	+
727	Dimethyldisulfide	+	+	+
739	1,1,2-Trichloroethane	++	+	+
749	Toluene	++	+	+
772 (A)	1-Pentanol	++		+
778 (B)	Hexanal	++	+	+
796	Tetrachoroethylene	++	+	+
800	Octane	++	+	
827	Ethyl cyclohexane	+	+	
878	2-*n*-Butyl furan	+	+	
884 (C)	1-Hexanol	++	+	
978	2-Pentyl furan	++	+	
1021 (E)	Limonene	++	+	+
1086 (H)	Nonanal	++	+	+

[a] K_1 index and peak identification corresponding to Fig. 2.
[b] + = Matches published mass spectra (MS) value and has appropriate K_1; ++ = matches authentic compound K_1 and MS value.
[c] + = Present in the isolated volatiles from this isolation procedure.

during the initial production drying process. More severe heat treatment would be expected to produce larger amounts of these "cooked" flavor components.

The presence of relatively large amounts of (E,E)- and (Z,E)-3,5-octadien-2-one (peaks F and G, Fig. 1) is of interest from the standpoint of mechanism of formation. They appear as minor components in the dry isolate but are quite abundant in the isolate from the hydrated system. These dienones have been found in red beans (Buttery et al, 1975), green beans (Murray et al, 1976), cooked asparagus (Tressel et al, 1977), oxidized linolenate (Badings, 1970), and autooxidized methyldocosahexaenoate (Noble and Nawar, 1975). All evidence indicates these compounds originate from a lipid oxidation reaction that is either enzymatic or chemical. Why they appear specifically in oats is unknown, but oat lipids do contain significant amounts of linolenic acid. Badings (1970) measured a flavor threshold of 0.30 ppm in paraffin oil for the dienone and described its flavor as fruity and fatty. Synthesis of this dienone and isolation of the (E,E) and (Z,E) isomers were undertaken by Heydanek and McGorrin (1981b). The odor appears to be more grassy and strawlike than fatty when diluted in air, and it resembles, to an extent, the grassy-straw character of oats. Further study is needed to determine the significance of these components to oat flavor.

Inherent flavor components in oat groats are present in very low levels and are not expected to be major factors in the flavor of processed oat products. The presence of latent enzyme activity at high moisture levels is important to understand because of the potential for flavor component development by these enzymes. The studies on oat groat volatiles lead to the expectation that further heat treatments are responsible for the development of true oat flavor.

C. Oatmeal Flavor

The transformation of dried oat groats into oatmeal involves several steps that are expected to be significant in further flavor development. The groats are treated with steam before flaking to inactivate enzymes and increase the moisture content. After flaking, the oats are cooled and packaged. The consumption of oatmeal occurs after a boiling-water treatment lasting anywhere from 1 to 10 min depending on flake thickness. Each heat step is expected to change and develop flavor components. The research focused on attempts to correlate sensory flavor impressions with volatile component changes during each heat treatment. Again, this problem was attacked by comparison of total volatiles isolated by vacuum steam distillation and headspace purge-trap techniques.

Oatmeal volatiles isolated by these techniques were separated into basic and acid/neutral fractions as previously described. Cooked oatmeal was the "old-fashioned" variety prepared by boiling for 5 min. The isolated and separated volatile fractions were compared on a sensory basis (Table II) and via GC-MS techniques for volatile identification. In an effort to describe the observed progressive appearance of oaty-nutty flavor in the cooked oatmeal system, the basic fraction was compared first.

Data collected from GC-MS analysis on basic fractions from oat groats and uncooked and cooked oatmeal are summarized in Table V. Corresponding to an increase in the oaty-nutty flavor development is an increase in the complexity and concentration of nitrogen heterocycles. Thiazoles are only detected after the final 5-min cooking in boiling water. The relative amount of these nitrogen heterocycles increases qualitatively; rigorous quantitative measurements have not been completed. Interestingly, the most abundant pyrazine and thiazole components are the more highly substituted isomers, which also have the lowest flavor thresholds (Maga, 1974). This suggests that compounds like 2-ethyl-3,5-dimethylpyrazine, 3-ethyl-2,5-dimethylpyrazine, 2-methyl-4-ethylthiazole, and 2,4-dimethyl-5-ethylthiazole are important to the nutty flavor developed in cooked oatmeal. All the alkyl pyridines identified to date are trace components that are probably below their respective flavor thresholds. Since there is a demonstrated development of new basic-fraction material in the 5-min cooking step, our isolation technique is appropriate in that the oatmeal to be distilled is never heated over 55° C after cooking. Basic fractions from volatiles isolated by atmospheric Nickerson-Likens distillations had higher levels of N-heterocycles and a different, more burnt-flavor character. Until further studies are completed, these must be considered artifacts because of the more severe conditions (2 hr in boiling water) of isolation. It does point out that oatmeal has the potential for further flavor development under these conditions.

The changes that occur in the neutral fraction of oat flavor isolates across the stages of oatmeal preparation are most dramatic. As was pointed out in the oat groat characterizations, little flavor contribution is expected from the inherent components present. After the rolling and flaking operation with its associated heating steps, certain volatiles began to appear. These were detected with GC-MS analysis but from a concentration standpoint were not much more abundant than the compounds naturally present in oat groats (i.e., terpene hydrocarbons and other alkyl/aryl compounds). Figure 3 and Table VI show several oxygenated volatiles appearing in oatmeal. None of these, except hexanal, was detected in dry oat groats.

TABLE V
Basic Fraction Composition of Oatmeal as a Function of Process Conditions

Compound	Groats	Uncooked Oatmeal	Cooked Oatmeal
Pyridines			
H-	+[a]	+	+
2-Methyl			+
3/4-Methyl			+
2,6-Dimethyl		+	+
2-Ethyl		+	+
3-Ethyl		+	+
2-Propyl-			+
Pyrazines			
2-Methyl	+	+	+
2,5-Dimethyl	+	+	+
2,6-Dimethyl		+	+
2,3-Dimethyl		+	+
Trimethyl		+	+
2-Ethyl, 5/6-methyl		+	+
2-Ethyl, 3,5-dimethyl	+	+	+
Tetramethyl		+	+
Thiazoles			
H-			+
2-Methyl, 4-ethyl			+
2,4-Dimethyl, 5-ethyl			+

[a] Denotes presence in volatile isolates from the named source.

In cooked oatmeal, the most abundant neutrals are oxygenated components (Figure 4 and Table VI). Most of the aldehydes are products of reactions logically expected to take place in the boiling-water heat treatment. The appearance of major amounts of 2,4-dienals is surprising, since they have not previously been implicated in grain flavors. However, their relatively high concentration suggests that they play a role in the composite flavor of oatmeal. The flavor components listed in Table VI are mostly derived from lipid oxidation processes. Furfural, benzaldehyde, and phenyl acetaldehyde probably arise from interactions of reducing sugars and amino acids. These flavorful aldehydes can contribute the backbone of grainy flavor impressions to cooked oatmeal, but all of the minor constituents have not been identified, and some of these may contribute further to the flavor. It is relatively obvious from these data that most of the flavor in cooked oatmeal arises in the final preparation step. Oat oil, reducing sugars, and free amino acid appear to be the most important precursors of flavor volatiles. No single component with "oaty" flavor identity has been found to date: the flavor of cooked oatmeal appears to be a blend of impressions contributed by both the basic nitrogen heterocyclic fraction and the neutral components.

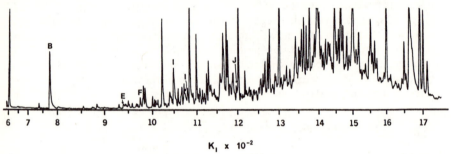

$$K_1 \times 10^{-2}$$

Fig. 3. SE-30 capillary gas chromatography-flame ionization detector responses of oatmeal total volatiles isolated by vaccuum steam distillation as a function of Kavat's retention indexes. Peak identifications correspond to those listed in Table VI.

TABLE VI
Oxygenated Neutrals in Oatmeal and Cooked Oatmeal

Compound	Peak[a]
Pentanal	A
Hexanal	B
Furfural	C
Heptanal	D
Benzaldehyde	E
1-Octen-3-ol	F
2,4-Heptadien-1-al	G
Phenylacetaldehyde	H
3,5-Octadien-2-one (two isomers)	I
2,4-Nonadien-1-al (two isomers)	J
γ-Octalactone	K
2,4-Decadien-1-al (two isomers)	L
γ-Nonalactone	M

[a] Letter code correlates with peaks labeled in Figs. 3 and 4.

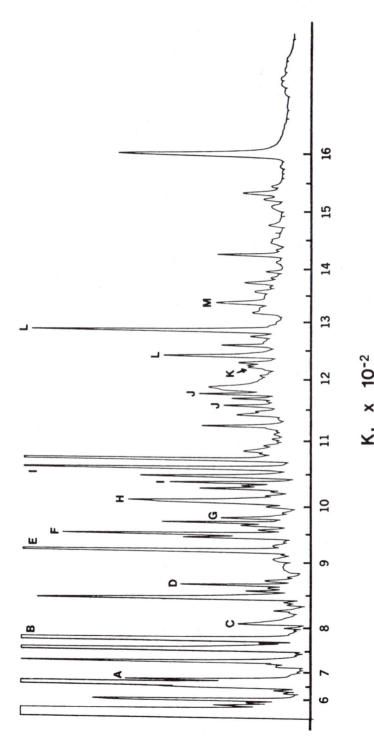

Fig. 4. SE-30 capillary gas chromatography-flame ionization detector responses of the neutral volatile fraction from cooked oatmeal isolated by vacuum steam distillation. Peak identifications correspond to those listed in Table VI.

D. Toasted Oats

Another end use and typical flavor characteristic of oats is developed by toasting them to a nutty, browned flavor either in the flaked or groat form. Toasted oat flakes and groats are commonly used in granola-type products and specialty breads. The toasting process develops a uniquely different flavor in oats, one that has nutty, browned characteristics. Therefore, additional flavor chemistry must be taking place to account for this difference.

To investigate the flavor volatile composition relative to toasting, oat groats were toasted in a forced-air oven. These groats had a strong browned, toasted, nutty flavor with a slight burnt aroma. A standard vacuum steam distillation was performed on the toasted groats, and the organic volatiles were isolated by CH_2Cl_2 extraction of the distillate water, similar to that used for whole oat groats. The distillate water had a strong nutty, burnt grain, and slightly oat character, as did the CH_2Cl_2 extract of the water. The organic extract was separated into basic and neutral/acid fractions. Sensory attributes were found to separate into two distinct types, as with previous oat isolates. The basic fraction had strong nutty, earthy, toasted, slightly harsh and burnt flavors, whereas the neutral fraction was described as burnt grain, browned, heavy, and not oaty.

GC analysis compared the basic fraction with that obtained from oatmeal. An immediate observation was that the toasted groats contained relatively large amounts of and many more types of material in the basic fraction. Figure 5 illustrates the GC pattern obtained from the toasted oat groat (basic fraction). GC-MS analysis of this fraction revealed an exceedingly complex mixture of various nitrogen heterocycles. As expected, alkyl-substituted pyrazines are the most abundant materials, and many isomers are present. Table VII details the analytical data and the mass spectral interpretations. The major pyrazines are those commonly found in roasted or toasted foodstuffs, mainly C_1-C_4 substituted alkyl pyrazines with methyl and/or ethyl substituents only. Although somewhat minor in concentration, alkyl thiazoles, pyridines, and oxazoles are found in this fraction. These components make an as yet unknown contribution to the flavor of toasted oat groats. Of considerable interest is the presence of fairly large amounts of alkyl/alkenyl substituted pyrazines. These compounds have significantly lower flavor thresholds than their saturated counterparts (Maga, 1974) and might logically be expected to contribute to the overall flavor. The relationship between the identified components and the flavor of toasted oats remains unknown because of the complexity of the mixture. (An estimate of relative concentration was obtained by comparison with an internal standard, 2-methoxy-3-methylpyrazine added to the distillation mixture at 0.01 ppm.) The major alkyl-substituted pyrazines are present at concentrations ranging from 1 to 5 ppm; alkyl/alkenyl-substituted pyrazines ranged from 0.01 to 0.5 ppm. The longer-chain alkyl-substituted pyrazines such as butyl and pentyl pyrazine are of some interest because they appear to be formed in a slightly different manner. These compounds and their methyl-substituted analogues are easily recognized by mass spectrometry because of their abundant McLafferty rearrangement ions. Table VII shows the series of compounds with alkyl substitution of butyl or pentyl only as tentatively identified because authentic standards are not available for the multiplicity of isomers possible. Masuda et al (1980) has shown that 5-substituted 2,3-dimethylpyrazines can arise from the reaction of 2,3-dimethyl-

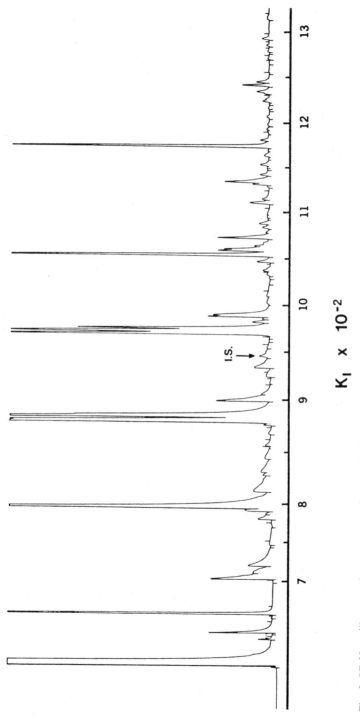

Fig. 5. SE-30 capillary gas chromatography-flame ionization detector responses to the basic fraction of vacuum-steam distilled toasted oat groats. Peak correlations are detailed in Table VII.

TABLE VII
Gas Chromatography-Mass Spectrometry Identifications of the Basic Fraction
of Toasted Oat Groats

Kovat's Retention Index[a]	Ions[b]	Compounds[c]
708	*80*,53,51,52	Pyrazine
711	*85*,58,57,84	Thiazole
719	*79*,52,51,50	Pyridine
782	58,*99*,57,59	2-Methyl thiazole
794	*99*,71,72,45	4-Methyl thiazole
799	*93*,92,78,66	2-Methyl pyridine
799	*94*,67,39,53	Methyl pyrazine
826	*111*,55,43,68	2,4,5-Trimethyl oxazole
829	*99*,71,72,45	5-Methyl thiazole
836	*93*,65,39,66	3/4-Methyl pyridine
860	*113*,72,71,40	2,4-Dimethyl thiazole
879	106,*107*,79,78	2-Ethyl pyridine
884	*108*,42,39,81	2,5-Dimethyl pyrazine
884	*108*,42,39,40	2,6-Dimethyl pyrazine
888	107,*108*,80,52	Ethyl pyrazine
891	*108*,67,42,80	2,3-Dimethyl pyrazine
902	*106*,52,53,79	Vinyl pyrazine
907	*113*,71,85,45	4,5-Dimethyl thiazole
940	98,*113*,71,45	5-Ethyl thiazole
947	92,*107*,106,65	2,4-Dimethyl pyridine
950	*124*,68,109,93	2-Methyloxy,3-methyl pyrazine (I.S. @ 0.01 ppm)
956	124,*139*,96,55	2-i-Propyl,4,5-dimethyl oxazole*[d]
974	121,*122*,39,94	2-Ethyl,6-methyl pyrazine
976	*122*,121,42,39	Trimethyl and 2-ethyl,5-methyl pyrazine
980	121,*122*,67,80	2-Ethyl,3-methyl pyrazine
983	94,*122*,121,107	*n*-Propyl pyrazine
985	*127*,112,71,85	2-Methyl,5-ethyl thiazole
989	*120*,52,54,51	2-Methyl,5-vinyl pyrazine
993	119,*120*,67,79	i-Propenyl pyrazine
1002	119,*120*	2-Methyl,3-vinyl pyrazine
1002	43,*122*,79,80	Acetyl pyrazine
1008	78,79,*121*,77	2-Propyl pyridine
1030	126,73,*141*,72	2-i-Propyl,4-methyl thiazole
1034	92,*121*,65,49	C$_3$ Pyridine*
1038	94,121,*136*,43	*n*-Butyl pyrazine
1050	119,*120*,67,39	2-(t,1-Propenyl) pyrazine
1057	135,*136*,42,39	2,5-Dimethyl,3-ethyl pyrazine
1063	*136*,135,121,56	2,3-Diethyl pyrazine
1064	135,*136*,54,42	2,3-Dimethyl,5-ethyl pyrazine
1068	121,135,*136*,56	2,5-Diethyl pyrazine
1068	108,*136*,40,67	2-Methyl,6-propyl pyrazine
1078	133,*134*,93,42	C$_4$-Monoene pyrazine*
1084	133,54,*134*	C$_4$-Monoene pyrazine*
1084	*141*,126,113,72	C$_4$-Sub thiazole*
1090	*134*,52,133,93	C$_4$-Monoene pyrazine*
1094	*133*,64,63,104	2-Me benzoxazole*
1119	108,135,*150*	2-Butyl,5-methyl pyrazine
1119	119,*134*,133	C$_4$-Monoene pyrazine* (labile Me)
1123	108,107,67,66,*150*	2-Butyl,6-methyl pyrazine
1140	*150*,135,56,149	2,3-Diethyl,5-methyl pyrazine
1140	149,*150*,135,39	2,5-Diethyl,3-methyl pyrazine

(*continued on next page*)

TABLE VII (continued)

$K_I{}^a$	Ions[b]	Compounds[c]
1142	122,121,42,107	Dimethyl propyl pyrazine*
1151	*150*,135,149,122	? Pyrazine
1163	94,147,*148*,107	Pentenyl pyrazine*
1163	*148*,52,147,53	Trimethyl, vinyl pyrazine*
1173	134,133,66,39	? Pyrazine
1179	122,121,45,123	Dimethyl,i-butyl pyrazine*
1184	93,106,120,65,*149*	2-Pentyl pyridine
1192	122,121,149,53,*164*	2,3 Dimethyl,5-butyl pyrazine
1219	149,133,164,148	?
1237	108,121,39,66,163	2-Methyl,3-pentyl pyrazine
1239	108,41,121,136,*162*	2-Methyl,?-pentenyl pyrazine*
1246	*146*,63,64,61	2-(2^1-Furyl) pyrazine
1248	93,92,*149*,106	3/4-Pentyl pyridine
1256	117,132,118,*147*	Methyl 1,2,3,5-tetrahydroquinoline*
1259	118,117,*147*,91	Methyl 1,2,3,5-tetrahydroquinoline*
1287	93,106,134,65,149	C_5/C_6 Mono-sub pyridine*
1294	107,120,134,106	Methyl, C_5 alkyl pyridine*
1296	122,121,135,42,(*178*)	Dimethyl, pentyl pyrazine*
1315	*160*,92,63	2-(2^1-Furyl), methyl pyrazine*

[a] K_I index corresponding to Fig. 5.
[b] Four major ions in decreasing abundance, with M^+ in italics.
[c] Identification based on authentic K_I and mass spectra (MS) or literature MS and appropriate K_I unless noted.
[d] * = Tentative identification, interpretative only.

5,6-dihydropyrazine and aldehydes or ketones in sodium ethoxide/ethanol solutions. Severe heat treatment processes, such as toasting, could possibly cause the aldehydes and ketones resulting from oat lipid degradation to react with dihydropyrazine intermediates and introduce longer-chain alkyl substituents.

As a model to test the validity of this hypothesis, we attempted to show that the most abundant aldehyde in oatmeal, hexanal, could be incorporated into pyrazine compounds under conditions in which Maillard reactions commonly occur. In a solution of 10% water in diethylene glycol, 10 mmol each of D-glucose, L-glutamic acid, ammonium hydroxide, and hexanal were stirred under reflux in a 130°C silicone oil bath for 18 hr. The reaction mixture was diluted with water and worked up as an isolation of a basic fraction from an aqueous flavor distillate. GC-MS analysis identified the expected lower alkyl pyrazines, 2-methyl, 2-ethyl, 2,3-dimethyl, trimethyl, and 2,5-dimethyl. Also, at longer retention times ($K_I > 1,300$), a series of pyrazines with M^+ ions at 164 (hexyl) and 178 (hexyl, methyl) were found, with three isomers of the latter present. All contained base peaks of M/e 94 or, in the latter set of isomers, 108. From characteristic pyrazine fragmentation patterns, it is reasonable to assume that hexylpyrazine and the three possible hexylmethylpyrazine isomers were formed in the model system. These preliminary experiments suggest that in the course of toasting, Maillard intermediates can react with longer-chain alkyl aldehydes from lipid oxidation and result in the formation of corresponding long-chain pyrazines.

The relatively large amount of 2-pentylpyridine further substantiates that these processes are taking place during toasting. Buttery et al (1977) identified 12 pyridines in roasted lamb volatiles. They postulated that 2,4-decadienal could condense with ammonia to form an aldimine. This could be followed by ring closure and oxidation to pyridines through 1,2-dihydropyridine intermediates. One of the components isolated from lamb was 2-pentylpyridine. As already mentioned in the discussion of oatmeal flavor and toasted oat neutrals, 2,4-decadienal isomers are formed in relatively large quantities. This would suggest a similar mode of alkyl pyridine formation in oat systems.

These data imply that in oats, browning reactions that form toasted, nutty flavor are not developed exclusively from a classic Maillard system utilizing amino acids and reducing sugars. The reaction of Maillard intermediates with lipid oxidation products also potentially plays a role in the formation of flavor compounds in oats under toasting conditions.

The neutral flavor fraction of toasted oat groats, representing brown, burnt-grain sensory impressions, contains even more components than the basic fraction. More than 150 components were noted in high-resolution capillary GC separations. GC-MS studies of the major components have been completed (Table VIII). These determinations show a major shift in the type of components present after toasting compared with the components (neutral volatiles) present in untoasted oat groats and oatmeal. After toasting, reducing sugar degradation

TABLE VIII
Gas Chromatography-Mass Spectrometry Identifications of the Major Neutrals
from Toasted Oat Groats

Kovat's Retention Index (K_I)	Ions[a]	Compounds[b]	Percentage[c]
631	44,43,41,58,*86*	Isopentanal	0.16
669	44,56,41,29	Pentanal	0.99
756	55,42,70,41	1-Pentanol	9.38
774	56,44,57,41	Hexanal	5.90
814	*96*,95,39,38	Furfural	21.05
860	*98*,53,81,97	Furfuryl alcohol	0.32
867	43,58,71,59,*114*	2-Heptanone	1.54
878	70,44,43,57,*114*	Heptanal	1.60
885	95,*110*,39,43	Acetyl furan	0.38
932	*106*,77,105,51	Benzaldehyde	6.40
938	*110*,109,53,51	5-Methyl furfural	9.19
970	81,98,43,*140*,	Furfuryl acetate	1.10
979	81,82,*138*,53	2-Pentyl furan	0.47
981	95,91,*124*,60	Propionyl furan	0.86
984	81,*110*,79,67	2,4-Heptadien-1-al	0.37
1002	119,91,63,64,*120*	*p*-Tolualdehyde	0.20
1013	91,92,65,*120*	Phenylacetaldehyde	0.62
1018	55,43,111,41,*126*	3-Octen-2-one	0.53
1083	57,41,56,55	Nonanal	0.83
1085	81,67,53,41,*124*	2,4-Octadien-1-al	0.37
1115	55,43,125,97,*140*	Nona-3-ene-2-one	0.23
1135	70,41,55,53,*138*	(E,Z)-2,6-Nonadiene-1-al	0.52
1185	57,43,55,41,*156*	Decanal	0.40
1272	81,67,68,152	(Z,Z)-2,4-Decadien-1-al	0.88
1290	81,67,68,*152*	(E,Z)-2,4-Decadien-1-al	2.58

[a] Largest mass spectra (MS) ions in decreasing abundance, with $M^{+\bullet}$ in italics.
[b] All identifications based on matching authentic compound MS and K_I.
[c] Percentage total area of isolated neutral volatiles.

products (i.e., furans) become much more abundant. Additionally, an increase in 2,4-decadienals can be noted. The data show that in the more severe heat treatment of toasting groats, Maillard browning reaction, Amadori rearrangements, and extensive lipid oxidation take place to provide many of the volatiles present in the neutral flavor fraction. No one compound is considered to be the most important flavor contributor; rather, the combined impact of several is suspected to be critical to toasted oat flavor. Other, as yet unidentified, minor components may also be quite important.

IV. OAT FLAVOR STABILITY STUDIES

Oat groats and oatmeal are generally accepted as stable food systems from an oxidative standpoint. Conversely, processed oat products often develop rancidity quite quickly, as previously observed by Martin (1958) in toasted oat flakes. Fritsch and Gale (1977) noted the presence of rancid odors in 12-week-old, oat-based, ready-to-eat cereals and correlated their appearance with hexanal concentrations of 5–10 ppm. In prior sections we have reported that carbonyls including hexanal, pentanal, 2,4-decadienal, and benzaldehyde are present in the volatiles of cooked oatmeal. The appearance of these oxygenated constituents in heat-processed oat products is consistent with oat lipid oxidation, and the highly unsaturated fat content (4–11%) of oats is usually cited as the potential source of rancidity (Sahasrabudhe, 1979). The contribution from enzymatic activity is expected to be negligible because of the very low moisture content and the heat treatments received by these cereal systems.

More recently, Heydanek and McGorrin (1981a) described the volatile components in noticeably rancid oat groats. The groats were prepared by boiling for 30 min in distilled water, draining, and freeze-drying overnight, followed by storage in polyethylene bags at ambient temperatures. Within one week, detectable rancidity odors were observed, and, when analyzed at three weeks of age, the oats had a pronounced "old oil, rancid, old chicken fat" aroma.

To define the volatile components causing rancidity, the rancid oat groats were vacuum-steam distilled. Figure 6 shows the capillary SE-30 separation obtained from this volatile isolation. As detailed in Table IX, a total of 45 compounds were identified, including 24 aldehydes, ketones, and alcohols. The most abundant volatiles found were hexanal, pentanal, 1-pentanol, and 3,5-octadien-2-one. All of the major components have been reported to occur in some form of oxidized lipid system (Forss, 1972). Consistent with the observation of Fritsch and Gale (1977), hexanal is the most abundant volatile produced; its concentration was estimated at 10–15 ppm in the rancid groat sample. However, the sensory characterization of the rancid odor as "old chicken fat" would suggest that the fatty, deep-fried odors of the 2,4-dienals are also important. In comparison, volatile isolations of sound oat groats provided compositionally different GC profiles.

Normally stable dried oat groats were also observed to generate an extremely rancid sensory character under a different set of processing conditions. Oat groats heated in a stream of air at 218° C for contact periods up to 2 min were found to produce an extremely green, rancid, old-oil aroma character after only five days of storage at ambient temperatures. Overall, the GC profile for rancid heated oat groats shows unique and higher levels of oxygenated volatiles, which

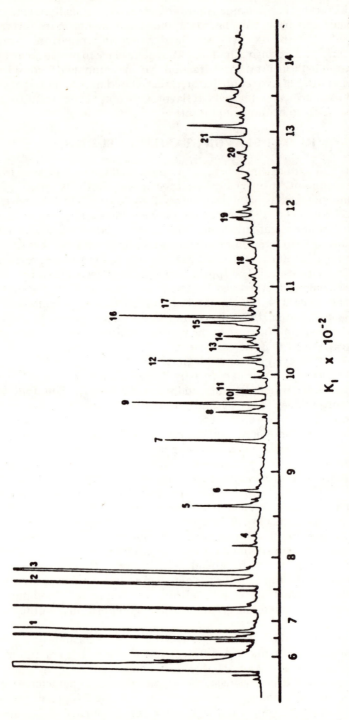

Fig. 6. SE-30 capillary gas chromatography-flame ionization detector responses of rancid oat groat volatiles. Peak identifications correspond to those listed in Table IX.

$$K_I \times 10^{-2}$$

TABLE IX
Gas Chromatography-Mass Spectrometry Identifications of Volatiles
Obtained from Vacuum Steam Distillation of Rancid Oat Groats

Peak[a]	Kovat's Retention Index (K_I)[b]	Compound[c]	Autooxidation Source[d]
	631	2-Me-butanal	
	642	Benzene	
1	669	Pentanal	L, A
	710	Pyrazine	
	718	Pyridine	
	748	Toluene	
2	756	1-Pentanol	L
3	776	Hexanal	L, Ln, A
	813	Furfural	
4	825	(E)-2-Hexen-1-al	Ln
5	859	1-Hexanol	L
	867	2-Heptanone	
6	876	2,4-Hexadiene-1-al	Unknown
6	877	Heptanal	O, L
	886	2,5-deMe pyrazine	
	889	2-Heptanol	
7	929	(E)-2-Heptene-1-al	L, A
7	931	Benzaldehyde	
8	960	1-Heptanol	L, O
9	970	1-Octen-3-ol	L
9	972	2-Octanone	
10	974	(Z,E)-2,4-Heptadien-1-al	Ln
11	981	Octanal	O, L
11	983	(E,E)-2,4-Heptadien-1-al	Ln
12	1015	3-Octene-2-one	A
13	1033	(E)-2-Octene-1-al	L, A
	1038	o-Tolualdehyde	
	1039	Acetophenone	
	1040	m-Tolualdehyde	
14	1045	(E,E)-3,5-Octadien-2-one	Ln
	1058	o-Cresol	
15	1061	1-Octanol	O
16	1068	(Z,E)-3,5-Octadien-2-one	Ln
17	1083	1-Nonanal	O
	1094	Isophorone	
18	1135	(E)-2-Nonen-1-al	
	1160	1-Nonaol	
	1062	(Z,E)-2,4-Nonadien-1-al	Unknown
	1163	Napthalene	
	1185	Decanal	
19	1188	(E,E)-2,4-Nonadien-1-al	L
20	1270	(Z,E)-2,4-Decadien-1-al	L, A
	1288	2-Me Naphthalene	
21	1291	(E,E)-2,4-Decadien-1-al	L, A
	1324	γ-Nonalactone	

[a] See corresponding number in Fig. 6.
[b] K_I for the SE-30 capillary system (Heydanek and McGorrin, 1981b).
[c] All compounds listed have K_I values and mass spectra identical to authentic standards.
[d] Per Murray et al (1976) or Forss (1972); O = oleic, L = linoleic, Ln = Linolenic, and A = archadonic.

were not present in the control sample. High-resolution SE-30 capillary GC-MS headspace analysis of 18-day-old groats that were heated at 218°C for 120 sec gave an abundance of aldehydes and alcohols, of which hexanal, pentanal, 3-methyl-1-butanol, and pentyl furan comprised 85–90% of the total GC area (Table X). Hexanal was again found to be the dominant volatile at a 5-ppm concentration. Two unusual compounds were identified in the headspace and can be attributed to a large excess of hexanal. The aldol self-condensation product, 2-butyl-2-octenal, was identified, as well as the corresponding hexanal-pentanal aldol condensation product, 2-propyl-2-octenal (Fig. 7). These aldehyde-derived products, as well as the pentanal aldol, 2-propyl-2-heptenal, and the other possible hexanal-pentanal aldol cross-condensation product, 2-butyl-2-heptenal, have also been observed in soy bran flour (R. J. McGorrin, *unpublished observation*, 1982). These unusual compounds are identified here for the first time, and their mass spectral fragmentation data are outlined in Table XI. In the context of oat rancidity, these components would be expected to contribute additional odors to the composite sensory impact. Additionally, oxygenated components such as (E)- and (Z)-3-octene-2-one, (E,E)- and (E,Z)-3,5-octadien-2-one are associated with lipid oxidation and can influence the rancid "old chicken fat" identity. Two compounds were tentatively identified on the basis of mass spectral fragmentation patterns as the epoxides 1,2-epoxydecane and 1,2-epoxyoctane. These epoxides are indicators of oleic and linoleic acid oxidation.

The oil composition of several common strains of oat groats has been determined by Sahasrabudhe (1979). The fatty acid composition of oat lipids is highly unsaturated, being comprised of an average of 36% oleic (18:1), 37% linoleic (18:2), 2.6% linolenic (18:3), and 2.2% C_{20}/C_{22} acids. The observed aldehyde, unsaturated aldehyde, and ketone volatiles, which play a dominant role in the odor and flavor of rancid oat groats, can be derived from the major unsaturated fatty acids as random oxidative breakdown products (Murray et al, 1976). Figure 8 summarizes the oxidative pathways and representative rancid oat volatiles formed from linoleic, linolenic, and oleic acids. In both cases where oat groats were subjected to boiling/freeze-drying or dry heat treatment, rancidity only appeared after several days of storage, which is consistent with an oxidative mechanism. Additionally, oat groats that were boiled for 30 min would not be expected to have residual enzyme activity.

Having established the rancid behavior of heated oat groats, we undertook a 23-day storage study to correlate the amount of headspace hexanal and pentanal with the degree of rancidity relative to fresh groats as determined by sensory evaluation. Samples of fresh oat groats were heated in a stream of hot air at 218°C for 0–2 min at 0.5-min intervals. The slightly expanded groats were then stored in glass jars at 35°C.

Odor descriptions by a 10-member trained laboratory panel were obtained simultaneously with GC analyses at three- to four-day intervals. The oat groat processing variables and a blind control were graded against a control sample of unheated groats. Raw data were analyzed by Wilcoxon Signed Rank Tests for nonparametric differences to yield an average flavor score as well as statistical confidence limits (Heydanek et al, 1979). As storage time increased, treated oat groats became typical of a highly rancid oil system. The onset of rancidity was related to the length of heat treatment.

TABLE X

Gas Chromatography-Mass Spectrometry Identifications of Headspace Volatiles from Rancid (218°C) Dry-Heated Oat Groats 18 Days Old

Kovat's Retention Index $(K_1)^a$	MS Ions[b]	Component	Probable Aroma Significance[c]
500	43,42,41,*72*	Pentane	
538	*72*,43,41,57	Isobutanal	
580	57,56,41,*86*	3-Methylpentane	
623	62,49,64,*98*	1,2-Dichloroethane	
633	41,44,39,58,*86*	Isopentanal	+
643	57,41,58,39,*86*	3-Pentanone	
644	*78*,77,51,39	Benzene	
674	44,41,58,57,*86*	Pentanal	+++
689	81,*96*,53,67	Ethyl furan	
700	43,57,71,*100*	(Heptane)	
750	91,*92*,65,63,39	Toluene	
770	70,55,42,41 (88)	3-Methyl-1-butanol	+
782	56,44,41,57 (100)	Hexanal	+++
800	43,57,71,85,*114*	Octane	
818	55,70,42,57 (88)	1-Pentanol	
843	55,41,69,83,98	Alcohol ?	
876	58,43,71,59,*114*	2-Heptanone	
883	56,81,43,69,*124*	2-*n*-Butyl furan	+
886	70,44,55,57 (114)	Heptanal	
892	71,41,55,58,56 (128)	1,2-Epoxyoctane	+
900	57,43,71,85,*128*	Nonane	
935	*106*,77,105,51	Benzaldehyde	+
943	83,55,41,39,*112*	Ethyl cyclohexane	
976	43,58,57,71,*128*	2-Octanone	
979	81,82,53,*138*	Pentyl furan	+
985	57,55,43,41,44 (128)	Octanal	+
1015	55,41,111,97,*126*	E-3-Octene-2-one	+
1045	95,41,39,*124*,109	E,E-3,5-Octadien-2-one	+
1050	55,43,111,97,*126*	Z-3-Octene-2-one	+
1061	70,55,41,57 (130)	1-Octanol	
1068	95,41,39,*124*	Z,E-3,5-Octadien-2-one	+
1078	135,*136*,108	3,5-DiMe-2-Et pyrazine	+
1087	71,41,55,57 (156)	1,2-Epoxydecane	
1098	55,56,69,41,*154*	1-Undecene	
1128	55,70,39,41 (144)	1-Nonanol	
1163	*128*,127,129,126	Napthalene	
1188	81,41,39,67,*138*	2,4-Nonadienal	+
1239	70,41,55,43 (158)	1-Decanol	
1258	55,111,83,139,*168*	2-Propyl-2-octenal	++
1288	*142*,141	1-Methylnapthalene	
1291	81,41,95,*152*	2,4-Decadienal	
1352	55,41,111,83,139,*182*	2-Butyl-2-octenal	++
1387	119,*162*,91,147,105	*p*-Diisopropylbenzene	

[a] SE-30 capillary retention index.

[b] M^+ in italics if present and in parentheses if absent.

[c] Estimated on the basis of threshold and relative abundance. + = Above threshold, ++ = probably significant, +++ = very significant.

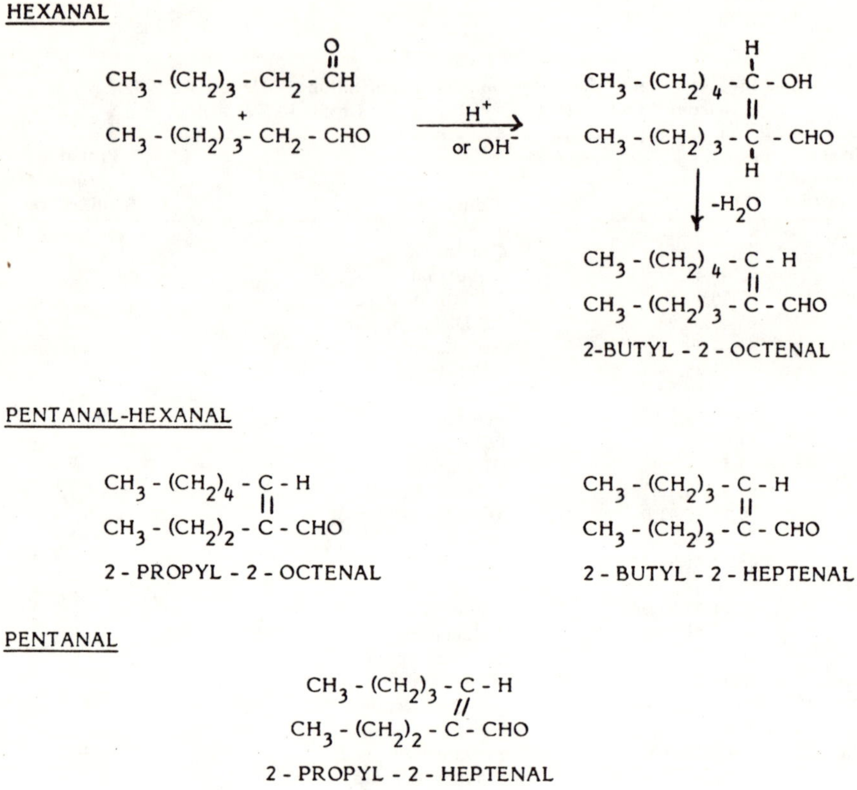

HEXANAL

$$CH_3 - (CH_2)_3 - CH_2 - \overset{\overset{O}{\|}}{C}H$$

$$\underset{+}{CH_3 - (CH_2)_3 - CH_2 - CHO} \xrightarrow[\text{or } OH^-]{H^+} \quad CH_3 - (CH_2)_4 - \overset{\overset{H}{|}}{\underset{\|}{C}} - OH$$

$$CH_3 - (CH_2)_3 - \overset{}{\underset{H}{C}} - CHO$$

$$\downarrow -H_2O$$

$$CH_3 - (CH_2)_4 - \overset{}{\underset{\|}{C}} - H$$

$$CH_3 - (CH_2)_3 - C - CHO$$

2-BUTYL - 2 - OCTENAL

PENTANAL-HEXANAL

$$CH_3 - (CH_2)_4 - \overset{}{\underset{\|}{C}} - H$$
$$CH_3 - (CH_2)_2 - C - CHO$$

2 - PROPYL - 2 - OCTENAL

$$CH_3 - (CH_2)_3 - \overset{}{\underset{\|}{C}} - H$$
$$CH_3 - (CH_2)_3 - C - CHO$$

2 - BUTYL - 2 - HEPTENAL

PENTANAL

$$CH_3 - (CH_2)_3 - \overset{}{\underset{/\!/}{C}} - H$$
$$CH_3 - (CH_2)_2 - C - CHO$$

2 - PROPYL - 2 - HEPTENAL

Fig. 7. Aldehyde self-condensation products identified in rancid dry-heated oat groats.

TABLE XI
Mass Spectral (MS) Data for Aldol Condensation Products

Compound	Kovat's Retention Index $(K_1)^a$	MS Ions[b]
2-Propyl-2-heptenal	1169	55 (100), 41 (86), *154* (64), 83 (60), 125 (58), 69 (47), 81 (47), 43 (44)
2-Butyl-2-heptenal	1266	55 (100), 41 (88), 125 (78), 83 (58), *168* (49), 43 (41), 67 (39), 97 (38), 139 (9)
2-Propyl-2-octenal	1258	55 (100), 111 (93), 41 (84), *168* (55), 43 (55), 83 (54), 81 (51), 139 (37), 125 (19)
2-Butyl-2-octenal	1352	55 (100), 41 (87), 111 (62), *182* (52), 83 (50), 43 (46), 139 (41), 125 (31)

[a] Using SE-30 headspace-purging gas chromatography conditions.
[b] Relative intensity shown in parentheses; M^+ in italics.

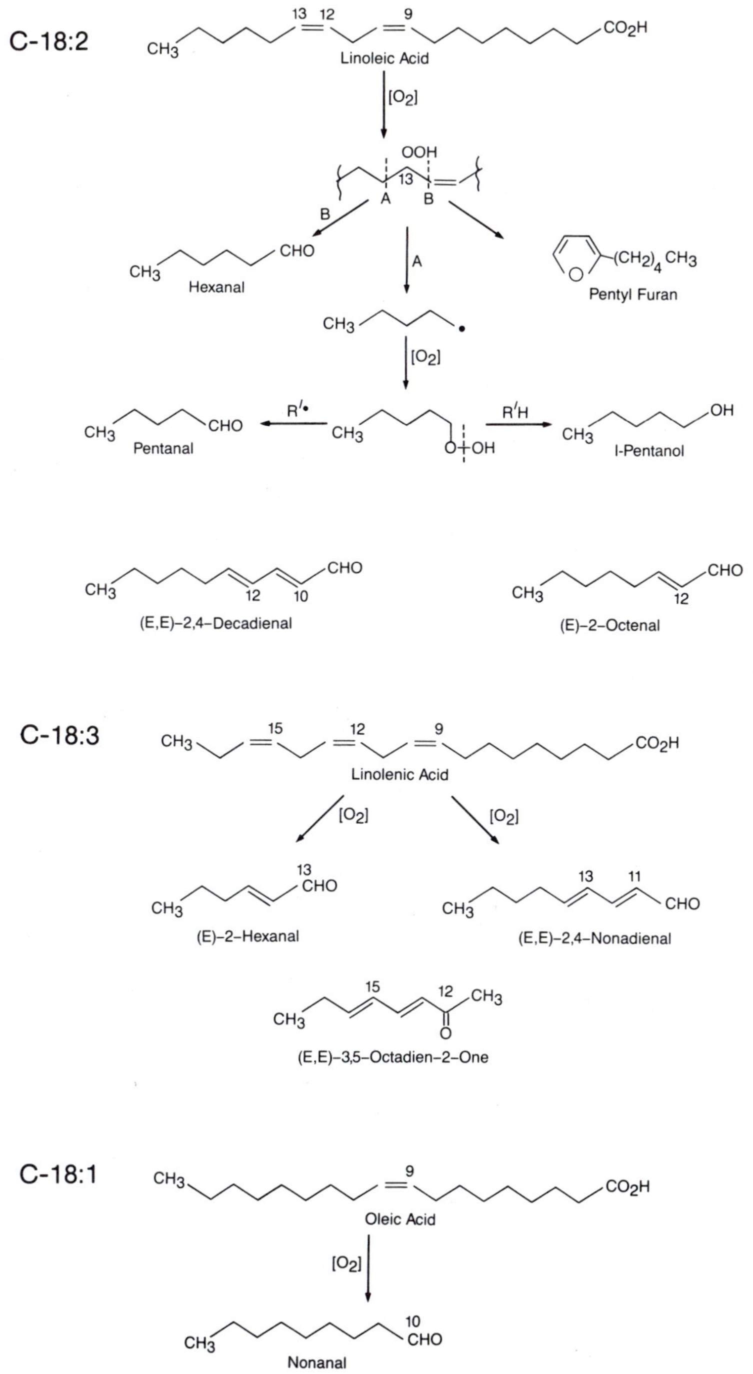

Fig. 8. Oxidative pathways for rancid oat groat volatiles derived from fatty acids.

It is readily apparent from Fig. 9 that a significant amount of hexanal was generated after the first day of storage for the 120-sec dry-heat treatment sample. This is consistent with rapid oxidative breakdown of linoleic acid hydroperoxides in surface oat lipid, which is accelerated by greater heat exposure. Overall, a greater number of new oxygenated volatiles appeared in the 120-sec sample during the 23-day storage period, indicating that extensive oxidation was occurring. Table XII shows the relative abundance of the nine

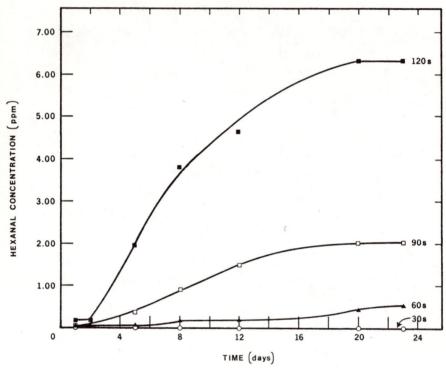

Fig. 9. Hexanal concentration in the headspace of dry-heated oat groats as a function of storage time at 25° C after heating at 218° C for indicated time, O − 30 sec, ▲ − 60 sec, □ − 90 sec, and ■ − 120 sec.

TABLE XII
Flavor-Significant Components Generated in Rancid Oat Groats (218°C for 120 Sec.)

Compound	Abundance (ppm)			Odor Threshold (ppm)[a]
	Day 1	Day 12	Day 23	
Hexanal	0.1	4.6	6.3	0.005
Pentanal	0.01	0.5	0.6	0.012
3-Methyl-1-butanol	···	0.1	0.13	0.04
Pentyl furan	···	0.06	0.06	1.0
1-Octanol	···	0.04	0.04	0.13
3-Octene-2-one	···	0.03	0.05	···
1-Pentanol	···	0.02	0.08	0.120
2-Butyl-2-octenal	···	0.02	0.06	···
3,5-Octadien-2-one	Trace	0.01	0.04	···

[a] Data from Fazzalari (1978).

major headspace volatiles and their corresponding odor thresholds. Hexanal is present well above threshold and is expected to provide a major impact. Although pentanal is also present in high concentrations, its sensory effect is lower. However, since the rancid aroma is not green but resembles old oil and old chicken fat, the other oxygenates must also provide significant odor contributions.

A correlation was made between the levels of hexanal and pentanal and the progression of perceived rancidity, as determined by a sensory panel. The levels of hexanal for the four process variables over 20 days of storage are displayed graphically in Fig. 9. Overall, hexanal headspace levels increased with storage time, and the processing time had a pronounced effect on the oxidation rate. In the 90- and 120-sec samples, a two-day induction period is apparent, followed by a very rapid rate of hexanal production, which achieves a maximum rate at eight days and then reaches a plateau. Pentanal levels also increased relative to the control but at a slower rate. However, the same trends are also evident.

The headspace GC analysis is supported by sensory data (Table XIII) that show increasingly negative flavor scores. An attempt to plot the aldehyde concentration versus flavor score, either as a linear or logarithmic plot, did not provide a linear relationship, indicating that other chemical factors are also important.

To understand more about the mechanism of oat rancidity, model systems were used to examine factors affecting oat oxidation off-flavor generation. The protocol for these experiments is as follows: (1) the control was oat groats heated for 120 sec at 218°C; (2) variable A was oat groats, extracted 18 hr with *n*-hexane-ethanol azeotrope, vacuum-dried, then heated as in (1); (3) variable B was oat groats heated as in (1) and then extracted 18 hr under conditions in (2) and vacuum-dried. The three samples were stored in glass jars at ambient temperature and evaluated after 40 days.

As expected, heated oat groats stored 40 days at ambient temperature were considerably rancid. Similarly, oat groats defatted before heat treatment were just as rancid after 40 days because only 30% of the total extractable fat was removed before processing, leaving a substantial amount of the internal lipid available for oxidation after heat expansion of the groat. The most significant finding was that the stability of groats defatted after heat treatment was greatly increased. This is consistent with the removal of all available surface lipid that participates in oxidative fragmentation to generate aldehydes and other volatiles associated with rancidity. High-resolution GC-MS of the headspace over oat groats defatted after heat processing shows a major reduction in overall

TABLE XIII
Sensory Scores of Dry-Heated Groats as a Function of Age

Treatment Time at 218°C (sec)	Flavor Score after Days at 35°C[a]			
	1	5	8	15
30	0	0	0	0
60	−1	−2	−1	−9
90	−4	−11	−12	−19
120	−13	−14	−17	−24

[a] Flavor score difference from control on a 30-point scale. Scores of −10 or worse indicate unacceptable magnitude of off-flavors present.

oxygenated volatiles (Table XIV), as compared with those compounds found in Table X. Of significance is the presence of hexanal at less than 2 ppm in variable B after 40 days. This finding provides additional evidence that oat lipid oxidation is a major factor in volatile production in rancid oats and provides an important mechanistic clue for further research on oat stability.

As evidenced by the extreme oxidative deterioration of oat lipids, a high turnover of linoleic and other unsaturated fatty acid peroxides occurs in heated oat groats; this is also supported by the relatively high pentanal and hexanal concentrations. Oxidative processes are generally recognized as surface phenomena, and the surface of the oat groat can provide a matrix in which molecular oxygen can interact with unsaturated lipids. Thus, it is plausible to assume that when the structure of the oat groat has been physically altered by expansion, the increased area of exposure apparently results in destabilization of oat lipid. Further, one can also postulate that the destruction or immobilization of natural antioxidants during severe heat treatment also contributes to instability.

The preceding studies were undertaken to define the causes and possible mechanism of rancidity in a noticeably oxidized oat groat system. Hexanal, pentanal, 3,5-octadien-2-one, and 2,4-decadienal would seem to be good indicators for the development of oxidative rancidity from oat oil fatty acids. Furthermore, rancidity in oat-containing foods is a function of the concentrations of these oxidative components rather than their absolute presence. Knowledge of the production mechanism of rancid volatiles and its relationship to the time and temperature of processing establishes a baseline of potential stability problems as well as flavor chemical interactions that may occur in oat-based products.

TABLE XIV
Gas Chromatography-Mass Spectrometry Identifications
of Headspace Volatiles from Dry-Heated Oat Groats
Extracted After Processing

Kovat's Retention Index (K_I)	Component
671	Cyclohexene
683	Trichloroethylene
774	Hexanal
794	Tetrachoroethylene
947	m-Et toluene
949	p-Et toluene
955	1,3,5-Me_3-benzene
964	o-Et toluene
979	1,2,4-Me_3-benzene
1017	C_4-Benzene
1063	C_4-Benzene
1100	Undecane
1200	Dodecane
1216	Branched alkane
1276	1-Me-napthalene
1300	Tridecane
1400	Tetradecane

V. FUTURE OAT FLAVOR RESEARCH

The work presented in this chapter represents the groundwork in terms of the chemical components of oat flavor from various sources; the next steps in oat flavor research, however, are more difficult. The composition of the volatile flavor fractions does give some clue to their origin. Little inherent flavor is present in oat groats. Flavor volatile precursors, and the routes to their specific formation in the oat flavor complex, are subjects for further study. Changes in fatty acid composition and content and lipid organization as a function of processing and flavor character need to be studied before we can understand fully the obviously complex mechanisms involved in neutral flavor complex formation, and much the same can be said of the free amino acids and reducing sugars of oats. The basic fraction obviously derives from Maillard reactions, and the importance of these precursor components is well documented.

In addition, once the volatiles are formed, specific correlations must be derived between composition and concentration of individual compounds relative to their effect on overall flavor perception. This area of objective-subjective flavor component correlation is a difficult one to research and one that requires extensive analytical and statistical data development. However, certain generalizations can be made from the data at hand, such as the correlation of off-flavor in stored oat systems with elevated concentrations of hexanal and 2,4-decadien-1-al. Even more specific is the direct correlation of the toasted flavor of oats with the amount and type of nitrogen heterocycles present. We are certainly closer to understanding their role in oat flavors, but a long way from being able to control their formation for specific flavor contribution.

Since this is an opportunity to take a somewhat futuristic look at oat flavor, we would mention that another possibility exists in terms of breeding oats for certain flavor attributes. It is known that lipid content and protein are heritable factors. If one examines more closely the actual fatty acid, amino acid, and reducing sugar composition, it may be possible to select certain flavor potential characteristics based on these heredity factors. With the current information on oat flavor, this is only in the realm of speculation, but it could become a reality if further research effort is made in this direction.

VI. SUMMARY

Oat flavor has been shown to be a complex, precursor-dependent, heat-induced collection of volatile flavor components. Oat groats, representing the precursors of commercial oat food products, are devoid of inherent flavor contributors. The well-known flavor of oatmeal and toasted oat flakes has been shown to be composed of both neutral and basic fraction flavor compounds. Nitrogen heterocycles, presumably formed from Maillard reactions, and lipid oxidation products are the key compositional types of flavor volatiles. Heat-induced reactions of precursors native to the oat groat are primarily responsible for the development of oat flavor during its normal processing into commercial food products.

Oat product flavor stability is dependent upon lipid composition and resistance to oxidation. Under certain processing conditions, oats are dramatically unstable, in contrast to the normally expected good stability of oatmeal, oat flour, and dried groats. Flavor instability of oats correlates directly with the appearance of low-molecular-weight lipid oxidation products and specifically with pentanal, hexanal, 2,4-decadien-1-als, and 3,5-octadien-2-ones.

LITERATURE CITED

AOKI, M., KOIZUMI, N., OGAWA, G., and YOSHIZAKI, J. 1981. Identification of the volatile components of buckwheat flour and their distributions of milling fractions. Nippon Skokuhin Kogyo Gakkai-Shi 24:476-481.

BADINGS, H. J. 1970. Cold-storage defects in butter and their relation to the autoxidation of unsaturated fatty acids (a thesis). Ned. Melk-Zuiveltijdschr. 24:147-257.

BIERMANN, U., and GROSCH, W. 1979. Bitter tasting monoglycerides from stored oat flour. Z. Lebensm. Unters. Forsch. 169:22-23.

BULLARD, R. W., and HOLGUIN, G. 1977. Volatile components of unprocessed rice (*Oryza sativa* L.). J. Agric. Food Chem. 25:99-103.

BUTTERY, R. G., LING, L. C., and CHAN, B. G. 1978. Volatiles of corn kernels and husks: Possible corn ear worm attractants. J. Agric. Food Chem. 26:866-869.

BUTTERY, R. G., LING, L. C., JULIANO, B. O., and TURNBAUGH, J. G. 1983. Cooked rice aroma and 2-acetyl-1-pyrroline. J. Agric. Food. Chem. 31:823-826.

BUTTERY, R. G., LING, L. C., and TERANISHI, R. 1980. Volatiles of corn tassels: Possible corn ear worm attractants. J. Agric. Food Chem. 28:771-774.

BUTTERY, R. G., LING, L. C., TERANISHI, R., and MON, T. R. 1977. Roasted lamb fat—Basic volatile components. J. Agric. Food Chem. 25:1227-1230.

BUTTERY, R. G., SEIFERT, R. M., and LING, L. C. 1975. Characterization of some volatile constituents of dry red beans. J. Agric. Food Chem. 23:516-520.

BUTTERY, R. G., XU, C., and LING, L. C. 1985. Volatile components of wheat leaves (and stems): Possible insect attractants. J. Agric. Food Chem. 33:115-117.

DANIELS, D. G. H., and MARTIN, H. F. 1961. Isolation of a new antioxidant from oats. Nature (London) 191:1302-1304.

DANIELS, D. G. H., and MARTIN, H. F. 1965. Antioxidants in oats: Diferulates of long-chain diols. Chem. Ind.:1763-1767.

DE LUMEN, B. O., STONE, E. J., KAZNIAC, S. J., and FORSYTHE, R. J. 1978. Formation of volatile compounds in green beans from linolenic and linoleic acids. J. Food Sci. 43:698-701.

ESSELMAN, W. J., and CLAGETT, C. O. 1974. Products of a linoleic hydroperoxide-decomposing enzyme of alfalfa seed. J. Lipid Res. 15:173-177.

FAZZALARI, F. A. 1978. Compilation of Odor and Taste Threshold Values Data. Am. Soc. for Testing and Methods, Philadelphia, PA.

FORSS, D. A. 1972. Odor and flavor compounds from lipids. Pages 181-258 in: Progress in Chemistry of Fats and Oils. R. T. Holman, ed. Pergamon Press, New York.

FRITSCH, C. W., and GALE, J. A. 1977. Hexanal as a measure of rancidity in low fat foods. J. Am. Oil Chem. Soc. 54:225-231.

HEIMANN, W., FRANZEN, K. H., RAPP, A., and ULLEMEYER, H. 1975. Radio gas chromatographic analysis of volatile aldehydes originating from soybean and oat lipoxygenase linoleic acid oxidation. Z. Lebensm. Unters. Forsch. 159:1-5.

HEIMANN, W., and SCHREIER, P. 1970. Formation of 9-hydroxy-10, 12- and 13-hydroxy-9, 11-octadecadienoic acid by lipoperoxidase from oats. Helv. Chim. Acta 53:2296-2304.

HEYDANEK, M. G., and McGORRIN, R. J. 1980. Oat flavor chemistry—Sensory properties of flavor isolates and composition of the basic fraction. AGFD No. 74. Abstr. of Papers, N. Am. Chem. Congr., 2d, Las Vegas, August. American Chemical Society, Washington, DC.

HEYDANEK, M. G., and McGORRIN, R. J. 1981a. Gas chromatography-mass spectroscopy investigations on the flavor chemistry of oat groats. J. Agric. Food Chem. 29:950-954.

HEYDANEK, M. G., and McGORRIN, R. J. 1981b. Gas chromatography-mass spectroscopy identification of volatiles from rancid oat groats. J. Agric. Chem. 29:1093-1096.

HEYDANEK, M. G., WOOLFORD, G., and BAUGH, L. D. 1979. Premiums and coupons as a potential source of objectionable flavor in cereal products. J. Food Sci. 44:850-855.

HRDLICKA, J., and JANICEK, G. 1964a. Carbonyl compounds in toasted oat flakes. Nature (London) 201:1223-1228.

HRDLICKA, J., and JANICEK, G. 1964b. Volatile amines as components of toasted oat flakes. Nature (London) 204:1201-1202.

KAMINSKI, E., WASOWICZ, E., and

PRZYBYLSKI, R. 1964. Volatile compounds in cereals: Separation and condensation for chromatographic analysis and description of their aromatic characteristics. Acta Aliment. Polonica 7:59-72.

LEU, K. 1974. Analysis of volatile compounds produced in linoleic acid oxidation catalyzed by lipoxygenase from peas, soy beans, and corn germs. Lebensm. Wiss. Technol. 7:98-100.

MAGA, J. A. 1974. Bread flavor. Crit. Rev. Food Technol. 5(1):55-147.

MAGA, J. A. 1978. Cereal volatiles: A review. J. Agric. Food Chem. 26:175-178.

MAGA, J. A. 1984. Rice product volatiles: A review. J. Agric. Food Chem. 32:964-970.

MARTIN, H. F. 1958. Factors in the development of oxidative rancidity in ready-to-eat crisp oatflakes. J. Sci. Food Agric. 12:817-820.

MASUDA, H., TANAKA, M., AKIYAMA, T., and SHIBAMOTO, T. 1980. Preparation of 5-substituted 2,3-dimethylpyrazines from the reaction of 2,3-dimethyl-5,6-dihydropyrazine and aldehydes or ketones. J. Agric. Food Chem. 28:244-246.

MURRAY, K. E., SHIPTON, J., WHITFIELD, F. B., and LAST, J. H. 1976. The volatiles of off-flavoured unblanched green peas (*Pisum sativum*). J. Sci. Food Agric. 27:1093-1099.

NOBLE, A. C., and NAWAR, W. W. 1975. Identification of decomposition products from autoxidation of methyl 4, 7, 10, 13, 16, 19-docosahexaenoate. J. Am. Oil Chem. Soc. 52:92-95.

SAHASRABUDHE, M. R. 1979. Lipid-composition of oats (*Avena sativa* L.). J. Am. Oil Chem. Soc. 56:80-85.

SHUKLA, T. P. 1975. Cereal proteins: Chemistry and food applications. Crit. Rev. Food Sci. Nutr. 6:383-428.

TRESSEL, R., BAHRI, D., HOLZER, M., and KOSSA, T. 1977. Formation of flavor components in asparagus. 2. Formation of flavor components in cooked asparagus. J. Agric. Food Chem. 25:459-466.

WANG, P., KATO, H., and FUJIMAKI, M. 1969. Roasted barley flavor—Basic fraction. Agric. Biol. Chem. 34:561-566.

WITHYCOMBE, D. A., LINDSAY, R. C., and STUIBER, D. A. 1978. Isolation and identification of volatile components from wild rice grain (*Zizania aquatica*). J. Agric. Food Chem. 26:816-822.

YAJIMA, I., YANAI, T., NAKAMURA, M., SAKAKIBARA, H., and HABU, T. 1978. Volatile flavor components of cooked rice. Agric. Biol. Chem. 42:1229-1233.

YAJIMA, I., YANAI, T., NAKAMURA, M., SAKAKIBARA, H., UCHIDA, H., and HAYASHI, K. 1983. Volatile compounds of boiled buckwheat flour. Agric. Biol. Chem. 42:729-738.

CHAPTER 13

OAT CLEANING AND PROCESSING

DONALD DEANE
EDWARD COMMERS
Carter-Day Company
Minneapolis, Minnesota

I. INTRODUCTION

The milling of cereal grains is civilization's oldest industry. Primitive people pounded grain upon flat stone or boulder, and for centuries developments were surprisingly slow. The rock on which the grain was crushed tended to become cup shaped, and the pounding stone oval or conical. These gradual changes foreshadowed the mortar and pestle, which were used in ancient Britain and by the aborigines of North America; they were also used in ancient Greece and Rome. The saddlestone, which followed, represented a distinct change from pounding to grinding: kneeling in front of a slab whose shallow concavity suggested the shape of a saddle, the operator, grasping each end of an oval stone having a flat undersurface, would, with a forward and backward rowing movement, grind the grain. Burr mills or stone mills gradually evolved in which the grain was ground between a stationary, circular, flat stone and a rotating stone, the distance between the stones determining the degree of grind.

The history of grain milling from the beginning of time until the last decades of the 19th century is surprisingly monotonous. Such changes as occurred in the industry were in motive power, or method of turning the grinding stones. The progression from human energy provided by slaves and others to energy provided by animals, the use of water and wind power, and finally the steam engine centered, apparently forever, around stone grinding mills. However, during the last decades of the 19th century, roller mills, using either iron or porcelain rolls, very gradually replaced stone mills; they were more efficient and produced more refined and better quality meals and flours.

Inventiveness was not confined to the milling of wheat; as early as 1840, a new device for hulling oats was produced in Camden, Maine. It was not a particularly ingenious machine, being similar to the hullers used in Scotland, where oatmeal in the form of porridge had been a staple of the diet for decades previously. However, its interest lies in the fact that it appears to be the first device intended to improve the oat milling process that was patented in the United States. Other patented developments for improving the cleaning of oats before milling

371

followed, but the first epochal invention in the field of oat milling did not appear until 1875, when Ferdinand Schumacher in Akron, Ohio, patented a cutting machine. Following testing and development, the outcome was a groat-cutting device that instead of converting the hulled kernels of oats (known as groats) into a coarse meal by crushing the grain between rollers or grinding them with burrs or millstones, sliced the kernels into three or four pieces with a minimum production of fine meal or flour. The slicing was achieved by feeding the groats into a hopper with a perforated bottom; the hopper had a free motion above a series of horizontal knives, so arranged that the groats falling endwise through the perforated bottom were sheared off as the hopper moved across them. *Steel-cut oatmeal*, as the product of the new machine was called, quickly became the favorite form of milled oats. Although requiring a longer time to cook than ground meal, it was less exposed (in those days of the open barrel for storage) to the action of the air and was less liable to become rancid.

The ability to cut or slice the oat kernels, rather than crushing them, eventually resulted in the use of roller mills to roll the cut pieces into flakes, generally known now as *rolled oats*. The practice of producing flakes or rolled oats rather than meal evolved on a serious commercial scale during the latter half of the 1800s and had the advantage of converting a much larger portion of the oat kernels into a more valuable product as compared with crushing, which resulted in the formation of a large percentage of fine flour. Also, rolled oats required less cooking time than steel-cut oatmeal, which had practically disappeared for table use by the start of this century. Steam was used as a "binder" to hold the groats intact while being subjected to the rolling pressure, and the application of steam immediately before flaking imparted the added advantage of partially precooking the oats, evolving into the "instant" oatmeal products. Apparently the first registration of these so-called instant cereal products occurred in 1877, with the symbol "3 minute" for application to "steam cooked cereal" (Thornton, 1933).

We have briefly traced the development of milling from the earliest times. During the last century, machine and process design advancements have progressed at a faster pace with the degree of improvement resulting in greater efficiency and better quality products. It is the purpose of this chapter to describe present-day practices and methods of oat cleaning and processing in which some of the basic principles developed by the early pioneers are still applied in a more sophisticated form.

II. OAT CLEANING AND PROCESSING—GENERAL STEPS

The general steps in oat processing are cleaning, hulling, steaming, and flaking (Fig. 1). In the cleaning step, dust, chaff, weed seeds, coarse grains, and other impurities are removed. In the hulling step, oats are graded by size and dried to permit efficient removal of the hull. The hulls are then abraded or knocked off the oats, the dehulled oat (groat) representing about 75% of the kernel. Then the groats are cut into two to four uniform pieces, steamed, and flaked for packaging.

The standard weight per bushel for oats is 32 lb (one Winchester or U.S. bushel equals 2,150.42 in.^{3}, or 35.24 L); the weight can vary from 22 to over 40 lb/bu. Oats fairly free of barley and other grains and weighing about 35–38 lb/bu are best for milling. Depending on the quality and test weight of the oats, 10–16 bu of oats is required to produce a "barrel" of rolled or flaked oats. A standard oat barrel in the United States is 180 lb (81.63 kg). A fair average test weight runs about 36 lb/bu for milling oats from the Central Plains to 40 lb/bu for oats grown in the western states and Canada. Shrinkage amounting to 4–6% of the purchased oat weight results from drying preparatory to milling. Field oats may contain 12% moisture and are dried down to 6.5%, or less, moisture. Rolled oats contain about 10% moisture and the residue oatfeed by-product about 7% (Loufek, 1944).

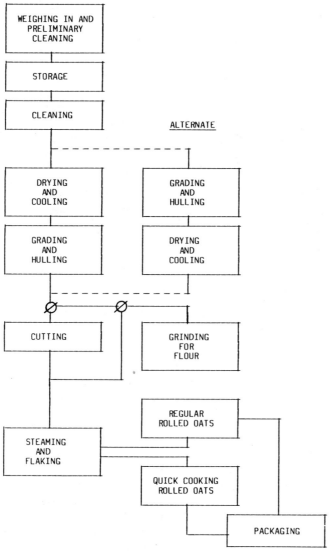

Fig. 1. Flow diagram of steps in oat cleaning and processing.

The following products typically result from milling oats, the percentage breakdown being dependent on the quality of the oats and the overall efficiency of the process: rolled oats, 45–60%; oat hulls, 24–27%; feed oats (doubles, thins, light), 10–20%; mixed grains and seeds (wheat, corn, barley, etc.), 2–3%; and fines (middlings), 2–5%.

III. INTAKE OF FIELD OATS AND PRELIMINARY CLEANING

The purpose of preliminary cleaning is to remove coarse field trash and objects that could damage conveying equipment and to remove dust, loose chaff, and other light impurities before storage.

Field impurities in oats vary widely depending on the location of the processing facility and where the oats are purchased. Some plants encounter impurities that other plants never see, and the preliminary cleaning is, therefore, undertaken on a local basis to suit the prevailing conditions.

When the oats are delivered to the mill, mechanical or electronic scales record the weight for record and inventory purposes. A rough, preliminary cleaning follows to remove coarse objects that could damage subsequent conveying equipment and to remove some of the dust, loose chaff, and other light impurities that cause dust hazards in the storage area and infestation during the storage period.

A receiving separator is used for this initial cleaning step. Generally, the receiving separators use one of two methods to remove the coarse and light impurities. The first method uses slightly inclined wire mesh or perforated sheet metal screens that are given a reciprocating or rotary motion. The perforation openings are selected to let the oats fall through while the coarse impurities are overtailed. The second method uses horizontal, slowly rotating, coarse wire-mesh reels or cylinders. The oats are either fed into the inside of the reel, where the oats fall through while the coarse objects are overtailed, or the oats are fed onto the outside of the rotating reel and pass through while the coarse objects are carried over and evacuated from the machine. The advantage of the latter method of feeding the reel is that the coarse material does not enter the inside of the scalping reel, so there is less chance of it becoming entangled and caught up inside the machine.

Most receiving separators, regardless of model type, incorporate an aspiration channel to remove light impurities from the oats before they leave the machine. Intake rates of the field oats arriving at the plant vary widely depending on the size and production output level of the plant; these can range from a low of 1,000 bu/hr at small mills to over 10,000 bu/hr at large facilities.

Figure 2 shows a receiving separator that is typical of screen-type models. Oats are fed onto the top coarse-screen deck, which is fitted with a perforated metal sheet that is 1 in. long by 1/2 in. wide (25.4 × 12.7 mm). Oats fall through these perforations onto the lower fine-screen layer, which is fitted with a triangular perforated sheet having sides measuring 9/64 in. (3.5 mm). Some fine impurities pass through the lower sieve deck while the oats overtail and pass into the aspiration channel for removal of light impurities, such as loose hulls and dust.

Figure 3 illustrates the working principles of a typical scalper-reel type receiving separator. Oats are feed along the full length of the horizontal main scalper reel, which is fitted with a $3/4 \times 3/4$ in. (19.05×19.05 mm) wire mesh screen rotating at 20 rpm. Coarse objects and long straws are carried over the top of the reel, and the oats pass through the wire mesh to an aspirating channel where light impurities are removed and conveyed to an integral dust-settling chamber. A conveyor then removes these light impurities out of the side of the machine. Coarse material and objects carried over the main scalper reel fall onto a rotating rescalper reel fitted with a $5/8 \times 5/8$ in. (15.9×15.9 mm) wire mesh to recover any oats riding on top of flat material such as paper. The coarse, rejected material falls into a receptacle on the floor adjacent to the machine for subsequent disposal.

The air volume required to aspirate the oats on receiving separators varies greatly according to the oat receiving rate and the size of separator used; it can range from a low of 1,000 ft^3/min (28.3 m^3/min) to a high of over 8,000 ft^3/min

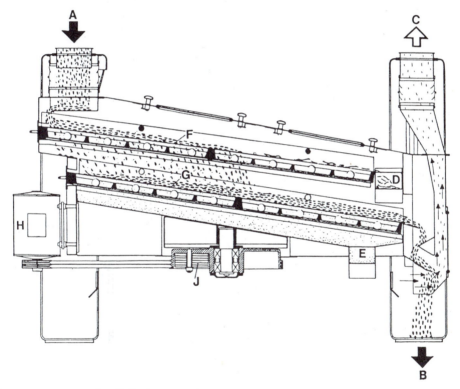

Longitudinal cross-section:

A Inlet of grain to be cleaned
B Outlet for cleaned product
C Connection for the aspiration of the machine
D Lateral outlet for coarse impurities (large kernels, strings, straw, etc.)
E Lateral outlet for fine screenings (light and broken kernels, sand, etc.)
F Coarse sieve
G Sand sieve
H Motor
J Flywheel

Fig. 2. Schematic of receiving separator. (Courtesy Buhler-Miag, Minneapolis, MN)

(226 m^3/min). Because of present-day high energy costs, many processing companies are now using closed-circuit receiving separators that, by using an integral fan, recycle the aspirating air and drop the removed light material into an internal expansion chamber. This eliminates the need for high-air-volume dust-collecting filter systems and conserves substantially on power costs, as well as reducing equipment capital investment costs.

Incoming oats are evaluated for moisture, test weight, insect contamination, groat discoloration, and milling yield. The rough yield determination is made by screening, hulling, and grading a small sample. The lower the moisture and the higher the test weight of the oats, the better the milling yield (i.e., a greater percentage weight of rolled oats can be produced from a given volume of oats). Insect contamination must be determined before storage so that the oats can be rejected if the contamination level is too high or stored in separate bins to prevent contamination from spreading and to permit the oats to be treated with approved chemicals to kill the insects. After initial testing, the oats are usually binned by yield, degree of contamination, and moisture. Blends are then delivered to the mill to reduce variations in yield and quality of end products.

The benefits of grading oats by milling yield, groat discoloration, moisture, and infestation and of then blending the grades for milling can be considered on the basis of "averaging out" the variations. For example, if those separate grades

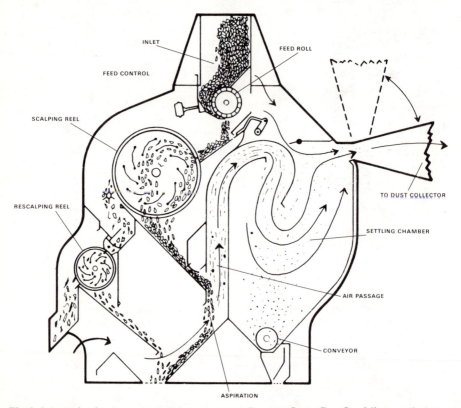

Fig. 3. Schematic of scalper-reel receiving separator. (Courtesy Carter Day Co., Minneapolis, MN)

of oats having test weights of 30, 33, and 38 lb/bu are blended in equal proportions (33-1/3% per grade), the blend for milling will have a test weight of 33.6 lb/bu. By blending the oat grades in suitable proportions, a satisfactory test weight average for milling can be obtained, even though the blend contains a percentage of oats that, on their own, would not give good milling yields. Blending enables commercial advantage to be obtained from the market conditions prevailing at the time the oats are purchased. By maintaining as constant as possible the test weight, moisture, etc., for milling, the end product quality remains consistent and deviations are minimized. Furthermore, process machinery does not have to be readjusted continually to obtain maximum efficiency when the oat quality average remains consistent.

Similarly, some groats are perfectly good but show signs of slight discoloration, resulting in a rolled oat product deviating from the generally accepted and usual flake color. These become considerably less noticeable when blended in small proportions with groats having normal color.

Preliminary cleaning of field oats before storage is only an initial, rough cleaning step, which is not meant to replace the thorough cleaning before milling. It is usually done at comparatively high rates corresponding to the unloading rates from trucks or rail cars. This necessitates the use of coarse screen meshes to achieve the throughput rates, and therefore only coarse objects are removed to protect subsequent conveying equipment and some of the dust, light impurities, and hollow and infested kernels are removed before storage. The total quantity of impurities removed in the initial cleaning stage (coarse trash, coarse and fine impurities, and light screenings) seldom exceeds 0.5% by weight of the incoming oats and more commonly averages only about 0.25%.

IV. OAT CLEANING—SPECIALIZED MACHINES

The initial processing step is the oat cleaning system, the purpose of which is to remove all grain and seed impurities from the oats and also to ensure that all unwanted extraneous material, such as stones, metal particles, sticks, etc., are extracted before milling. Removal of oats that are not suitable for milling is also a purpose of the cleaning stage.

A typical oat cleaning system is illustrated in Fig. 4. The foreign materials removed during cleaning are corn, seeds, sticks, soybeans, barley, wheat, loose hulls, stones, and dust. The contaminants usually become mixed with the oats in the field and during handling in various grain elevators. Oats that are not suitable for milling and that are removed include the following: (i) Double oats (bosom). The hull of the primary kernel envelops the second grain. Normally, groats in both kernels are poorly developed, resulting in a high percentage of hull. (ii) Pin oats. These are usually very thin and short and very poor yielding, with little or no groat inside. (iii) Light oats. Although generally equal in size to normal oats, light oats contain small groats in comparison to the hull; they are separated by aspiration. (iv) Other types of oats. These consist of twins and discolored, green, and hulless kernels, which may or may not be removed in the cleaning plant depending on their size (Salisbury and Wichser, 1971).

The first machine in the cleaning flow is a milling separator combining coarse and fine screening with an efficient aspiration. Different sieve deck motions are available depending on the manufacturer and design concept, and they fall into one of the following categories:

1. Rotary motion in which the sieve moves in a circle in a near horizontal plane. The speed and radius of rotation are interrelated and vary according to the design; with a radius of 1 in., experience has shown that the best speed is around 300 rpm, whereas with a radius of 1-3/4 in., the best speed is about 195 rpm. A rotary motion has a slight tendency to separate products by length as they lie flat on the screen in their passage down the sieve, and the shorter grains pass through the perforations while the longer pieces cannot pass through and are overtailed.

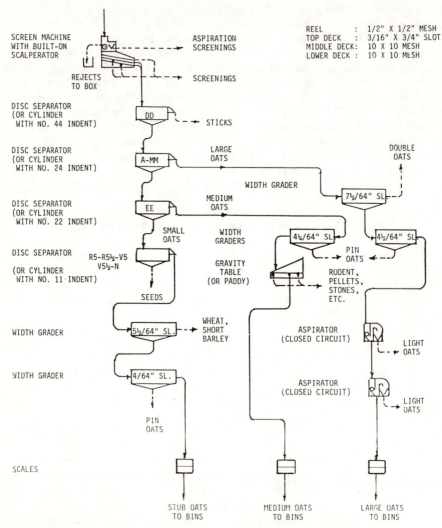

Fig. 4. Typical oat cleaning flow. (Courtesy Carter-Day Co., Minneapolis, MN)

2. Oscillating or reciprocating motion in which the sieves slope slightly downward. The speed and stroke of the sieve deck is, again, interrelated for optimum results. For oat cleaning, the range of speed is about 300–450 strokes per minute and the range of stroke about 1/2–1-1/4 in. (12.7–31.7 mm). In general, the higher the speed, the shorter the stroke should be: for example, at 300 strokes per minute the greatest capacity is achieved with a stroke of 1-3/8 in., whereas at 450 strokes per minute the best stroke is 7/8 in. (22.2 mm). The slope of the sieves varies depending on design between 1 in 6 and 1 in 12. Because a reciprocating motion has a tendency to upend grains and material on the sieve, products separate on a width or thickness basis as the kernels are presented longitudinally to the perforations or wire mesh openings when they upend (Lockwood, 1960).

3. Combined head-end rotary motion and tail-end reciprocating motion. This sieve movement combines the advantages of both previously described sieve motions. The rotary, or gyratory, motion has no vertical component, and this allows the product to lie flat and be in contact with the screen at all times. In a reciprocating machine, with each stroke the product is momentarily suspended off the screen, which lowers screening efficiency. With this combination movement, the tail end, or lower part, of the sieves actually reciprocate, which helps move the product down and off the screens and improves product throughput. The gyratory motion is concentrated at the head-end section of the machine where 70–90% of the screening is performed.

In a milling separator, the top sieve deck is clothed with screen material (either perforated sheet metal or wire mesh) to provide a close scalping separation. The oats and fine material fall through the top sieve layer onto the lower sieve layer (or layers) clothed with finer screens for fines removal. Research over the last few years (Carter-Day Company, *unpublished data*) has shown that, for maximum separation efficiency, twice as much fine-screen area as coarse-screen area should be provided because the separation required in the coarse-scalping separation is not normally a near-size one. Therefore, the throughput per square foot of screen is much greater, requiring less screen area for a given capacity rate. In the fine-screening area of the separator, a near-size separation is required, which means that additional screening area is needed to provide time for the fine material and particles to migrate to the bottom of the oat layer, come into contact with the screen mesh, and pass through the openings.

Sieve throughput in a milling separator varies depending on such design parameters as sieve motion, speed, angle of inclination, and depth of product on the sieve. Accordingly, throughput can range from 15 to 27 lb/ft^2·min on coarse-scalping screens and from 12 to 17 lb/ft^2·min on fine screens.

Most milling separators incorporate an aspiration to remove dust and light material from the oats before leaving the machine. Depending on type of separator used, the aspiration is on the oat stream entering the machine on the theory that screening efficiency is improved with preremoval of fines and light material, or else on the oat stream leaving the machine after screening on the theory that the aspiration is more efficient with some impurities removed. Maximum air-volume requirement is about 6.5 ft^3/min per bushel of oats and maximum air velocity in the aspiration channel is about 1,400 ft/min, although these parameters vary according to the machine design.

Figure 5 shows a typical milling separator with built-on scalper-aspirator unit for preremoval of light impurities. This machine is equipped with three screen decks—one coarse and two fine layers, thus incorporating twice the fine-screen to coarse-screen area. Sieve deck assembly motion is the combination head-end rotary and tail-end reciprocating type, and the separations performed result in the following fractions:

1. Scalper reel (wire mesh measuring 1/2 × 1/2 in. [12.7 × 12.7 mm]). Generally, reel rejects consist of straws and coarse foreign objects that may have inadvertently been mixed with the oats during storage or subsequent handling.
2. Scalper-aspirator aspiration. This stream consists of dust, loose hulls, light trash, and poor oats with little or no groats inside.
3. Coarse top screen (over slotted perforation measuring 3/16 × 3/4 in. [3.4 × 19.0 mm]). These overtails usually consist of corn, soybeans, sticks, large weed seeds, and larger stones.
4. Middle and lower fine screens (over 10 wire mesh). These main streams consist mostly of good oats.
5. Middle and lower fine screens (through 10 wire mesh). These fines contain sand, groat chips, and many small seeds.

The next stage of the cleaning process utilizes a series of specialized cleaning machines that selectively remove weed seeds, double oats, any remaining stones or sticks, and low-quality oats such as pin oats. We will first describe the operating principles of these cleaning machines to show how certain types of separations are made, what the limitations are, and how, when operating as an integrated system, they achieve the desired cleaning results.

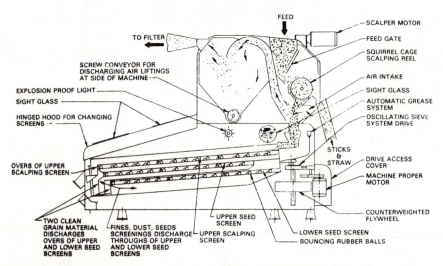

Fig. 5. Schematic of typical milling separator with top-mounted scalper-aspirator. (Courtesy Carter-Day Co., Minneapolis, MN)

A. Disk Separator

Some impurities are the same width or thickness as the oats but differ in length. For example, a round vetch may have the same general width or "plumpness" as an oat kernel, but it is always shorter in overall length. A flat screen removes impurities when the overall dimensions are quite different, but it does not make a precise separation when the dimensions are similar.

A disk separator is specifically designed to perform a separation by differentiating between the difference in length between the desired and unwanted products (Fig. 6). The separator shown consists of a horizontal shaft on which is mounted a series of vertical cast iron disks of 25- or 18-in. (635- or 457-mm) diameter with indented pockets cast in on each side. The disks revolve at 56–58 rpm, which experience has shown to be the best speed to permit grains or seeds to enter the indented pockets and be thrown out of the pockets by centrifugal force when the pockets have rotated past the top center position. A wide range of different indented shapes and sizes is available according to the separation required, and Figs. 7–9 illustrate three of the 52 types of indents that are available. The oats are fed in at one end of the series of disks, and all particles that fit into the indented pockets are lifted out of the mass as it is conveyed through the machine by inclined blades on the spokes of each disk, which attach

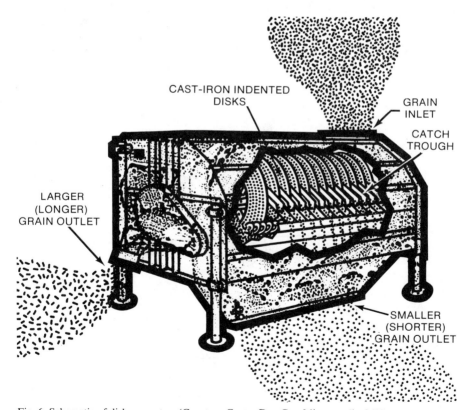

Fig. 6. Schematic of disk separator. (Courtesy Carter-Day Co., Minneapolis, MN)

the indented pocket section to the central shaft; the product mass literally passes down the machine between the spokes from one disk to the next. The larger particles, or grains, do not fit into the indented pockets and are conveyed down the machine, from disk to disk, by the inclined blades until they reach the last

Fig. 7. Disk pockets—sticks being removed from oats (sticks longest).

Fig. 8. Disk pockets—oats being removed from wheat (oats longest).

disk and are overtailed from the disk machine through an outlet at the tail end. Small particles that fit into the indented pockets and are lifted out of the mass are thrown out onto catch troughs between the disks when the indented pocket has passed top center position on the disk rotation (much as grain is thrown out of bucket elevator cups when the cups have passed over the top of the head pulley). The lifted product slides off the catch troughs into a collecting hopper running the entire length of the machine.

In a disk separator, the shortest product is always lifted by the indented pockets and the longest product is rejected by the indents and overtailed. Hence, large indented pocket disks lift oats out of larger impurities such as sticks, and small seeds are lifted out of oats with smaller indented pocket disks. To make a good separation, there must be at least 1/16-in. (1.6 mm) difference in length between the products to be separated.

In a typical disk separator (27 disks, each with a 25-in. [635-mm] diameter) used for separating seeds out of oats using R5-1/2 size indented pockets, there are 8,596 indented pockets on both sides of each disk, or a total of 232,092 indented pockets in the machine. The number of indented pockets coming into

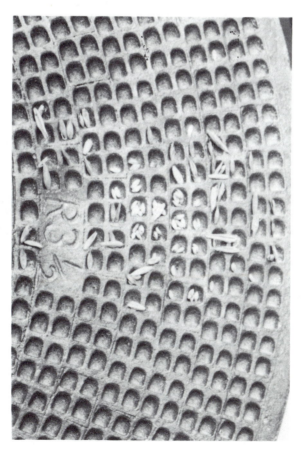

Fig. 9. Disk pockets—seeds being removed from oats (seed shortest).

contact with the oats during each minute of operation at 56 rpm disk speed is 12,997,152. This size of machine has a maximum capacity of 200 bushels of oats per hour.

B. Indented Cylinder

A second type of separator is an indented cylinder separator that performs the same length separation of particles as a disk separator (Fig. 10). This machine has a slowly rotating (56 rpm), horizontal cylinder with indents punched in the inner surface. The indents lift the shorter particles out of the mass and drop them into a catch trough with a screw conveyor in the bottom, positioned down the center of the cylinder, from where they are conveyed out of the machine; Fig. 11 shows the indented cylinder drum surface. The longer particles, which do not fit into the indents, overtail out of the end of the cylinder. Hence, small seeds can be lifted out of oats with smaller indents and oats can be lifted out of longer sticks with larger indents. The angle of the catch trough in relationship to the rising side of the cylinder can be adjusted with the machine in operation, and thus the "cut point" of the separation can be selected to provide the separation required. The mass of product entering the indented cylinder tends to ride up on the rising side of the cylinder because of friction, but any particle that does not fit into the

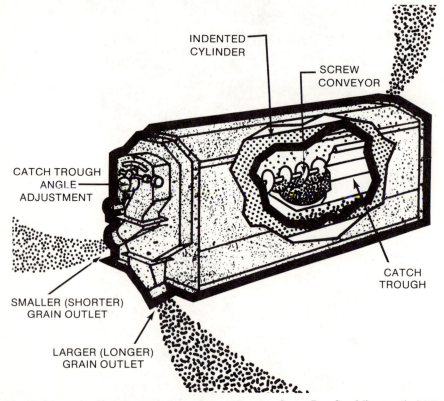

Fig. 10. Schematic of indented cylinder separator. (Courtesy Carter-Day Co., Minneapolis, MN)

indents eventually falls back into the mass as a result of gravity. By positioning the edge of the catch trough lower down the rising side of the cylinder, more longer particles have the opportunity to ride up and fall into the catch trough. By raising the edge of the catch trough higher up the rising side, only shorter particles that fit snugly into the indents are able to fall into the catch trough and be separated from the mass.

This flexibility of separation adjustment is helpful for more difficult separations, such as lifting large barley out of short oats when the length of the two grains is similar but not identical. Indented cylinder separators have lower capacities than disk separators, and hence more machines are required for a given flow rate. Capacity is reduced because there are fewer indents in a single cylinder than there are indented pockets in the series of disks in a disk separator. For example, the total number of comparable indents to the disk pockets in the Model 2527 disk separator discussed previously is 89,100 on a No. 3 Carter Uniflow Cylinder Separator; this compares with a total of 232,092 indent pockets on the disk machine. In general terms, the capacity of an indented cylinder separator is about one-quarter that of a disk separator on similar applications.

Fig. 11. Indented cylinder drum surface.

C. Width Sizer

The width or thickness sizer (Fig. 12) separates grains when their length is the same but their width or thickness (plumpness) is different. It consists of a slowly rotating (56 rpm), horizontal cylinder with round hole perforations (Fig. 13) or rectangular slots (Fig. 14) cut in. In cylinder shells with round hole perforations, the small perforation sizes (3/64–3/16 in. [1.2–4.8 mm] diameter) have internal ribs placed lengthwise down the cylinder between each several rows of perforations, and the larger sizes (13/64–27/64 in. [5.2–10.7 mm] diameter) are recessed (i.e., the hole is at the bottom of an indent). Slotted cylinder shells have the rectangular slots in rows, with the long side of the slot running around the shell circumference (Fig. 15). Perforated slot sizes are available in a range from 2.5/64 to 20/64 in. [1.0 to 7.9 mm] wide.

When slotted perforations are used, channels or grooves upend and align the seed or other grains presenting the *thickness* of the seed to the opening. If

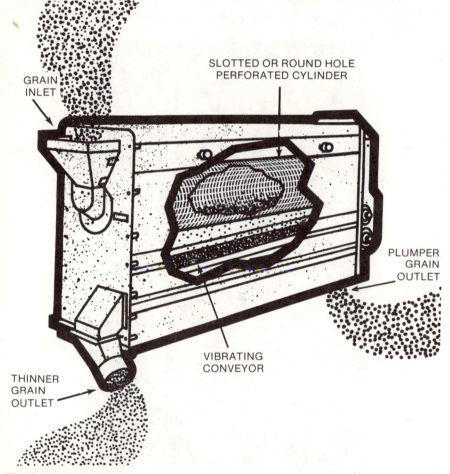

Fig. 12. Schematic of width or thickness sizer. (Courtesy Carter-Day Co., Minneapolis, MN)

sufficiently thin, they then fall through the slotted perforation at the bottom of the groove while oversize grains, which are unable to fall through, pass down the inside of the cylinder shell and overtail through an outlet at the tail end of the machine. The thinner products passing through the cylinder slots are conveyed by a vibrating conveyor under the cylinder to an outlet at the end of the machine.

Fig. 13. Round-hole ribbed cylinder shell.

Fig. 14. Slotted cylinder shell.

Cylinder shells with recessed round perforations (13/64–27/64 in. [5.2–10.7 mm] diameter) are used for *width* sizing (Fig. 16). The recess or indent at the top of the perforations causes seed or other grain to upend, presenting the diameter or width of the seed or grain to the opening that enables seeds, etc., to fall through the perforations.

The ribs in the small round-hole perforations (3/64–12.5/64 in. [1.2–5.0 mm] diameter) agitate the materials and upend them to present the product diameter to the perforation in much the same fashion as the recessed indent does in the larger size perforations (Fig. 17).

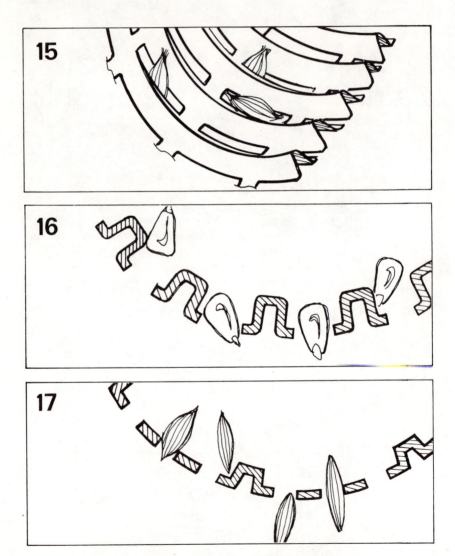

Figs. 15–17. Schematic of (15) slotted cylinder shell, (16) round-hole recessed cylinder shell, and (17) round-hole ribbed cylinder shell. (Courtesy Carter-Day Co., Minneapolis, MN)

D. Gravity Separator

Some impurities have very similar length and width characteristics to oats and, therefore, they cannot be accurately separated by the length- or width-sizing equipment just described. Such impurities as oat kernels slightly hollowed because of insect damage and sticks the same length and width as an oat kernel come into this category. Many of these impurities are, however, lighter or heavier than oats and thus have a different specific gravity. The term *gravity separator* is a contraction of the proper name of this machine, *specific gravity separator*, which means a separator of particles differing in their specific gravities.

About 250 B.C., Archimedes discovered the law of specific gravity, which states that all bodies floating on or submerged in a liquid are buoyed up by a force exactly equal to the weight of the liquid they displace. The specific gravity of a particle is the ratio of its density to some standard substance, the standard usually being water with a specific gravity of one. Particles having a specific gravity of less than one will float, and those with a specific gravity of greater than one will sink. All gravity separators use air as a standard rather than water; since air is lighter than water, the relative difference between particles of differing weights is increased. For this reason, the gravity separator is a very sensitive machine that, when operated correctly, can produce a very precise separation.

Air is used as the separating standard through the process of stratification (Fig. 18). Stratification occurs by forcing air through the particle mixture so that the particles rise or fall by their relative weight to the air. Figure 18A represents a cross section of a gravity separator directly over the fan; a particle mixture has been introduced on top of the screen deck with the fans off. In Fig. 18B, the fan has been turned on and adjusted so that the heaviest particles rest on the surface of the deck and the lightest particles are completely free of the surface of the deck. Proper regulation of the airflow at this time is critical or, as shown in Fig. 18C, all particles are lifted free of the separating deck surface by excess air. Figure 19 shown the ideal situation in the operation of a gravity separator.

The particle mixture, oats containing lighter and heavier impurities (similar to the situation shown in Fig. 18A), falls from the feeder onto the deck. The area

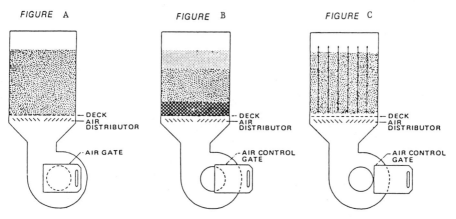

Fig. 18. Principle of stratification. (Reprinted, with permission, from Oliver Manufacturing Co., undated)

immediately around the feeder is called the stratifying area, where the vibration of the deck and the lifting action of the air combine to stratify the material into layers, with the heavier layers, such as stones, on the bottom and the lighter layers, such as hollow weed seeds or hollow oats, on the top (Fig. 18B). Separation cannot occur until the material becomes stratified. The size of the stratification area depends on the difficulty of the separation and on the capacity at which the machine is processing. Once the material becomes stratified, the vibrating action of the deck begins pushing the heavier layers in contact with the deck toward its high side. At the same time, the lighter layers, which are at the top of the material bed and do not touch the vibrating deck, float downhill toward the low side of the slightly sloping deck.

As the material flows downhill from the feed end to the discharge end of the deck, the vibrating action gradually converts the layers of vertical stratification to a horizontal separation. By the time the material reaches the discharge end of the deck, the separation is complete. Heavier materials, such as rocks and stones, are concentrated at the high side of the gently sloping deck. Light materials, such as hollowed weed seeds, are at the low side of the deck, and intermediate materials, the good oats, are in between. Movable vanes at the discharge end of the deck are adjusted to give the desired cut point between the three fractions (Oliver Manufacturing Co., undated).

E. Paddy Separator

When known mechanical methods such as sieves, aspirators, and length, width, and thickness grading fail to separate effectively, different materials of nearly equal size, shape, and almost equal specific gravity can be separated by a paddy separator. Gravity separators are limited in their ability to make a precise separation when the specific gravity of the particles to be separated is almost identical. The concept of the paddy separator was based on a discovery, made by Frederic H. Schule in Germany in 1892, proving that grains, although almost

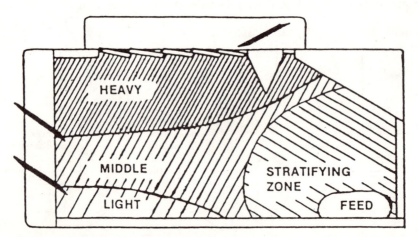

Fig. 19. Ideal situation in operation of gravity separator. (Reprinted, with permission, from Oliver Manufacturing Co., undated)

equal in size and shape, may vary with regard to specific gravity, surface texture, and effect of gravitation.

The tendency to settle depends on the difference in specific gravity (however small), surface smoothness, and grain shape. Thus, hulled oats are smoother and weigh more, so they travel down the table while the grains with husks still on them, which are rougher in surface texture, travel upward (Fig. 20). The table, which is made to pivot around its longitudinal axis, rests upon rubber tires. Sorting takes place within the compartments, being affected by a series of zigzags arranged at right angles to the direction of motion.

The oats to be treated accumulate in the separating compartments, forming a mound of a certain height above the channel bottom with the layer diminishing in thickness toward the discharge areas. If the compartments are correctly fed, the shaking motion of the table causes the formation of layers in a vertical direction according to the physical properties of the different grains (hulled and unhulled oats), with the top layer consisting (to a large extent) of the larger and lighter kernels (unhulled) and the bottom layer consisting of smaller, yet specifically heavier grains (hulled).

Apart from the formation of layers and the preseparation, the varying surface texture of the kernels affects the separating process. Following the table slope and gravity, the grains tend to travel toward the lower side (discharge of heavy kernels). The impact imparted to the grains by being thrown against the channel flanks, which are arranged at an angle of 28° to the direction of table movement, causes the grains, especially those of the top layer (unhulled), to travel toward the upper side (discharge of light kernels) (Fig. 21). The knee in the channel is indispensable for an efficient separation. Width of channel, length of stroke, and acceleration of the deck have to be precisely adjusted to achieve optimum separating efficiency and peak capacity (F. H. Schule GMBH, undated).

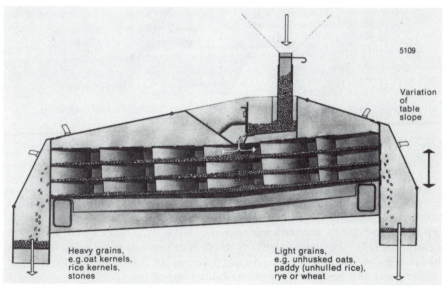

Fig. 20. Sectional view of separating table. (Reprinted, with permission, from F. H. Schule GMBH, undated)

Fig. 21. Channel flanks. (Reprinted, with permission, from F. H. Schule GMBH, undated)

V. FURTHER OAT CLEANING AND GRADING

We now return to the initial aspirating and screening performed on the milling separator and continue with the oat cleaning and grading process that incorporates some, or all, of these specialized machines. After the initial cleaning step (refer to Fig. 4), the oats are routed to disk separators where sticks are removed by DD type disk pockets (or to cylinder separators with No. 44 indent shells—e.g., 44/64-in. [17.5 mm] diameter across the top of the indent) that lift out the good oats and reject, or overtail, long impurities.

Following stick removal, the aim is to classify the oats into the three fractions of stub (short) oats, medium-size oats, and large oats so that each fraction, which contains its own particular types and sizes of impurities, can receive individual cleaning treatment more selectively and more efficiently than if all sizes were kept together as one mixed stream. The oats go to another disk separator with A and MM pockets (or cylinders with No. 24 indents) that separate large oats from the shorter medium-size and small oats; the last two fractions contain all the weed seeds present because these seeds are considerably smaller, or shorter, than the large oats. The pockets lift out all the seeds with the medium-size and short oats and reject and overtail the large oats. The larger or longer oats rejected by the A and MM pockets (or No. 24 indents) next pass to a width sizer fitted with a 7.5/64-in. (3.0-mm) slotted cylinder that permits good oats to fall through and overtails the double oats not suitable for processing. Pin (thin) oats are then removed by a slot that is 4.5/64 in. (1.8 mm) wide on a second width sizer. The good, large oats then receive a two-stage aspiration on multilouver or on closed- or open-circuit aspirators to remove light oats. Following aspiration, the large oat stream is weighed and binned. Some processors prefer to use gravity separators at this point instead of aspirators to remove light oats and also to ensure removal of any heavy impurities, such as mud balls or stones, that may be present and that aspirators will not separate.

The mixture of medium-size and small oats lifted by the A and MM disk pockets (or No. 24 cylinder indents) flows to another disk separator fitted with EE type disk pockets (or a cylinder with No. 22 indents), which reject or overtail the medium-size oats and lift out the small oats containing weed seeds, some groats, wheat, and short barley. The seeds are then removed from the small oats by another disk machine equipped with a variety of R5, R5-1/2, V5, and N pockets (or a cylinder with No. 11 indents). The small oats from this separator pass to a width sizer with a slotted cylinder shell 5.5/64 in. (2.2 mm) wide that permits the oats to pass through and overtails any wheat or barley present. A final width grading is made on a 4/64-in. (1.6-mm) slotted cylinder shell to drop pin (thin) oats through before the good, small, or stub oats are weighed and binned.

The medium-size oats separated earlier by the EE disk pockets (or No. 22 indents) are routed to a width sizer with a slotted cylinder shell 4.25/64 in. (1.7 mm) wide to remove any pin (thin) oats present. The medium-size oats then proceed to a gravity separator that removes light rodent pellets and any other light impurities present, such as similar-size weed seeds, the lighter of which have been hollowed out by insect damage or lack of development; this separator also separates out any heavier stones. A specific gravity separation is used at this point because the impurities to be removed are generally of similar length and width to the oats. They are, however, lighter or heavier, which permits a well-adjusted gravity separator to make a good separation. As described in the section on specialized machines, a combination of sieve deck inclination, speed, throw (amount of horizontal movement), and a high volume of air passing through the screen deck to achieve stratification is used to achieve the separation. A very careful adjustment of all these settings is required to obtain effective results on a specific gravity separator. The medium-size oats from the gravity separator are weighed and binned.

At this point, the oats—which have been cleaned, sized, and put into storage bins—are ready for hulling. The percentage breakdown of each size naturally varies depending on the place of origin of the oats and the growing-season conditions, but a typical breakdown would be 30–35% large oats, 45–50% medium-size oats, and 20–25% stub oats.

It is customary to have magnetic separations in the cleaning and sizing steps to remove ferrous metal particles from the oats. Such particles, consisting of nuts, bolts, and rivets, etc., may be present in the raw oats delivered to the mill or have become detached from machines during processing. Magnetic protection is provided by permanent spout magnets, electromagnets, or rotating drum magnets positioned so that the stream of oats entering and leaving the cleaning system passes over them.

In a commercial operation, the machinery used varies depending on the capacity or flow rate of the system. Larger plants contain several machines operating in parallel on each particular separation, whereas smaller plants may have only one or two machines. Also, the disk pocket selections, cylinder indents, and width-sizer slotted cylinder shell sizes given above are intended to be typical; different processors may use variations, as well as different approaches to the actual cleaning flow. Indented cylinders may be used instead of disk machines on some or all of the length separations (see Fig. 4). Regardless of the approach taken, the desired end result remains the same—namely, to produce well-cleaned oats graded according to size for hulling, which, as the name implies, entails removal of the hull or husk. Before hulling, however, the oats are dried.

VI. DRYING AND COOLING

Hulled oats, or groats, contain about 6.5% fat, a quantity that is not found in any other cereal grain. During normal storage of oats, for example, at 13% moisture and up to 65° F (18° C), free fatty acid content increases only very slowly; but if the oats are crushed or milled into meal, the production of free fatty acid is considerable in a matter of two or three days. The catalyst in the production of free fatty acid is an enzyme of the lipase type. Free fatty acid causes the product to become rancid and unpalatable; however, if the oats are subjected to heat, the lipase is inactivated within a few minutes provided the moisture does not fall below 12% and the oats are maintained at a temperature of 194–212° F (90–100° C).

The next step in the oat processing system is, therefore, drying and cooling. The objectives of this stage are to inactivate the lipase or fat-splitting enzymes sufficiently to prevent the development of undesirable flavors during processing and to prevent rapid rancidity in the end product (Hutchinson et al, 1951); to develop a slightly roasted flavor, which is considered desirable by most processors; and to make the oat hulls more friable, or brittle, to facilitate their removal during the subsequent dehulling stage.

Different approaches to drying and cooling are taken by processors according to their individual preferences and market requirements. One older method is to use pan dryers, which are normally about 12 ft (3.6 m) in diameter and placed one above the other in stacks of 7–14 pans depending on the capacity and retention time required; each pan is steam jacketed and open on the top. A retention time in the stack of at least 1 hr of drying is usually required to achieve the desired degree of flavor development. The oats are moved from the inside to the outside of each pan by sweep arms, and they move through the system by dropping from the outside of the top pan to the inside of the pan below (Fig. 22).

Temperature of the oats usually ranges from 190 to 200° F (88 to 93° C) during drying, and normally only 3–5% moisture is removed. Oats entering the dryer average around 12% moisture content and are in the 7–10% range leaving the dryer. During the drying process, by the time the oats reach a temperature of 190–200° F (88–93° C), the moisture content has fallen below 12%; when the oats are cooled, they still show some 20–40% of their original lipase activity. The oat steaming process, which is described in a later section, completes the inactivation of the lipase.

Another form of oat dryer is the radiator column type, in which a vertical column has banks of horizontal radiators arranged down the height of the column in a staggered fashion so that all the oats come into contact with the steam-heated surfaces in their slow passages down and through the radiators.

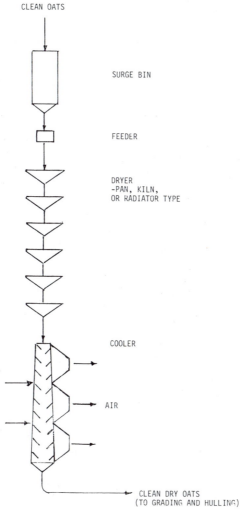

Fig. 22. Flow of oat drying and cooling system. (Courtesy Carter-Day Co., Minneapolis, MN)

Smaller-capacity mills sometimes use rotary steam dryers, but generally the flavor development is considered lower than in pan or radiator column dryers because the retention time is much reduced. Many European processors use charcoal-fired kilns, in which the combustion gases passing through the oats impart a distinctive roasted flavor to the oatmeal.

Some mills are now drying (with kilns, pan, or radiator type dryers) later in the process after the hulls have been removed from the oats (i.e., groat drying). It is claimed that the groats are tougher because they have not been dried at the dehulling stage and, therefore, that the tips do not suffer as much damage in the hullers. Consequently, a higher yield is claimed (about 1–5% more), and the husk is at a higher moisture level since it has not been dried out, giving a higher weight by-product with a greater commercial value.

Following drying, the oats are cooled for further processing, usually on multilouver-type cooling columns or in a cooling section on the bottom of radiator dryers, in which ambient air is drawn through the oats to reduce temperature. Leaving the cooler, the oats have a moisture content of about 8–11% and a temperature in the 100–120° F (38–49° C) range.

VII. GRADING AND HULLING

After the cooling stage the oats are ready for hulling, which, as the name implies, separates the outside hulls from the groats inside the kernels. (Where groat drying is practiced, the hulls have been removed following cleaning.) The hulls must be removed because they are tough, lack flavor, and are unpalatable for human consumption. Hulling efficiency is improved by prior grading or sizing of the oats so that each hulling machine is fed a limited, rather than wide, range of kernel sizes (i.e., either longer or shorter oats, but not a mixture of both sizes). In the system previously described, the sizing was carried out in the cleaning process, which resulted in large, medium-size, and small (stub) grades; some processors, however, prefer to clean, dry, and cool the oats as one stream and then grade the oats by size on a separate processing system just before hulling. Figure 23 shows a typical grading flow for hulling when this approach is taken, which essentially constitutes the head-end grading section of the comprehensive cleaning and grading flow shown in Fig. 4. Impurity removal by screening, aspiration, length and width grading, and specific gravity separations are performed in the oat cleaning section, but the grading by kernel size for hulling is performed separately.

Some processors, in addition to performing separate grading for hulling, also remove the smaller weed seeds in this separate section as shown, rather than in the oat cleaning section. The size of the oat processing facility frequently determines which approach is taken. In small-capacity plants, it is not feasible to grade the oats by size and then clean each grade separately because the quantity of each grade is too small to warrant separate cleaning flows.

Regardless of which method is used, the oats should be graded into at least two sizes by length; plants with very large capacity may grade into three sizes for hulling, as the flow rate of each fraction is sufficiently high to feed the hullers adequately and permit them to be adjusted precisely to suit the size of oats fed to them, thus obtaining maximum hulling effect with minimum tip breakage.

The impact huller shown in Fig. 24 has completely replaced the stone huller; it is much more efficient, produces a better yield, and requires less horsepower per unit of capacity. The large/medium-size oat hulling and final grading stage and the short (stub) oat hulling system are shown in Fig. 25. The oats enter the center of a high-speed rotor that is fitted with blades or fins that centrifugally throw the oats against a Carborundum ring fixed to the machine housing, where the hulls are detached from the groats by impact and abrasion. The speed of the rotor is adjusted by a variable speed control, according to the type and condition of the oats being processed, to obtain maximum hulling efficiency with minimum breakage of groats and production of fines and chips. Normally, the rotor speed is between 1,400 and 2,200 rpm depending on the size and condition of the oats.

Some processors prefer impact hullers that are fitted with a hard rubber ring, as opposed to a Carborundum ring insert, claiming that the rubber is not as severe and results in less breakage of the groats and less formation of fines. The

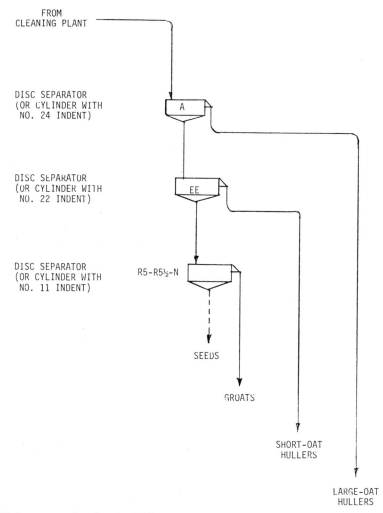

Fig. 23. Separate grading flow for hulling. (Courtesy Carter-Day Co., Minneapolis, MN)

huller produces a mixture of free groats, free hulls, groat chips, fines, and some unhulled oats. A good huller that is well adjusted for the size of oats being fed to it has an efficiency of around 90–95% (i.e., that percentage of the oats is dehulled in one pass through the machine). Badly maintained hullers, or incorrect speed selection according to the size, type, and condition of the oats being processed, can result in an efficiency as low as 55%.

Next comes separation of the mixture of free groats, groat chips, fines, and unhulled oats leaving the huller so that the groats can be isolated for further processing and the unhulled kernels subjected to further hulling. The fines and detached hulls are lifted out of the stream by aspiration; their terminal velocities are sufficiently lower than those of groats and unhulled oats to permit an effective separation on aspirators in which the product mixture falls down an upward-rising air column that lifts out the light particles. Extreme care is required in adjusting air volume and velocity in the air column to avoid excessive loss of small groats and chips with the hull by-product, which has a lower commercial value.

The hulled oats (or groats) are then polished on scourers that subject the groats to a mild surface-abrading action. A scourer consists of a horizontal cylinder, the

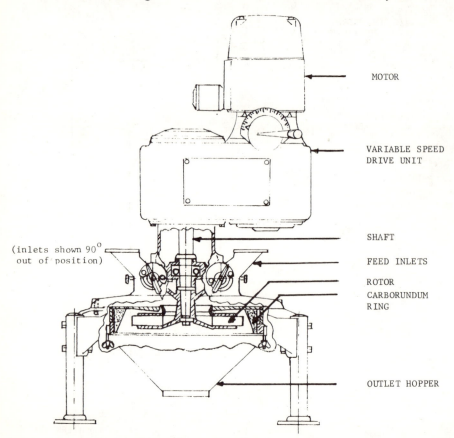

Fig. 24. Schematic of impact huller. (Courtesy Entoleter Inc., New Haven, CT)

inner surface of which is lined with a moderately rough material such as a heavy, intercrimped wire mesh; a horizontal shaft runs through the center of the cylinder, on which are mounted inclined blades that convey the groats against and along the rough cylinder surface. The moderately mild abrading action removes any small pieces of husk still adhering to the groats, and these husk

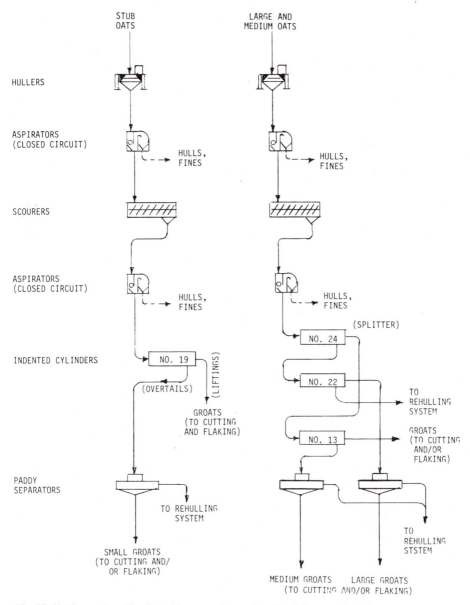

Fig. 25. Hulling and grading flow. (Courtesy Carter-Day Co., Minneapolis, MN)

pieces are then removed by a second aspirator, along with any loose hulls that may have been missed by the initial aspirator.

In the next step, the unhulled oats are separated from the groats. Most of the groats are sufficiently shorter than the unhulled oats to enable a practical separation to be made by disk separators or indented cylinder separators. However, this separation is made less effective by some oats whose groats are as long as the oat and by the huller "clipping" or damaging the tips of some oats that are not hulled in the first pass. The final, difficult separation between groats and "clipped" oats, where little difference in length or specific gravity exists, is made on paddy separators (or gravity separators). The paddy separator is, as previously described, a very sensitive machine that is used in cases where sieving or air separation does not give enough separation; it permits a good separation to be made on products of the same size and shape, based on their slight differences in specific weight, behavior of impact, and floatability. The sensitivity of paddy separators makes them more accurate than gravity separators for the difficult final separation of similar-size groats and unhulled oats. The paddy separator is, however, a rather low-capacity machine, and consequently it is preferable to perform the bulk of the separation, where the size difference between the groats and the unclipped, unhulled oats is not similar, on disk machines or indented cylinder separators and to finish the similar-size separation on paddy separators.

The premium (large and medium-size) oat stream passes from the hullers, aspirators, and scourers to indented cylinder separators fitted with No. 22 and No. 13 indent shells; these separators grade the groats into medium-size and large fractions to facilitate the final separation on the paddy separators (or gravity separators). The initial No. 22 indent cylinder separator splits the stream, lifting out the shorter, medium-size groats mixed with clipped, unhulled oats and overtails the longer, large groats mixed with whole unhulled oats and longer clipped oats. This longer fraction then goes to a second No. 22 indent cylinder separator that overtails the large groats and lifts out the slightly shorter groats mixed with unhulled oats and longer clipped oats, which then flow to a large-fraction paddy separator for the final separation of the groats from the unhulled oats. The groats separated out by the second No. 22 indent cylinder separator and by the paddy separator are ready for further processing on the next stage, whereas the unhulled oats from the paddy separator require further hulling. The shorter, medium-size groats mixed with clipped, unhulled oats lifted out by the initial No. 22 indent cylinder separator go to a third No. 13 indent cylinder separator that lifts out the medium-size groats and overtails the slightly larger groats with the unhulled, clipped oats. This stream then flows to the medium-fraction paddy separator for the final separation of the groats from unhulled, clipped oats that require further hulling action.

Like a conventional gravity separator, a paddy separator works more efficiently when the kernel sizes fed to it are within a small range—hence the need to grade the groats and unhulled oats carefully by size at this stage of the process.

The stub oats are hulled and graded basically in the same manner as the large and medium-size fractions. Following hulling, aspiration, and scouring, the small groats are lifted out from the smaller unhulled oats in this fraction by a No. 19 indent cylinder separator while the cylinder overtails go to a paddy separator for the final separation.

The recovered unhulled oats require a second pass on a huller to detach the hulls. In smaller mills, these unhulled oats are recycled back to the initial hullers on each respective grade size, and they then pass through the grading system once again. In larger plants, where the quantity of unhulled oats requiring further treatment is sufficiently large to feed hulling and grading equipment, a separate rehulling system is frequently used (see Fig. 26). Since the fraction for rehulling has already been subjected to impact abrasion on the initial hullers, and some of the unhulled oats present have already been clipped, the production of fines and chips tends to be higher on a rehulling system or second pass through the hullers. Consequently, the groats tend to be slightly smaller, and a rehulling system is basically a repeat of the stub-oat hulling and grading system.

The groats separated out in hulling and rehulling systems flow to the cutting and flaking plant for production of rolled oats or to a milling unit to be ground into oat flour, depending on each processor's market requirements.

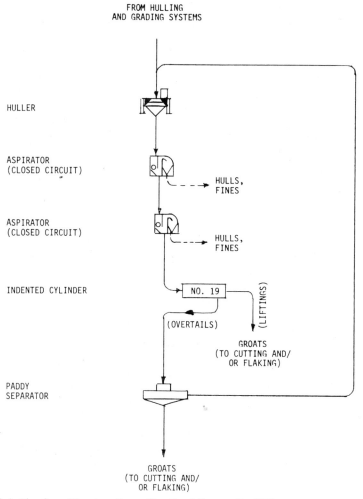

Fig. 26. Rehulling flow. (Courtesy Carter-Day Co., Minneapolis, MN)

VIII. CUTTING AND FLAKING

The groats are now ready for processing into the end products. Broadly speaking, oats for human consumption are sold either as oatmeal or rolled (flaked) oats, or to a lesser extent, as oat flour. Rolled oats constitute the main end product, and, as the name implies, these are produced by flattening the groats between rolls under heavy pressure. To prevent disintegration of the groats and production of fines under heavy pressure, a binding agent must be used to hold the groat particles together. This is achieved by subjecting the groats to steaming immediately ahead of rolling, the heat and moisture serving to bind the flattened groat together. Whole groats produce extremely large flakes that are difficult to handle, store, and package; consequently, the groats are generally cut into uniform pieces, two to four per groat, immediately before steaming. This results, after rolling, in a flake size that is more compatible with packaging requirements and consumer acceptance. Figure 27 illustrates a typical groat-cutting flow.

Groats are cut with rotary granulators, machines consisting of rotating drums perforated with countersunk round holes through which the groats align themselves endwise and fall against stationary knives that are arranged around the bottom and outside of the drum (Fig. 28); there are usually four or six individual drums per machine. The groats are fed into the inside of each drum through chutes at the side of the drums, which are rotating at about 37–40 rpm. The groats fall into the countersunk perforations and come endwise against the series of stationary knives on the outside and bottom of the drums. As the tip of the groat protrudes through the perforation, it is sheared off by the first knife it encounters and, as the groat continues to fall through the perforation, it encounters the next knife as the drum is rotating. This action continues until the entire groat has passed through the perforation and finishes up in two to four pieces. The rotation speed of the drums obviously has an influence on the number of pieces that are produced from a single groat: the slower the speed of rotation, the fewer the groat pieces that will be produced; the faster the speed, the greater the number of pieces. The initial size of the groat also influences the number of pieces of oat groat produced.

The aim is to cut each groat into the minimum number of pieces consistent with the end product requirement because some fines are inevitably produced each time a groat is cut. The condition (sharpness) of the cutting knives has a large influence on the amount of fines: a well-maintained rotary granulator produces about 1.2–1.3% fines, and seldom above 2%; a badly maintained granulator with blunt or incorrectly positioned knives, or with the wrong drum speed for the size and condition of the groats being processed, can produce up to 10% fines, thereby drastically reducing overall process efficiency. Some processors prefer to steam the groats immediately ahead of cutting, rather than after cutting and just before flaking, claiming that the binding action of the heat and moisture reduces the amount of fines produced by the cutting knives. In other operations, groats are conditioned for cutting by adding a small amount of water, rather than steam, about 15–30 min before cutting. The cut groats are then steamed in the conventional way immediately ahead of flaking.

The fines produced by cutting, consisting of oat middlings and flour, are then removed by a shaker screen (or small sifter) fitted with a 22 mesh tin-mill screen

(0.032 in. [0.8 mm] opening), although different mesh screens may be used by various processors. The oat middlings and flour are frequently used as a high-quality, high-protein animal feed; again, though, end use depends on each processor's market requirements.

The cut groats are separated from any uncut groats by length-sizing separation on a disk separator using V5-1/2 indent pockets (or a No. 13 indented cylinder separator) (Fig. 27) that lift out the shorter cut groat pieces and reject, or overtail, any longer, uncut oats. The uncut groat fraction is recycled back to the rotary granulator, and some processors prefer to aspirate this recycled fraction to remove any hull pieces before they are returned to the cutters.

The cut groats are now ready for the flaking stage shown in Fig. 29. Conditioning the groats for flaking is accomplished by steaming them with live

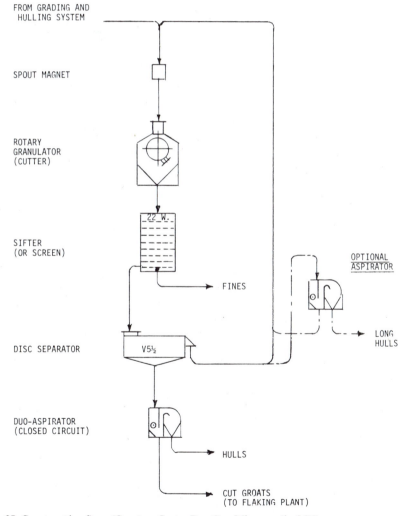

Fig. 27. Groat-cutting flow. (Courtesy Carter-Day Co., Minneapolis, MN)

steam at atmospheric pressure immediately ahead of the flaking rolls. The steaming softens the groats and permits flakes to be formed with a minimum of breakage, and the heat and moisture also complete the inactivation of the enzyme systems that would otherwise cause rancidity and undesirable flavors in the oatmeal.

Although different types of steamers are used, a common one consists of a gradually expanding, cylindrical or rectangular column in which steam nozzles are carefully arranged to ensure that all areas of the slowly descending mass of

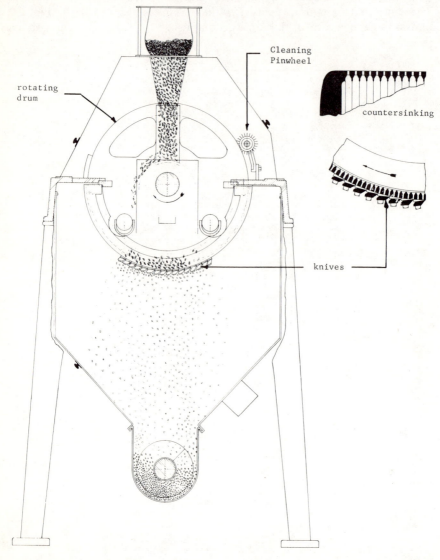

Fig. 28. Schematic of rotary granulator. (Courtesy Kipp-Kelly Ltd., Winnipeg, Manitoba, Canada)

cut groats are subjected to the same amount of steam injection (Fig. 30). Consistent and homogeneous heat and moisture distribution are essential to ensure good quality flakes and yield; too much moisture results in a high percentage of agglomerates, whereas too little produces excessive small flakes and fines. Ideally, the steamer is equipped with automatic controls whereby temperature probes control the amount of steam injected according to predialed and continuously recorded settings. Some processors use horizontal steamers in the form of two- or three-tier screw conveyors with mixing flights and with steam nozzles located in the troughs. This type, however, requires more frequent cleaning to remove the residue that accumulates in the bottom of the trough and

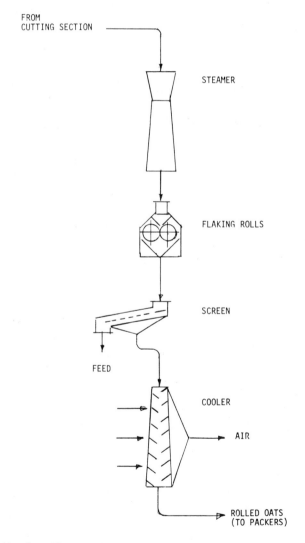

Fig. 29. Oat-flaking flow. (Courtesy Carter-Day Co., Minneapolis, MN)

on the screw flighting; the expanding vertical-column steamer tends to be more self-cleaning because of the continuous downward movement of the cut groat mass.

Retention time in the steamer is about 12–15 min, during which time the cut groat temperature is increased from ambient to 210–220°F (99–104°C). The

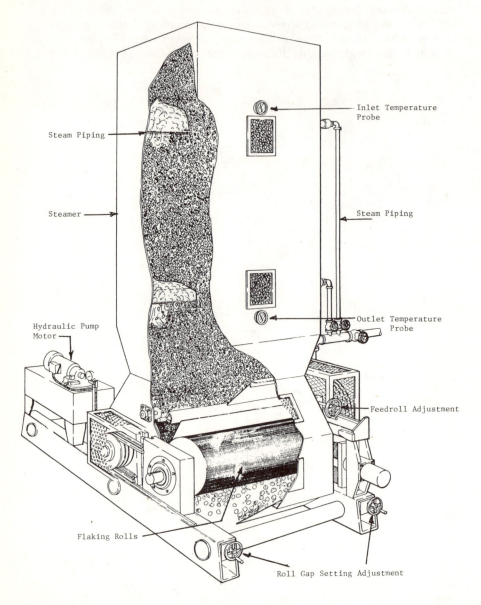

Fig. 30. Schematic of vertical steamer and flaking roll. (Courtesy Roskamp Mfg., Inc., Cedar Falls, IA)

steam increases the groat moisture from about 8–10% when entering to about 10–12% at the steamer outlet.

We noted earlier that some processors prefer to steam the groats before cutting, claiming that the steam conditioning toughens the groats and results in fewer fines during cutting. When this approach is taken, the steamed and cut groats from the rotary granulator pass to the flaking roll.

Regardless of the cutting approach taken, the steamed, cut groats pass directly into the rolls from the steamer. The flaking roll consists of two rigid end frames supporting two cast iron rolls that measure, depending on capacity requirements, from 12 in. diameter × 30 in. long (30.5 × 76.2 cm) to 28 in. diameter × 52 in. long (71.1 × 132.1 cm). A feed gate above the rolls spreads the groats from the steamer along the full length of the roll nip area. The rolls run at zero differential (i.e., the same speed) and rotate at from 250 to 450 rpm, depending on roll diameter and processor preference.

The cut groats are rolled into relatively thin flakes for instant or quick-cooking oatmeal; these flakes have an average thickness of around 0.010–0.015 in. (0.25–0.38 mm). Uncut groats are flaked about 50–75% thicker, or to a thickness of about 0.020–0.025 in. (0.5–0.63 mm) and marketed for regular oatmeal. The moisture content of rolled oats after flaking is about 10–12%. The roll gap distance and nip pressure, which on modern flaking rolls are hydraulically controlled by oil or compressed-air pressure, are adjusted to produce uniform flakes of the thickness desired. The roll gap distance controls the flake thickness and the hydraulic system exerts sufficient pressure to keep the two rolls in the set position.

Following the rolls, the flakes pass directly to a shaker screen (or sifter) where any fines produced in the flaking process are removed. Overcooked pieces, which are generally agglomerates of several flakes, are also scalped off. Finally, the flakes go to a cooler to reduce moisture and temperature to ensure acceptable shelf life. The most common form of cooler is the multilouver type, in which the flakes pass slowly down vertical, rectangular columns through which ambient air is being drawn. Some processors prefer band-type coolers, in which a slow-moving mesh band conveys the flakes from one end of an enclosed housing to the other while ambient air is drawn through the thin, moving bed of flakes.

Flakes enter the cooler, regardless of type, with a moisture content of around 10–12% and a temperature of about 200° F (93° C) and leave the cooling process at 9–11.5% moisture and a temperature of around 110° F (43° C), ready for packing.

The cooled flakes, which represent from 50 to 60% by weight of the oats processed, depending on the quality of the oats and the general efficiency of all steps in the processing system, are next conveyed to the packaging system. Because of the fragile nature of the quick-cooking, thinner flakes in particular, handling and conveying should be kept to a minimum to reduce breakage. Ideally, the flaking system should be located directly above the packaging equipment. Packaging usually includes weighing the contents of each container to safeguard against any inadvertent density variation or change in flake size, which could result in underweight product content. It is also a good idea to have metal detection devices monitoring the packed products before shipment to safeguard against ferrous metal particles becoming mixed in as a result of fragments becoming detached from machines or packaging equipment.

IX. OTHER OAT PRODUCTS AND BY-PRODUCTS

A. Oat Products

Although the major end product is regular rolled oats or quick-cooking (instant) oats for preparation of oatmeal, some processors also manufacture oat flour. Oat flour is primarily used in baby food cereals and ready-to-eat breakfast cereals, providing a good source of high-protein nutrition and flavor compared with flours produced from most cereal grains. It is generally not of the fineness associated with, say, wheat flour, and it has a coarser granulation; the granulation varies depending on the specifications of the end user and the type of product to be manufactured—breakfast cereal, baby food, etc. Oat flour specifications, for example, can range from 100% through 38 wire mesh (0.0198 in. [505 μm] opening) to 75% through 60 wire mesh (0.0127 in. [316 μm] opening).

In the preparation of oat flour, the oats are thoroughly cleaned and hulled as previously described. The hull content must generally be reduced to a very low level, particularly for the preparation of baby foods. The groats are steamed before milling to improve the stability and complete the inactivation of the enzyme system to prevent rancidity. After steaming, the groats are cut or sheared open on a rotary granulator; some processors use a roller mill with grooves or corrugations cut in the rolls for this initial shearing step. Next, the cut groats pass to an impact-type hammer mill where a substantial reduction in particle size takes place, and the ground product is then sifted on a gyratory sifter equipped with wire meshes to correspond to the granulation requirement of the oat flour. Particles that do not pass through the wire mesh are recycled back to the hammer mill in small plants or routed to another, separate hammer mill for further grinding in larger systems. Some processors use a stone burr mill for this final reduction step, reducing the particle size of the cut groats until all the product sifts through the specific wire mesh.

Because of the comparatively high fat content of the groats, it is necessary to have a large volume of exhaust air drawn through impact-type hammer mills to keep the screen perforations open, to prevent fine product from sticking inside the screen and on the beaters, and to prevent heat buildup in the machine.

A few processors produce oat flour on a series of roller mills and sifters. In this type of system, a series of three or four break roller mills gradually reduce the particle size of the groats on grooved or corrugated rolls, the fines produced being sifted out after each grinding passage. Since the groats have a high fat content and have been steamed, only a very gradual reduction of the groats is carried out at each passage to prevent product from sticking in the roll corrugations. From the sifters, particles that are not sufficiently fine pass to roller mills equipped with smooth rolls, which reduce the flour down to the required granulation.

B. By-Products

OAT HULLS

At least 25% of oats is hulls, and consequently a large volume of hulls must be ground and sold. Impact-type hammer mills are used for this purpose. Most hulls are used as a high-fiber animal feed ingredient or for the production of industrial solvents.

FEED OATS

Light, double, and thin oats that are removed during the cleaning operation are generally used in animal feeds. Although undesirable for milling, these oats are almost equal to normal oats nutritionally for livestock feeding purposes. Typically the feed oats represent about 8–11% of the original oats, although the quantity can vary considerably depending on the quality of the oats purchased and the region of origin.

MIXED GRAINS AND SEEDS

About 2–3% of mixed grains, comprised of corn, wheat, barley, soybeans, sunflower seeds, and weed seeds, are removed during the oat cleaning process. These mixed grains are generally ground on impact hammer mills and sold as an animal feed ingredient.

OAT MIDDLINGS

The fines produced from cutting and flaking normally represent about 3–5% in an efficient plant. Depending on market conditions prevailing for the individual processor, the fines are either sold as a high-grade animal feed supplement or blended into the oat flour milling system.

X. OAT CONVEYING AND EXHAUST SYSTEMS

As these descriptions of the various stages of oat processing suggest, an oat processing plant requires a good deal of both horizontal and vertical conveying of the product from one machine to another, or from one stage to the next. The continuous movement of oats within machines and in conveying equipment creates a considerable amount of dust composed of very small husk fragments and field dirt, which must be carefully controlled for reasons of safety and operator comfort.

For vertical and horizontal conveyance of oats in the intake and cleaning systems, bucket elevators, screw conveyors, or chain conveyors are preferred over positive- or negative-pressure pneumatic conveying lines; oat hulls are most abrasive, and they wear out the pneumatic conveying components (rotary air locks, pipe bends, and cyclone receivers) very quickly, resulting in frequent repair or replacement and high maintenance costs. In addition, pneumatic conveying of oats requires higher power consumption than mechanical conveying equipment, and this represents an ongoing and ever-increasing operating cost. Pneumatic conveying generally takes from 5 to 10 times more power to convey vertically a given weight of oats the height of an average processing facility. Lining the bucket elevator heads with polyurethane, or other abrasion-resistant material, helps to reduce wear and tear.

Gravity spouting between oat cleaning and sizing equipment should be arranged so as to avoid long, straight runs that cause high oat kernel velocities and resultant impact wear on machine inlets, as well as kernel breakage. Self-cleaning checks or "dead" boxes should be used to break long spout runs at changes in spout direction so that oat kernels impinge on other kernels rather than on the spouting. When the flow of oats stops, the check boxes clean themselves out, preventing stagnant pockets that could become a source of infestation.

Areas on or within cleaning and sizing machines that come into contact with the mass of moving or sliding oats should also be protected with such materials as urethane liner plates, abrasion-resistant steel, or ceramic tiles. Machine inlet and discharge chutes and pans underneath sieve screens come into this category of areas subject to rapid wear, and they can benefit from suitable protection. Disk pockets, indented cylinder shells, and slotted sizing shells can be supplied with electroless nickel-plated surfaces to better withstand abrasion from the oat hulls during the separating process.

Adequate protection of conveying equipment and machines, though entailing higher initial costs, drastically reduces long-term repair, replacement, and maintenance costs, and most oat processors consider the expenditure to be a worthwhile investment.

Air contamination can be a major problem in oat cleaning and hulling systems. Small oat-hull sliver fragments and dust escaping to the mill atmosphere can cause considerable operator discomfort in the form of eye and skin irritations. To ensure general plant cleanliness, oat cleaning and conveying equipment must be efficiently exhausted by a well-designed, central system that exhausts air and dust at all points. A central exhaust system consists of an exhaust fan to move the air and create suction, a fabric-filter dust collector, and a network of piping or trunking running from the filter collector to all machines and points to be exhausted. Fabric-filter dust collectors should be selected to operate at liberal air-to-cloth ratios to permit efficient self-cleaning of the fabric filtering tubes and to avoid a buildup of resistance in the system from dust clogging or blinding the tubes. Air-to-cloth ratio is, as the name implies, the ratio of the air volume (cubic feet per minute) entering the filter collector to the fabric cloth area (square feet) inside the filter. Depending on the efficiency of the type of filter collector used, an air-to-cloth ratio of 8:1 (8 ft^3/min per square foot of cloth area), and an absolute maximum of 10:1, is preferred because oat hulls and slivers tend to cling to the filtering cloth media, making self-cleaning of the tubes more difficult than with exhaust systems operating on most other cereal grains.

The cloth tubes inside a fabric filter are usually automatically cleaned by periodic injection of a blast of air from a positive-displacement pump or a high-pressure centrifugal fan onto the inside surface of each tube. The blast of air creates a shock wave down the fabric, loosening the dust that has settled on the outside surface of the tubes as the dust-laden air of the exhaust system passes through the fabric. The loosened dust falls down into a hopper located under the series of cloth tubes and leaves the filter collector through a rotary air lock on the hopper outlet. After a period of time (several months or longer, depending on air-to-cloth ratio and filter efficiency), the filtering media eventually become partially blocked or blinded with dust; the tubes must then be removed and cleaned, or the increased resistance to the air being drawn through the cloth by the exhaust fan drastically reduces the exhaust air on the machinery and conveying equipment, which in turn leads to dust in the plant atmosphere. Fabric filter collectors that enable the cloth tubes to be removed and replaced from the top are desirable because the operator is able to work on the clean-air side of the unit rather than having to enter the dirty side to perform maintenance.

XI. SUMMARY

Previous sections have described the various steps in the oat milling process in general terms. Throughout the industry, each processor has preferred methods and approaches that are suited to particular market requirements; therefore, the process flows and machine details stated are meant to be typical and not necessarily applicable in all cases. Regardless of the individual approaches taken, the overall processing requirements are as described, and variations on them are a means to an end and not another form of oat processing. Whether groats are steamed directly ahead of flaking or directly ahead of groat cutting is a preference based on the experience and requirements of the processor for maximum efficiency. Steaming of the groats remains an essential part of the process regardless of the stage at which it is performed. Similarly, whether oats are graded according to size for hulling during the cleaning process or directly ahead of the hullers is a preference based on results obtained from experience; the fact remains that the oats must be graded for maximum efficiency on the hullers. Each processing stage is an essential part of the overall oat-milling process. Methods may vary, but no step can be eliminated, nor can shortcuts be introduced without sacrificing either quality or efficiency, or both.

Looking to the future, we see no radical changes in the fundamental steps of converting raw oats into rolled oats or oat flour that are followed today. The oats will still have to be cleaned, the hulls removed, and the groats cut, steamed, and rolled or ground. The methods used in achieving each of the processing steps could be changed and improved in future years as technology advances.

The present-day range of oat cleaning equipment, covering screening, grading, sizing, aspiration, and specific-gravity separation functions, is generally efficient, but we expect that refinements and improvements will evolve over the years ahead. The separations may be monitored and controlled by sensors to remove the human adjustment element and to achieve peak operating efficiency at all times. Such factors as the speed, throw, and inclination of screening sieves, the intensity of an aspiration, and the speed of sizing reels in relationship to feed rate may one day be sensor controlled to ensure that no good product is lost with the by-products of less commercial value. Drying of oats may, in the future, be accomplished by infrared heat systems when solar energy has been developed to the point of economic feasibility. Removal of oat hulls may be achieved by ultrasonics with the advancement of sound technology and cheaper natural energy. Small laser light beams may cut the groats for flaking at unimaginable speed and with no fines production, replacing the mechanical method that has been in existence for over a century with little improvement in efficiency.

The roll pressure and speed of flaking rolls may be automatically controlled by sensor and computer to ensure an absolutely consistent flaked end product, regardless of input product variations. As the technological advances over the last 50 years have clearly demonstrated, the limits to advancement in processing technology are far distant; what today seems unlikely or unfeasible may well become a reality tomorrow.

In spite of the potential for technical advancement over the next century, it is necessary to be realistic and to appreciate that oat processing, like other cereal milling processes, is more of an art than a science. The raw material varies in characteristics and quality from one crop year to the next, and the processing requirements are not labor-intensive. Reducing processing costs by replacing manual labor with automated systems is, therefore, difficult to achieve; each advancement in process technology must, of necessity, be financially justifiable in relationship to savings in labor, improved efficiency, or both. With this in mind, we believe that the oat processing technology described in this chapter is likely to change and advance, but only over a long period of time.

LITERATURE CITED

F. H. SCHULE GMBH. Undated. High Capacity Paddy Separator Bulletin. No. 1730E-4712. F. H. Schule GMBH, Hamburg, West Germany.

HUTCHINSON, J. B., MARTIN, H. F., and MORAN, T. 1951. Location and destruction of lipase in oats. Nature (London) 167:758-759.

LOCKWOOD, J. F. 1960. Flour Milling, 4th ed. Northern Publishing, Liverpool.

LOUFEK, J. 1944. Milling of rolled oats. Milling Prod. 9:18-22.

OLIVER MANUFACTURING CO., INC. Undated. Gravity Separator Operating Instructions Manual. Oliver Mfg. Co., Rocky Ford, CO.

SALISBURY, D. K., and WICHSER, W. R. 1971. Oatmilling—systems and products. Bull. Assoc. Oper. Millers May:3242-3247.

THORNTON, H. J. 1933. The History of the Quaker Oats Company. University of Chicago Press, Chicago.

CHAPTER 14

OAT UTILIZATION: PAST, PRESENT, AND FUTURE

FRANCIS H. WEBSTER
John Stuart Research Laboratory
Quaker Oats Company
Barrington, Illinois

I. INTRODUCTION

Oats have been grown for thousands of years; their primary use, however, has been as a forage crop and a feed grain rather than a food for people. The first reference to human consumption is by Pliny in his *Natural History*. He reported that the Germanic tribes of the first century knew oats well and made "their porridge of nothing else." Despite evidence of early oat consumption, the grain found long-term acceptance only in Ireland and Scotland, where it was used for a variety of porridges. Oatmeal was brought to the New World by Scottish colonists but was not widely accepted as part of the diet. Samuel Johnson said, "Oats—a grain which in England is generally given to horses, but in Scotland supports the people." To which, according to Sir Walter Scott, Lord Elibank is reputed to have replied: "True, but where will you find such horses, where such men?"

During the early 19th century, almost all oatmeal available in this country was imported from Scotland and Canada and sold almost exclusively in pharmacies. Most 19th-century U.S. cookbooks are reported to have either completely omitted recipes for oatmeal or, at best, suggested it as a food for the infirm. *Cookery As It Should Be,* published in 1859, suggested oatmeal for the sickroom and recommended the addition of a huge spoonful of brandy "if the patient could bear it"! Based on this beginning, oats have developed an image as a wholesome and nourishing, hot breakfast cereal. Oats are generally recognized as having the highest protein and lipid content of any cereal grain, and they are also known to be a rich source of dietary fiber (see Chapters 10 and 11).

Three developments in the late 19th century helped establish oatmeal as a staple breakfast item: the large-scale domestic milling of oats; movement of oatmeal from the druggist's shelf to the grocery; and development of packaging, brand names, and promotion. These innovations, pioneered by Ferdinand

413

Schumacher, a German immigrant, have developed oatmeal into the largest selling cereal in the United States (hot or cold). Current volume (1985) for the hot cereal industry (which includes oat, wheat, and multigrain cereals) is about 360 million pounds (164 million kilograms). Oatmeal-based products are the biggest portion of the hot cereal industry, contributing 66% of all hot cereal tonnage.

This chapter focuses on commercial oat products and their utilization as food products and in specialty application. Finally, potential new applications for oats or oat fractions described in the preceding chapters are briefly evaluated.

II. PRODUCTS AND APPLICATIONS IN FOODS

Oats are unique in their uses and attributes in comparison to most other cereal grains. First, they are used with rare exception as a whole-grain flour or flake; in contrast, the germ and significant portions of the bran are generally removed from other grains before they are introduced into food systems. Second, oats are heat processed to develop the characteristic toasted-oaty sensory notes. Enzyme inactivation is another objective of the heat treatment. Untreated oat flours rapidly develop soapy and bitter flavors because of the action of lipase, lipoxygenase, and peroxidase (Hutchinson et al, 1951; Kazi and Cahill, 1969; Biermann and Grosch, 1979). This heat treatment also has a major impact upon the functional characteristics of oat products; for example, protein solubility is dramatically reduced. Commercial oat product heat treatments are regulated by residual enzyme measurements. Oat product specifications typically call for materials to be either tyrosinase negative or peroxidase negative. Tyrosinase activity measurements are used as an indicator of lipase inactivation. This enzyme is more heat stable than lipase and provides a rapid analytical test for oat processors. Peroxidase is the most heat stable of these enzymes, and it must be monitored directly to assure elimination.

A. Commercial Oat Products

The commercial oat processor purchases and mills about 175 kg of oats to produce 100 kg of food-grade oat products; Table I lists typical yields of products and by-products. This presentation assumes that the mill is operating to produce only two products: quick and regular oat flakes. In practice, a variety of flake types or oat flour could be produced while maintaining the same yield figures.

TABLE I
Typical Yields of Oat Mill Products and By-Products[a]

Product/By-Product	Yield (lb)	Percentage of Total
Feed-oat extraction	15	8.6
Mixed-grain extraction	5	2.9
Hulls and screening	45	25.7
Oat middlings	10	5.7
Quick-cooking flakes	80	45.7
Regular flakes	20	11.4
Total	175	100.0

[a] Source: Youngs et al (1982); used by permission.

Mill yields vary depending upon the milling quality of the oats and mill operating efficiency. The oat berry or groat is encased in a hard, fibrous hull that requires rigorous processing to remove. The groat-to-hull ratio, which varies with oat variety and growth conditions, can have a significant impact upon yield. Oats are generally purchased on the basis of milling quality rather than chemical composition or functional attributes. Some processors in the United States specify a minimum protein content; however, the U.S. oat crop typically exceeds this minimum value.

Typical commercial oat products are rolled oats, steel-cut groats, quick oats, baby oat flakes, oat bran, and oat flour. Many intermediate flakes and flour are produced by oat processors to meet individual customer specifications. Descriptions of typical products follow:

1. Rolled oats, which are produced by flaking whole groats, are the thickest of the standard oat-flake products; flake thickness varies from 0.020 to 0.030 in. (0.508–0.762 mm) depending upon the intended end use. The thicker flakes require longer cooking periods and maintain flake integrity for extended time periods.
2. Steel-cut groats are produced by the sectioning of groats into several pieces; they are used in the preparation of flakes and flour and as a specialty ingredient.
3. Quick oats are flakes produced from steel-cut groats. In this process, oat groats are typically cut into three to four pieces before the final steaming and flaking process. Quick oats, which are usually 0.014–0.018 in. (0.356–0.457 mm) thick, require less cooking time than the whole-oat flakes.
4. Baby oats are also produced from steel-cut groats, but the flakes are thinner and have a finer granulation than quick oats. These smaller, thinner flakes cook more rapidly than quick oats and have a smoother texture.
5. Instant oat flakes are produced from "instantized" steel-cut groats. Before cutting, the groats have been subjected to a special proprietary commercial process that results in rapid-cooking flakes. The flakes are typically 0.011–0.013 in. (0.279–0.330 mm) thick.
6. Oat flour is produced by grinding flakes or groats into flour for use as an ingredient in a wide variety of food products.
7. Oat bran is a bran-rich fraction produced by sieving coarsely ground oat flour.

B. Food Applications

In comparison to the literature on other cereal grains, there is a paucity of published material on oat applications and functionality. The premier use of oats is in hot breakfast cereals; most other uses could be classified as specialty applications that result from the unique flavor and texture, moisture retention, or stabilizing properties of oats. The two factors that differentiate oats from most other cereal ingredients, its whole-grain identity and its thermal processing requirement, also serve to limit its application potential. Nevertheless, oats have a substantial number of applications and are used throughout the food industry in products that benefit from the unique characteristics of oats.

HOT CEREALS

Hot breakfast cereals are the number-one application for oat flakes. In fact, oatmeal (all types combined), according to one source, is the number-one dry cereal in the U.S. market (Anonymous, 1980). Products vary from rolled oats (whole-oat flakes), which require five or more minutes to prepare on the stove top, to prepare-in-the-bowl (instant) products made from instantized cut groats. The instant products are generally fortified and are sold in a wide variety of flavors. In addition to convenience, the prepared products differ considerably in other characteristics, especially texture. The larger, longer-cooking flakes retain flake identity and texture, and they exhibit the inherent gum (β-glucan) characteristics; the smaller, instant flakes tend to lack flake identity and thus are smoother in texture. Instant products also do not exhibit as much of the gum characteristic as the larger flakes because the smaller, instantized flakes have more readily rehydratable starch exposed, which can compete effectively for the available water.

Oat bran is a recent addition to the hot oat cereal market. This bran-enriched product contains about twice the total plant fiber as rolled oats (see Chapter 11). The product can be prepared as a hot cereal or substituted for oat flakes in other applications. The apparent health benefits of this β-glucan-rich fraction make it an attractive ingredient for health-related products.

Manufacturers have developed a number of process variations in attempts to gain a competitive edge in this market. Details of the quick-cooking oats patent literature have been summarized by Daniels (1974).

COLD CEREALS

Oat flour is a frequently used ingredient in cold cereals. Youngs et al (1982) conducted a survey of cold cereals and found that oats were the major ingredient in 14.6% of 82 cold cereals surveyed; wheat and corn were the major ingredients in 36.6 and 26.8%, respectively. The high fat and fiber contents of oat flour tend to restrict its utilization in highly expanded extruded cereal products.

Production of oat-based cereals requires the processor to exercise control over process temperature and final product moisture. Martin (1958) demonstrated that the stability of ready-to-eat, fully gelatinized crisp oat flakes was dependent upon duration of cooking, drying method, and final product moisture content. Reduction of the final moisture content below critical values caused the rapid onset of oxidative rancidity. Puffed oat products are also highly susceptible to oxidative rancidity (Martin, 1958). Although oats have a potent natural antioxidant system, it can be rendered ineffective by overprocessing. Thermal breakdown is one factor that may be responsible for oat antioxidant destruction (see Chapter 9); however, specific investigations need to be conducted to determine the fate of oat antioxidants in processed cereals.

BREAD PRODUCTS

Oat products are ingredients in many bakery items, to which they add important benefits. First, oats have excellent moisture-retention properties that keep breads fresher for longer periods of time (McKechnie, 1983). Dodok et al (1982a) concluded that up to 15% of the wheat flour in biscuit formulations could be replaced with oat flour, which improved product consistency and shelf life. In a related study, Dodok et al (1982b) concluded that, in addition to the favorable physical effects, oat flour was able to stabilize the fat component. This phenomenon may be related to oat antioxidant properties. Oomah (1983) investigated composite blends of wheat and oat flour in bread making. Dough development time, dough strength, centrifuge water retention, and loaf volume decreased and mixing tolerance increased as the oat flour level increased. D'Appolonia and Youngs (1978) reported that oat bran increased farinograph absorption and produced doughs with good stability but decreased loaf volume. These observations are explained by the basic differences in the protein composition of oats and wheat (see Chapter 7). A study on the effects of cereal malts on bread making indicated that oat malts were inferior to other cereal malts (Finney et al, 1972).

Oat products are used in some breads to appeal to the health-conscious market. Oats provide protein, dietary fiber, and the whole-grain image. Bakers are able to produce breads with a wide variety of textures by varying the oat product type, level, and method of introduction. McKechnie (1983) provided the following description of oat product application to breads. Rolled oats are typically added to variety breads when some flake identity and a chewy texture are required in the final product. Quick or baby oats provide good moisture retention, less flake identity, and a smoother texture, and steel-cut oats add a seed or nutlike texture to variety breads. Oat flour at levels up to 30% can be used to replace wheat flour, with the primary benefit being moisture retention and freshness. Oat flour alone may not be used to make bread because it lacks the gluten component.

COOKIES

Oats are a major ingredient in many cookie mixes (Smith, 1973; McKechnie, 1983). Inclusion of oat ingredients in cookie mixes affects the absorption of dough water and the flavor and texture of the final product. The choice of flake type is influenced by the desired end result. Since oat flake types differ in flake size, thickness, and amount of fines, each produces corresponding differences in the finished cookie texture profile. Larger flakes retain flake identity, whereas baby oats provide a smoother texture. Steel-cut oats impart a nutlike flavor and texture.

Water absorption is crucial in cookie doughs (Smith, 1973; McKechnie, 1983). Oat product absorption rates vary with the size and thickness of the oat ingredient particle and temperature; to produce the same type of cookie consistently, it is thus necessary to maintain control of oat ingredient particle size distribution and moisture absorption characteristics. The flake moisture should also be maintained within a narrow range. Control of these factors, plus dough temperature and lay time, are essential in oatmeal cookie manufacture.

Smith (1973) stated, "The inclusion of an oat product confers crispness to a cookie and, some bakers allege, allows a reduction of the shortening. Conversely, because of oats' water absorption potential, it allows use of increased shortening in the richer types of cookies." Unfortunately, very little information has been published on the impact of oat ingredients on cookie doughs. Oomah (1983) reported that cookie spread factors increased with increased amounts of roller-milled oat flour, whereas those of a commercial oat flour remained constant. These differences may have been influenced by differences in bran level, but the data were not clear on this issue. Jeltama et al (1983) investigated the effect of dietary fiber components upon cookie quality. Various cereal (including oats) and legume brans were added to sugar cookies. In general, their results indicated that increased fiber levels were associated with an increase in tenderness and moistness and with decreased spread.

INFANT FOODS

Oatmeal or flour is a major component of many infant foods. Drum-dried oatmeal is typically an infant's first introduction to solid food in many parts of the Western world, and oat flour is used as a thickener in many commercial infant foods. Important considerations in the selection of oats for this use include allergenicity, flavor compatibility, nutritive value, shelf life and stability, economics, and availability (Shukla, 1975). Numerous formulations have been developed for infant foods based on these criteria. Other ingredients are often mixed with oats to achieve a particular nutritional objective, including milk solids, soy flour, cottonseed flour, wheat germ, sugar, vegetable oil, malt, dried yeast, and vitamin and mineral mix. For example, a soy-oat infant formula was developed in Mexico to combat malnutrition (Table II). This product is nutritionally equivalent to commercial milk-based formulas; it is suitable for feeding lactose-intolerant infants, and it uses inexpensive, locally grown ingredients (Mermelstein, 1983).

OTHER OAT APPLICATIONS

Oatmeal is a major ingredient in granola cereals and bars. Rolled oats are used as a thickener in soups, gravies, and sauces, as a meat extender in meat loaf and meat patties, and in many home-cooked baked goods. Specialty flours have been developed that can stabilize dispersions or add viscosity to a wide variety of food products (Shukla, 1975); the stabilizing effect is attributed to the oat β-glucan. Oat flour is the base for *frescavena*, a popular South American beverage; oat-based soups are also quite popular in these countries. Oats have also been proposed as a main-meal side dish (Webster, 1983). Rashbaum et al (1983) reported on a fermentation process to produce a natural red color using oats as a substrate.

In earlier chapters in this monograph, the authors have described research conducted on fractionation of oats into their component parts. The potential

TABLE II
Soyaven (Soy-Oat) Formulation Developed
for Use by Infants in Mexico[a]

Ingredient	Percentage
Dehulled soybeans	32.1
Pearled oats	25.6
Sucrose	34.1
Vegetable oil	5.8
Tricalcium phosphate	1.2
NaCl	0.5
DL-methionine	0.2
Vitamins and minerals	0.4
Flavoring	0.1

[a]Source: Mermelstein (1983); used by permission.

applications of oat protein, starch, β-glucan, and oil are outlined in the respective chapters. Interested readers should refer there for complete details. Although a great deal of interest has been created in the attributes of specific components, none of the processes has been developed commercially.

C. Oat Antioxidants

Oats usage as an antioxidant was proposed by Peters and Musher (1937). Since that time, it has received considerable attention in laboratories and in the scientific trade literature. A specially fine-ground oat flour was marketed for antioxidant purposes and found to be effective to some extent in milk and milk powder, butter, ice cream, fish, bacon, frozen fish or sausage, cereals, and other products that are sensitive to fat oxidation during storage. The oat flour antioxidant was added directly to the product or, in some instances, to the packaging material (Bedford and Joslyn, 1937; Tribold, 1938; Atkinson, 1947; Janecke, 1954).

In the past, several theories were advanced on the nature of the oat antioxidant system. Proteins, phospholipids, tocopherols, and phytic acid were proposed as being responsible for the antioxidant activity (Taufel and Rothe, 1944; Evans et al, 1953). However, more recent research has identified what appears to be the major antioxidant system present in oats. An extensive investigation of oat antioxidants resulted in the isolation of 24 different phenolic compounds that had antioxidant activity. The major components are all derivatives of hydroxycinnamic acid (Daniels and Martin, 1964a, 1968; King, 1968), and several types have been identified. Eight were esters of long chain monoalcohols, diols, or α-hydroxyacids, and others were found to be glycerol esters (Daniels and Martin, 1964b, 1965, 1967, 1968; King 1968). The antioxidant activity is attributed to the caffeic and ferulic acid contents (Daniels and Martin, 1961). Structurally, caffeic and ferulic are very closely related to the common commercial antioxidants butylated hydroxyanisole and butylated hydroxytoluene (Fig. 1). Daniels and Martin (1961) demonstrated experimentally the antioxidant potential of oat antioxidants (Table III). The test, based on the increase in the induction period of oat oil oxidation, showed that on an equal weight basis the purified oat antioxidant had effectiveness equal to that of commonly used commercial antioxidants.

Oat antioxidants have been used to only a limited extent. The availability and effectiveness of synthetic antioxidants have led to their widespread usage, displacing oat-based and other natural antioxidants. Widespread health concerns, however, have resulted in the total ban or restricted usage of synthetic antioxidants by some countries, and manufacturers may have to look elsewhere for solutions to food stability problems. Natural antioxidants provide an obvious and readily implementable solution to some of these problems. Oat-based antioxidant systems are an alternative that food manufacturers could use to stabilize their products without raising concern among health-conscious consumers. Further, natural antioxidants can be used to provide protection where usage of synthetic antioxidants is restricted by food type or fat level. Natural antioxidants such as oat flour or its extract will not solve or be appropriate for all applications, but they have a place in food quality preservation.

III. INDUSTRIAL AND OTHER NONFOOD USES

Several unusual applications have developed for oats and oat fractions over the years. Interestingly, some of these uses stem from the early implication of oats as having medicinal properties, whereas others were developed to utilize the by-products of the oat milling operation. In some cases, little published information is available to provide a clear understanding of the observed functionality.

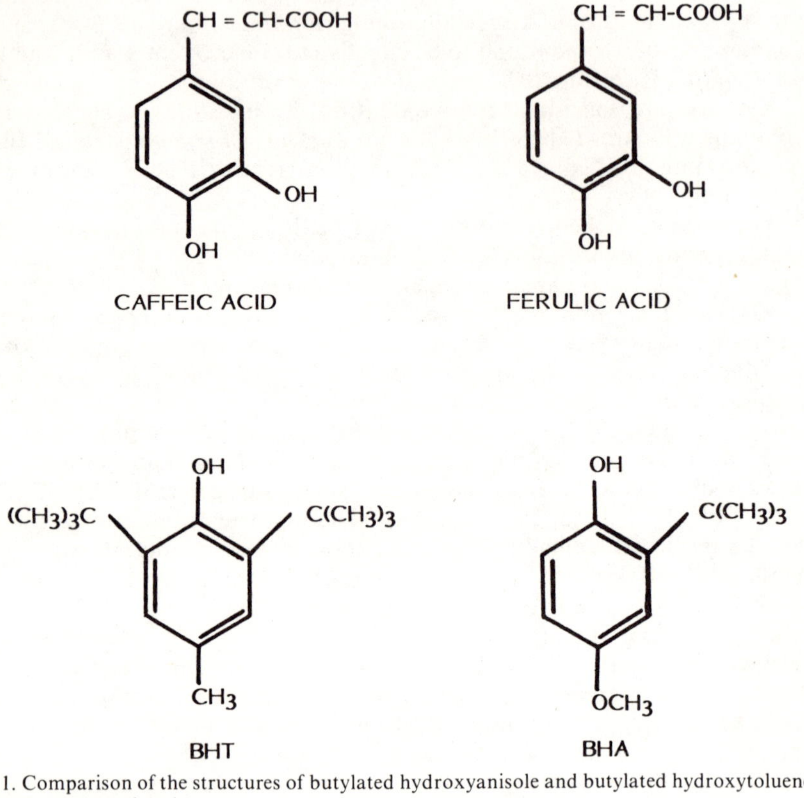

Fig. 1. Comparison of the structures of butylated hydroxyanisole and butylated hydroxytoluene to those of oat phenolic acids.

TABLE III
Relative Antioxidant Activity of Oat Antioxidant
and Its Constituent Structures[a]

Substance	Relative Activity Units
Purified oat antioxidant	30
Caffeic acid	90
Ferulic acid	14
Neutral component	0
Propyl gallate	30
Butylated hydroxytoluene	30

[a]Source: Daniels and Martin (1961). Copyright Macmillan Journals Ltd.; used by permission.

A. Adhesives

Several investigators have proposed the use of oat starch in adhesive applications (Kessler and Hicks, 1949; Secondi and Secondi, 1955; Waggle, 1968). The ready availability of wheat and corn starch, however, has prevented attempts to commercialize this concept. Caldwell and Pomeranz (1973) suggested that oat starch might find application in cosmetic or pharmacological preparations because of its unique granular size.

B. Cosmetic Products

Cosmetic manufacture is the only significant nonfood, nonfeed utilization of oatmeal or flour. Oatmeal has been used in facial masks and bath oil as a soap replacement in order to obtain relief from itching (Brownlee and Gunderson, 1938). The cleansing action of oatmeal preparations has been related to its slightly higher acidity relative to the skin, which reportedly enhances its ability to absorb dirt and sebaceous secretions (Caldwell and Pomeranz, 1973); the oatmeal preparation is thought to leave a protective film that is attributed to the colloidal properties of the β-glucan (Stanton, 1951). Oat oil has also been suggested as playing a role in the cosmetic properties of oatmeal (Jainstyn, 1957). Some cosmetic claims suggest that oatmeal has hypoallergenic properties. Meister (1963) proposed an oatmeal preparation as a soothing treatment for relief from irritation caused by contact or geriatric dermatoses, eczema, measles, chicken pox, pityriasis rosea, and sunburn. Early patents (Musher, 1944, 1948) made similar claims about oatmeal usage in facial packs and cosmetics. Commercial products include bars for normal cleansing, for dry skin, for acne and oily skin, a shampoo, and a bath preparation. The benefits of oat flour in cosmetics were reviewed by Miller (1977).

C. Cariostatic Properties

Oat hulls contain a factor that has cariostatic activity. A study on the cariogenic potential of cereals and the relationship to the degree of refining showed that whole-grain cereals were less cariogenic than refined or processed ones (Constant et al, 1952). Subsequent studies demonstrated that the anticaries fraction was located in the hulls (Constant et al, 1954; Buttner and Muhler, 1959). The oat-hull anticaries factor does not appear to be related to the ash, protein, fat, fiber, or total phosphorus content (McClure, 1964). Organic phosphates (McClure, 1964), polyphenols (Vogel et al, 1962), and a nitrogenous base conjugated with palmitic acid (Buttner and Muhler, 1959) have been implicated as possible cariostatic components. The cariostatic agent can be extracted with acidified ethanol or other polar organic solvents (Taketa and Phillips, 1957; Thompson, 1963). The activity in the ethanol extract was not only effective in inhibition of dental caries but also inhibited growth of cariogenic microorganisms. The anticariogenic factors have not been identified at this time.

Utilization of oat hulls as a cariostatic agent has never passed beyond the experimental stage. Chewing gum containing an oat extract has been tested experimentally, but the development of clinically verifiable evidence on the effectiveness of products containing oat hulls remains as a major hurdle for commercialization. Readers desiring a more complete discussion on anticariogenic factors in cereals should consult recent reviews by Madsen (1981) and Lorenz (1984).

D. Furfural Synthesis

Furfural and furan production from oat hulls represent the successful commercialization of a research program directed at by-product utilization. Furfural is formed by a series of consecutive reactions that are initiated by the hydrolysis of pentosans and completed by a complex cyclodehydration to furfural. (For details of this process, see Chapter 4 in this volume; Dunlop and Peters, 1953; and Dunlop, 1973.) Oat hulls are an excellent source of pentosans. Since the demand for furfurals exceeds the availability of oat hulls, other raw materials are necessary to fulfill industry needs. This need has provided a market for other agricultural by-products: Table IV lists the furfural potential of the commonly utilized pentosan sources. Most vegetable materials contain pentosans, but not all raw materials offer equivalent economic opportunity. Furfural and its derivatives are currently used in petroleum extractions, resins for manufacture of foundry cores and molds, vinyl resin plasticizers, chemical intermediates for insecticides and pharmaceuticals, and formation of polytetramethylene ether glycol (Dunlop, 1973).

E. Other Oat Hull Usage Opportunities

Many other uses have been suggested for oat hulls. Shukla (1975) discussed the potential use of oat hull pentosans (xylans) for the production of xylitol; this concept has been investigated by Buchala et al (1972) and Jaffee et al (1974). Xylose yield from acid hydrolysis of oat hemicelluloses was about 14%. Other proposed uses of oat hulls include an antiskid tread component (Boyle, 1952), a brewery filter aid (Myasnikova et al, 1969), in linoleum (Kent, 1966), a hydrogen peroxide explosive component (Baker, 1962), and as a component of plywood glue extenders (Stone and Robitschek, 1978). Some work has also been conducted on fermentation of oat hulls. Rosenberg et al (1978) studied the production of fungal protein from oat hulls.

F. Oat Enzymes

Some authors have suggested that oats might be utilized as an industrial source of lipase (Caldwell and Pomeranz, 1973; Shukla, 1975). Oats contain significant

TABLE IV
Furfural Potential of Selected Agricultural By-Products[a]

Material	Furfural Potential (%)[b]
Corncobs	23
Oat hulls	22
Bagasse	17
Rice hulls	12

[a] Source: Dunlop (1973); used by permission.
[b] AOAC method, dry basis.

levels of lipase, especially in the bran and aleurone layers. However, no visible attempt has been undertaken to develop a commercial oat-based enzyme preparation.

G. Pharmaceutical Products from Oats

Several pharmaceutical applications have been identified for portions of the oat plant. Because details on these applications are beyond the scope of this article, they will be mentioned only briefly. The most interesting application is in the use of an oat-leaf extract to reduce the craving for tobacco or opium (Anand, 1971). Connor et al (1975) determined that an extract from green plants contained a substance that had morphine-antagonistic properties. Other pharmacological activities include a substance that has properties that stimulate "lutenizing" hormones (Shukla, 1975) and a stimulant for excretion of follicle-stimulating hormones from the cerebral cortex, hypothalamus, and pituitary (Sugaware, 1970). The cholesterol- and insulin-lowering properties of oats are discussed in Chapter 11.

IV. SUMMARY

From their auspicious beginning as a nutritious food, oats have developed into a staple food item, especially at the breakfast table. In the future, oats will continue to occupy this position: the hot cereal market continues to grow, fueled by new products that are designed to appeal to a wide variety of tastes. The nutritional benefits of oats will appeal to health-conscious consumers and strengthen this market position. These events alone, however, will not bring about a dramatic increase in oat utilization.

We will only see major increases in oat utilization as a result of new products that are designed to meet specific market needs. Newly realized nutritional attributes of oats provide one such opportunity. The cholesterol-lowering properties of oats (Chapter 11) should certainly appeal to a society where coronary heart disease is a leading factor in the mortality rate; the insulin-lowering property is another key health benefit. Additional clinical evidence is needed on the effectiveness of oat-based diets in these cases.

The protein, starch, β-glucan, and oil fractions all have potential as food ingredients. In this monograph, the authors of chapters on these subjects have discussed processes developed to isolate each component. Subsequent evaluations demonstrated that each component had useful properties for food and specialty applications. However, none of these processes has been commercialized to date. The major roadblock is by-product utilization. For example, a process was developed to produce a useful oat protein concentrate, but the protein fraction only accounts for about 20% of the starting raw material. The value added is not sufficient for the protein fraction to carry the economic load. Thus, the processor must simultaneously develop a market for the large by-product fraction. Oat breeding directed at increasing the content of specific components—such as protein or β-glucan—might be an approach to improving the economic picture for fractionation processes.

An opportunity exists to create renewed interest in oat antioxidants. Recent changes and potential changes in regulatory attitudes regarding synthetic antioxidants in many countries have caused food processors to evaluate other mechanisms of ensuring product oxidative stability. Natural antioxidants, like those in oats, certainly warrant investigation.

Oats are a cereal with unrealized potential. Their unique composition and multiple nutritional attributes provide a basis for the development of new oat products and ingredients. However, very little information exists in published literature on the functional attributes of oats and oat products. Establishment of a sound technical data base on oat functionality would be a major step toward expanding oat utilization. Completion of these activities could allow oats to compete more successfully with wheat, rice, and potatoes as a component of meals other than breakfast.

LITERATURE CITED

ANAND, C. L. 1971. Effect of *Avena sativa* on cigarette smoking. Nature (London) 233:496-497.

ANONYMOUS. 1980. Food in the A.M. time. Time 115:53.

ATKINSON, I. 1947. Controlling rancidity in frozen sausage meat. Food Ind. 19(Oct.):104-105.

BAKER, A. W. 1962. Hydrogen peroxide explosives. U.S. patent 3,047,441.

BEDFORD, C. L., and JOSLYN, M. A. 1937. Oat flour and hexane extract of oat flour as antioxidants for shelled walnuts and walnut oil. Food Res. 2:455-469.

BIERMANN, U., and GROSCH, W. 1979. Bitter tasting monoglycerides from stored oat flour. Z. Lebensm. Unters. Forsch. 1969:22-26.

BOYLE, T. E. 1952. Antiskid composition. U.S. patent 2,585,219.

BROWNLEE, H. J., and GUNDERSON, F. L. 1938. Oats and oat products. Cereal Chem. 15:257-272.

BUCHALA, A. J., FRAZER, C. G., and WILKIE, K. C. B. 1972. Extraction of hemicelluloses from oat tissues during the process of delignification. Phytochemistry 11:1249-1254.

BUTTNER, W., and MUHLER, J. C. 1959. The effect of oat hulls on the dental caries experience in rats. J. Dent. Res. 35:823-824.

CALDWELL, E. F., and POMERANZ, Y. 1973. Oats. Pages 393-411 in: Industrial Uses of Cereals. Y. Pomeranz, ed. Am. Assoc. Cereal Chem., St. Paul, MN.

CONNOR, J., CONNOR, T., MARSHALL, P. B., REID, A., and TURNBULL, M. J. 1975. The pharmacology of *Avena sativa*. J. Pharm. Pharmacol. 27:92-98.

CONSTANT, M. A., PHILLIPS, P. H. and ELVEHJEM, C. A. 1952. Dental caries in the cotton rat. XIII. The effect of whole grain and processed cereals on dental caries production.

J. Nutr. 46:271-280.

CONSTANT, M. A., SIEVERT, H. W., PHILLIPS, P. H., and ELVEHJEM, C. A. 1954. Dental caries in the cotton rat. XIV. Further study of caries production by natural diets with especial reference to the role of minerals, fats, and the stage of refinement of cereals. J. Nutr. 53:17-27.

DANIELS, D. G. H., KING, H. C. C., and MARTIN, H. F. 1963. Antioxidants in oats: Esters of phenolic acids. J. Sci. Food Agric. 14:385-390.

DANIELS, D. G. H., and MARTIN, H. F. 1961. Isolation of a new antioxidant from oats. Nature (London) 191:1302.

DANIELS, D. G. H., and MARTIN, H. F. 1964a. Antioxidants in oats: Light induced isomerization. Nature (London) 203:299.

DANIELS, D. G. H., and MARTIN, H. F. 1964b. Structures of two antioxidants from oats. Chem. Ind. London. p. 2058.

DANIELS, D. G. H., and MARTIN, H. F. 1965. Antioxidants in oats: Diferulates and long chain diols. Chem. Ind. London. p. 1763.

DANIELS, D. G. H., and MARTIN, H. F. 1967. Antioxidants in oats: Mono-esters of caffeic and ferulic acids. J. Sci. Food Agric. 18:589-595.

DANIELS, D. G. H., and MARTIN, H. F. 1968. Antioxidants in oats: Glyceryl esters of caffeic and ferulic acids. J. Sci. Food Agric. 19:710-712.

DANIELS, R. 1974. Quick cooking oats. In: Breakfast Cereal Technology. Noyes Data Corp., Rev. 11. Parkridge, NJ.

D'APPOLONIA, B. L., and YOUNGS, V. L. 1978. Effect of bran and high-protein concentrate from oats on dough properties and bread quality. Cereal Chem. 55:736-743.

DODOK, L., MOROVA, E., and GALLOVA-ADASZOVA, M. 1982a. Influence of inactivated oat flour on gluten, dough and biscuit quality. Bull. Potravinarskeho Vyskumu 21:45-48.

DODOK, L., MOROVA, E., and SUROVA, E. 1982b. Experience with storage of oat biscuits. Bull. Potravinarskeho Vyskumu 21:49-52.

DUNLOP, A. P. 1973. The furfural industry. Pages 229-236 in: Industrial Uses of Cereals. Y. Pomeranz, ed. Am. Assoc. Cereal Chem., St. Paul, MN.

DUNLOP, A. P., and PETERS, F. J., Jr. 1953. The Furans. Reinhold Publishing, New York.

EVANS, C. D., COONEY, P. M., and MOSER, H. A. 1953. The flavor problem of a soybean meal. XI. Phytic acid as an inactivating agent for trace metals. J. Am. Oil Chem. Soc. 30:143-147.

FINNEY, K. F., SHOGREN, M. D., POMERANZ, Y., and BOLTE, L. C. 1972. Cereal malts in breadmaking. Baker's Dig. 46:36-38.

HUTCHINSON, J. B., MARTIN, H. F., and MORAN, T. 1951. Location and destruction of lipase in oats. Nature (London) 167:758-759.

JAFFEE, J. M., SZKRYBALO, W., and WEINERT, H. 1974. Xylose. U.S. patent 3,784,408.

JAINSTYN, H. 1957. Cosmetic preparation. Ger. patent 957,597.

JANECKE, H. 1954. Uber das vorkomen von taklonen die oxydation von l-ascorbinsaurells hemmen. Hafer Dtsch. Lebensm. Radsch. 65:69-71.

JELTAMA, M. A., ZABIK, M. E., and THEIL, L. J. 1983. Prediction of cookie quality from dietary fiber components. Cereal Chem. 60:227-230.

KAZI, T., and CAHILL, T. J. 1969. A rapid method for the detection of residual lipase. Analyst (London) 94:417.

KENT, N. L. 1966. Technology of Cereals with Special Reference to Wheat. Pergamon Press, Oxford.

KESSLER, C. C., and HICKS, W. L. 1949. Adhesive from cereal flour. U.S. patent 2,466,172.

KING, H. G. C. 1968. New antioxidant glycerides. Chem. Ind. London. p. 1468.

LORENZ, K. 1984. Cereal and dental caries. Pages 83-137 in: Advances in Cereal Science and Technology, Vol. VI. Y. Pomeranz, ed. Am. Assoc. Cereal Chem., St. Paul, MN.

MADSEN, K. O. 1981. The anticaries potential of seeds. Cereal Foods World 26:19-25.

MARTIN, H. F. 1958. Factors in the development of oxidative rancidity in ready-to-eat crisp oat flakes. J. Sci. Food Agric. 9:817-824.

McCLURE, F. J. 1964. Inhibition of experi-mental caries by oat hulls. Arch. Oral Biol. 9:219-221.

McKECHNIE, R. 1983. Oat products in bakery foods. Cereal Foods World 28:635-637.

MEISTER, B. 1963. A bath oil high in phospholipid content. J. Am. Geriatr. Soc. 11:789-791.

MERMELSTEIN, N. H. 1983. Soy-oats infant formula helps fight malnutrition in Mexico. Food Technol. 37:64-72.

MILLER, A. 1977. Cosmetic ingredients. Soap Deterg. Toiletries Rev. 7:21-25.

MUSHER, S. 1944. Oat products suitable for stabilizing and thickening various cosmetics, inks, etc. U.S. patent 2,355,028.

MUSHER, S. 1948. Cosmetic preparation. U.S. patent 2,436,818.

MYASNIKOVA, A. V., RALL, Y. S., TRISVYATSKII, L. A., and SHATILOV, I. S. 1969. Handbook of Food Products: Grain and Its Products. (Transl. from Russian) Israel Program for Scientific Translations, IPST Cat. No. 5168. U.S. Dep. Agric. and Natl. Sci. Found.

OOMAH, B. D. 1983. Baking and related properties of wheat-oat composite flours. Cereal Chem. 60:220-225.

PETERS, F. N., Jr., and MUSHER, S. 1937. Oat flour as an antioxidant. Ind. Eng. Chem. 29:146-151.

RASHBAUM, S. A., and YUEH, M. 1983. Natural red coloring prepared from an oat substrate. U.S. patent 4,418,081.

ROSENBERG, H., OBRIST, J., and STOHS, S. J. 1978. Production of fungal protein from oat hulls. Econ. Bot. 32:413-417.

SECONDI, A., and SECONDI, A. 1955. Nonflammable materials of construction. Ital. patent 509,408.

SHUKLA, T. P. 1975. Chemistry of oats: Protein foods and other industrial products. Crit. Rev. Food Sci. Nutr. 6:383-431.

SMITH, W. H. 1973. Oats can really 'beef up' cookies. I. Snack Food 62:33-35.

STANTON, T. R. 1951. New products from old crops. Pages 341-344 in: USDA Yearbook of Agriculture. U.S. Govt. Printing Office, Washington, DC.

STONE, J. B., and ROBITSCHEK, P. 1978. Factors affecting the performance of plywood glue extenders. For. Prod. J. 28:32-35.

SUGAWARE, K. 1970. Effects of various drugs on FSH secretion. Nippon Funin Gakkai Zasshi 15:126-131.

TAKETA, F., and PHILLIPS, P. H. 1957. Oat hull fractions and the development of dental caries. J. Am. Diet. Assoc. 33:575-578.

TAUFEL, K., and ROTHE, H. 1944. Chemistry of fat spoilage. XX. Characteristics of fat antioxidant complexes in oats. Chem. Abstr. 42:9203.

THOMPSON, D. J. 1963. Dietary factors affecting dental caries. Ph.D. thesis, Univ. of Wisconsin. (Diss. Abstr. 24:3089.)

TRIBOLD, H. O. 1938. Keeping quality of crackers improved by antioxygen. Food Ind. 10(Feb.):71.

VOGEL, J. J., THOMPSON, D. J., and PHILLIPS, P. H. 1962. Studies on the anticariogenic activity of oat hulls. J. Dent. Res. 41:707-712.

WAGGLE, D. H. 1968. Dextrin adhesive composition. U.S. patent 3,565,651.

WEBSTER, F. H. 1983. Method for preparing whole grain oat product. U.S. patent 4,413,018.

YOUNGS, V. L., PETERSON, D. M., and BROWN, C. M. 1982. Oats. Pages 49-105 in: Advances in Cereal Science and Technology, Vol. V. Y. Pomeranz, ed. Am. Assoc. Cereal Chem., St. Paul, MN.

INDEX